Principles of Microbial Metabolism and Metabolic Ecology

Alfred M. Spormann

Principles of Microbial Metabolism and Metabolic Ecology

Springer

Alfred M. Spormann
Departments of Chemical Engineering, and of Civil
and Environmental Engineering
Stanford University
Stanford, CA, USA

ISBN 978-3-031-28217-1 ISBN 978-3-031-28218-8 (eBook)
https://doi.org/10.1007/978-3-031-28218-8

This Springer imprint is published by the registered company Springer Nature Switzerland AG
The registered company address is: Gewerbestrasse 11, 6330 Cham, Switzerland

For Anthea and Samantha

Preface

This textbook is about unifying principles and variations of metabolism and the relationship to a microbe's ecophysiology. Historically, the field of microbial metabolism and its diversity emerged out of the quest of chemists and biochemists to understand fundamental chemistries of life, often with a focus on human health and disease. Still today, microbial metabolism is often introduced from an anthropocentric rather than an evolutionary or ecological perspective: Statements such as "*The main catabolism of a microbial cell is glycolysis and the TCA cycle; other pathways are exceptions or 'alternatives'*..." are surviving in frequently used phrases such as "alternative electron acceptor" or "carbon and energy source." Such limited view, for example of glycolysis, misses that there are at least three different versions of the glycolytic pathway, where each version confers a different fitness to a microbe! How can the "same" pathway confer different fitness? Understanding metabolism as a result of evolution of metabolic modules and their recombination into pathways provides conceptual flexibility to not only recognize diverse, extant metabolism but also to make powerful predictions on ecological implications, including from genomic data.

Most of the diversity of microbial metabolism can be explained in light of a few unifying principles. These principles are introduced here using examples primarily of anaerobic catabolism (energy metabolism) of chemoorganotrophic microbes, as the finite number of principles is best illustrated in this environmentally, biotechnologically, and medically important group of microbes. These principles can easily be transferred to lithotrophic microbial life.

In the first part (Chaps. 2–5), selected ecological, energetic, kinetic, and chemical principles are explained to set the organismal and ecological context for microbial metabolism and its ecophysiological consequences. In the second part (Chaps. 6–8), seven prototypic and six enzymatic auxiliary reactions are introduced in detail and shown to recur in common metabolic modules, which themselves are recurring and recombine into diverse, major metabolic pathways. The third part (Chaps. 9 and 10) shows how the organizing principles can explain metabolism, and how environmental constraints elicit pathway variations or different metabolic strategies.

This book makes frequent references to *E. coli* and its metabolism of glucose. It seems counterintuitive to use a single microbe as a reference. The reference is nevertheless useful to expand the "*E. coli*-centric" view of metabolism and to show both the unity and diversity of metabolism. The unprecedented diversity of prokaryotic metabolism, however, can be better understood knowing that *E. coli* is just a very special case of a microbe adapted to a specific niche and that many other metabolic strategies exist and are pursued by other microbes.

Each chapter is internally consistent, where the concepts relevant for each chapter are either introduced or, if previously discussed, summarized briefly. Therefore, each chapter contains summaries and redundancies of previously discussed concepts. Other redundancies were introduced for didactic reasons to emphasize important prevailing concepts and observations.

"Metabolism" and "catabolism" are often used interchangeably with the understanding that the principles developed here apply primarily to microbial catabolism.

Prokaryotes, i.e., archaea and bacteria, are frequently addressed here collectively as "microbes" with the understanding that "microbes" do not include, in this context, eukaryotes and phages. The choice of the word "microbe" is deliberate to emphasize the microbial cell as the selectable unit of microbial biology and is not limited to a taxonomic descriptor. Enzyme names are written in italics to emphasize their specificity.

This book is intended to help the reader organize the seemingly unranked details and "exceptions" of diverse metabolisms from the perspective of simple, recurring metabolic modules. Such tools allow us to easily deconvolute new metabolism in light of and to predict a microbe's ecophysiology. This book is written as an introduction to microbial metabolism for students with some background in biology and chemistry, for graduate students and postdocs, who seek to obtain more detailed and conceptional insights into microbial metabolism, and for professionals teaching or working on microbial metabolism, who look for a reference book on microbial catabolism, or for inspiration.

The idea and material for this book emerged out of teaching microbial metabolism at Stanford University for 28 years as well as directing summer courses in microbial biology over 16 years (4 years Microbial Diversity Course, MBL, Woods Hole with Carrie Harwood as co-director, and 12 years Hopkins Microbiology Course, Stanford University, Pacific Grove with Paul Rainey, Chris Francis, and John Roth as co-directors). Being part of the NSF-funded Center for Dark Energy Biosphere Investigations, directed by Jan Amend and Julie Huber, provided a greater appreciation of deep-sea environments and understanding of "microbial life in the slow lane." My longer quest to understand organizing principles of microbial metabolism and to connect its molecular and cellular features with a microbe's ecology arose from my early studies and lifelong conversations with Rolf K. Thauer, and my other mentors, Ralph S. Wolfe and John R. Roth.

I am indebted to my colleagues, coworkers, and scientific and personal friends for constructive and lively discussions, for critical comments, and for important suggestions for improvement of this book, including (in alphabetical order): Harry Beller, Jonas Cremer, Joerg Deutzmann, Maeva Finker, Arthur Grossman, Wenyu Gu, Thomas Hansen, Susan Holmes, Ann Karim, Frauke Kracke, Bruce Levin, Ian Marshall, Nicole

Matis, Alexandria (Ali) McCully, Jack Meeks, Albert Müller, Volker Müller, Soeren Molin, John Roth, Bernhard Schink, Rolf K. Thauer, Julia Vorholt, Tom Wandless, and Gregg Whited.

My special thanks and gratitude go to Rolf K. Thauer for his generosity in sharing his time, his insights, expertise, and unrivaled, encyclopedic knowledge and understanding of microbial metabolism, and his invaluable critique, suggestions, and perspectives offered during the writing of this book. The book would not be the same without him!

Up and enjoy discovering the fascinating world of microbial metabolism!

Stanford, CA Alfred M. Spormann
June 2022

Contents

1 Microbial Life on Earth, Metabolism, and Metabolic Diversity

Contents

1.1 Introduction

Life arose on Earth ~3.8 billion years ago in form of single-celled prokaryotic microorganisms. Today, the collective global number of microbial cells is estimated to be ~5×10^{30}, constituting about 13% of the total biomass on Earth (Fig. 1.1a). The largest fraction of microbial biomass is present in complex and heterogeneous subsurface environments (~3.5×10^{30} cells in the terrestrial subsurface and ~$0.25–2 \times 10^{30}$ cells in marine deep subsurface). The remaining microbial biomass is in soil (~0.26×10^{30} cells) and the open ocean (~0.12×10^{30} cells) (Fig. 1.1b). For reference, the well-studied microbes of the gut of mammals account only for ~0.0001% of all microbes. Thus, most microbes on Earth live and metabolize in complex, heterogeneous environments, far different from simple and well-understood laboratory environments.

On a global scale, archaeal biomass accounts for roughly 10% of the total prokaryotic biomass. There are notable differences in relative abundance of archaea in different environments: Archaea are more prevalent in the marine subsurface and marine planktonic environments relative to soil and the terrestrial subsurface.

Life requires metabolism, reproduction, and heredity as well as evolution by natural selection. The most noticeable impact of microbial life on its environment is via the microbe's energy metabolism, required for providing energy from environmental resources. Chemotrophic microbes that use chemical energy have been shaping our physical-chemical environment because of the rates of substrate consumption and release of catabolic end products. Photoautotrophic microbes and plants profoundly shaped Earth's geochemistry because of the need for electrons as reductant in CO_2 fixation. Arguably the most impressive manifestation of microbial metabolism is the unprecedented metabolic and physiological diversity, in particular with respect to carbon degradative metabolism. Annually, ~100 Gt carbon is mineralized via microbial metabolism, which corresponds approximately to 90% of all CO_2 fixed by photosynthesis.

A. M. Spormann, *Principles of Microbial Metabolism and Metabolic Ecology*,
https://doi.org/10.1007/978-3-031-28218-8_1

Microbial metabolism can be reduced to a limited set of seven prototypic and six auxiliary enzyme reactions, consisting of oxidoreductase and carbon–carbon cleavage reactions, that combine into a limited set of diverse metabolic modules. These metabolic modules are recurrent in the majority of the diverse catabolic pathways. Natural selection shaped the kinetics of enzymes with respect to K_M and v_{max} to mediate a coordinated metabolic flux in response to the flux of the primary energy substrate in a microbe's environment. These reactions, modules, and pathways are introduced in Chaps. 6–8. Understanding principles of metabolism is much easier when understanding first how the drivers of metabolism such as a microbe's physical-chemical environment, thermodynamics and kinetics including fluxes of energy, and reactivities of organic molecules, confer fitness. These principles are introduced first in Chaps. 2–5. Chapters 9 and 10 provide key examples for the link between pathway architecture and fitness based on the principles discussed (Chaps. 2–8). Such link is illustrated well in the following simple laboratory experiment, which will serve as a "red thread" throughout this textbook.

1.2 Principles and Patterns in Metabolic Ecology; the Julian Adams Experiment

A profound experiment conducted by R. B. Helling and C. N. Vargas in the laboratory of Julian Adams revealed some insights into the link between molecular metabolism and a microbe's ecology. This experiment and subsequent analyses are referred to in this book as the "Julian Adams" experiment with the understanding that many coworkers over time contributed significantly to the understanding presented here. The experimental system consisted of a chemostat (Fig. 1.2, more details in Chap. 5) which is a constant-volume reactor where the rate of continuous inflow of nutrients (dilution rate = flow rate divided by reactor volume) is equal to the rate of outflow of newly synthesized cells (growth rate), of unused

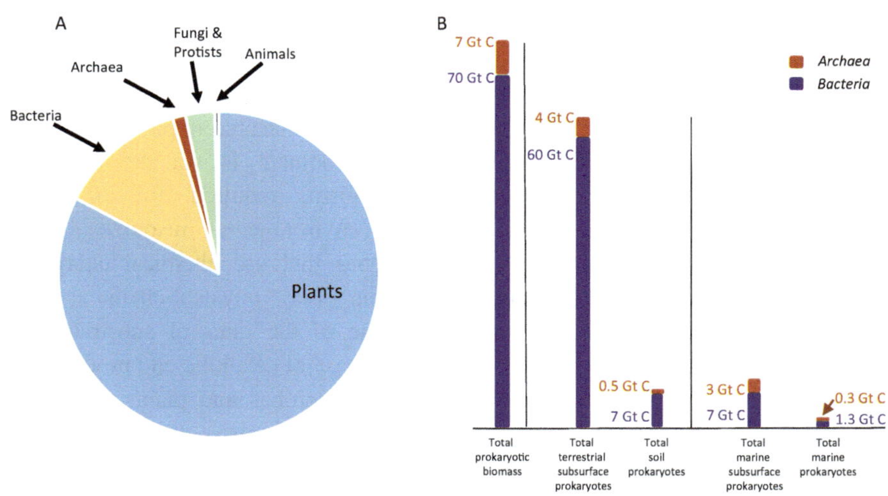

Fig. 1.1 Estimated distribution of the microbial biomass on Earth. Numbers given are rounded estimates. (**a**) Terrestrial biomass is composed of and dominated by 450 Gt (Giga = 10^9) C (carbon) of plant biomass, 70 Gt C Bacteria, and 7 Gt C Archaea, 12 Gt C fungi, and 2 Gt C Animal. Terrestrial biomass is 2 orders of magnitude higher than marine biomass. (**b**) Most of the ~90% of the total prokaryotic biomass on Earth is in the terrestrial subsurface and is bacterial; ~10% is archaeal. Soil and marine plankton contain prokaryotic carbon in the same order of magnitude, with a relative enrichment in Archaea in marine environments. Numbers do not add up because of uncertainties in measurements and estimations and should be used to gauge orders of magnitude and relative composition rather than as accurate numbers. Modified after Bar-on et al. (2018)

nutrients, and of metabolic end products. In an ideal chemostat, the cell density stays constant while microbes grow continuously at a fixed rate determined by the dilution rate. In other words, the chemostat provides the most simple, constant, and predictable environment of a microbe. In the Julian Adams experiment, the catabolic substrates were glucose and O_2, where glucose was the growth-limiting substrate.

The glucose-limited chemostat was seeded with a single clonal population of *E. coli* (JA122, derived from MG1622) and grown and maintained at a dilution rate of 0.2 h^{-1} (doubling time of 3.5 h). During the adaptive evolution experiment and after about 800 generations, samples of the chemostat were plated on tryptone agar plates and colony morphologies were recorded (Fig. 1.2). The striking observation was that most colonies exhibited a morphology that was different from the seeding (ancestral) *E. coli* strain; the relative frequencies were dynamic over the time course of the experiment. Among the numerically-dominant colonies were small colony variants (SCV), whose diameter on tryptone agar plates was smaller than that of wild type (Fig. 1.2). The frequency of SCV in advanced chemostat cultures dynamically fluctuated between 50 and 90% of the total population (i.e., colonies). Interestingly, also a few larger-sized colonies (large colony variants, LCV) were observed with a colony size larger than wild type. The colony morphology phenotypes were stable and inheritable, indicating that these variants had evolved from the ancestral *E. coli* strain within the approximately 800 generations in the chemostat environment. The parental wild-type strain was largely absent in the advanced stage of the experiment.

Notably, cells of SCV and LCV exhibited metabolic and physiological traits that were different from those of the ancestral strain. In order to test whether the difference in colony morphology correlated with a phenotype that was beneficial in this particular chemostat environment or had accrued simply by chance, Adams and coworkers seeded a new glucose-limited chemostat with defined mixtures of evolved and ancestral stains and operated the chemostat under the same conditions, including the same dilution rate (Fig. 1.3). Interestingly, the different variants were stably maintained and co-existed in that chemostat in roughly the ratio as they were found in the original chemostat. This observation showed that the evolved variants (SCV, LCV) had an increased fitness in the chemostat environment relative to the ancestral wild type because

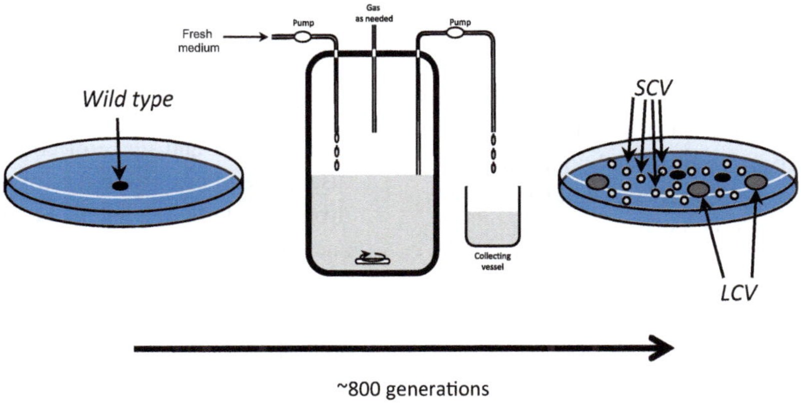

Fig. 1.2 Simplified summary of the adaptive laboratory evolution experiment by Julian Adams and coworkers. *E. coli* (JA122) was grown for 800 generations in an aerobic, glucose-limited chemostat at a dilution rate equivalent to a 3.5 h doubling time. Periodically, liquid samples were removed and plated on tryptone agar plates, and morphologies of the colonies were examined. The relative frequency of colony variants was dynamic and fluctuated over the course of the experiment. Chemostat systems are described in detail in Chap. 5. Abbreviations: SCV, Small Colony Variants; LCV, Large Colony Variants. See text for details

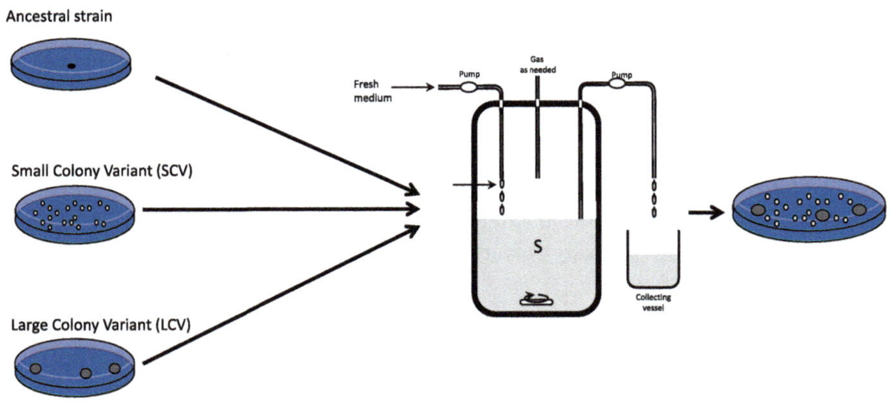

Fig. 1.3 Defined mixtures of evolved variants co-existed stably in a reconstructed chemostat, operated under same conditions. See text for details

wild type was largely outcompeted by the variants. The experiment showed further that the different variants could stably co-exist in this environment because no single variant had taken over the entire chemostat culture.

Since growth of any strain in the chemostat is associated with utilization of the only energy resource in the chemostat, glucose (O_2 was not limiting), wild type and evolved variants were tested for their efficiency of glucose utilization. The fact that SCV and LCV did stably co-exit suggested that both variants do not occupy the same niche because of the niche exclusion principle. This begged the question of whether any of the strains could grow on an excreted metabolite that is derived via another strain from the only substrate glucose. To test this hypothesis, cross-wise growth experiments with wild type, SCV, and LCV (tester strains) were conducted to examine whether any strain can grow on spent supernatant from glucose-grown batch cultures of the other monocultures (Fig. 1.4). It turned out that only significant growth was observed when LCV was inoculated into the spent supernatant of SCV (Fig. 1.4). SCV grew neither on supernatant of LCV nor of wild type. Further analyses revealed that the cross-feeding metabolite was acetate that was released by SCV in large amounts and utilized by LCV. Apparently, wild-type strain had speciated into two strains: SCV metabolized glucose incompletely and inefficiently to acetate, and LCV adapted to the newly constructed niche acetate by specializing on acetate in lieu of competing for glucose. Moreover, SCV outcompeted the ancestral wild type for the resource glucose.

> Not all small colony variants isolated in this experiment after the original 800 generation chemostat experiment were identical. Some SCV also released glycerol into the medium and other colony variants not described here consumed glycerol, just like LCV consumed acetate. For the sake of simplicity, the partitioning of glucose into glycerol and glycerol consumption is disregarded here and in the following. The drivers underlying that niche partitioning and speciation of pathways are very similar to that what will be discussed here extensively for acetate. Analyzing both acetate and glycerol cross-feeding in parallel is more confusing and, therefore, eliminated here. However, the experiment revealed that not only acetate, but also glycerol is cross-fed.

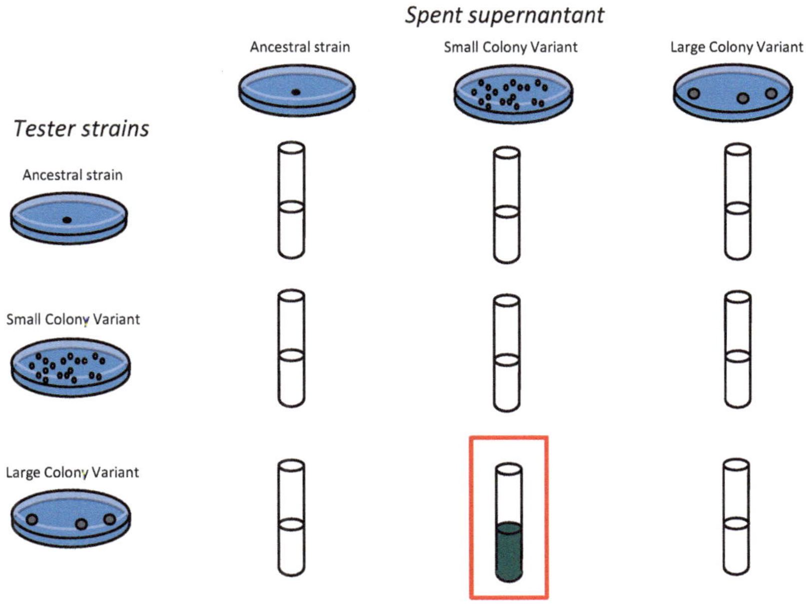

Fig. 1.4 Schematic representation of the design and outcome of cross-feeding via culture supernatant. Filter-sterilized supernatant was prepared from batch monocultures of either ancestral wild type, small colony variant, or large colony variant, grown aerobically in glucose minimal medium. Filter-sterilized supernatant was then inoculated with either ancestral wild type, small colony variant, or large colony variant (tester strains) and growth was monitored. Growth was observed when supernatant of small colony variant was inoculated with a large colony variant tester strain, indicating that small colony variant produced a metabolizable compound in quantities that allowed growth of the large colony variant. Outcomes are pictorially emphasized for didactic reasons. See text for details

These experiments revealed several important findings for a link between metabolism and ecological success:

1. Apparently, a constant environment with high flux of resources can drive metabolic speciation. The resource glucose became partitioned into glucose and acetate.
2. SCV can exist despite using the resource glucose inefficiently and outcompete, at the same time, the efficient ancestral strain.
3. A food chain had evolved where LCV could feed off a metabolic end product (acetate) secreted by the small colony variant.

1.3 An Initial Metabolic View of the Julian Adams Experiment

Under the chemostat conditions, *E. coli* wild type oxidizes glucose mostly to CO_2 via glycolysis, citric acid cycle, and reduces molecular oxygen as terminal electron acceptor via a respiratory chain (see the following Chapters). These pathways, abbreviated for the following as "respiratory" metabolism, collectively represent the energy-conserving pathways of *E. coli* wild type, making complete use of the available resource glucose + O_2. Figure 1.5 provides the metabolic overview of this metabolism and metabolic speciation by SCV and LCV. SCV became better adapted in acquiring glucose from the chemostat environment: SCV evolved to have a higher affinity for glucose, a more efficient uptake of glucose, and a faster consumption (higher rate of glucose metabolism) in the chemostat. At the same time, utilization of glucose was incomplete and less efficient. It is important to note that acetate is a metabolic end product of the energy metabolism of the SCV and not some intermediate that leaked out of a cell. Paradoxically, this inefficient use of glucose enabled SCV to compete better against

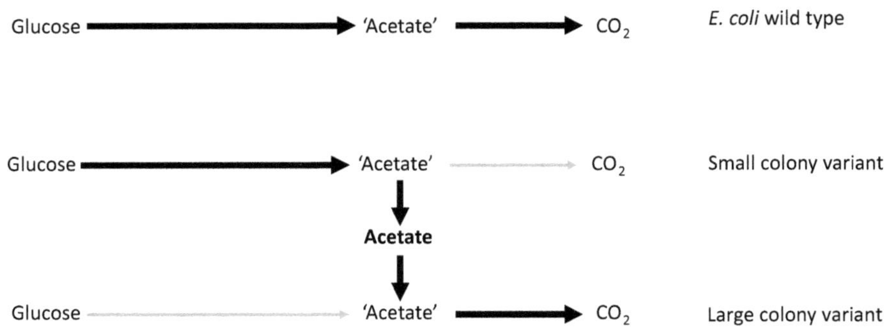

Fig. 1.5 Partitioning of resources in the aerobic, glucose-limited chemostat. While *E. coli* wild type uses the Embden–Meyerhof pathway, the TCA cycle, and the respiratory chain as catabolic pathways, the small colony variant became specialized to operate the Embden–Meyerhof pathway at higher rate at the trade-off of employing the TCA cycle and respiratory chain. The large colony variant is maladapted to using glucose but gained the capacity to uptake and utilize acetate efficiently via the TCA cycle and respiratory chain. Molecular oxygen (not shown here for simplification) is the electron acceptor for the three strains for oxidation of glucose or acetate. "Acetate" in parenthesis is to indicate that intracellular acetate in form of acetyl-CoA is produced and consumed in the metabolic pathways listed above. See text for details

and outcompete wild type in the chemostat environment, although the overall metabolic pathway in SCV became shorter. As a consequence of this particular evolutionary outcome, a new niche was formed: acetate in the environment. This new niche selected for the LCV, which adapted to utilizing acetate as substrate. This specialization of the large colony variant to acetate came with the trade-off of LCV becoming maladapted to accessing glucose. However, this new metabolic trait allowed the LCV to escape competition with the SCV and wild type.

This profound outcome of the Julian Adams experiment raises some key questions of microbial metabolism:

- Why and when are microbes with shorter pathways more fit than strains with longer pathways and a more efficient use of a resource?
- What are the drivers of speciation?
- What determines the specific architecture of the (evolved) metabolic pathways?
- What are the ecological and environmental consequences of having short or long pathways?
- Why are there trade-offs?
- What would have been the outcome of such long-term evolution experiment if a different limiting substrate or a different microbe was used?
- What are the rules of metabolism?

This Julian Adams experiment is used as a "red thread" in this textbook for understanding the links between metabolism, pathway architecture, and fitness. Chapters 2–5 first introduce some foundational thermodynamic, kinetic, and environmental concepts; Chaps. 6–8 introduce the metabolic principles that govern microbial metabolism, and Chaps. 9 and 10 provide ample examples, in particular of anaerobic metabolism of organic compounds that mirror the outcome of the Julian Adams experiment on a global scale.

Further Reading

Bar-on YM, Phillips R, Milo R (2018) The biomass distribution on earth. Proc Natl Acad Sci U S A 115(25): 6506–6511

Helling RB, Vargas CN, Adams J (1987) Evolution of *Escherichia coli* during growth in a constant environment. Genetics 116(3):349–358

Kinnersley M, Holben WE, Rosenzweig F (2009) E unibus Plurum: genomics analysis of an experimental evolved polymorphism in *Escherichia coli*. PLoS Genetics. https://doi.org/10.1371/journal.pgen.1000713

Kinnersley M, Wagner J, Kroll E, Adams J, Sherlock G, Rosenzweig F (2014) Ex Uno Plures: clonal reinforcement drives evolution of a simple microbial community. PLOS Genetics 10(6):e1004430. https://doi.org/10.1371/journal.pgen.1004430

Rosenzweig RF, Sharp RR, Treves DS, Adams J (1994) Microbial evolution in a simple unstructured environment: genetic differentiation in *Escherichia coli*. Genetics 137(4):903–917

A Microbe's Environment and Natural Selection

2

Contents

2.1 Introduction

A famous essay by the evolutionary geneticist Theodosius Dobzhansky is entitled "Nothing in Biology Makes Sense Except in Light of Evolution," and nothing is truer in microbial metabolism than that. The interplay between metabolism and evolution is reciprocal as a physical-chemical

A. M. Spormann, *Principles of Microbial Metabolism and Metabolic Ecology*,
https://doi.org/10.1007/978-3-031-28218-8_2

environment is altered as a consequence of microbial metabolism, setting thereby new selective forces for new metabolic and physiological traits. Local environments are shaped by fluxes of resources that determine ecological success. This chapter introduces some organizing principles on the intricate relationship between the physical-chemical characteristics of a microbe's external and internal environment, the fitness of a microbe in its immediate and global environment, and selection for and against specific metabolic traits.

This chapter uses the terms and concepts of "catabolism" and "anabolism," although they are introduced in more detail in Chap. 4. Briefly, catabolism is equivalent to energy metabolism and includes all metabolic reactions that are directly or indirectly involved in ATP synthesis. Anabolism is often used synonymously to biosynthesis and includes all metabolic reactions that are involved in the synthesis of a living cell. Repair and metabolism required for maintenance under nongrowing conditions is broadly described as maintenance metabolism. Catabolic reactions are those involved in catabolism and anabolic ones in anabolism.

2.2 Microbes and the Physics in Their Environment

At the μm scale, the effects of mass, force, time, and mass transport on microbes are different from what they are in macroscopic environments on macroscopic organisms. This is due to the small size of microbes in conjunction with the viscosity of water at the μm scale. The length size of prokaryotes varies enormously between 0.17 μm (*Thermophilum* sp.) and 100–300 μm (*Thiomargarita namibiensis*, some up to 750 μm). Cell volume spans a range of over 10 orders of magnitude (<0.01 μm^3 to 200×10^6 μm^3). In many microbes, such as *E. coli*, cell size varies with growth rate, where the cell volume of slow growing cells is smaller by an order of magnitude than of fast-growing cells. For simplification of the following discussion, however, let's assume a spherical microbe of 1 μm in diameter.

To illustrate the difference in prevailing physical forces, compare a person and a microbe swimming in water. Inertia can move a swimmer by a single stroke by a meter or two. On a microbial scale, there is no significant inertial force. The viscosity of water acting on a microbe when swimming is equivalent to that of a human swimming through molasses: in other words, the inertial motion of an *E. coli* cell, which stopped swimming, ceases in less than the diameter of a hydrogen atom and in less than a millisecond!

2.2.1 Maintaining Osmotic Pressure within a Prokaryotic Cell

All living cells have in common that they are surrounded by a semipermeable lipid bilayer membrane. Semipermeable indicates that the membrane is readily permeable for water but not for most solutes. While prokaryotes generally only have one cell compartment surrounded by a semipermeable membrane, eukaryotes have several such compartments (e.g., mitochondria, chloroplasts, endoplasmic reticulum, etc.).

Due to the semipermeability, solutes inside a microbial cell exert an osmotic pressure on the cytoplasmic membrane that is proportional to the difference in solute concentration between inside and outside. At 25 °C, a 1 M concentration difference of solutes exerts an osmotic pressure of 22.4 bar, which is the osmotic pressure in many Gram-positive bacteria, in which the semipermeable cytoplasmic membrane is surrounded by a multilayered murein sacculus that prevents the bacteria from bursting. Osmotic pressure is similarly high in some archaea that are enclosed by a multilayered pseudomurein sacculus. Osmotic pressure in Gram-negative bacteria, which are enclosed by only a monolayer of murein, is generally only near 5 bar. As a rule of thumb, prokaryotes living in environments with large fluctuations in water activity, such as soil, have a higher internal osmotic pressure than those living in environments with a more constant high water activity, such as lakes and oceans. Water activity at a given temperature is P/P_o, where P is the partial pressure of water with solutes and P_o the partial pressure of water without solutes. The water activity of distilled water is 1.0.

When the concentration of solutes outside a cell exceeds that of the solutes inside, the cytoplasm, but not the rigid cell wall, shrinks, a phenomenon called plasmolysis. Under these conditions a cell no longer divides, however can increase the intracellular solute concentration, e.g., by synthesizing compatible solutes, such as glutamic acid, and/or by importing potassium cations to neutralize glutamic acid. Compatible solutes are compounds that, at high concentration, do not affect the activity of intracellular enzymes. Best known compatible solutes are glycine betaine and ectoine.

2.2.2 Diffusion into and Membrane Permeability of the Prokaryotic Cell

Transport of dissolved resources to a microbial cell at the µm scale is not controlled by mixing or stirring, because the viscosity of water makes local turbulences of less than 1–6 mm insignificant. Mass transport to and from, and within, a microbe is determined by molecular diffusion. The flux of resources via diffusional transport is given by Fick's Law:

$$J = D\frac{dC}{dx},$$

where dC is the concentration difference between two points, dx the distance between the two points, and D the diffusion coefficient [cm^2/s]. Flux J has the dimension of mmol s^{-1} cm^{-2} (flux of molecules through an area of 1 cm^2). The diffusion coefficient D of a molecule in water is dependent on the size and shape of the molecule as well as its ability to coordinate H_2O. D increases with increasing temperature. Of the molecules listed in Table 2.1, protons diffuse in water at 25 °C most rapidly and serum albumin most slowly. Compared to macromolecules such as the protein serum albumin, diffusion of metabolites in water is faster by more than an order of magnitude (Table 2.1). However, even for small molecules mass transfer is only rapid when the distance is short.

Assuming that the intracellular concentration of a catabolic substrate is close to zero and the distance across the cell membrane is constant, the rate of diffusion depends only on the substrate concentration in the boundary layer of a cell and on D (Fig. 2.1). As alluded to above, mixing at a scale below 1 mm is insignificant for mass transport. Thus, a microbe is surrounded by a "diffusional halo" that is unaffected by turbulence (Fig. 2.1). In contrast to a grazing cow, for example, a microbe cannot increase the flux of nutrients for uptake by swimming. A microbe such as *E. coli*, swimming at ~30 µm/s, cannot outswim its diffusional halo; the cell moves only between local environments with different substrate concentration to enhance diffusional transport by increasing the concentration gradient (see below). (A cow can increase the flux of grass intake by moving and eating faster). For a cell larger than 10 µm in length, swimming motility can theoretically enable to escape the diffusion halo, which would result in uptake of resources directly coupled to movement. Although diffusional transport to a cell increases linearly with cell size, the volumetric rate of substrate utilization decreases with increasing cell size. Thus, it is advantageous for microbes to be small when utilizing substrates at low concentration to avoid limitation by diffusion. This is observed in *E. coli*, for example, when growing slowly at low substrate concentration. Extracellular enzymes (e.g., *hydrolases*) that aid in accessing substrates, which cannot be transported into a cell, are most effective if they are present in a sphere of less than 10 µm around a cell as diffusional transport is most effective at that scale and

Table 2.1 Diffusion coefficients *D* in water and artificial lipid membrane permeability *P* at 25 °C. The size of *P* depends considerably on the lipid composition of the membrane. *P* is generally smaller when determined for cytoplasmic membranes. Therefore, the values given are more to show relative differences in permeability between the different compounds. Diffusion coefficients are given in the literature also as $\mu m^2/ms = 10^{-3}\ mm^2/s = 10^{-9}\ m^2/s = 10^{-5}\ cm^2/s$

Compound	Diffusion coefficient *D* $[10^{-5}\ cm^2\ s^{-1}]$	Membrane permeability *P* $[cm\ s^{-1}]$
H_2O	2.3	10^{-3}
H^+ (H_3O^+)	9.1	10^{-6}
H_2	5.1	10^{0}
O_2	2.3 $0.2\ cm^2\ s^{-1}$ air	10^{0}
CH_4	1.8	10^{-1}
CO_2	1.9	10^{-1}
Formic acid	1.4	10^{-3}
Formate$^-$	1.5	10^{-6}
Acetic acid	1.2	10^{-2}
Acetate$^-$	1.1	10^{-6}
Lactic acid	0.9	10^{-4}
Lactate$^-$	1.0	10^{-7}
Na^+	1.3	10^{-14}
K^+	1.7	10^{-14}
Cl^-	2.0	10^{-10}
Glucose	0.6	10^{-10}
Saccharose	0.5	$<10^{-10}$
Lysozyme (14.3 kDa)	0.1	$<10^{-10}$
Serum albumin (66 kDa)	0.06	$<10^{-10}$

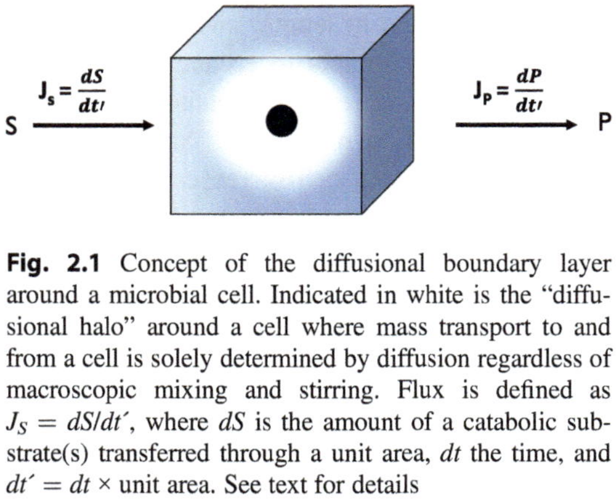

Fig. 2.1 Concept of the diffusional boundary layer around a microbial cell. Indicated in white is the "diffusional halo" around a cell where mass transport to and from a cell is solely determined by diffusion regardless of macroscopic mixing and stirring. Flux is defined as $J_S = dS/dt'$, where *dS* is the amount of a catabolic substrate(s) transferred through a unit area, *dt* the time, and $dt' = dt \times$ unit area. See text for details

diffusional loss of extracellular enzymes is minimal.

In densely packed microbial biofilms, competition for substrates occurs within the diffusional halo. Moreover, when accessing insoluble substrates, such as cellulose or other complex carbohydrates, microbes colonize such surfaces as biofilms and excrete respective *hydrolases* that stay confined to that ~10 μm^3 space.

The cytoplasmic membrane is a diffusional barrier preventing the loss of cytoplasm content, including enzymes, other macromolecules, coenzymes, and metabolic intermediates, as well as gating compounds entering and exiting a cell. The outer membrane of Gram-negative bacteria is generally not a diffusional barrier for solutes with a molecular mass of up to 500 Da. Diffusion through the lipid bilayer of the cytoplasmic membrane is primarily determined by a molecule's hydrophobicity and given by the permeability coefficient *P* (Table 2.1). In contrast to diffusion coefficients, permeability coefficients vary by several orders of magnitude. For example, the gases H_2, O_2, CO_2, and CH_4 rapidly diffuse through the cytoplasmic membrane, indicating that an active transport for obtaining a higher intracellular concentration of a gas would be futile, because a "concentrated" gas would immediately diffuse out of the cell. On the other hand, glucose has a low permeability, and uptake into a cell requires enzymatic transport systems. Ions, such as Na^+ and Cl^-, have a very low permeability due to their charge properties, and active transport is required for uptake. Interestingly, sodium ions have a lower permeability than protons, which may be one reason for why some microbes use a sodium motive rather than a proton motive force to conserve energy (see Chap. 6). The permeability constant of H_2O (10^{-3} cm s^{-1}) is lower than that of gases (Table 2.1). However, the concentration difference of water between inside and outside is in the order of 0.2 M and can be as high as 1 M in Gram-positive bacteria. This allows H_2O to almost freely diffuse through the membrane considering that the diffusion rate is directly proportional to the concentration difference.

Membrane permeability for short-chain fatty acids is of considerable importance. While protonated acids have a relatively high membrane permeability, the deprotonated anionic forms, e.g., lactate$^-$, butyrate$^-$, or acetate$^-$, have a low and differential permeability (Table 2.1). p*Ka* is that pH value where half of the acid is in the deprotonated and the other half in the protonated form. The p*Ka* of lactic acid is 3.8, which makes it a stronger acid than acetic acid (p*Ka* = 4.7). Therefore, at pH 7 more than 99.94% of lactic acid is present as lactate$^-$, which permeates through the cytoplasmic membrane by three orders of magnitude slower than lactic acid. On the other hand, at the same pH 0.5% of the total concentration of acetate is protonated; this significant fraction of acetic acid is membrane permeable. If the pH outside a cell is identical to that inside, the internal pH is not affected by the diffusional transport of a proton via acetic acid across the membrane because acetic acid dissociates inside to acetate and a proton until the acetate and proton concentrations inside and outside are identical. However, when the pH outside is, e.g., 5, acetic acid will diffuse into the cell until the internal pH is also pH 5. Such diffusional acidification of a cell's interior is limited only by the intracellular buffering capacity for protons which is estimated to be near 50 mM. Generally, growth of microbes producing short-chain fatty acid seems to cease if the cytoplasmic pH decreases to below 6. Such toxicity effect of acetic acid is more pronounced in the presence of lactic acid. Lactic acid is less permeable than acetic acid, however, is a much stronger acid. On the other hand, if a growth medium is acidified with H_2SO_4 (sulfuric acid) rather than acetic acid, the situation is quite different. H_2SO_4 at pH 5 is completely dissociated into SO_4^{2-} and 2 H^+, and neither ion easily permeates the membrane. Thus, no proton enters a cell via diffusional transport and the internal pH is not affected as long as a cell is intact. This explains why chemolithotrophic sulfur-oxidizing microbes can exist in low pH (pH 1–2) hot sulfur springs or solfatara fields.

2.2.3 Diffusion of O_2

The time of diffusion to homogeneous distribution of small molecules inside a microbial cell is rapid and in the order of msec. For example, driven by a concentration difference of 0.1 mM (0.1 atto-mol/μm^3), one atto-mol O_2 (10^{-18} mol) ($D = 2.3\ 10^{-5}$ cm^2 s^{-1} = 2.3 μm^2/ms) requires only 0.43 ms to diffuse a distance of 1 μm (e.g., across the length of a cell) at 25 °C; however, it will take 0.43 s to diffuse 1 mm. Such rapid intracellular diffusion eliminates any spatial concentration gradient or partitioning of metabolism inside a μm-size microbial cell.

The diffusion coefficient of O_2 is 10,000 times higher in air than that in water at 25 °C (Table 2.1). On the other hand, the diffusion rate of O_2 in water is frequently lower than the respiration rate of aerobic microbes, which can be in the order of μmol/(min × cm^3) assuming biomass concentration in the range of mg/cm^3. In equilibrium with air, O_2 concentration in water is 240 μM (0.24 μmol/cm^3) at 25 °C. Thus, the diffusion rate of O_2 to an aerobic microbial cell is 0.33 nmol cm^{-2} min^{-1}, which is lower by roughly three orders of magnitude than its above calculated respiratory rate, assuming a concentration difference of 0.24 μmol cm^{-3} per 1 cm distance (from completely oxic to completely anoxic). A consequence of this relatively slow O_2 diffusion rate is that aqueous environments containing metabolizable compounds very rapidly become anoxic even close to the air–liquid interface based on the activity of aerobic and facultative microbes.

With increasing temperature, respiration rate increases more than the rate of diffusion. In addition, the solubility of O_2 in water decreases with increasing temperature. A mass transfer for O_2 between air and water can be increased by increasing the surface area of O_2 exchange and by reducing the diffusion distance via mixing. The former can be achieved, e.g., by bubbling air as small bubbles (higher surface area) into the aqueous solution, or by mixing, which are basic requirements for growing aerobic microbes in the laboratory.

Although O_2 inside an aerobic microbe is not partitioned due to rapid intracellular diffusion (see above), O_2-consuming metabolic processes generate stable gradients. O_2 is consumed mainly at the site of the cytoplasmic membrane and at rates that exceed the diffusion rate. As a consequence, when O_2 concentration in the surrounding aqueous environment is in the range of μM, the cytoplasm of actively respiring, aerobic microbes is almost always completely anoxic.

2.2.4 Intracellular Compartmentalization via Proteins and the Constraint of Protein Space

In a μm-sized cell, an enzyme encounters its cognate substrate within 1 ms by diffusion (see above). Enzymes have a catalytic efficiency $\left(\frac{k_{cat}}{K_M}\right)$, which is, for most enzymes, 2–3 orders of magnitude lower than the diffusion limit (~10^8–10^9 M^{-1} s^{-1}).

The diffusion coefficient for small proteins such as lysozyme (Table 2.1) is smaller than that of O_2 by more than an order of magnitude.

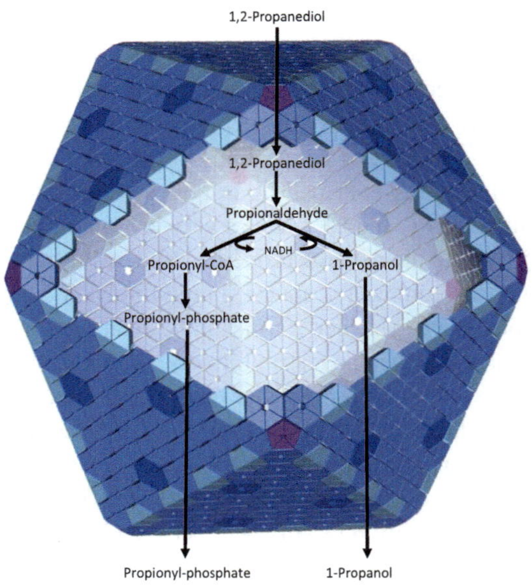

Fig. 2.2 Schematic representation of a microbial microcompartment. Microcompartments kinetically isolate metabolic reaction sequences that involve toxic or volatile intermediates, such as acetaldehyde or propionaldehyde. Blue hexagonals indicate shell proteins. See text for details. From Chowdhury et al. (2014) with permission

Despite this fact, the diffusion rate of enzymes is still sufficient to eliminate any spatial concentration gradient or partitioning of metabolism inside a microbial cell. However, in prokaryotes some proteins form with other proteins functional macro-complexes (molecular masses >1000 kDa); other enzymes can be concentrated in microcompartments (Fig. 2.2) which contain a permeable protein "membrane," both leading to a kinetic compartmentation of metabolites. Metabolic intermediates that are formed in such complexes or compartments are prevented from mixing with other cytoplasm content. An important example is acetaldehyde metabolism. Acetaldehyde is toxic and its metabolism is separated from the cytoplasm by containing this metabolite either into a multi-enzyme complex composed of *ethanol dehydrogenase* and *acetaldehyde dehydrogenase* (AdhE) or a micro-compartment.

Such physical compartmentalization allows prokaryotes to kinetically compartmentalize metabolism in the absence of organelles, such as

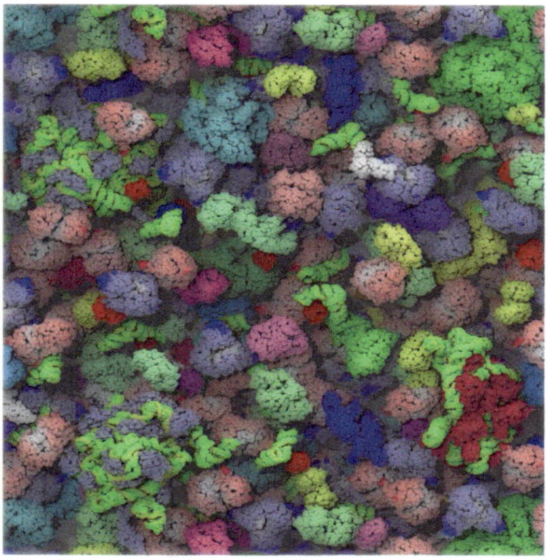

Fig. 2.3 Cytoplasmic model of the macromolecular content and crowding of an *E. coli* cell. RNA is shown in green and yellow. See text for details. With permission from McGuffee and Elcock (2009)

mitochondria or chloroplasts, which require a lipid bilayer membrane and membrane trafficking.

The biomass-to-volume ratio of microbes is remarkably constant and invariant of the growth rate, and so is the surface areal density. The mass and internal volume of a prokaryotic cell is dominated by some 2–4 million protein molecules and ribosomes. The entire set of proteins is defined as the proteome. The cytoplasm is a crowded protein space, containing about 70% of all proteins at a concentration of ~200 mg protein/ml as the computational model shows in Fig. 2.3. Cytosolic as well as membrane-associated or membrane-integrated enzymes involved in energy conservation comprise the catabolic proteome, while those involved in biosynthesis, including in transcription, translation, and DNA polymerization, constitute the anabolic proteome. The collective total proteome space is limited and confined by the fixed volume of a cell. This intrinsic constraint causes important trade-offs in allocation of proteome space to specific cellular functions, and, consequently, proteome allocation is under natural selection. The membrane proteome is even more constrained because of the restriction of membrane proteins to a 2D layer rather than a volume as in the cytoplasm. Thus, there is even stronger natural selection on the specific allocation of proteins with specific functions that occupy a limited membrane proteome space.

Assuming the mass of an average protein of 30,000 Da, the cytoplasmic protein concentration is equivalent to ~10 mM protein, which is ~100,000 times higher than the concentration of free intracellular protons at pH 7 (~10 H^+ per cell)! Since most proteins are enzymes, and since the rate of an enzyme-catalyzed reaction is dependent on the total enzyme concentration ($v_{max} = k_{cat}\,[E_T]$, see Chap. 5), the space that is allocated to specific proteins profoundly determines the kinetic coordination of catabolism, anabolism, and with that growth and fitness. For example, in fast-growing *E. coli*, up to 80% of the total intracellular space can be taken up by ribosomes. This significant demand for space has consequences for the total space that is available for a catabolic proteome (see Chaps. 4 and 5). In other microbes, such as *Bacillus subtilis*, *Lactococcus lactis*, and the archaeon *Methanococcus maripaludis*, the ribosome content is relatively invariant with the growth rate in catabolism-limited cells.

2.2.5 Facilitated Diffusion and Active Transport of Catabolic Substrates

Transport of low molecular mass substrates into a cell for energy conservation via specific transporters is often limiting the rate of microbial growth. On the other hand, diffusion of catabolic substrates to the cytoplasmic membrane is only rate-limiting at very low substrate concentrations. Thus, the catabolic substrate concentration around the cells is often that of the growth medium.

The semipermeable membrane is composed of lipids and proteins, the latter being involved in transport of molecules into and out of the cells, in biosynthesis of, e.g., cell wall constituents and in energy conservation. Generally, only relatively small molecules (<2000 Da) can be transported from the external environment into a microbial cell, which is why proteins, polysaccharides, and nucleic acids are partially hydrolyzed to oligopeptides, oligosaccharide, or oligonucleotide, respectively, before their transport.

In microbes, there are three types of enzymatic systems known for transport of low molecular mass compounds: a) **facilitated diffusion**, e.g., for glycerol or, in some microbes, glucose uptake, which increases selectively the permeability across the cytoplasmic membrane for a specific compound by *permeases*, equilibrating concentration inside and outside. The driving force for facilitated "transport" is simply the concentration difference between outside and inside, where the substrate is consumed; b) **active transport** leads to a higher concentration inside a cell relative to outside and, therefore, requires metabolic energy. The energy can be derived from ATP hydrolysis, such as in ABC transporters, or from $\Delta\mu H^+$ or $\Delta\mu Na^+$ as in sym-/antiporters. An important variation of active transport is c) **group translocation** as found in uptake of carbohydrates in some enteric bacteria or Clostridia, where uptake is coupled to a simultaneous covalent activation of the carbohydrate. In case of glucose uptake by the phosphoenolpyruvate (PEP) transferase system, glucose transport is coupled to transfer of a phosphoryl group from PEP to glucose yielding glucose-6-phosphate (see Chap. 9).

One way a microbe can increase the rate of uptake is to increase the number of transporter proteins in the cytoplasmic membrane. However, an increase in transporter proteins comes at a significant cost as it infringes on the membrane proteome space by competing with other membrane-associated functions, such as respiration in respiring microbes.

2.3 Microbes and the Chemistry of Their Environment

Many of Earth's environments, that are inhabited by microbes, have strikingly different total compositions of chemical elements resulting in significant metabolic consequences. For example, some soils are generally deficient in copper, cobalt, molybdenum, or selenium; parts of the open ocean are relatively deficient in iron (Pacific) or phosphate (Atlantic), and hot springs are deficient in bioavailable molybdenum resulting in the prevalence of tungsten-containing enzymes instead. There are environments without O_2 and others without light. Also, environments can vary greatly in pH and temperature. Thus, the elemental composition and properties of the environment shape the composition and function of a microbial cell.

2.3.1 Chemical Composition of the Prokaryotic Cell and Trace Element Requirements

Table 2.2 provides an approximated elemental composition of a prokaryotic cell. About half of a cell's dry mass is carbon, followed by ~20% O, ~8% H, and ~14% N. The percentages vary to some extent between cells growing at different rates and in different growth phases, the largest differences being between cells with storage materials (starch, polyhydroxybutyrate, or

Table 2.2 Approximate composition of chemical elements and their oxidation state of a prokaryotic cell. The values can vary depending on growth conditions. See text for details

Element	% dry weight	Oxidation state
C	50	0, variable
O	20	−2
H	8	+1
N	14	−3
P	3	+5
S	1	−2
K	1	+1
Na	1	+1
Mg	1	+2
Ca	Not always present	+2
Trace elements <1		

polyphosphates) and those without. The values in Table 2.2 are useful when designing a growth medium to obtain a desired amount of cell mass. While O and H are acquired mostly from H_2O, other elements need to be supplied in a growth medium at appropriate levels.

The relative elemental composition of a dry microbial cell can be simplified and approximated to a theoretical chemical formula: $\mathbf{H_8O_{1.25}C_{4.17}N \approx H_7O_2C_5N}$. The apparent molecular mass of the latter formula is 113 g/ "mol". Considering the invariant oxidation state in biomass of H, O, and N, the average carbon in cell biomass is slightly negative (i.e., reduced, ~−0.6). Carbon in lipids is more reduced relative to carbon in nucleotides. In the following discussion, the abbreviation $\mathbf{<CH_2O>}$ is used to indicate that the oxidation state of carbon in biomass is simplified as zero and, therefore, formally at the level of the carbon of formaldehyde (thus the parenthesis).

Most prokaryotes also contain Na^+ and/or Ca^{2+}, however, these elements may not be essential for growth. All prokaryotes are dependent on the presence of the trace element (Table 2.3) zinc (Zn^{2+}), although the required zinc concentration

Table 2.3 Some functions of trace elements that are generally components of growth media. %, abundance of metals in the earth's crust

Trace Element(TE)	Function
Zinc (Zn) 0.0070%	Zinc is essential for all living organisms. The non-redox active transition metal is a cofactor of over 300 enzymes and other proteins
Iron (Fe) 5.63%, but not always bioavailable	Iron is essential for most living cells, exceptions being lactic acid bacteria and *Mycoplasma*. Most abundant iron proteins are heme proteins and iron-sulfur proteins
Cobalt (Co) 0.0025[a]	B_{12} and coenzyme-B_{12} dependent enzymes; *methionine-amino-peptidase*; *nitrilase*. *E. coli* can grow in the absence of Co and/or B_{12}
Nickel (Ni) 0.0084%	*Urease*; *[NiFe]-hydrogenases*; F_{430}-*hydrogenase*; *methyl-coenzyme M reductase*; Ni-*containing carbon monoxide dehydrogenase*; *acetyl-CoA synthase /decarbonylase*; *Ni-containing superoxide dismutase*. Ni has no apparent function in animals and many prokaryotes
Copper (Cu) 0.006[a]	Many oxidoreductases catalyzing reactions with O_2, e.g. *cytochrome oxidase*, *methane-*, *NH_3 monooxygenases*, *Cu-dependent superoxide dismutase*; also *NO- and N_2O reductases*. Cu is not required by strict anaerobes
Manganese (Mn) 0.095%	Oxygenic photosynthesis; *Mn-dependent superoxide dismutase*; *Mn-dependent catalase*. Mn probably not required by strict anaerobes
Vanadium (V) 0.012%	*V-nitrogenase*; *haloperoxidases* in fungi V only required for *nitrogenase* synthesis in the absence of Mo
Molybdenum 0.00012%[a]	*Mo-nitrogenase*; *nitrate reductase*; *formate dehydrogenase*; *formyl-methanofuran dehydrogenase*; *xanthine dehydrogenase/oxidase*; *Mo-dependent carbon monoxide dehydrogenase*. Mo not required by all prokaryotes
Tungsten (W) 0.000125%	W-enzymes instead of Mo-enzymes in thermophiles and hyperthermophiles but no tungsten *nitrogenase* or *nitrate reductase*
Lanthanum (La) 0.0039	Cofactor of *pyrroloquinoline-dependent alcohol dehydrogenases* including *methanol dehydrogenase*, in methylotrophs and even non-methylotrophs
Boron (B) 0.001%	Quorum-sensing signal containing boron in *Vibrio harveyi;* antibiotic boromycin formed by *Streptomyces antibioticus*. No B-dependent enzyme is known. Most prokaryotes do not require B for growth
Selenium (Se) 0.000005% [a]	Some *[NiFe]-hydrogenases*; some *formate dehydrogenases*; *glycine-reductase* *E. coli* contains a Se-dependent *formate dehydrogenase*. In the absence of Se, *E. coli* no longer converts formate to H_2 and CO_2

[a] Areas of deficiency in some parts of the world

in growth medium is so low that zinc contamination in other chemicals or glassware is often sufficient to enable growth. Almost all prokaryotes also require iron and that at relatively high amounts. The only prokaryotes known not to require iron are the lactic acid bacteria and most *Mycoplasma* species. Copper appears to be essential only in aerobic microbes and in denitrifying microbes when growing anaerobically. Requirement for selenium appears to be restricted to some anaerobically growing prokaryotes. Also, aerobically growing eukaryotes contain selenoenzymes.

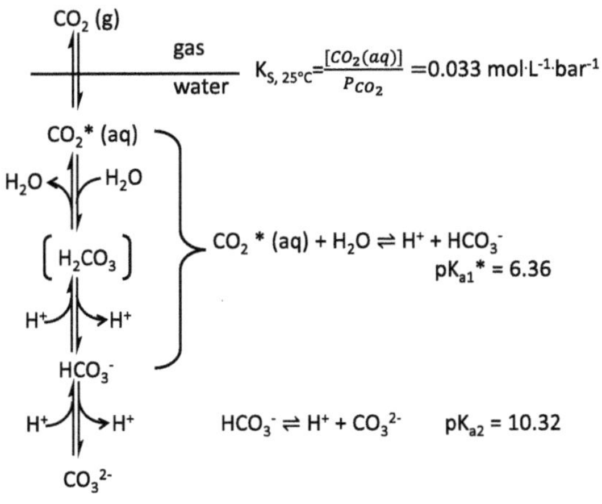

Fig. 2.4 CO_2/bicarbonate buffer in microbial ecosystems. In contrast to other aqueous buffers such as inorganic phosphate, the CO_2/bicarbonate buffer system extends from the aqueous to the gaseous phase. $pK_{a1}*$ is the apparent pKa of the overall reaction and dissociation of aqueous CO_2^* + H_2O to bicarbonate + H^+. Values for Henry's constant (Ks) and $pK_{a1}*$, pK_{a2} are at 25 °C. The CO_2/bicarbonate buffer is the relevant buffer in most microbial ecosystems, in particular in anoxic, organic rich environments. See text for details

2.3.2 CO_2 and pH

CO_2 is more than 30-fold more soluble in water compared to other gasses of biological importance such O_2, N_2, and H_2 (Henry constant, i.e., the gas concentration in solution at 1 bar gas partial pressure, for CO_s is 0.033 M bar^{-1} and in the order of 0.001 M bar^{-1} for the other gases at 25 °C). CO_2 concentration in the atmosphere is presently 0.042% and continuously increasing. Until 1760 when industrialization began with the use of fossil fuels, the steady-state CO_2 concentration in the atmosphere has been 0.028% for thousands of years. Annually, in the order of 260 Gt CO_2 is fixed by photosynthesis, and the anthropogenic release of CO_2 in 2021 was ~36 Gt. CO_2 is net assimilated by autotrophs and net produced by organotrophs, as well as used as electron acceptor in hydrogenotrophic methanogens and acetogens (see Chap. 10).

When dissolved in water, CO_2 partially reacts reversibly with H_2O to form carbonic acid, H_2CO_2, which reversibly dissociates to H^+ and HCO_3^-. An "apparent dissociation constant," $pK_{a1}*$, describes the overall aqueous dissociation:

$$CO_{2^*} + H_2O \rightleftharpoons HCO_3{}^- + H^+$$

where CO_2* = dissolved CO_2 + H_2CO_3 (Fig. 2.4). Because $pK_{a1}* = 6.36$, the HCO_3^-/CO_2* couple is an effective buffer at circumneutral pH. In particular, in anaerobic organic-rich environments, such as the mammalian gut, rumen of the cow, anaerobic digesters, or some sediments, CO_2 is released as catabolic end product and contained in an often closed gas atmosphere, so that the aqueous combined dissolved inorganic carbon (DIC = CO_2 + HCO_3^- + CO_3^{2-}) concentration can reach tens to hundreds of mM. Up to 40% of the off-gas of an anaerobic digester can be CO_2. The dissociation of HCO_3^- to CO_3^{2-}, $pK_{a2} = 10.32$, becomes relevant for buffering in alkaline environments.

Distilled water in equilibrium with the 0.042% CO_2 in the atmosphere has a DIC concentration of 13.9 μM and a pH = 5.7. The percentage of CO_2* (= dissolved CO_2 + H_2CO_3) dissociated in water into H^+ and HCO_3^- can be calculated via the Henderson–Hasselbalch equation:

$pH = pK_{a1}* + \log [HCO_3^-]/[CO_2*]$.

With $pK_{a1}*$ at 25 °C being 6.36, the numerical value of $\log([HCO_3^-]/[CO_2*])$ is - 0.66 or $[HCO_3^-]/[CO_2*] = 1/4.57$ for the pH to be 5.7.

Adjusting pH by a CO_2/Bicarbonate Buffer

How much pCO_2 and how much $NaHCO_3$ are required to adjust the pH, for example of an anoxic growth medium, to 7.2? The aqueous CO_2 concentration is determined by pCO_2 through the Henry constant K_S, ($[CO_2\ aq] = K_S \cdot P_{CO_2}$) as well as by the aqueous $NaHCO_3$ and bicarbonate concentration, $[CO_2\ aq] = [NaHCO_3] - [HCO_3^-]$. Using the Henderson–Hasselbalch equation, $[HCO_3^-] = \frac{Ka_1 * [CO_{2aq}]}{[H^+]}$ ($K_{a1}* = 4.45 \times 10^{-7}$ at 25 °C). It follows that $[CO_2\ aq] = [NaHCO_3] - \frac{Ka_1 * [CO_{2aq}]}{[H^+]}$ and $[CO_2\ aq][H^+] = [NaHCO_3][H^+] - Ka_1 * [CO_{2aq}]$. Thus, $[CO_2\ aq] = \frac{[NaHCO3][H^+]}{Ka_1 * + [H^+]}$, which is equal to $KS.P_{CO_2} = \frac{[NaHCO_3][H^+]}{Ka_1 * + [H^+]}$. So, for a desired pH = 7.2 and a pCO_2 of 0.1 bar, e.g., 1 bar pressure of 10% CO_2 in 90% N_2, or 2 bar pressure of 5% CO_2 in 90% N_2), $[NaHCO_3] = \frac{P_{CO2} * K_S (Ka_{1*} + [H^+])}{[H^+]}$, and, thus, 23.3 mM $[NaHCO_3]$ are required.

Thus, when aqueous CO_2* concentration is in equilibrium with atmospheric CO_2, 22% of the 13.9 µM CO_2* are dissociated into H^+ and HCO_3^-.

Prior to the industrial revolution, the average pH of the ocean was about 8.2. Today, the average ocean pH is about 8.1. This lower pH is not only due to the increase in atmospheric CO_2 concentration within the last 260 years but also because of acid rain and organic runoffs from continents that are converted to CO_2. The present pH of ocean water would even be lower if not for alkaline runoffs from the continents due to weathering of minerals. Oceans will remain a sink for atmospheric CO_2 until a pH of below 5.7 is reached (because with increasing atmospheric CO_2 the pH at equilibrium will decrease).

2.3.2.1 Other Environmental Parameter Affecting Microbial Fitness

Temperature: The average temperature of the oceans is 2 °C, and 18 °C in ocean surface water. The average body temperature of humans is near 37 °C. The temperatures in hot springs and deep-sea hydrothermal vents are 100 °C and higher. In all these environments microbes are found that are adapted to these different temperatures. Within limits, growth rates increase with increasing temperature until a maximum is reached and then decrease. There are microbes with growth temperature optima between 0 °C and 25 °C (psychrophiles), between 25 °C and 45 °C (mesophiles), between 45 °C and 80 °C (thermophiles), and with optima between 80 °C and 120 °C (hyperthermophiles). In general, at their optimum growth temperature, psychrophiles grow slower than mesophiles which grow slower than thermophiles which grow slower than hyperthermophiles; however, there are many overlaps. Below the optimum growth temperature, growth rate mostly decreases with a Q_{10} of slightly above 2, so that a decrease in temperature by 10 °C leads to a decrease in growth rate by a factor of slightly above 2. In case of hyperthermophiles, Q_{10} can be as high as 3. With a Q_{10} of 2.2, growth rate of a thermophile with a temperature optimum near 70 °C is predicted to decrease by a factor of $2.2^4 = 23.4$ when growing at 30 °C. Growth rate of a mesophile with a temperature optimum near 37 °C decreases by a factor of 10.6 at 7 °C and a factor of 23.4 at −3 °C, i.e., in the refrigerator.

Pressure: It was long assumed that life only occurs at atmospheric pressure (~1 bar). However, microbes living in the depth of oceans at 11 km or as low as 5 km below the continental surface are adapted to very high pressures. At 11 km deep in the oceanic water column, the pressure is 1100 bar (1 bar per 10 m). Interestingly, some microbes are dependent on the presence of high pressures (barophiles) for growth while others are not and can simply tolerate high pressure. For example, the doubling time of *Methanococcus jannaschii,* which was isolated from a deep-sea hydrothermal vent, decreases from 83 min at 7.8 bar to 18 min at 750 bar at the same temperature. In contrast, the hyperthermophile *Pyrolobus fumarii,* which was isolated from a hydrothermally heated black smoker at the Mid Atlantic Ridge at 3650 m depth showed no growth enhancement when incubated at 250 bar relative to 3 bar.

pH: Microbes also have an optimum pH for growth that can be as low as pH = 1 for some acidophiles, and as high as pH = 11 for some alkaliphiles. Independent of the pH optimum, cytoplasmic pH in different microbes remains near 7 in growing cells. Under growth conditions, cytoplasmic pH can deviate from 7 by one pH unit when extracellular pH is lower or higher from pH 7 by several pH units. The effect of external pH on internal pH is also dependent on the type of acids or bases present and was discussed above and in detail in Chap. 9. Lowering external pH, for example, by acetic acid, propionic acid, or butyric acid with p*K*a near 4.7 is more effective in lowering the cytoplasmic pH than HCl with a p*K*a near 0. At pH 5 these three carboxylic acids are mainly undissociated and can freely diffuse through the cytoplasmic membrane; HCL at pH 5, on the other hand, is fully dissociated and cannot as such diffuse through the cytoplasmic membrane. Thus, short-chain carboxylic acids can collapse the internal pH when the external pH is 5 and growth ceases; this cannot occur if the acid is HCl. This has an applied aspect in food biotechnology. Fermentation of food by microbes to acetic acid and lactic acid leads to collapse of the intracellular pH of spoilage microbes and, thus, prevents their growth. Stabilization of cabbage in form of Sauerkraut or Kimchi are other examples. Just acidification by HCl would not do the same job.

2.3.2.2 Water Activity

Water activity (a_w) is the partial vapor pressure of water containing solutes divided by the vapor pressure of distilled water at the same temperature. In distilled water, a_w is 1. Water activity decreases with increasing concentration of solutes. At very low solute concentration, the decrease is independent of the nature of the solute and only dependent on the number of molecules present. At high concentration, intermolecular interactions become important. For example, a 1 M NaCl solution has a water activity of 0.965 and a 1 M sucrose solution has a water activity of 0.955. At very low solute concentration, water activity is almost temperature independent. At high concentration, water activity increases with temperature because of a loss of interactions at higher temperatures.

Most microbes grow optimally at water activities above 0.98. For halophiles, growth optimum is at water activities as low as 0.95. Growth rate decreases with decreasing water activity, and growth completely ceases at a water activity in the range between 0.9 and 0.8 in prokaryotes and in fungi in a range between 0.7 and 0.6. This is one explanation for why, e.g., at a sufficiently low water activity in bread, fungi but not bacteria can grow easily. When growth ceases at decreasing water activity, cells typically do not die but continue to metabolize albeit at a very low rate. This is important because food that is preserved with high salt or sugar concentration is not sterile but contains transiently inactive microbes. This is also true for dried soil. As soon as the water activity is increased, e.g., by rain in soil or by increasing temperature in case of food, or both, microbes start growing again.

2.4 Environment-Specific Catabolic Fluxes

Energy resources are available to microbes in form of fluxes which are subject to environment-specific changes. Flux is defined as $J_S = \frac{dS}{dt'}$ where dS is the amount of a catabolic substrate(s) transferred through a unit area, dt the time, and $dt' = \text{dt} \times \text{unit area}$ (Fig. 2.1). The catabolic substrate is in most cases two catabolic compounds, an electron donor and an electron

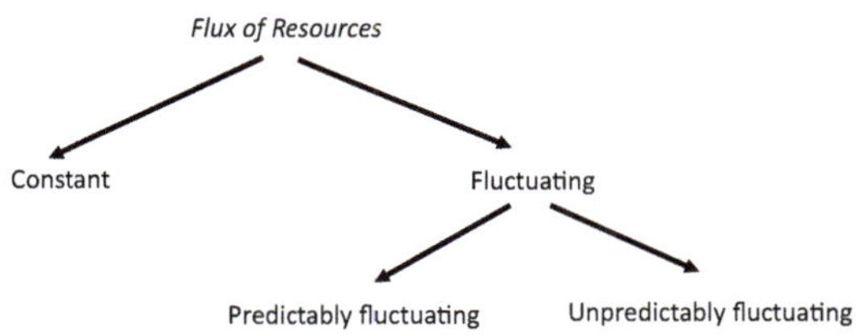

Fig. 2.5 Types of resource fluctuations in environments. See text for details

acceptor. For simplicity of the following discussion, such two substrates are treated as only one "S" to avoid confusion arising from stoichiometries. Also, the unit area through which the mass of *dS* flows is neglected at this point as it is always the same unit area of a cell. Fluxes and rates are discussed in Chap. 5 in more detail. Fluxes of catabolic substrates vary greatly in diverse natural and engineered environments, and the type of flux determines major ecophysiological characteristics, including metabolic pathway architecture and energy utilization.

2.4.1 Constant and Fluctuating Environments

Based on the change of fluxes of resources, microbial environments can be categorized into constant or fluctuating (Fig. 2.5). Fluctuating environments are distinguished further into predictably fluctuating and unpredictably fluctuating environments. In a chemostat environment for example, there is a constant flux of resources and a constant (ef)flux of new cells and metabolic end products (see Chap. 5 for more details). Under steady-state conditions, neither the concentration of resources nor of end products changes, and cellular metabolism simply transforms resources under constant flux conditions. Such chemostat environment was the basis of the Julian Adams experiment (Chap. 1). Some natural environments can be approximated as a "constant" or fed-batch, high flux environment such as the rumen of the cow with a semi-constant influx of metabolizable substrates (celluloses, hemicelluloses, pectins, and starch from grass) or a sewage treatment plant. In general, microbes that are adapted to constant, high flux environments have metabolic and physiological traits that are different from those found in microbes inhabiting constant, low flux environments, such as deep-sea sediments. High flux environments with competition between microbes for energy resources select for microbes with high metabolic rates and short, specialized metabolic pathways (see Chap. 5).

Biological Interactions in an Environment

The biological environment of a microbe includes all organisms, including phages that, in one way or another, affect the flux of resources toward a microbe via interactions. Common types of biological interactions are (1) mutualistic interactions, where two microbes benefit from an interaction with each other; (2) commensal, where only one microbe benefits from the activity of the other; (3) competitive, where both microbes compete for the same resource; and (4) predatory, where one microbe exploits the biomass of the other microbe as resource. In all cases, these biological interactions have a metabolic basis, and the outcome of these interactions affects microbial community structure. Microbiota (microbial community) in conjunction with metabolic fluxes to and within the community are often referred to as the "microbiome." Fast-growing microbes metabolizing at high rates are often found in environments with high fluxes. As a consequence, their biomass is increasing, which further increases the microbial rate of resource utilization and increases competition among kin, and, therefore, can alter the type of interaction prevalent in a microbial ecosystem.

Most natural environments are fluctuating, and the frequency and extent of fluctuation matters. If resources (and end product accumulation) change slowly, microbes tend to acclimate to a new flux based on their metabolic capacity and genomic potential. For example, phototrophic microbes in cyanobacterial mats in salt marshes acclimate to natural changes of light intensity. The microbial community composition and the populations of the individual microbes remain more or less unaltered during those fluctuations. On the other hand, if environmental resources change more rapidly than microbes can acclimate to, metabolic niches open or close leading to changes in microbial community composition based on selection, a process also termed "habitat filtering."

Adaptation Versus Acclimation

"Adaptation" and "acclimation" are often used synonymously, which is incorrect. In particular, "adaptation" is unfortunately often used to describe differential gene expression induced upon changes in an environment. Adaptation is defined as a permanent, genetic (mutational) change resulting in a new phenotype relative to that of the ancestor. For example, a loss-of-function mutation in a repressor controlling expression of a transporter gene may lead to an increased level of transporter protein resulting in an increased uptake of new resources and enable a cell to metabolize a substrate faster. This is an *adaptation* to a new environment. Acclimation is a temporary change in physiology, making use of the extant genotype and cellular wiring, and results in a different gene expression profile; it does NOT involve a genetic change. For example, up-regulation of the expression of degradative enzymes or a transporter protein in response to newly available environmental resources is an acclimation but not an adaptive response.

Fluctuations of resources can be predictable, such as the circadian light–dark cycle or the tidal cycle in the oceans, or unpredictable, such as surface soil being subjected to random episodes of rain, flooding, desiccation, and temperature fluctuations (Fig. 2.5). Other examples of unpredictable fluctuations include mechanical perturbations e.g., of soil such as mixing, plowing, or earthworm migration. For phototrophic microbes, light is the prime environmental energy resource, and the availability and intensity of light matter as much as the concentration of a chemical compound for a chemotrophic microbe. Fluctuations in light level can be predictable (circadian) as well as unpredictable (clouds). Temperature, although not an environmental resource but a key environmental parameter, controls fitness as well. In constant temperature environments, such as in Hot Springs, the physiology and metabolism of inhabiting microbes is adapted to prevailing temperature. This is different for soil or cyanobacterial mat microbes from which microbes cannot escape, and summer temperatures during the day can lead to well over 40°C, whereas the same microbe might be subjected in soil to sub-zero temperatures in the winter, and survive.

Adaptive evolution experiments show that fluctuating environments select for variant with deregulated catabolic pathway(s). This enables microbes to have simultaneously expressed catabolic pathways for several substrates regardless of their absence or presence. This bet-hedging trait, however, comes at the cost of a lower growth rate to accommodate the larger fraction of catabolic proteins in the proteome (see Chap. 5).

2.4.2 Unstructured and Structured Environments

An important physical aspect of an environment is whether it is physically structured or unstructured. Unstructured environments are typically aqueous environments where there is no physical support to allow for stable local interactions between microbes; resources are available to all microbes based on complete mixing of resources. This is not a contradiction to the above-discussed primary access to resources by diffusion because mixing here is macroscopic and includes microbes with their diffusional halo. Open oceans and other bodies of water are, in the first approximation, relatively mixed, except on a μm scale as discussed above. In unstructured environments, competition for environmental resources is global throughout the entire environment, because all microbes have access to all resources. Even in the presence of multiple energy resources (e.g., in the cecum of mammalian gut), diverse microbes compete among each other for the same resources that are accessible to all microbes via diffusion and mixing. If a finite amount of resources is present, an unstructured environment will be dominated in the end only by a few, well-adapted microbes. Once abundant resources are consumed, the overall environment contains only low genetic diversity. When the ecosystem is operated under constant and predictable conditions for a long period of time at relatively high-energy flux rates, metabolic speciation is likely to occur (see Julian Adams experiment).

In structured environments, such as soil or biofilms, microbes are physically "fixed" relative to each other, and global mixing is prevented. Resources are "privatized" from competitive use by other microbes. In other words, competition between microbes in a structured environment is local, because access to resources is local. This typically results in multiple, diverse, local gradients, and metabolically diverse microbes, that are adapted to multiple energy fluxes through such gradients, co-exist. Consequently, a structured environment maintains a higher genetic and metabolic diversity. If formation of structure is allowed in a system, this will select for metabolic interactions, such as in nitrification or in naturally occurring biofilms and flocs in sewage treatment plants and trickling filters. While microbes in unstructured environments may be well adapted to rapidly recruit resources, a structured environment contains a metabolically diverse range of microbes including specialists and generalists as well as fast and slow-growing microbes, including those adapted to a narrow range of gradients and low concentrations of resources.

Microbes, adapted to an environment with high, constant fluxes tend to be specialists, using a rather limited spectrum of catabolic substrates in short metabolic pathways and small genomes (e.g., rumen of the cow, *Lactobacillus*, *Propionibacterium*, syntrophs, methanogens). Microbes in unpredictably fluctuating, low flux environments such as in soil (e.g., *Pseudomonas*, *Nocardia*, *Flavobacteria*, *Mycobacteria* species) tend to have a larger number of extensive pathways for accessing diverse organic resources (electron donors, acceptors) and larger genomes.

2.4.3 Microbial Biofilms

Formation of biofilms is initiated by a single microbial cell or cluster of cells attaching to an organic or inorganic surface. Attached cells divide, driven by nutrient flux across the surface, and cells are retained locally in a structured environment.

In biofilms, microbes are spatially immobilized relative to a surface and to each other by an extracellular matrix consisting of polysaccharides, proteins, nucleic acids, as well as cell structures such as pili. The thickness of biofilms can range from a single cell layer to hundreds of μm. Catabolic substrates are often derived from soluble compounds in an aqueous phase around a biofilm. In other biofilms, microbes metabolize the substratum surface such as chitin of crustacea, or plant material, Fe^0 surfaces, or electrons from electrodes in engineered systems. In biofilms, mass transfer and microbial substrate consumption rates are different relative to an unstructured planktonic system as substrate concentrations are highly stratified in a biofilm resulting in metabolism-induced gradients and heterogeneity. For example, if the resource is a soluble compound in the aqueous phase, biofilm cells close to the substratum surface will be more substrate-limited than cells at the immediate interface to the bulk liquid. Thus, growth rate of cells deeper in the biofilm is lower, or even zero, contributing, e. g., to a persistence phenotype by some microbes in response to antibiotic treatment, which typically targets growing cells. In biofilms that formed on the surface of a catabolic substratum, growth activity can be higher for cells directly at the substratum surface.

Competition for a substrate in a biofilm is local and between cell populations in the immediate vicinity of each other. This results in steep local gradients of substrates and end products. Therefore, many different selective forces for diverse metabolic traits act on biofilm microbes. In an unstructured planktonic environment such as in a batch media test tube, all cells are mixed and competition is global, and slow-growing cells will be outcompeted by faster growing cells and lost. In biofilms, slow-growing cells are retained. Therefore, biofilms maintain and contain more genetic and phenotypic diversity because there are more diverse niches available.

Selection for biofilm formation is strong and often driven by adaptive mutations if the size of the seeding population is large. Multiple molecular mechanisms, such as secreted polymers (polysaccharides, protein, DNA), pili and other cellular appendages, lead to surface-specific or -unspecific, tight association of a cell to a surface. In laboratory environments, variants with increased adherence to a surface are rapidly selected for from microbial populations of a reasonable size. The strong selection for biofilm formation results from the fitness gain of cells obtaining constant exposure to a continuous flux of energy resources. In biofilms, even a (long-lasting) flux of very low, e.g., µM, concentration of a catabolic substrate can lead to slow but sustained substrate consumption and growth, while consumption of an equivalent substrate concentration in a batch culture would not result in any significant cell growth. This is the reason why trickling filters are used for removing of trace contaminants in treatment of wastewater to low drinking water level.

2.5 Fitness

The above discussed physical and chemical parameters of a microbial environment introduced some of the constraints on fitness to inhabiting microbes. Microbial fitness is the ability of a microbe to leave viable offsprings. Other traits such as to persist during resource limitation or extreme physical/chemical perturbations are important as well; however, microbes ultimately need to grow and generate viable offsprings. Fitness is effectively illustrated by the analogy of a "fitness landscape" to describe the relationship between a microbe's genotype or phenotype and its relative fitness in a given environment (Fig. 2.6). A microbe's genotype is the sum of all genes in its genome as they encode functions that, when coordinately expressed, result in an observable fitness phenotype. A single microbial genotype can be expressed as a number of (related) phenotypes, which enable a microbe to acclimate to changes in substrate fluxes.

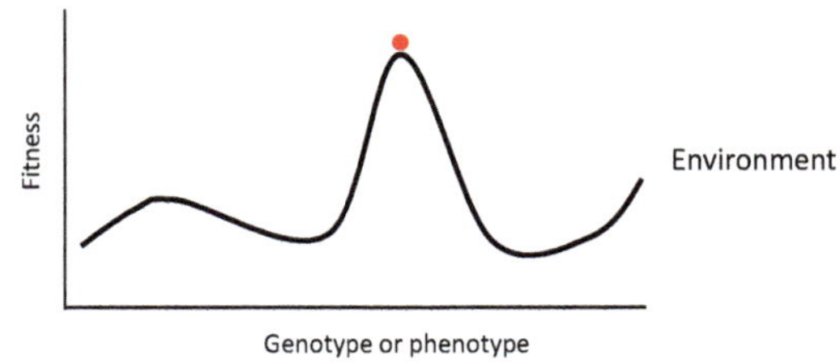

Fig. 2.6 The fitness landscape as a visual metaphor for microbial fitness and natural selection. For a given environment with a given flux of catabolic substrate(s), each microbial genotype expresses a phenotype that has a certain fitness relative to another genotype. The red dot indicates a genotype or phenotype, whose fitness is optimally adapted to environment 1. See text for details

Borrowed from concepts of macro-ecology, the landscape analogy is to indicate that for a given microbial environment (here, a specific flux of substrate(s)), there is a defined and distinct phenotype conferring a maximal fitness (Fig. 2.6). Fitness and genotype scales are relative, indicating a fitness with respect to related, i.e., "neighboring" genotypes/phenotypes.

It is generally assumed that extant microbes are well adapted to and populate a fitness peak in their respective natural environment (Fig. 2.6). During million years of evolution, fitness peaks have been "climbed" via adaptive evolution. Most catabolic pathways are well-fixed in microbes, because energy resources are mostly recurrent.

The fitness landscape analogy is useful to describe that fitness differences exist (1) between two closely related microbes in a single environment and/or (2) in a single microbe when in two different environments. For example, the facultative pathogen *Vibrio cholerae* has both a host-associated lifestyle during acute infection of the human gut and a lifestyle metabolizing crustacea-derived chitin. Both habitats collectively lead to the microbe's lifestyle to produce viable offsprings, and there is a fitness landscape for each environment. *V. cholerae* navigates multiple environments so that the overall ecological success depends on each environment and the time spent in each.

In addition to the flux of energy/substrate, dispersal and population size contribute to fitness. How dispersal can shape the metabolic and physiological strategy of a microbe is particularly apparent, for example, in pathogenic microbes causing transmissible infectious diseases, where a microbe's virulence is often directly correlated to the rate of its transmission between hosts. A strain of *Vibrio cholerae*, causing cholera, evolves to be more virulent when the transmission rate is high. A high virulence induces the host to produce more resources for the pathogen at the trade-off of its own viability (and death). This is beneficial to the pathogen when transmission rate is high because by the time a host is severely diseased, the pathogen has shed from the old, and migrated to a new, uninfected host. Conversely, low transmission rates tend to select for less virulent pathogens as it is more advantageous for the pathogen to only slightly compromise the host, so that the host is infectious for a longer period of time when transmission is uncertain. Strains adapted to a high transmission rate have faster and less efficient metabolism than those adapted to slower transmission rates. The density of the host population (e.g., cities vs rural areas) and the population size of the pathogenic microbe are important factors. For example, large, undirected dispersals are offset by a large microbial population size and by a finite range of metabolizable compounds in the environment. Moreover, many well-adapted microbes have mechanisms, such as adhesion, that aid in persisting in the environment and enable access to continuous resources (see Biofilms Sect. 2.4.3 above). Finally, the ability to survive under conditions of starvation is an important aspect as well, although this capacity does not directly generate viable offsprings; it prevents cells from dying so that they will be able to multiply again when resources become available again. Thus, microbes do not need to be constantly at the peak of each fitness landscape in their respective diverse environments, but in at least one, where they reproduce to a population size that is greater than the loss in others.

2.6 Selection and Drift

Natural selection is the systematic sorting through microbial genotypes/phenotypes and biasing a microbe's survival based on its fitness to a particular environment. Positive selection increases the frequency of microbes metabolizing a resource; the metabolic phenotype of the microbe being selected for reflects competitive traits of how a resource (electron donor, electron acceptor) is used considering the dynamics of its flux in the microbe's physical-chemical environment. Natural selection inherently involves competition between individuals for a resource.

Selection can be strong or weak. In the fitness landscape analogy, weak selection favors broad fitness peaks, whereas sharp peaks represent strong selection (Fig. 2.7). In most natural settings, selection is weak and maintains closely related microbes of similar fitness traits at a broad peak. On the other hand, in laboratory environments, in particular in "unnatural"

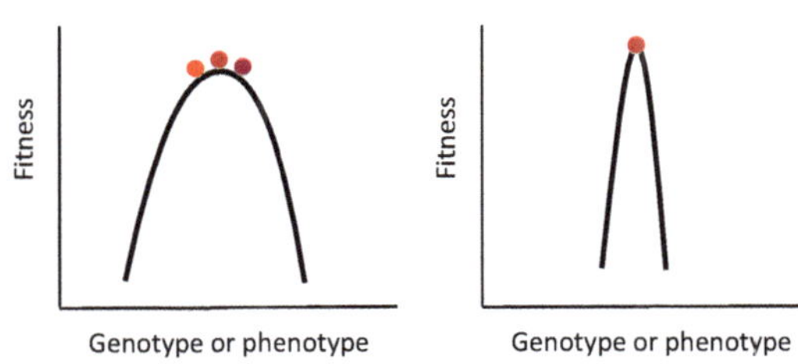

Fig. 2.7 Weak and strong natural selection. Weak selection (left) maintains multiple genetically closely related microbes with similar fitness at close to the fitness peak, whereas strong selection (right) results in loss of populations with only slightly different fitness

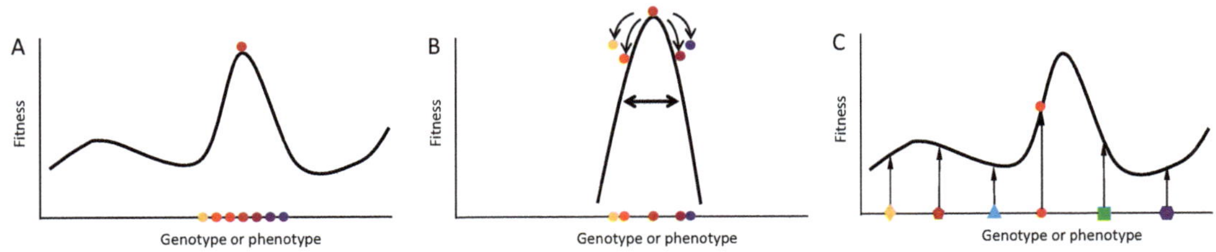

Fig. 2.8 Natural selection. (**a**) Evolutionary selection "searches" through naturally-generated, random mutants in a microbial population (same symbol but different color shades), that is well-adapted to a fitness peak (maximal fitness, strain as red circle). (**b**) Negative selection as a force for sharpening the genetic coherence of a population adapted to a given environment. Most point mutations are neutral or detrimental, leading to a decreased fitness phenotype of such variants in a particular environment. Less fit lineages are lost over time by competition. (**c**) Ecological selection (habitat filtering) selects from a pool of existing, genetically diverse microbial strains/species for those that have the highest fitness in a given environments. See text for details

environments created by experimentalist, selection for novel phenotypes can be very strong. For example, strong selection for antibiotic resistance or horizontal gene transfer in the laboratory results in a few variants while closely related genotypes without such specific trait are lost (Fig. 2.7). As outlined below, the arising of genetic variants in the Julian Adams experiment is an example for rare, but strong (and highly visible) positive selection.

In addition to selection, drift also biases the survival of members of a population, however, in the opposite way as selection. Drift is the change in gene frequency due to a random process, often by chance. Drift is inversely proportional to the population size N ($D \propto \frac{1}{N}$), meaning that drift is maximal when the population size is small, and lowest if the population size is 1 individual. In other words, when processes affect very few cells, for example by a sampling procedure in the laboratory, drift rather than selection determines a cell's survival. The transfer of colonies on agar plates as discussed below is such example (see below, Fig. 2.8b). Also, a single microbe (i.e., genotype) settling randomly and by chance onto an environmental surface in the absence of other microbes (i.e., competition) and colonizing this surface to form a biofilm is an example of survival based on drift. The random event explains the survival of this particular genotype in this environment. The metabolic and physiological traits of these microbes do not need to be anywhere close to a fitness peak (i.e., microbes do not need to be particularly well adapted); cells survive and grow simply because of the absence of better competitor! Once this single cell grows and produces viable offsprings that remain physically close to the parental cell, competition and, therefore, positive selection kick in and shape the composition of the microbial genotype/phenotypes based on the initial random event.

2.6.1 Evolutionary Selection

Evolutionary selection acts on individual cells in a microbial population containing naturally (or artificially) generated genetic variants (Fig. 2.8a). "Clonal" populations contain cells with point mutations at a frequency of about one in 10^{-6}–10^{-8} per gene and per generation. Thus, a clonal population of 10^9 cells contains a few hundred cells with random single point mutations. Because most microbes are adapted to and exist in a specific environment, most mutations have a neutral or slightly deleterious phenotype. Natural selection in natural environments is relatively weak but omnipresent, because the parental strain is already at the fitness peak. The arising of a large gain in phenotype is rare.

Extant microbes are well adapted to and populate a fitness peak of their respective environment (Fig. 2.8a). The reason for this is that catabolic resources have been largely the same over the last several hundred million years. Catabolic pathways for using this finite spectrum of naturally occurring organic and inorganic substrates have been selected for and maintained. Microbes adapted to some of the most impactful "catastrophic" long-term biogeochemical events, such as the arising of the oxidized inorganic compounds SO_4^{2-}, Fe^{3+}, NO_3^-, and O_2 from photosynthetic processes. These inorganic compounds became major new electron acceptors, and with that, new fitness landscapes and peaks emerged that are well occupied by now. For microbes at the fitness peak, there is no mutational benefit for increasing fitness further as microbes are already well adapted, and acclimation is sufficient to mitigate environmental fluctuations. Thus, strong selection plays only a minor part in shaping and maintaining the genotype/phenotype of extant microbes in nature. Anthropogenic compounds and pollutants, on the other hand, were only recently introduced into the environment, and strong selection shapes the catabolic pathways for their degradation. This may only be true partially, as a surprising plethora of organic anthropogenic "pollutants," such as the chlorinated ethenes, tetrachlorethene, or vinyl chloride, are formed also naturally, however at very low levels. Thus, an apparent de novo evolution for degradation of an anthropogenic organic compound can very well reflect the presence of already evolved metabolism that has been unknown or is present only at very low frequency.

Population Dynamics

Population dynamics is based on fitness differences within a population. How long, i.e., how many generations, does it take, for example, for a beneficial mutation to sweep through a population and come to fixation. The following equation (courtesey Bruce Levin) for a population calculates the number of generations (t) required for a specific change in gene (mutant) frequency.

$$t = \frac{1}{s} \ln \frac{p_t(1 - p_0)}{p_0(1 - p_t)},$$

where s is the selection coefficient, p_0 the frequency of the beneficial allele in the population at the beginning, p_t its final frequency. If the selection coefficient is 0.10, for example equivalent to a fitness gain of 10% (which is high), and the frequency of this allele at t_0 is 10^{-4} and the final frequency is 0.99, then 138 generations are required.

2.6.2 Negative, or Purifying, Selection

Negative selection reduces the frequency of microbes carrying a deleterious allele. In a microbial population that is well-adapted to a fitness peak, purifying selection sharpens the fitness peak by eliminating the (slightly) deleterious variants from the population, resulting in a genetic cohesion of the most fit genotype in a particular environment (Fig. 2.8b). Microbial species can, therefore, be defined as a group of genetically related (clonal) microbes that conserve genome information to occupy the same fitness peak. The arising of microbial species is a consequence of natural selection, and purifying selection is primarily a force for conservation of a species. While purifying selection is abundant, adaptive evolution by positive selection is rare.

2.6.3 Ecological Selection

Selection in an ecological context, sometimes also referred to as habitat filtering, "searches" through the fitness space of microbes; however, the pool of microbes being "searched through" consists of genetically diverse microbial species. Consider, for example, the structured environment of soil with its thousands of local fitness peaks, most of which are populated by metabolically diverse microbes containing diverse genotypes. When using a gram of soil to isolate, let's say, an aerobic glucose-degrading microbe using laboratory-based liquid isolation techniques (see below), ecological selection "searches" through the diverse genotype space of the microbial community of the soil sample for that microbe that has the highest fitness to the laboratory (aerobic glucose utilization) environment (Fig. 2.8c). This genotype does not have to be at the fitness peak of this environment; it just needs to be more fit relative to all other glucose-degrading microbes in the soil and enrichment sample!

Microbial Populations Versus Microbial Communities

The terms microbial "population" and "community" are often used interchangeably at the expense of an accurate description of the underlying biology. Microbial populations described genetically closely related, "clonal" microbes where individual cells may carry, on rare occasions, a few mutations relative to each other. On the other hand, microbial communities contain genetically different microbes, including their populations, with clearly distinguishable metabolism, physiologies, and phenotypes and niche. Most environmental samples will contain a community based on the presence of genetically and phenotypically different microorganisms, as well as populations within each of these phylogenetic units.

2.7 Examples of Selection and Drift

Natural selection, when in action in natural environments, is largely invisible because positive selection is weak and the loss of lineages via purifying selection remains mostly unnoticed. Nevertheless, this is where most of selection occurs in Nature. Laboratory experiments with strong selection can, however, make natural selection visible. The following examples illustrate the omnipresence of strong selection in laboratory experiments.

2.7.1 Selection in the Julian Adams Chemostat Experiment

The Julian Adams experiment was introduced in Chap. 1 to illustrate speciation and metabolic specialization as a function of environmental flux of resources to explain general patterns of metabolism in extant microbes. It also illustrates how strong laboratory selection can result in metabolic speciation. *E. coli* is believed to occupy a niche in the mammalian gut for tens of thousands of years and is presumably well adapted to that (fluctuating) environment. The Small Colony Variant (SCV) and the Large Colony Variant (LCV) arose by mutation(s) from the parental population; the fitness landscapes of the evolved genotypes and steps of adaptive evolution are displayed conceptually in Fig. 2.9. With its single peak, the first landscape (A) represents the fitness of *E. coli* in the organic-rich environment in the mammalian gut (Environment 1, fluctuations in electron acceptor and organic electron donors). Landscape B represents the glucose-limited chemostat environment 2, and landscape C is the fitness landscape of the glucose-limited chemostat environment containing additionally the newly developed niche acetate (Environment 3).

While the ancestral strain is well adapted to the "natural" environment (Fig. 2.9a), it is not at the fitness peak in the chemostat environment (Fig. 2.9b) (otherwise it would not evolve). Since this Environment 2 existed for 800 generations, the small colony variant was positively selected for in this environment because the variant is slightly more fit in Environment 2 than wild type. This initial fitness advantage was further amplified by accumulation of additional beneficial mutations, making this variant lineage "climbing up" to the peak (Fig. 2.9b). Ancestral wild-type genotype was lost from the population in the chemostat environment because the evolved metabolic features of SCV enabled it to outcompete wild type for glucose. From a

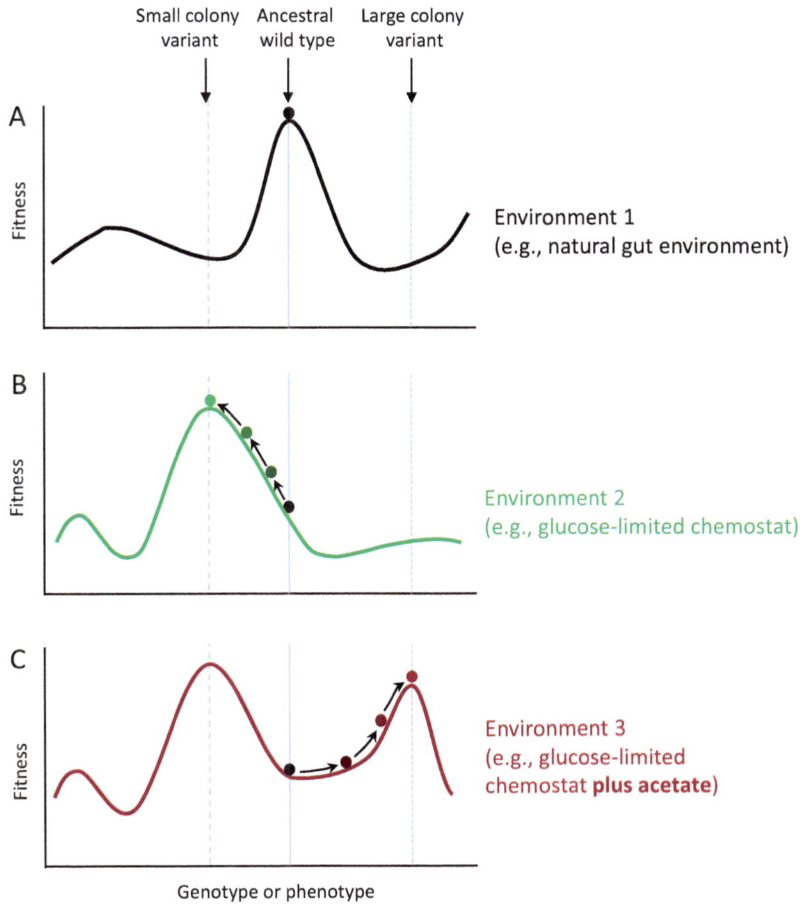

Fig. 2.9 Simplified fitness landscape for visualizing selection and adaptive evolution in the Julian Adams experiment (actual genealogy is a bit more complex). (**a**) Theoretical fitness landscape of the ancestral *E. coli* "wild-type" strain with a maximal fitness in the natural gut environment. (**b**) The glucose-limited chemostat environment (Environment 2) is different from the gut environment with respect to physical-chemical composition, available metabolic substrate, and fluxes thereof, resulting in a different fitness landscape including another fitness peak, which is not occupied by wild type. In Environment 2, the Small Colony Variant (SCV) arose by mutation and positive selection for utilizing glucose more rapidly and at lower concentration (i.e., climbing up and occupying the fitness peak of environment 2). Multiple sequential mutations were acquired by SVC to move up toward the fitness peak (green trajectory). As a consequence of the altered metabolism of SCV enabling this variant to climb the unoccupied peak, SCV released acetate into the chemostat environment, thereby creating Environment 3 (i.e., niche creation). (**c**) In Environment 3, the new acetate niche and new fitness peak is climbed and occupied by Large Colony Variant (LCV), another variant that evolved via multiple mutations to gradually occupy the new acetate peak. SCV and LCV co-exist as they do not compete for the same substrate, but ancestral *E. coli* wild type is outcompeted by SCV for glucose. For simplification, it is assumed that SCV and LCV are at the fitness peak in their respective environments in which wild type is less fit and outcompeted. See text for details

metabolic point of view, the adaptive changes resulting in this fitness increase were due to (1) a faster glucose metabolism, (2) utilization of glucose at lower concentration, and (3) an incomplete utilization of the resource glucose and concomitant efflux of acetate, which became available as the new resource in Environment 3. The latter environment contains an additional fitness maximum (niche) that was subsequently occupied by the LCV also via adaptive evolution. By metabolic specialization to acetate, LCV escaped competition with wild type and SCV.

Acquisition of this new metabolic trait was traded off in LCV with a loss of ancestral glucose utilization (Fig. 2.9c). Often, an increase in fitness in one environment is associated with a fitness loss (cost) in a different environment. As Fig. 2.9 indicates, the ancestral genotype is maladapted to Environments 2 and 3, and the SCV genotype is less fit in Environment 1. It seems reasonable to assume that SCV and LCV would be less competitive in the gut environment relative to the ancestral strain.

2.7.2 Selection During Isolation of Microbes in Pure Culture

Isolation of microbes in the laboratory into pure, axenic culture has been key to our current understanding of microbial biology including metabolism and physiology. "Model" microbes, such as *E. coli*, *Pseudomonas sp.*, *Vibrio sp.*, *Clostridia sp.*, or methanogens have led to foundational discoveries in metabolism, biochemistry, molecular biology, signal transduction, population genetics, and ecology. Most of these microbes were isolated into pure culture and studied because they grew rapidly and robustly in laboratory growth media. Thus, most science advances are primarily based on studying fast-growing microbes. This raises the questions: For what specific traits did the employed isolation methods actually select for, and why have we been obtaining predominantly microbes with fast-growing phenotypes?

Two strategies involving different selections have been in use for isolating aerobic and, with some modification, anaerobic prokaryotes in pure culture: growth in liquid enrichment cultures or on agar plates. The following example illustrates the intrinsic selection bias associated with each platform. Let's assume we want to isolate an aerobic, glucose-oxidizing microbe from soil. A 5 ml liquid enrichment culture contains a sterile minimal medium with an excess of the catabolic resource glucose (e.g., 10 mM); anabolic resources are not limiting. Composition of such growth medium with high substrate concentration is typically guided by the desire of the experimentalist to observe visible growth as turbidity. At a cell density of about 10^6–10^7 cells per ml, growth-dependent turbidity is observable. The test tube is well aerated to introduce the electron acceptor O_2. When such liquid enrichment is inoculated with a soil sample, ecological selection "searches" through the genotype/phenotype space of the inoculum for that microbe that has the best fitness in this particular (laboratory) environment (Fig. 2.10). Because the glucose concentration chosen is nonphysiologically high and because the 5 ml liquid test tube is an unstructured environment, competition for glucose is global. Such enrichment scheme inherently selects for those microbes that can recruit the resource glucose the fastest (Fig. 2.10). Growth typically ceases if a key nutrient becomes limiting or because of accumulation of inhibitory end products, including pH, in the batch culture. Once growth is observed in the first liquid culture (let's assume to a density of ~10^8 cells/ml), a sample of that enrichment is microscopically inspected and transferred (5–10% (V/V) inoculum) into a second sterile enrichment test tube.

It may take several serial transfers until the culture is sufficiently enriched and dominated by one or two species. With increasing liquid transfers, which translates into an increasing population size as well as number of generations grown, the resulting microbial population is increasingly shaped by evolutionary selection (adaptive evolution) and less by the initial ecological selection. Each liquid transfer with, e.g., a 2% inoculum and growth is equivalent to ~5.6 generations. Thus, in a prolonged isolation procedure, you might evolve the microbes that you want to study rather than isolate such strain as present in nature!

Single clonal colony isolations by direct agar plate isolation are usually conducted for aerobic isolates from a liquid enrichment (Fig. 2.10b) or as a strategy for direct isolation of microbes (Fig. 2.10c). An agar plate is a structured environment in the sense that it prevents mixing and global competition. For isolations from a highly-enriched liquid culture (Fig. 2.10b), a small droplet containing, e.g., less than 50 viable cells is plated on the agar plate. This is a bottleneck step

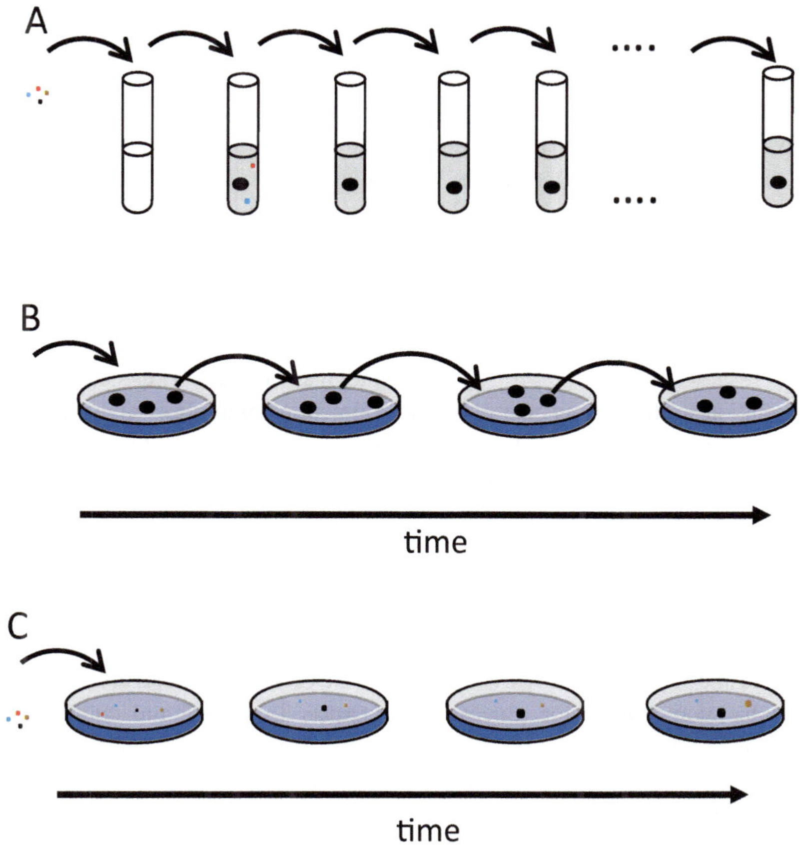

Fig. 2.10 Ecological and evolutionary selection during isolations of microbial strains in the laboratory by (**a**) liquid enrichment or (**b**, **c**) agar plates. An example inoculum contains numerous different microbial species, indicated as dots of different color (**a** & **c** left edge; dot size and color indicate abundance and diversity of different genotypes.) In (**a**), the environment of the serial liquid dilution positively selects for the faster growing, black-dotted microbes and counterselects against the other colors; those lineages disappear because competition for resources is global. With increasing number of serial transfers, ecological selection becomes less relevant in lieu of evolutionary selection in terms of shaping the genotype of variants of "the isolate." (**b**) Serial agar plating and isolations of individual microbial strains as colonies from restreaked colonies. Evolutionary selection and drift shape the outcome of this isolation protocol. (**c**) Direct isolations of single cells (colonies) from environmental samples (multiple-colored dots). The sample is plated directly at low cell density to allow non-competitive, clonal growth on an agar plate, if cells can grow at an agar–air interface. Ecological selection initially selects locally for growth, and because competition is not global, even slower growing colonies can co-exist on the plate and appear after some period of time. As colonies grow to higher cell density, the increasing population in each colony is subjected to evolutionary selection, which selects for variants of "the isolate" with a better fit to the environment. See text for details

with significant drift because only a very small population (50 out of 10^8 cells) was selected randomly to survive. The agar plate is incubated, and, once visible, a single colony is picked, randomly again (drift is back!!), and streaked out aseptically onto a fresh agar plate. A colony is assumed to comprise a clonal population of ~10^8 cells (*E. coli*), seeded by a single cell. During growth from this single cell, local competition increases rapidly as sibling cells accumulate around the seeding cell in this structured environment, sharpening the fitness peak of the agar plate fitness landscape. The population structure of the colony is governed by selection although the single cell that seeded that colony originated by drift. Typically, a colony is picked and streaked again for single colonies on a fresh plate. Here again, the combination of drift (which clone is picked) and evolutionary selection (for the fittest genotype in the growing colony) controls the outcome. Continued propagation by colony transfers includes intrinsically a strong bottlenecking of a culture. Because of this, laboratory studies with microbes need to start from freezer rather than from continuously propagated cultures.

A second isolation strategy using agar plates is direct isolation, where an environmental sample is plated directly onto the surface of a selective agar medium (Fig. 2.10d). High glucose concentration as sole catabolic substrate in the agar plate favors rapidly growing microbes, which become quickly visible as colonies (because the experimentalist may be too impatient to wait). However and in contrast to the liquid culture enrichment, there is no global competition for glucose as mixing is prevented and competition on the agar surface is local. Microbes with the same catabolic property but different substrate utilization kinetics, for example, can co-exist on the agar surface if they are prevented from spatially interacting competitively with each other. Colonies of fast-growing strains will be visible after a few days, whereas slow-growing colonies might appear after weeks. Thus, an experimentalist may miss slow-growing microbial strains, if the plates are discarded after the first colonies appeared!

2.8 Other Selection Acting During Microbial Isolations

In addition to the strong selection for utilizing a catabolic substrate, other important selective forces act during the liquid → plate transfer sequence and shape the metabolic features of the isolated microbe:

Selection of aerobic or aero-tolerant glucose-fermenting microbe. Although the liquid test tube enrichment discussed above was designed for selecting for aerobic glucose-oxidizing microbes, solubility of oxygen from air at 25 °C is only 240 μM O_2 (with the higher solubility in a low salt, freshwater medium). As introduced in Sect. 2.2.3, the poor solubility and diffusion of O_2 in aqueous solution relative to high specific rates of microbial O_2 consumption renders many aerobic cultures O_2-limited. Even with optimum aeration, catabolic oxidation of glucose will be limiting considering the stoichiometry of $C_6H_{12}O_6 + 6O_2 \rightarrow 6CO_2 + 6H_2O$, low solubility of O_2, and the slow mass transfer from the gas phase. Thus, aero-tolerant glucose-fermenting microbes might also be selected for (by ecological selection) in the liquid enrichments. The air–liquid interface, where O_2 concentration is higher than inside the bulk liquid, might positively select for aerobic glucose-oxidizing variants that form some floating pellicle to maintain such variant population close to the O_2 at the air–liquid interface.

Selection for substrate and end product tolerance. In both liquid culture and agar plates, there is an intrinsic selection against microbes that do not tolerate the nonphysiologically high substrate concentration of a laboratory medium. In most natural environments, transportable catabolic substrates are present only at very low μM steady-state concentration, because compounds are rapidly consumed, while the flux through this low concentration might be very high. Thus, in those environments, there may never have been any positive selection for tolerance of high substrate concentration. Similarly, in most natural environments end products often do not accumulate. End product accumulation and potential inhibitory effect in this particular enrichment environment (see Sect. 2.3.2 discussion on short-chain fatty acids acetate, butyrate, lactate) may select for those few microbes that are able to tolerate.

Selection for stationary phase metabolism and rapid exiting from lag phase. Since the first steps in the above liquid enrichment experiment were serial transfers, each growth cycle in a test tube ends in a non-growth (stationary) phase where cells spend some time under non-growth conditions including at high concentration of potentially inhibitory end products, etc. Within a population or microbial community, there will be some variation in how well a lineage tolerates stationary phase. Positive selection will favor those strains where exposure to those conditions is less detrimental than to others. Moreover, the timing of the transfers chosen by the experimentalist (periodically, e.g., every 24 or 48 h, or unpredictably after a long weekend away from the lab) determines the tolerance of the resulting isolate to stationary phase conditions. Such selection shapes the physiological profile of the finally microbial isolate.

Upon transfer to a new test tube, stationary phase-adapted cells need to resume growth, and the time period (lag phase, see Chap. 5) that cells need before growth resumes matters significantly. A "head start" of a subpopulation initiating growth by escaping faster from stationary phase provides a selective advantage to this lineage: there is less competition for the high substrate concentration in the new test tube, because other cells are slower to acclimate.

Selection for resistance to surface tension, aerotolerance, and to plating. Plating of cells on an agar surface creates surface tension stress for cells at the air/water interface. Thus, the plating procedure inherently selects for those microbes that can tolerate the experimentally-induced surface tension stress and counterselects for those that can't. To bypass such negative selection, cells are sometimes diluted into a "soft agar" overlay or in anaerobic agar-shake tube cultures. Similarly, single cells on an aerobic agar surface are exposed to high pO_2 (20% O_2) and may not tolerate such oxidative environment which they would rarely encounter in their natural environment. Finally, the procedure of plating, which involves a Drigalski spatula, glass beads, a pipette tip, or some other device to disperse a volume of 50–100 μl sample evenly on the plate may be counterselective as cells might adsorb onto the material. Alternatively, traces of the ethanolic disinfectant or other means to sterilize the device might be inhibitory to growth.

Selection for temperature tolerance. Temperature is highly selective/counterselective both in liquid enrichment cultures and agar plate enrichments. This is particularly important when isolating microbes from environments with fluctuating temperatures, such as soil, but less so from relatively constant temperature environments, such as hot springs or deep-sea sediments.

Obviously, there is much more to known and unknown selective forces acting during isolations. The discussion above, however, illustrates how pervasively present natural selection is during isolations and in seemingly mundane microbiological procedures. Moreover, it shows how ecological and evolutionary selection act in concert in shaping the metabolic and physiological phenotype of the microbe being isolated. A common "wisdom" among microbiologists (and geneticists) is that "... you get what you ask for, but sometimes you do not know what you ask for." It might also explain how we arrived at the current "model microbes" that have been well studied subsequently, and what that "model" stands for.

Further Reading

Bergthorsson U, Andersson DI, Roth JR (2007) Ohno's dilemma: evolution of new genes under continuous selection. Proc Natl Acad Sci U S A 104(43): 17004–17009

Copley S (2015) An evolutionary Biochemist's perspective on promiscuity. Trends Biochem Sci 40(2)

McGuffee SR, Elcock AH (2010) Diffusion, crowding & protein stability in a dynamic molecular model of the bacterial cytoplasm. PLoS Comput Biol 6(3): e1000694

Purcell EM (1977) Life at low Reynolds number. Am J Phys 45:3–11

Schulz HN, Jørgensen BB (2001) Big bacteria. Annu Rev Microbiol 55:105–137

3 Microbial Energetics

Contents

The original version of the chapter has been revised. A correction to this chapter can be found at https://doi.org/10.1007/978-3-031-28218-8_11

A. M. Spormann, *Principles of Microbial Metabolism and Metabolic Ecology*, https://doi.org/10.1007/978-3-031-28218-8_3

3.1 Introduction

Resources that microbes use from the environment are chemicals (matter) derived from the elements C, N, S, P, H, O, relevant metals and non-metals, and energy. Energy is used in metabolic processes to produce viable offsprings. The energy can be chemical energy as in chemotrophs (Fig. 3.1) or light energy as in phototrophs. How cells access and utilize available energy and how this energy is used to produce new cells define a microbe's physiology and ecology. Available useful energy is conserved and stored via **catabolic pathways** in form of adenosine triphosphate (ATP), while **anabolic** (= biosynthetic) metabolism and macromolecular syntheses as well as maintenance metabolism consume ATP (Fig. 3.1). ATP is the key cellular energy carrier and couples catabolic and anabolic pathways.

This chapter provides some very specific examples, which are introduced in more detail in later chapters. The intent of using these examples here is to emphasize the importance and show early the pervasive implications of thermodynamics on metabolism and ecophysiology. Thus, examples that appear to be too specific and seem to lack background might need to be re-read after a more systematic introduction in later chapters.

3.2 Change in Gibbs Free Energy, (ΔG); in Brief

Biologically useful energy is released or consumed in chemical reactions. The useful chemical energy is quantified as the change in **Gibbs free energy, ΔG**, which is the focus of this section. In equilibrium thermodynamics of chemical reactions operating under reversible conditions, Gibbs free energy is that fraction of energy that can be used maximally for work. Although most living systems are not in a state of equilibrium and proceed irreversibly (life is irreversible!), the use of ΔG to describe a cell's energy metabolism nevertheless enables to make quantitative predictions about energy flow at the level of

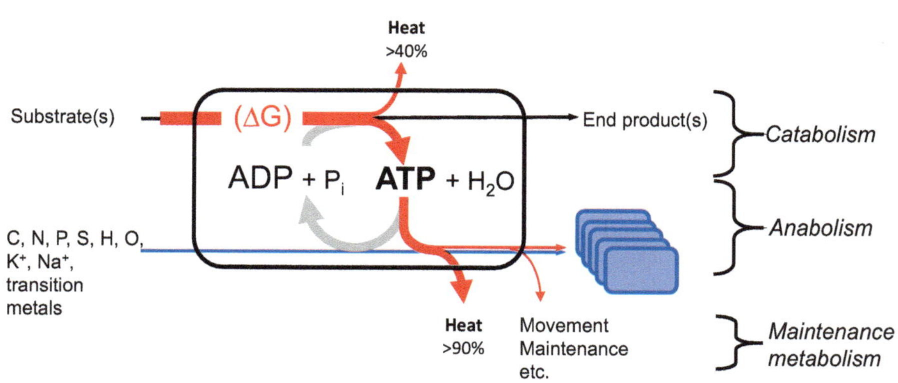

Fig. 3.1 Microbial use of chemical energy during energy-limited growth. Red lines show the approximate flow of energy (ΔG). Catabolic reactions and pathways convert environmentally available energy into ATP, while anabolic metabolism consumes ATP for assimilation of C, N, H, O, P, and other elements into new cells. The thermodynamic efficiency of converting available energy into ATP is variable and depends on the value of $\Delta G_{S \rightarrow P}$ of the actual catabolic reaction and is here assumed to be 60%. Thus, ~40% of the energy retrieved from the substrate(s) by catabolism escapes as heat. Notably, the thermodynamic efficiency of ATP synthesis is higher in very slowly growing, obligate syntrophic microbes (near 80%) and lower in very rapidly growing microbes (as low as 20%). Energy consumption in anabolism is only efficient by about 10% ($\Delta G_{ATP \rightarrow ADP} = -50$ kJ/mol). This low thermodynamic efficiency is mainly due to protein synthesis consuming most ATP in a growing cell. Four ATP equivalents are consumed per peptide bond ($\Delta G^0 = +15$ kJ/mol) formed, and protein constitutes about 50% of the microbial dried cell mass (see Chap. 5). As the environmental consequence of metabolism and growth, end products as well as new cells are released in the environment. When anabolism rather than catabolism is limiting, heat production increases, resulting in decreased growth yield (g cells (dried mass) per mol substrate). Abbreviation: Pi, inorganic phosphate. See text for details

individual reactions, cells, microbial communities, and ecosystems.

First and second law of thermodynamics: Equilibrium thermodynamics—for non-chemists generally a horror—when applied to microbial systems is much simplified because changes in temperature, pressure, and volume during metabolism can be neglected, because a microbial system essentially operates at isothermal, isobaric, and isochoric conditions. Microbes generally live in aqueous environments, in which these parameters are constant at the timescale of observation.

In thermodynamic terminology, microbes are "open systems" that can exchange with its environment energy in the form of either heat and/or work. If, however, the environment is part of and included in the entire thermodynamic system, then the system of microbes plus their environment can be treated as an "isolated" system, which does NOT exchange energy outside of that system. Therefore, the thermodynamic laws apply, namely that **energy (work + heat) in the system remains constant** (1. Law of thermodynamics) and that **work in the system can be converted to heat**; however, **heat cannot to be converted to work** (2. Law of thermodynamics). As an important consequence of the second Law and the isothermal conditions of microbial life, **the only energy that is useful to a microbe is work, ΔG.**

Exergonic and endergonic reactions, ΔG^0, and $\Delta G^{0\prime}$. Let's assume a simple reversible aqueous reaction where a single substrate is converted to a single product: A $\leftrightarrows$ B. In equilibrium thermodynamics, the useful energy (work) released in a chemical reaction under reversible conditions is referred to as the change in Gibbs free energy, ΔG, with the unit kJ per mol substrate (A) converted to product (B). Reactions that release useful energy under reversible conditions have a negative ΔG and are referred to as **exergonic**. When the change in Gibbs free energy of the reaction is positive ($\Delta G > 0$), then the reaction is **endergonic** and requires the input of useful energy to proceed from the left to the right. This applies both to single reactions and to entire metabolic pathways. In an overall pathway converting substrate(s) S in a sequence of reactions to product(s) P, the ΔG values of the individual pathway reactions are **additive** and equal to the $\Delta G_{S \rightarrow P}$ of the overall pathway reaction. In order for a pathway to be catabolic, $\Delta G_{S \rightarrow P}$ needs to be sufficiently negative (exergonic) to drive the endergonic synthesis of ATP from ADP and P_i (see below).

The free energy change of the aqueous reaction A $\leftrightarrows$ B depends on the concentration of A, [A], and B, [B], but not on the concentration of water, if H_2O is a reactant or product, which remains essentially constant near 55 M. As intuitively expected, a reaction is more exergonic (ΔG is more negative) if [A] is high and [B] is low and vice versa. [A] and [B] at 1 M concentrations in 25 °C water, or, if A or B is a gas, at a partial pressure of 1 bar are defined as **standard conditions**. The free energy change when 1 mol of substrate is converted to 1 mol of product under standard conditions at 25 °C during this conversion is referred to as $\boldsymbol{\Delta G^0}$.

Many metabolic reactions involve protons and are pH dependent. The intracellular proton concentration in all living cells is generally near 10^{-7} M (pH ~ 7) at 25 °C rather than 1 M (pH = 0, i.e., standard conditions). Therefore, the free energy change under **"physiological standard conditions"** (all substrates and products 1 M **except** protons at 10^{-7} M) is given by $\boldsymbol{\Delta G^{0\prime}}$, where the "prime" indicates 10^{-7} M H^+ (see Sect. 3.3).

Concentration dependence. When the concentration of substrate and product is different from 1 M, the free energy change associated with the reaction under such non-standard conditions can be calculated by:

$$\Delta G = \Delta G^0 + RT \ln \frac{[B]}{[A]}.$$

where [B] and [A] are the prevailing, "actual" concentrations (more accurately "activities,"), R is the Gas constant (8.314 J/mol K), and T is the absolute temperature in Kelvin (298.15 K = 25 °C). The term $RT \ln \frac{[B]}{[A]}$ indicates the deviation in Gibbs free energy from standard condition. Thus, a reaction or a pathway can be exergonic at substrate and product concentrations prevailing in cells or the environment while under standard conditions they might be endergonic. For example, $\Delta G^{0'}$ of conversion of butyrate to two acetate and $2H_2$ by syntrophic bacteria is +48.3 kJ/mol (see Chap. 9). At very low in vivo H_2 concentration, $\Delta G'$ is negative by ~−15 kJ/mol, which is sufficient to drive the phosphorylation of 0.2 ATP (see below). However, these catabolic pathways are slow because of the endergonic $\Delta G^{0'}$ (see above).

ΔG^0 and the equilibrium constant K_{eq}. ΔG^0 can be calculated from the equilibrium constant K_{eq} according to

$$\Delta G^0 = -RT \ln K_{eq}$$

where R is again the Gas constant (8.314 J/(mol K), T the absolute temperature in Kelvin, and K_{eq} the equilibrium constant of the reaction, and the unit is kJ/mol. Substituting ln$X$ with 2.3 lg$X$, the equation simplifies to

$$\Delta G^0 = -5.7 \lg K_{eq}$$

at 25 °C = 298.15 K in mesophilic environments. The equilibrium constant at a given temperature is calculated from [A] and [B] in equilibrium: $K_{eq} = \frac{[B]_{eq}}{[A]_{eq}}$. If K_{eq} is >1, then ΔG^0 is negative; if K_{eq} is <1, then ΔG^0 is positive.

K_{eq} can also be calculated from the rate constants k_{+1} of the forward reaction divided by the rate constant k_{-1} of the reverse reaction: $K_{eq} = \frac{k_{+1}}{k_{-1}}$. The forward rate is $v_{+1} = k_{+1}[A]$, and the reverse rate is $v_{-1} = k_{-1}[B]$. Thus, **ΔG^0 and the rate (v) of a reaction are related** via the above equation and $\Delta G^0 = -RT\ln K_{eq}$ (see also Haldane equation, Sect. 5.4.2). When $\Delta G^0 = 0$, K_{eq} is equal to 1 and k_{+1} and k_{-1} are identical ($k_{+1} = k_{-1}$), the net rate of conversion of A to B under standard conditions is zero ($v_{+1} - v_{-1} = 0$). **Thus, in thermodynamic equilibrium, there is no net flux.** When $\Delta G^0 < 0$, K_{eq} is >1 and k_{+1} is $>k_{-1}$, the reaction proceeds under standard conditions from the left to the right at a rate determined by the ratio of k_{+1}/k_{-1}. In other words, the larger k_{+1} is relative to k_{-1}, the higher is the net forward rate of a reaction under standard conditions. This does not mean that the absolute rate is high, since the equilibrium constant only makes a prediction about the ratio of k_{+1}/k_{-1} rather than about the absolute value of k_{+1}. These relations hold true regardless of whether or not the reaction is enzyme-catalyzed (catalysts do not affect the equilibrium constant!).

Enthalpy and entropy. As discussed above, ΔG is the energy form work that is released or required, when 1 mol A is converted to 1 mol B under reversible conditions. To predict the total energy change (work + heat) associated with the reaction under **irreversible conditions** the following Gibbs–Helmholtz equation is used:

$$\Delta G = \Delta H - T\Delta S,$$

where ΔH is the enthalpy change in kJ/mol, T the temperature in Kelvin, and ΔS the change in entropy in kJ/mol. The change in enthalpy ΔH is the energy released as heat, when the reaction proceeds **under completely irreversible conditions**. It is equal to ΔG when $\Delta S = 0$. When $\Delta S \neq 0$, which in most chemical reactions is the case, then ΔH differs from ΔG. ΔS reflects the change in heat capacity, unit kJ/(mol K), when 1 mol A is converted to 1 mol B. When the number of products in a stoichiometrically balanced equation exceeds the number of substrates, ΔS generally increases and, vice versa, decreases when the number of products is smaller than that of the substrates. ΔG of such reactions is generally strongly temperature-dependent. This is of importance when ΔG values are extrapolated to thermophilic or psychrophilic environments from ΔG at 25 °C for metabolic predictions. Most ΔG^0 values reported in the literature are given for 25° C = 298.15 K.

In microbial systems, the energy that is not used for work is irreversibly lost as heat (Fig. 3.1) because under isothermal conditions, heat cannot be converted back to work (2nd Law of thermodynamics). This irreversibility, however, is biologically not "useless" since it provides a directionality of metabolic fluxes through pathways as outlined below. The energy form heat in the most naturally occurring microbial processes generally does not result in a temperature change, as almost all microbial transformations occur at dilute concentrations in water and low cell densities. Under circumstances, however, where this heat cannot be dissipated rapidly, development of reaction heat is significant. For example, during microbial composting of organic matter and in industrial microbial fermenters at high cell density and high substrate flux, significant heat is generated. Many industrial reactors need to be cooled rather than heated!

3.3 Calculating ΔG of Metabolic Reactions

Section 3.2 discussed how ΔG^0 and $\Delta G^{0\prime}$ can be obtained for a single-substrate-to-single-product reaction (e.g., catalyzed by *epimerases*, *isomerases*, *mutases*), which are relatively rare in metabolism. Reactions with more than one substrate and product are much more frequent and discussed below.

Let's first consider the following general multi-substrate and multi-product chemical reaction:

$$a\mathrm{A} + b\mathrm{B} \rightleftharpoons c\mathrm{C} + d\mathrm{D},$$

where A and B are the reactants, C and D are the products, and *a*, *b*, *c*, and *d* are the stoichiometric coefficients balancing the equation. The change in Gibbs free energy, ΔG, of the above reaction is given as:

$$\Delta G = \Delta G^0 + RT \ln \frac{[\mathrm{C}]^c[\mathrm{D}]^d}{[\mathrm{A}]^a[\mathrm{B}]^b}$$

The term $RT \ln \frac{[\mathrm{C}]^c[\mathrm{D}]^d}{[\mathrm{A}]^a[\mathrm{B}]^b}$ accounts for the actual concentration(s) (more accurately "activities,") of reactants [A], [B], and products [C], [D], representing deviations from standard conditions and can be simplified to $5.7 \lg \frac{[\mathrm{C}]^c[\mathrm{D}]^d}{[\mathrm{A}]^a[\mathrm{B}]^b}$ when $T = 298.15$ K.

ΔG^0 can be obtained in four ways: (1) from the concentration of reactants and products in equilibrium, (2) from the enzyme's v_{max} and K_M of the forward and the reverse reaction, (3) from the free energy change of formation of compounds from the chemical elements, and (4) from redox potential differences, ΔE^0.

The first two methods are generally used to determine ΔG^o of an individual metabolic reaction. The third method is used mostly for assessing ΔG^o of overall catabolic pathways in the context of (1) cellular energy allocation including growth yield, (2) thermodynamic efficiencies, and (3) predicting the extent of energy conservation (n_{ATP}). The fourth method is used to define the electron flow in catabolic pathways considering that most catabolic reactions are redox reactions. The difference in redox potential, ΔE^0, also allows (1) to predict electron fluxes in microbial ecosystems and the involvement of particular metabolic types of microbes, (2) to scrutinize individual redox reactions in the context of in situ concentrations, as well as (3) to assess cellular fluxes of electrons via cellular electron carriers.

3.3.1 From the Equilibrium Constant K_{eq}

As introduced above, ΔG^0 of a reaction can be calculated from the equilibrium constant K_{eq} ($\Delta G^0 = -RT \ln K_{eq}$), which is determined from the equilibrium concentrations $[\mathrm{A}]_{eq}$, $[\mathrm{B}]_{eq}$, $[\mathrm{C}]_{eq}$, and $[\mathrm{D}]_{eq}$. Calculating K_{eq} from equilibrium concentrations provides reliable values only if K_{eq} is between 100 (exergonic reactions) and 1/100 (endergonic reactions) because of the difficulty in measuring accurately substrate and product concentration, when their concentration ratios $\frac{P_{eq}}{S_{eq}}$ are higher or lower.

3.3.2 From Kinetic Constants

Another, less frequently used, method to determine ΔG° is by making use of the relationship between the rate of the forward and reverse reaction of enzyme catalyzed reactions by using the Haldane relationship:

$$K_{eq} = \frac{V_f K_{M_r}}{V_r K_{M_f}}$$

where V_f is the maximal rate in the forward direction, V_r, the maximal rate in the reverse direction, and K_{Mf} is the K_M in the forward direction, and K_{Mr} the K_M in the reverse direction (see Chap. 5 for details). By experimentally determining V_{max} and K_M values for both the forward and reverse of an enzyme-catalyzed reaction, K_{eq} can be calculated. This method can be used if K_{eq} cannot be determined from the substrate and product concentrations at equilibrium.

3.3.3 From the Free Energy Change of Formation

Calculating ΔG^o of the reaction

aA + bB $\rightleftharpoons$ cC + dD makes use of the relationship:

$$\Delta G^0 = \sum \Delta G^o_{(f\text{Products})} - \sum \Delta G^o_{(f\text{Substrates})}$$
$$= \left(c\Delta G^0_{\text{fC}} + d\Delta G^0_{\text{fD}}\right) - \left(a\Delta G^0_{\text{fA}} + b\Delta G^0_{\text{fB}}\right)$$

where $\Delta G_f{}^0$ is the free energy of formation of compounds from the elements under standard conditions and 25 °C. The relationship is a direct consequence of the first law of thermodynamics. Consider the example of the catabolic reactions:

$$\text{glucose} + 6O_2 \rightarrow 6CO_2 + 6H_2O$$

The free energy of formation under standard conditions ($\Delta G^0{}_f$ at 25 °C) of the substrates and products from the elements is:

Glucose, −971.22 kJ/mol;

O_2, 0 kJ/mol;

CO_2, −394.359 kJ/mol and

H_2O, −237.178 kJ/mol

(glucose is 1 M aqueous solution and CO_2 and O_2 in the gaseous state at a partial pressure of 1 bar). Thus at 25 °C, when glucose is respired aerobically according to the above equation, −2872 kJ ($\Delta G^{0\prime}$) are released per mol glucose.

This method is generally applied to obtain ΔG^0 for entire pathways.

Values of $\Delta G_f{}^0$ at 2 °C, 25 °C, and 100 °C can be obtained from various published tables and databases, including one in Table A.1 in the Appendix at the end of this book. These temperatures are relevant because 2 °C is the average temperature of ocean water at which many marine microbes live; 25 °C is the standard temperature; 70 °C is the near optimum growth temperature of many thermophiles; and 100 °C is the growth temperature optimum of many hyperthermophiles near the boiling point of water at 1 atm (≈1 bar). The importance of considering the free energies of formation at these different temperatures is shown in the following three examples:

$$4H_2 + CO_2 \rightarrow CH_4 + 2H_2O$$
$$\Delta G^{0\prime} = -140\text{kJ/mol at } 2\,^\circ\text{C};$$
$$\Delta G^{0\prime} = -131\text{kJ/mol at } 25\,^\circ\text{C};$$
$$\Delta G^{0\prime} = -100\text{kJ/mol at } 100\,^\circ\text{C}$$

$$4H_2 + 2CO_2 \rightarrow \text{acetate}^- + H^+ + 2H_2O$$
$$\Delta G^{0\prime} = -110\text{kJ/mol at } 2\,^\circ\text{C};$$
$$\Delta G^{0\prime} = -95\text{kJ/mol at } 25\,^\circ\text{C};$$
$$\Delta G^{0\prime} = -44\text{kJ/mol at } 100\,^\circ\text{C}$$

$$\text{Acetate-} + H^+ \rightarrow CO_2 + CH_4$$
$$\Delta G^{0\prime} = -30\text{kJ/mol at } 2\,^\circ\text{C};$$
$$\Delta G^{0\prime} = -35\text{kJ/mol at } 25\,^\circ\text{C};$$
$$\Delta G^{0\prime} = -56\text{kJ/mol at } 100\,^\circ\text{C}$$

The differences in $\Delta G^{0\prime}$ at different temperatures matter significantly because they are in the range of the energy required for ATP synthesis (see below). This temperature dependence can explain, for example, why acetogenesis from H_2 and CO_2 is less observed at temperatures above 70 °C (ΔG at pH_2 $<10^{-3}$ bar becomes positive! See Chap. 10) and why reversed acetogenesis is mainly found in habitats with temperatures above 70 °C (ΔG at pH_2 $<10^{-3}$ becomes negative). However, the temperature dependence does not explain why methanogenesis from acetate is not observed at temperatures above 70 °C although the reaction becomes more exergonic at increasing temperature. The most compelling explanation is that methanogenesis from acetate evolved only when the ambient temperature decreased to below 70 °C assuming that the first living cells on Earth were hyperthermophiles.

Calculations of ΔG^0 from the free energies of formation are unambiguous when the p*K*a of acids or bases involved is at least one pH unit different from the pH of interest. However, determining a ΔG^0 is more problematic when the actual pH is within the range of the respective p*K*a values. For example, at circumneutral pH, reactants such as inorganic phosphate, ATP, aqueous carbon dioxide, and others have p*K* values in the range of 7. Thus, at circumneutral pH these reactants exist as mixtures of chemical species, and the equilibrium of the reaction is not represented accurately by a single chemical reaction. Under those circumstances, standard transformed Gibbs energies have to be used (see further readings in this chapter) that are not discussed further here.

3.3.4 From Redox Potentials

Most catabolic reactions are oxidation-reduction (redox) reactions. For example, NADH is oxidized via the aerobic respiratory chain and O_2 is reduced:

$$\mathrm{NADH} + \mathrm{H}^+ + 0.5\,\mathrm{O}_2 \rightleftharpoons \mathrm{NAD}^+ + \mathrm{H_2O}$$
$$\Delta \mathrm{G}^{0\prime} = -220\ \mathrm{kJ/mol\ NADH}.$$

The $\Delta G^{0'}$ of this reaction is calculated from the **difference in redox potential, $\Delta E^{o\prime}$**, of the two redox couples ($NAD^+/NADH$; $E^{0\prime} = -320$ mV and $O_2/2H_2O$; $E^{0\prime} = +820$ mV) via the following equation:

$$\Delta G^{o\prime} = -nF\Delta E^{o\prime}$$

where *F* is the Faraday constant (96,485 coulombs per mol electrons or 96,485 J/mol V) and *n* is the number of electrons transferred per overall reaction (in the above example $n = 2$). $\Delta E^{o\prime}$ is the **difference in redox potentials, $\Delta E^{0\prime} = (E^{0\prime}_{Acceptor} - E^{0\prime}_{Donor})$** of the conjugated redox couples of each half-reaction under standard conditions and at pH 7. (Table A.2 in the Appendix contains $E^{o\prime}$ and E' values of microbially important conjugated redox couples).

When conditions deviate from standard, except for pH = 7, the redox potential E' of a conjugated redox couple (at pH 7) is:

$$E' = E^{o\prime} + \frac{RT}{nF}\ln\frac{[C_{\mathrm{ox}}]}{[C_{\mathrm{red}}]}$$

where $[c_{ox}]$ is the actual concentration of the oxidized form (in this example: NAD^+ for $E'_{NAD+/NADH}$ or O_2 for $E'_{O2/H2O}$), and $[c_{red}]$ the actual concentration of the reduced form (in this example, NADH or H_2O). Simplifying the above Nernst equation for conditions at 298 K, and including the Gas constant (8.3144621 J K^{-1} mol^{-1}) and Faraday constant (96,485 J mol^{-1} V^{-1}), and converting ln*X* −>2.3lg*X*, the redox potential of a general conjugated redox couple is

$$E' = E^{o\prime} + \frac{0.059}{n}\lg\frac{[C_{\mathrm{ox}}]}{[C_{\mathrm{red}}]}.$$

For simplification, RT2.3/F = 0.059 V is from here on rounded to 0.06 V.

For deducing an in situ redox potential, for example, of electron-carrying coenzymes or redox-active metabolites via the $\frac{[C_{ox}]}{[C_{red}]}$ ratio, it is useful to note a few "rules of thumb" that are based on metabolomics data and theoretical considerations (Table A.1 in the Appendix):

1. For intracellular electron- and hydride carriers, the sum of the concentrations of c_{ox} plus c_{red} is constant because of the confined intracellular pool (Table A.1 in the Appendix). Their intracellular $\frac{[C_{ox}]}{[C_{red}]}$ ratio deviates from 1 at most by a factor of $10^{\pm 2}$ (Change in redox potential of ±120 mV when $n = 1$ and ±60 mV when $n = 2$). Thus, E' of the Fd_{ox}/Fd_{red} couple at an $E^{0\prime}$ of −400 mV (Fd from *C. pasteurianum*) and $n = 1$ can be as negative as −520 mV, and E' of the $NAD^+/NADH$ couple at an $E^{0\prime}$of −320 mV and $n = 2$ was found to be as high as

−280 mV. E' of the NADP+/NADPH couple also at $E^0 = -320$ mV can be as low as − 380 mV in situ.

2. For redox-active metabolites, the sum of the concentrations of c_{ox} plus c_{red} is not constant. For metabolites undergoing a redox transition of $n = 2$, for example pyruvate ↔lactate at $E^{0\prime} = -190$ mV, the $\frac{[C_{ox}]}{[C_{red}]}$ ratio can deviate from 1 by a factor of $10^{\pm 3}$ (equivalent to ±90 mV). Thus, E′ of the pyruvate/lactate couple can be as low as −280 mV.
3. When the concentration of the oxidized form of redox-active metabolite (c_{ox}) remains constant and only c_{red} varies, the $\frac{[C_{ox}]}{[C_{red}]}$ ratio can deviate from 1 by a factor of $10^{-6 \text{ to } -8}$ (equivalent to +180 to +240 mV if $n = 2$). For example, as elemental sulfur (S^0) is essentially insoluble and, therefore, at a fixed "concentration," in situ E' of the S_o/H_2S couple ($E^{0\prime} = -$ 270 mV) can be as high as −90 mV (ratio = 10^{-6}). Similarly at a constant pH, E' of the $2H^+/H_2$ couple ($E^{0\prime} = -420$ mV) can be as high as −180 mV (ratio = 10^{-8}).

It is also important to note that the absolute intracellular concentration of redox active coenzymes involved in catabolism is, in general, higher than those used exclusively in anabolism. For example, the concentration of tetrahydromethanopterin (C_1 group carrier), methanofuran (formyl group carrier), and F_{420} (ox + red) is up to 100-fold higher in methanogens than in archaea using these coenzymes exclusively in biosynthesis (see Chaps. 6 and 10). The same is true for the C_1-carrier tetrahydrofolate (H_4F) (see Chaps. 6 and 10): its concentration is much higher in acetogenic bacteria and sulfate-reducers oxidizing acetate catabolically via the Wood–Ljungdahl pathway than in microbes that employ H_4F only in serine, purine, thymidylate, and methionine biosynthesis. In most microbes, NAD^+ is involved in both catabolism and anabolism, whereas $NADP^+$ is mainly involved in anabolism. Accordingly, in most microbes total concentration of NAD^+ plus NADH is higher than that of $NADP^+$ plus NADPH. Also, cellular content of ferredoxins and flavodoxins is much higher in microbes using these 1-electron redox carriers in catabolism than in those using them only anabolically.

An Intuitive Thinking About Reduction Potentials

One way to develop a quantitative intuition of "redox potential" is to think about it in terms of "affinity" of an atom or molecule toward electrons. If a redox couple has a low, **negative redox potential**, such as $E^{0\prime}_{H^+/H_2} = -420$ mV, this means that the reduced form, H_2, has a **high** tendency to give off its electron(s) and become oxidized (H^+); in other words, H_2 is a good electron donor. Conversely, H^+ at that potential has a very **low** tendency to accept electrons to become reduced. Similarly, a **positive redox potential** of $E^{0\prime}_{O_2/H_2O} = +820$ mV indicates that the oxidized form, O_2, has a **high** tendency to accept electron(s), so it becomes reduced (to H_2O); in other words, O_2 is a good electron acceptor. Thinking about $E^{o\prime}$ and $\Delta E^{o\prime}$ from such perspective is useful when predicting which half-reactions can be coupled by microbes to form an exergonic catabolic overall reaction. A positive value of $\Delta E^{o\prime}$ is equivalent to a negative $\Delta G^{0\prime}$ of an exergonic reaction (see Nernst equation). It is the **difference** between the redox potential of the electron acceptor and electron donor, $\Delta E^{o\prime}$, that is directly proportional to the driving force of the reaction.

Electrical Energy Conversion

With the increasing interest in microbial electrosynthesis, which is the use of cathodic electrons, either directly or indirectly, in microbial catabolism, it is useful to relate electrical units (Coulomb, Ampere, and Watt) to Joule. 96,485.34 C is the charge of 1 mol electron; 1 A = 1 Coulomb per second; 1 W = 1 Joule per second = 1 V × A. Thus, an electrical current of 1 A is equivalent to a flux of ~10.4 μmol electrons per second.

3.3.5 Calculating ΔG at Temperatures Other Than 25 °C

Because of the prevailing anthropocentric world view, standard conditions were set historically to a temperature of 25 °C and atmospheric pressure of 1 bar. However, microbes have been discovered that thrive outside such narrow range, for example in hot springs or the deep biosphere where the temperature can reach 120 °C and the pressure 1100 bar. Since the effect of temperature on $\Delta G^{0\prime}$ is much more pronounced than that of pressure, only the temperature dependence is discussed here.

For thermophilic (growth temperature optimum above 45 °C) and hyperthermophilic (growth temperature optimum above 80 °C) microbes, the free energy change associated with an overall catabolic reaction needs to be determined at temperatures above 25°C. In single-substrate to single-product reactions, the temperature dependence is given, in first approximation, by: $\Delta G^{o} = -RT \ln K_{eq}$, where T is the absolute temperature (at 25 °C, $T = 298$ K; at 100 °C, $T = 373$ K; at 2 °C, $T = 275$ K). When the number of substrates differs from the number of products, entropy changes need to be considered in addition. In the latter case, the free energy change has to be recalculated from free energies of formation. These are available for many (but not all) catabolic substrate and products at temperatures between 2 and 200 °C (see Table A.1 in the Appendix).

When dissociated acids are involved as substrate or product in catabolism, the temperature dependence of dissociation of water into H^+ and OH^- needs to be considered. Many bioenergetic calculations are performed at pH = 7 because this denotes neutrality at 25 °C and 1 bar. Neutrality is defined as that pH, where activities of H^+ and OH^- are equal. Because the dissociation constant for H_2O is temperature-dependent, pH representing neutrality also varies with temperature. Neutral pH decreases at P_{SAT} (saturation pressures for H_2O) from ~7.4 at 0°C to ~5.6 at 200 °C. Thus, the "prime" in $\Delta G^{0\prime}$ refers to a pH 7 only at 25 °C (Fig. 3.2)!

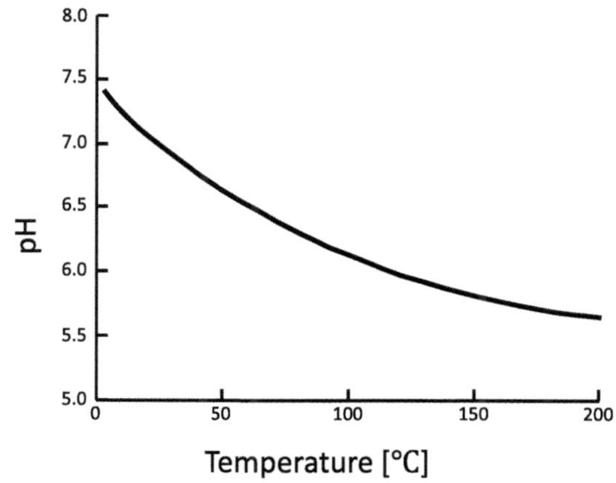

Fig. 3.2 Neutral pH as a function of temperature of the saturation pressures for H_2O. (From Amend and Schock 2001, with permission)

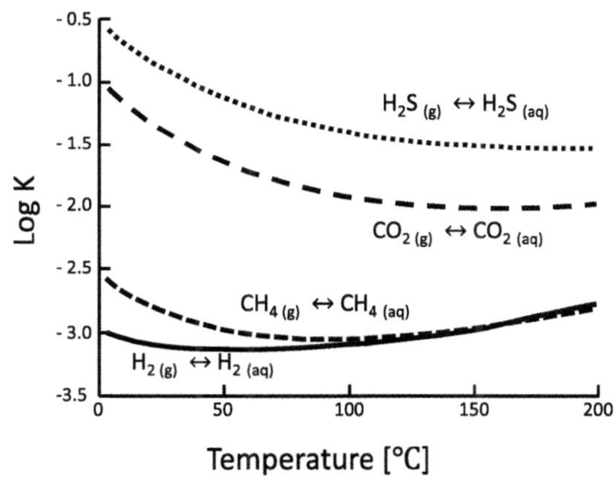

Fig. 3.3 Log K plotted against temperature at P_{SAT} for the solubility of gaseous H_2S, CO_2, CH_4, and H_2. P_{SAT} = saturation pressure for H_2O. K is the equilibrium constant for the gas in the gas phase and in solution. (From Amend and Schock 2001, with permission)

When gases are involved as substrates or products in catabolism, temperature dependence of their solubility in water needs to be considered as well. Only dissolved gases are the substrates of enzymes. Hydrogenotrophic methanogens grow on the gases H_2 and CO_2 and form as product the gas methane that is, in turn, substrate of methanotrophic archaea and bacteria. Temperature dependence of the solubility of the three gases versus temperature is displayed in Fig. 3.3. A key point to note is that solubility of H_2 gas decreases only moderately with temperature and increases again at higher temperatures. In fact, H_2 (g) is more than twice as soluble in water at 200°C than at 50 °C (at saturation pressure for H_2O at each temperature)!

3.3.6 Calculating ΔG of pH-Dependent Reactions at pH Other Than 7

Consider, for example, the ΔG of the H^+-dependent reaction of glucose conversion to lactate at pH 7:

$$1\ \text{Glucose} \rightleftharpoons 2\ \text{Lactate}^- + 2\text{H}^+$$

At circumneutral pH, lactic acid is deprotonated to lactate$^-$ + H^+ (p*K*a = 3.86). $\Delta G^{0\prime}$ of this reaction is calculated from $\Delta G^0{}_f$ as

$$\begin{aligned}\Delta G^{0'} &= \left(2\times\Delta G^0{}_{\text{f,lactate}} + 2\ \Delta G^{0'}{}_{\text{f,H}^+\text{at pH 7}}\right)\\ &\quad -\left(\Delta G^0{}_{\text{f,glucose}}\right)\\ &= [(2\times(-517.81)+2\times(-39.9)\\ &\quad -(-917.22)]\ \text{kJ/mol glucose}\\ &= -198.2\ \text{kJ/mol} \approx -200\ \text{kJ/mol}.\end{aligned}$$

If the above reaction proceeds at pH 8 rather than at pH 7, for example, and only the proton concentration is different, then the Gibbs free energy is $\Delta G = \Delta G^{0'} + 5.7\ \lg\ [H^+]^2$, where the $[H^+]$ concentration in the second term of the equation is now the **deviation (difference) from pH 7,** $H^+{}_{pH\ 7} \rightarrow H^+{}_{pH\ 8}$; $= 10^{-1}$. Consequently, at pH 8 $\Delta G^0{}_{pH8} = \Delta G^{0'} + 5.7\ \lg\ [10^{-1}]^2$. Since 2 protons are formed per mol glucose, the standard free energy of glucose fermentation to lactate at pH 8, $\Delta G^0{}_{pH8}$, $=-200$ kJ/mol + (-11.4 kJ/mol) = -211.4 kJ/mol.

If the above reaction proceeds at pH 3 rather than at pH 7, then the p*K*a of 3.8 of lactic acid has to be considered. At pH 3, most of the lactate is present in the protonated form:

pH = pK + log [Lactate-]/[Lactic acid], and the fermentation is

$$1\ \text{Glucose} \rightleftharpoons 2\ \text{Lactic Acid}$$

Therefore, in the calculation of ΔG^0 at pH 3, free energy change of formation of lactic acid rather than of lactate + H^+ is used.

pH dependencies are important in microbial processes because some microbes have their optimum growth pH at acidic pH (e.g., in acid mine drainage (pH 0.1 to 2) or in solfatara fields (pH ~2 to 3)) and some at alkaline pH (e.g., soda lakes, such as Mono Lake, CA or Lake Natron (pH 9 to 11)). Intracellular pH of the acidophiles and alkalophiles is, however, mostly near 7.

The redox potential of the pyridine nucleotide electron carriers (NAD^+, $NADP^+$) is also pH sensitive. If intracellular pH is ≠7, the redox potential of the NAD^+/NADH + H^+ couple, for example, changes by 30 mV per pH unit ($E^{0\prime}$of the NAD^+/NADH+H^+ couple is -0.320 V). E^0 at a pH other than 7 is equal to - 0.32 V + 0.03 V × ΔpH). The redox potentials of ferredoxins and cytochromes are theoretically pH independent.

3.3.7 Calculating Redox Potentials of the $2H^+/H_2$ Couple

The redox potential of the $2H^+/H_2$ couple is particularly important in many anoxic, organic-rich environments, in which H_2 carries electrons between species (interspecies H_2 transfer), resulting in metabolic networks between different microbes. For example, H_2 concentration in subseafloor sediments varies between 0.05 and 1000 nM, depending on in situ concentration of the predominant electron acceptor which is used by H_2-consuming microbes (Fig. 3.2; at 1 bar H_2 and 25 °C, H_2 solution concentration is 0.78 mM).

The redox potential of the reaction $H_2 \rightleftharpoons 2\ H^+ + 2\ e^-$ is

$$E = E^o + \frac{0.06}{2}\lg\frac{[\text{H}^+]^2}{[\text{H}_2]} = E^o + 0.03\lg\frac{[\text{H}^+]^2}{[\text{H}_2]}.$$

By definition, E^0 for $2H^+/H_2$ is zero under standard conditions (pH 0; $[H^+] = 10^0$ M and $pH_2 = 1$ bar), and the term $0.03\lg\frac{[H^+]^2}{[H_2]}$ accounts for the deviation from standard state. E^{0}'is the redox potential of $2H^+/H_2$ under physiological standard conditions (pH 7 ($[H^+] = 10^{-7}$ M) and pH $_2 = 1$ bar). $E^{o\prime} = E^{o} + 0.03 \lg \frac{[H^+]^2}{[H_2]} = 0 + 0.03\ (-14) = -0.420$ V.

The in situ redox potential of the $2H^+/H_2$ pair in anoxic environments, such as freshwater sediments or the large intestine of mammals, where pH_2 is near 10 Pa (10^{-4} bar) and pH = 7, is $E' = E^{o\prime} + 0.03 \lg \frac{1}{[10^{-4}]} = -300$ mV. At pH 8, the in situ redox potential is $E = E^{o\prime} + 0.03 \lg \frac{[10^{-1}]^2}{[10^{-4}]} = -360$ mV. At 0.05 nM H_2 (equal to a $pH_2 = 0.53 \times 10^{-7}$ bar at 25°C), as in nitrate-rich, anoxic sediments, the redox potential of the $2H^+/H_2$ couple at pH 7 is $E = -200$ mV and at pH 6 $E = -140$ mV. Thus, the redox potential of the $2H^+/H_2$ couple changes by 60 mV per pH unit and by 30 mV, when H_2 concentration changes by a factor of 10.

Δ*G* is highly concentration-dependent when the number of substrates and products differs. An entropic effect on the Δ*G* of a reaction becomes significant when the cumulative number of substrate and cumulative number of product molecules differs. Consider, for example, fermentation of glucose by *Moorella thermoacetica* (Fig. 10.14) to acetate and H^+ according to:

$$\text{Glucose} \rightleftharpoons 3\ \text{Acetate}^- + 3H^+$$

One mol glucose gives rise to 3 mol end products (protons are not considered in the concentration dependence here since the proton concentration is taken as constant). Under physiological standard conditions (all substrates and products 1 M, except protons 10^{-7} M), $\Delta G^{0\prime} = -304$ kJ/mol glucose. When concentrations are non-standard and in the range of 1 mM (10^{-3} M) as they are often in situ in anoxic environments, the free energy released in the above reaction is $\Delta G' = -304 + 5.7 \lg \frac{[10^{-3}]^3}{[10^{-3}]}$ kJ/mol $= -304 - 5.7 \times 6$ kJ/mol $= -338$ kJ/mol glucose, and at 0.1 mM concentrations (10^{-4} M) –350 kJ/mol glucose. This significant increase in free energy is available for microbial ATP synthesis (see below).

The entropic effect has the opposite consequence, e.g., in "chain elongation fermentation" of *Clostridium kluyveri* (see also Fig. 9.18), where the cumulative number of substrate molecules is larger than the cumulative number of product molecules:

$$6\,\text{Ethanol} + 4\,\text{Acetate}^- \rightleftharpoons 5\,\text{Butyrate}^- + 4H_2O + 2H_2 + H^+$$

Ten substrate molecules are converted to 7 product molecules (H_2O and protons are not considered in the concentration dependence because H_2O and proton concentrations are taken as constant). Thus, when concentrations of ethanol, acetate, and butyrate are 1 mM, free energy change decreases from –184 kJ/mol H^+ ($\Delta G^{0\prime}$) to –98.5 kJ/mol H^+ ($\Delta G'$). In the above calculation, H_2 partial pressure is considered to be 1 bar (i.e., standard condition), because H_2 gas escapes as bubbles from a *C. kluyveri* culture indicating that its partial pressure is near 1 bar. The difference in Δ*G* between standard and mM in situ conditions (85.5 kJ/mol H^+) is equivalent to a "loss" of approximately 1 ATP (see below).

When, however, the number of substrates and products in a reaction are equal, then $\Delta G^{0\prime}$ and $\Delta G'$ are identical when concentrations of substrates and products decrease from 1 M to 1 mM: e.g., Glucose + ATP $\rightleftharpoons$ Glucose-6-phosphate + ADP.

3.3.8 Accuracy vs Precision When Calculating ΔG Values and Redox Potentials

When calculating ΔG^0 and E^0 values, use of molar (mol/l solution) rather than molal concentrations (mol/kg solvent) is actually incorrect. Molal concentrations need to be employed because only molal concentrations are temperature- and pressure-independent. Molal and molar concentrations are nearly identical when the solvent is water because 1 l water solvent ≈1 kg solvent (in contrast, 1 l ethanol as solvent = 0. 78 kg solvent at 25 °C). While remaining, the difference between mol/g solvent and mol/l solution is very small at low concentrations. Nonetheless, use of molar concentrations to calculate ΔG and E, is incorrect. Molal activity is the molar concentration multiplied by the activity coefficient. The activity coefficient corrects for interactions between solute molecules that reduce the effective concentration of the solute. These interactions are zero only at infinite dilution (activity coefficient = 1). At molar (1 M) concentration, activity coefficient can be as low as 0.8 and even lower in the presence of salts. For example, the pKa of KH_2PO_4 dissociation to $KHPO_4^-$ + H^+ ($\Delta G^0 = -2.3$ lg Ka) decreases from 7.2 at infinite dilution to a pKa = 6.8 at 100 mM. The activity coefficient is also temperature-dependent. For example, the pKa = 8.2 of the buffer TRIS at 25 °C decreases to 7.9 at 35 °C. Therefore, use of concentrations instead of activity in the calculations of ΔG and E values is incorrect.

"Accuracy" refers to how truthfully a measurement reflects a true value, while "precision" indicates how well a measurement or calculation repeatedly generates the same result, which affects how many significant figures are appropriate to report. Biological variation between experiments and unknown intracellular concentrations, ionic strength, and pH result in inaccuracy, although a measurement technique itself may be very precise. For example, a ±5% error in measurement of pH can easily result in a variation of the measured redox potential by 5 mV, considering that redox potential changes by 30 mV per pH unit, when the redox reaction involves one proton and by 60 mV, when it involves 2 protons. Thus, literature values given in Table A.2 in the Appendix for redox potential were rounded (e.g., − 420 mV rather than − 414 mV in case of the $2H^+/H_2$ redox couple, and −820 mV rather than +818 mV in case of the O_2/H_2O redox couple). On the other hand, standard free energies of formations from the elements as given in Table A.1 in the Appendix have a relatively high precision and are, therefore, given up to three decimals.

Some of the redox potentials given in Table A.2 in the Appendix were fitted to conform with the free energy change of the overall catabolism as exemplified by methanogenesis from H_2 and CO_2 ($\Delta G^{0\prime} = -131$ kJ/mol, Chap. 10). Four redox reactions and 4 non-redox reactions are involved in this pathway. For two of the eight reactions, namely methyl-transfer to coenzyme M and methyl-coenzyme M reduction to methane, the $\Delta G^{0\prime}$ value has not been determined directly. Assuming $\Delta G^{0\prime}$ of the two latter reactions to be − 30 kJ/mol, then $E^{0\prime}$(methenyl-H_4MPT/methylene-H_4MPT) = −390 mV and $E^{0\prime}$(methylene-H_4MPT/methyl-H_4MPT) = −320 mV are the best fit (Table A.2 in the Appendix). Making slightly different assumptions, −360 and − 330 mV, respectively, fits better.

Finally, not all published ΔG^0 and E^0 values are correct. For example, $E^{0\prime}$of the crotonyl-CoA/butyryl-CoA couple has been reported in 1986 to be −10 mV by one laboratory and in the same year to be −126 mV by another laboratory. The correct standard redox potential is −10 mV because only with this potential all the standard free energy changes associated with the individual reactions involved in butyrate oxidation to 2 H_2, 2 acetate$^-$ and 2 H^+ add up to +48.3 kJ/mol. This is the standard free energy change calculated for fermentation of butyrate from the standard free energies of formation from the elements (Table A.2 in the Appendix).

Another example is the redox potential of the methylene-H_4F/methyl-H_4F couple that is given in Table A.2 in the Appendix as −200 mV. The first redox potential published was −100 mV, which in the meantime was shown to be incorrect.

In summary, thermodynamic data given in Tables A.1 and A.2 in the Appendix are only as accurate, precise, and correct as scientists have measured or calculated them.

3.4 ATP as Energy Carrier in all Living Cells

3.4.1 ΔG of ATP Formation

ATP (adenosine triphosphate) is the key coenzyme carrying usable cellular energy. Physiologically, it is the complex of ATP with the magnesium cation Mg^{2+} that is the energy carrier; however, this is—for simplification—neglected in the following. In ATP (Fig. 3.4), energy is stored in two anhydride bonds between two phosphoric acid groups. In most ATP-consuming or ATP-forming reactions, only one of these anhydride bonds is hydrolyzed or formed (ATP + H_2O ⇌ ADP + P_i; $\Delta G^{0\prime}$ = −32 kJ/mol). In some reactions, however, pyrophosphate (PP_i) is released (ATP + H_2O ⇌ AMP + PP_i; $\Delta G^{0\prime}$ = −42 kJ/mol), and this release is considered to be irreversible because pyrophosphate, in most cases (there are only a few exceptions), is irreversibly hydrolyzed by a cytoplasmic *pyrophosphatase* to two phosphates (PP_i + H_2O → 2 P_i; $\Delta G^{0\prime}$ = −22 kJ/mol). PP_i formation from ATP is of importance mainly in biosynthesis (DNA, RNA, protein, and isoprene biosynthesis; *ATP citrate lyase*) but also in some catabolic pathways (e.g., sulfate reduction; acetate activation via *acetyl-CoA synthetase*). ADP and AMP are equilibrated in the *adenylate kinase* (or *myokinase*) reaction (ATP + AMP ⇌ 2 ADP; $\Delta G^{0\prime}$ = 0 kJ/mol). The free energy changes of the above two ATP-hydrolyzing reactions have been experimentally determined from K_{eq} of ATP-dependent reactions.

Intracellular concentrations of ATP, ADP, AMP, and inorganic phosphate (Pi) are in the order of mM rather than 1 M. In growing cells, ATP/ADP ratio is generally 10/1. In aerobically growing *E. coli*, for example, the intracellular inorganic phosphate concentration is near 10 mM. Therefore, in this microbe approximately +50 kJ/mol is required to phosphorylate one ADP.

$$\begin{aligned}\Delta G' &= \Delta G^{0'} + 5.7\lg\frac{10/1}{[10^{-2}]}\\ &= +32 + 5.7\lg(10^3)\\ &= +32 + 5.7\cdot 3 \approx +50\ \text{kJ/mol}\end{aligned}$$

Intracellular P_i concentration can vary considerably between microbes and growth conditions. Given these variations, the Gibbs free energy

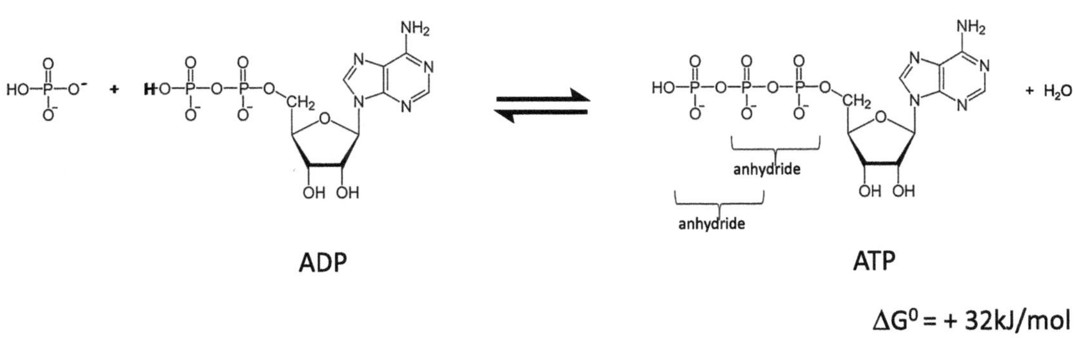

Fig. 3.4 Gibbs free energy of phosphorylation of ADP (adenosine diphosphate) to ATP (adenosine triphosphate). The anhydride bond between two phosphates is either formed to an anhydride or hydrolyzed. Under standard and reversible conditions, the formation of an anhydride bond in ATP requires +32 kJ/mol; the second one from AMP and PP_i requires +42 kJ/mol. See text for details

required under **reversible conditions** for synthesis of 1 mol ATP from ADP and P_i in vivo at pH 7 can be 5 kJ/mol higher or lower than +50 kJ/mol. The mean phosphorylation potential is slightly lower at an intracellular pH of 6 (48 kJ/mol) and slightly higher at an intracellular pH of 8.5 (57 kJ/mol).

3.4.2 The Adenylate Energy Charge

Many enzymes catalyzing ATP- or ADP-dependent reactions are regulated by the adenylate energy charge

$$= \frac{[\text{ATP}] + 0.5[\text{ADP}]}{[\text{AMP}] + [\text{ADP}] + [\text{ATP}]}.$$

Energy charge, as detailed in Sect. 3.4.2, is indicative of the energetic and, therefore, viability state of a cell as it regulates the activity of ATP- and ADP-dependent catabolic and anabolic enzymes.

Light as Energy Resource

Phototrophic microbes use the physical energy of light as their catabolic resource. While cyanobacteria utilize light in the range of 450–750 nm, purple sulfur bacteria are typically specialized on wavelengths in the range of 750–950 nm, and purple non-sulfur bacteria in the range of 800–1050 nm, reaching into the invisible infrared. How much energy is available in a mole of photons at these wavelengths? The energy of a photon is $E = h\nu$, where h is the Planck constant $h = 6.626 \times 10^{-34}$ Js and ν the frequency of light, which is related to the wavelength by $\nu = \frac{c}{\lambda}$, where c is the speed of light at $c = 3 \times 10^8$ ms^{-1}. Considering the Avogadro constant of $N_A = 6.023 \times 10^{23}$ mol^{-1}, the energy content of 1 mol of photons is $E = N_A\frac{hc}{\lambda}$, = 266 kJ/mol at 450 nm and 150 kJ at 800 nm. The energy per mol photons is equivalent to ΔG (free energy change per mol substrate) in chemical reactions.

3.4.3 n_{ATP} and Growth Yields Y_S and Y_{ATP}

Figure 3.5 depicts a simple catabolic pathway and associated Gibbs free energy change $\Delta G_{S \rightarrow P}$, where the Gibbs free energy released in the pathway reaction is conserved and used for ATP synthesis (more detail in Chap. 6).

$\boldsymbol{n_{ATP}}$ **is the number of mol of ATP** formed net per mol substrate in a specific catabolic pathway. $\boldsymbol{n_{ATP}}$ is also referred to as **ATP gain** (sometimes also called "**ATP yield**"). While n_{ATP} describes the number of ATP formed per pathway reaction, the terms ATP gain and ATP yield emerged from the experimental field as a specific, experimentally-determined value [mol ATP formed per mol substrate used or product formed]. Either term describes the same and is used here interchangeably.

Gibbs free energy can vary by more than 2 orders of magnitude between catabolic pathways. For example, aerobic glucose respiration to 6 CO_2 is associated with a $\Delta G^0 = -$2872 kJ/mol glucose, and a ΔG near −3000 kJ/mol at 1 mM glucose and 10^{-4} bar CO_2. On the other hand, $\Delta G^{0\prime}$of butyrate-forming fermentation of glucose to 2 acetate$^-$, 2 H_2, and 1 H^+ is +48.3 kJ/mol butyrate, and $\Delta G'$ of −14.4 kJ/mol when forming butyrate and acetate instead at 1 mM concentration, and H_2 at 10^{-4} bar. Not surprisingly, different catabolic pathways will have different n_{ATP} depending on the absolute free energy change.

n_{ATP} can be higher than 30 in aerobic glucose respiration and as low as 0.2 in syntrophic butyrate fermentation (for mechanisms, see Chaps. 6 and 9). In several pathways, such as in the Embden–Meyerhof pathway for glucose oxidation to pyruvate, ATP is not only formed but also consumed in reactions that are endergonic even at very high substrate and low product concentrations. One example is formation of glucose-6-phosphate from glucose and phosphate ($\Delta G^{0\prime} = +16.5$ kJ/mol). Thus, for an entire catabolic pathway, more ATP needs to be formed net than consumed. In such cases, n_{ATP} is the relevant

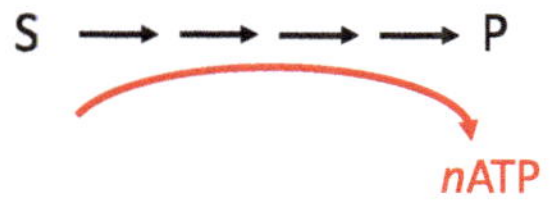

$\Delta G_{S \rightarrow P} = n_{ATP}\Delta G_{ATP}$ + Heat (ΔG_{unused})

Fig. 3.5 Energy conservation in a catabolic pathway converting substrate S to product P. The overall energy released in this transformation, $\Delta G_{S \rightarrow P}$, consists of a fraction that is used for ATP synthesis (n_{ATP} $\Delta G'_{ATP}$), where n_{ATP} is the number of mol ATP formed per mol substrate in the pathway, $\Delta G'_{ATP}$ = – 50 (±5) kJ/mol, and heat, which is the fraction of $\Delta G_{S \rightarrow P}$, which is unused for ATP synthesis. n can be as low as 0.2. See text for details

number of **net** mol ATP conserved per mol substrate transformed in a pathway.

Generally, n_{ATP} can be estimated by dividing $\Delta G_{S \rightarrow P}$ by −80 kJ/mol ATP, the latter being the energy change of ATP hydrolysis under the irreversible conditions in most living cells. However, in extreme cases, the latter can be as low as – 70 kJ/mol ATP as in very slowly growing anaerobic syntrophs and much higher than −90 kJ/mol ATP as in microbes with a high rate. The high rate is traded off in these microbes against a low yield. n_{ATP} can also be determined experimentally by dividing the growth yield (g cells/mol substrate) by Y_{ATP}, the theoretical growth yield (g cells/mol ATP) (see below).

Growth yields are given in g cells (dried cell mass) per mol of catabolic substrate. Y_{ATP} (growth yield per mol ATP) is obtained by dividing growth yield by n_{ATP}. Notably, Y_{ATP} for a given anabolic substrate is constant, e.g. ca.10 g/mol ATP if the anabolic carbon substrate is glucose, and ca. 5 g/mol ATP if it is CO_2 (see Chap. 4). Prerequisite is that growth was energy limited. Only under those conditions, the relation $Y_S = n_{ATP} \times Y_{ATP}$ is valid.

ATP yield and ATP growth yield are not the same. ATP yield or ATP gain = n_{ATP} and denotes the mol ATP gained per mol substrate catabolized and can be calculated from $\Delta G_{S \rightarrow P}$/80 kJ/mol ATP. On the other hand, the ATP growth yield, Y_{ATP}, is an empirical number describing how much biomass (in g dry mass) is theoretically synthesized per mol ATP.

3.4.4 Thermodynamic Efficiency of ATP Synthesis

The term $\frac{n_{ATP}\Delta G'_{ATP}}{\Delta G_{S \rightarrow P}} \times 100$ is the **thermodynamic efficiency** (in %) of a pathway, where $\Delta G'_{ATP}$ = – 50 kJ/mol. The thermodynamic efficiency of a catabolic pathway needs to be less than 100% in order for a pathway to have a net forward flux, because at equilibrium (100% efficiency) there is no net flux (see above $K_{eq} = k_{+1}/k_{-1}$, also Chap. 5). Consequently, release of heat (Fig. 3.6) is necessary for a pathway to proceed at a finite rate. On the other hand, thermodynamic efficiency cannot be too low, because at low thermodynamic efficiency growth yield (n_{ATP} Y_{ATP}) is also low. The observed "sweet spot" seems to be a thermodynamic efficiency near 60% (Fig. 3.1). Thermodynamic efficiency of ATP synthesis is highest in very, very slowly growing obligate syntrophs (near 100%) and lowest in very rapidly growing microbes (near 20%).

It is conceivable that ADP phosphorylation is being *pulled* by anabolism rather than being *pushed* by catabolism, since ATP consumption in anabolism is highly irreversible: up to 90% of the free energy being dissipated as heat (Fig. 3.1). As a consequence, steady-state ATP concentration, ATP phosphorylation potential, and adenylate energy charge would be lower than when ATP synthesis in catabolism is irreversible. The thermodynamic efficiency of ATP synthesis at $\Delta G'_{ATP} > -50$ kJ/mol would approach 100% and growth rate would be zero. There is experimental evidence that in microbes with a catabolism operating near thermodynamic equilibrium the phosphorylation potential can be considerably lower than −50 kJ/mol.

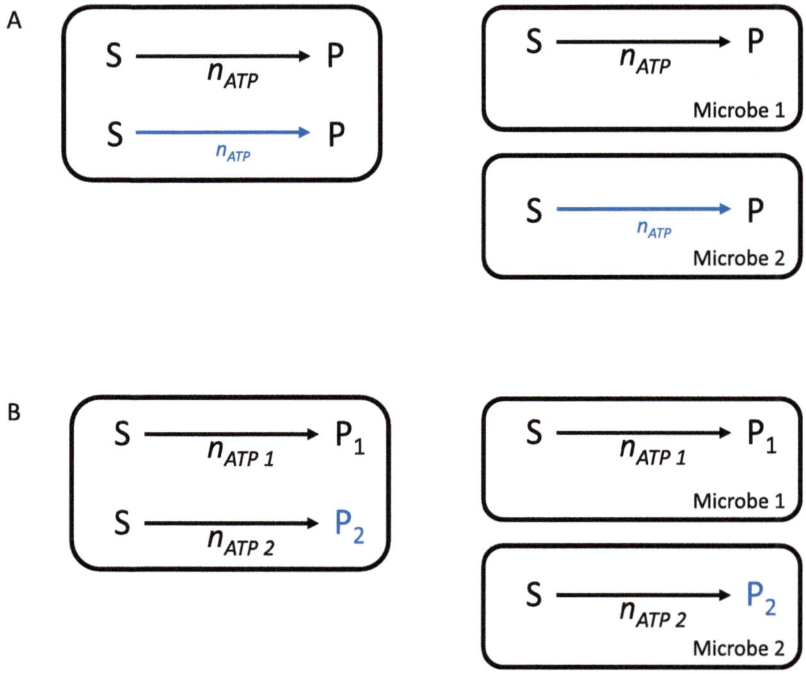

Fig. 3.6 Significance of n_{ATP} and thermodynamic efficiencies on metabolic traits and speciation. (**a**) A single microbe contains two pathways for the same overall reaction, however, each one with a different n_{ATP} (different arrow color), resulting in a high and a low substrate affinity pathway (large letter n_{ATP}, small letter n_{ATP}). Speciation of each pathway type into two different microbes. (**b**) A single microbe carries more than one pathway for utilization of a single substrate, where each pathway results in different end products and operates with different n_{ATP}. Speciation of the different pathway types into two different microbes. Chapters 9 and 10 provide examples for both scenarios. See text for details

3.5 Different Thermodynamic Efficiencies and Their Ecophysiological Consequences

The significance of n_{ATP} and associated thermodynamic efficiency for substrate utilization cannot be overemphasized as a key determinant for a microbe's metabolic strategy and ecological success. n_{ATP} is biologically controlled and under strong natural selection.

One and the same catabolic substrate can be transformed to the same end products by different pathways with different n_{ATP} (Fig. 3.6a). Some microbes switch to a pathway with low n_{ATP} to gain access of a catabolic substrate at low concentration. Pathways with different n_{ATP} can speciate into different microbes, resulting in two species (Fig. 3.6b): one with a high n_{ATP} which is adapted to using high substrate concentration, and one with a low n_{ATP} adapted to low substrate concentration. Also, there can be different pathways in a single microbe that converted a single substrate to different end products at different n_{ATP} and thermodynamic efficiencies (Fig. 3.6b). The following examples show also that thermodynamic efficiency of ATP synthesis correlates negatively with growth rates, the reasons for which are explained in Chap. 5.

3.5.1 Constant n_{ATP} in Anaerobic H_2 Oxidizers Over a Wide Range of H_2 Concentrations

Methanothermobacter marburgensis grows converting H_2 and CO_2 to CH_4 with a n_{ATP} of 0.5 and an apparent affinity constant Ks for H_2 of below 10^{-1} bar. n_{ATP} does not change when H_2 concentration decreases from 10^0 bar to 10^{-4} bar and when the free energy change associated with methane formation decreases from −131 kJ/mol to –40 kJ/mol. Thermodynamic efficiency of ATP synthesis is 20% at high H_2 concentration and slightly above 60% at the low H_2 concentration. With increasing thermodynamic efficiency, the rate of growth decreases almost linearly. Such n_{ATP} and thermodynamic efficiency management (metabolic "business plan") is also found in other H_2 oxidizing anaerobes such as acetogens and sulfate reducers.

3.5.2 Different n_{ATP} in Glucose Oxidation to the Same End Products

Aerobic growth on glucose: In growing *Pseudomonas* sp. cells, when respiring glucose aerobically to CO_2 ($\Delta G = -3000$ kJ/mol), 36 ATP are estimated to be formed from ADP and P_i ($n_{ATP} = 36$). A value near 35 is also found for mitochondria of most animal cells. This appears to be the maximum n_{ATP} for aerobic glucose respiration. Thus, the thermodynamic efficiency of the *Pseudomonas* sp. pathway is 62% ($\frac{36 \times 50}{3000}$ 100). In growing *E. coli* cells, for the same overall reaction of respiring glucose aerobically to CO_2, n_{ATP} is only 24, and the thermodynamic efficiency of ATP synthesis is only 40%. This correlates with the finding that *E. coli* can grow aerobically under optimal conditions at a faster rate than *Pseudomonas* (see Chap. 5 for more on rate-yield trade off). The energy of ≈-3000 kJ/mol glucose of the common overall reaction is available to both microbes but only used partially.

The different n_{ATP} in this example can be explained mainly by differences in composition of the electron transport chain from NADH to O_2 which is common to all aerobic glucose-oxidizing microbes:

$$\mathrm{NADH} + \mathrm{H}^+ + \tfrac{1}{2}\mathrm{O}_2 \rightleftharpoons \mathrm{NAD}^+ + \mathrm{H_2O} \qquad \Delta G^{0\prime} = -220\ \mathrm{kJ/mol\ NADH}$$

Energy is conserved via an electrochemical H^+ gradient across the cytoplasmic membrane (or inner mitochondrial membrane in eukaryotes) and converted to ATP via a membrane-bound *ATP synthetase* (see Chaps. 5 and 6 for more details). In the respiratory chain, three sites allow H^+ translocation across the cytoplasmic membrane for energy conservation at various stoichiometries (nH^+ translocated per $2e^-$ transported), which is equivalent to different n_{ATP} for the same overall reaction. Oxidation of NADH via the aerobic respiratory chain in *Paracoccus* sp. and *Pseudomonas* sp. involves the three proton-translocating sites (Fig. 3.7): *NADH dehydrogenase*, translocating four H^+/2 e^- for this particular *NADH dehydrogenase*, *ubiquinol dehydrogenase* (cyt *bc1* complex), translocating four H^+, and cytochrome *c* oxidase, translocating two H^+. Collectively, 10 H^+ are translocated per NADH oxidized with ½O_2, which is equivalent to ~2.5 ATP (4H^+/ATP stoichiometry, see Chap. 6). For the same overall reaction, *Helicobacter pylori*'s electron transport chain contains a cytochrome *c* oxidase that only translocates one H^+. *E. coli* does not contain cytochrome *c* and a cytochrome *bc1* complex (= testing negative in the cytochrome *c* oxidase test). Apparently, it is advantageous for *E. coli* to only translocate 6 H^+ per NADH using a cytochrome *bo oxidase* under fully oxic conditions. When oxygen-limited, *E. coli* switches from *cytochrome bo oxidase* to a *cytochrome bd oxidase* and from NDH-1 to NDH-2. Neither *cytochrome bd oxidase* nor NDH-2 translocates a proton (although energy is available) and O_2 is used as electron acceptor (Fig. 3.7). This trait enables *E. coli* to adjust rate and yield of ATP production to varying environmental conditions.

Both the Embden–Meyerhof and the Entner–Doudoroff pathway (see Chap. 8) are pathways catalyzing the identical overall reaction: the oxidation of 1 glucose to 2 pyruvate plus 2 NADH; yet, in the Embden–Meyerhof pathway n_{ATP} (=2) is twice the n_{ATP} of the Entner–Doudoroff pathway ($n_{ATP} = 1$). Most carbohydrate-metabolizing microbes contain the Embden–Meyerhof pathway; the Entner–Doudoroff pathway is found in only a few prokaryotes. *E. coli* contains both pathways, however, uses the Embden–Meyerhof pathway when metabolizing glucose, and the Entner–Doudoroff pathway when metabolizing gluconate. Gluconate is more oxidized than glucose by 2 electrons; therefore, gluconate

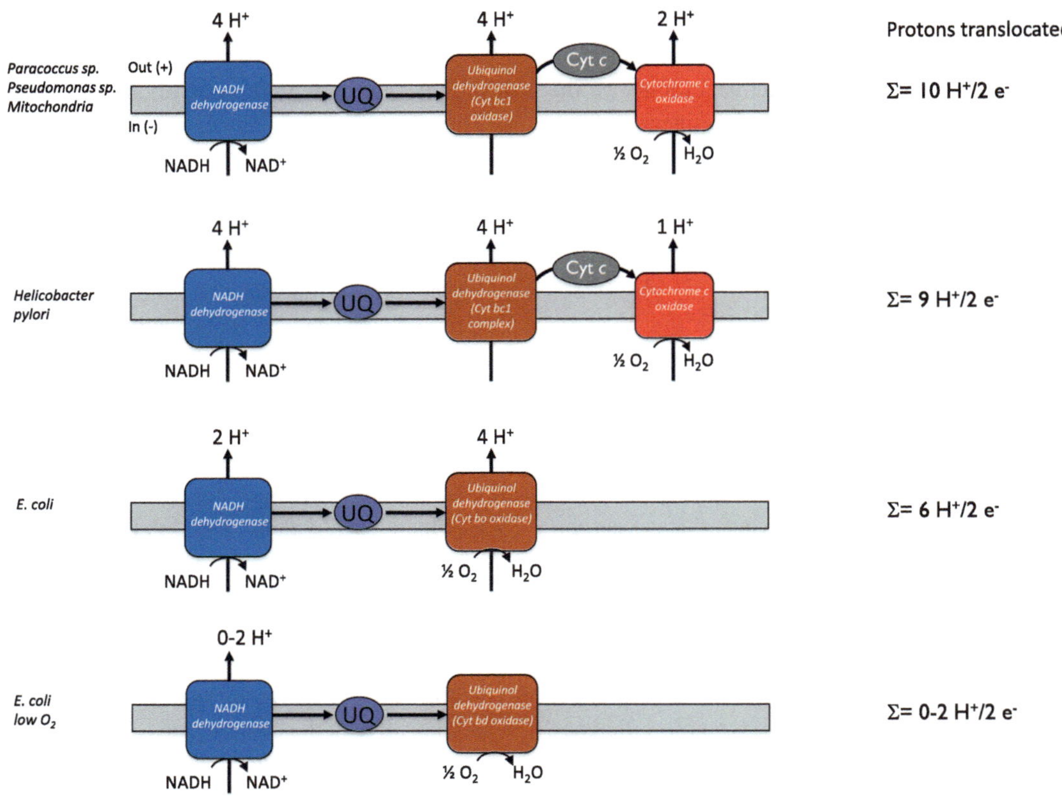

Fig. 3.7 Architecture of different aerobic, NADH-oxidizing respiratory chains. From the number of H^+ translocated per 2 e- (i.e., 1 NADH oxidized), n_{ATP} can be calculated assuming that in the microbes presented, four electrogenic protons drive the phosphorylation of 1 ADP. The respiratory chains of all four examples mediate the same overall reaction (see above); however, they differ in components and stoichiometry of H^+ translocation in *cytochrome c oxidase*, cytochrome *bo oxidase*, *cytochrome bd oxidase,* and *cytochrome bc1* complex. In addition, there are different ubiquinone-coupling *NADH dehydrogenases*, which also have variable stoichiometries of H^+ or Na^+ translocation. Such differences in pathway architecture result in different n_{ATP} which enables microbes to express different growth rate and metabolic phenotypes that affect their fitness in their respective environment. The color scheme is not to imply phylogenetic, but only functional relationships. See text for details

oxidation to two pyruvate produces only 1 NADH and, therefore, only 1 ATP in that pathway.

Anaerobic growth on glucose. Yeast ferment glucose to two ethanol ($\Delta G^0 = -235$ kJ/mol glucose) via the Embden–Meyerhof pathway, while *Zymomonas mobilis* uses the Entner–Doudoroff pathway for the same overall reaction. At high glucose concentration in their natural environment, growth yield and thermodynamic efficiency of ATP synthesis associated with ethanol formation in yeast is twice as high as in *Z. mobilis.* As predicted, *Z. mobilis* ferments glucose to ethanol at much higher rates and to much lower cell concentrations than yeast, which is taken advantage of in biotechnology.

3.5.3 Different n_{ATP} of Glucose Fermentation to Different End Products

E. coli ferments glucose anaerobically at pH 7 mainly by mixed acid fermentation to acetate, ethanol, succinate, formate, H_2, and CO_2 ($\Delta G^{0\prime} = -218$ kJ/mol glucose; $\Delta G' = -$

260 kJ/mol at 1 mM concentrations of substrate and products); $n_{ATP} = 3$ at a thermodynamic efficiency of 63%. Lactic acid bacteria ferment glucose to 2 lactate ($\Delta G^{0\prime} = -198$ kJ/mol glucose; $\Delta G' = -215$ kJ/mol at 1 mM concentrations); $n_{ATP} = 2$ at a thermodynamic efficiency of 51%.

These examples show different metabolic strategies with different thermodynamic efficiencies to utilize a single substrate (glucose) resulting in distinctive ecophysiological traits.

Another way to control thermodynamic efficiency is by varying the number of end products. In mixed acid fermentation of glucose by *E. coli*, lactate, acetate, ethanol, succinate, formate, H_2, and CO_2 are formed; however, relative stoichiometries can vary depending on growth conditions. For example, in the absence of selenium *E. coli* no longer converts formate to CO_2 and H_2 (*formate dehydrogenase* involved is a selenium-containing enzyme). Such variation modulates also thermodynamic efficiencies in particular at low substrate concentration as discussed above.

Another example for the dependence of product formation on available micronutrients is in *Clostridium pasteurianum* (Chap. 9). This Gram-positive anaerobe ferments glucose to acetate, butyrate, CO_2, and H_2 ($n_{ATP} = 3.3$), when growing on iron-sufficient media (Fig. 9.16) and to 2 lactate ($n_{ATP} = 2$) when growing on iron-deficient media although very slowly. Lactic acid bacteria, that do not contain iron-requiring enzymes, ferment glucose to 2 lactate rapidly in the absence or presence of iron. Iron is growth-limiting for microbes, for example in milk, blood, the human gut, or the body surface. This can explain why lactic acid bacteria may dominate in these environments, and why *Clostridia*, which contain Fe-containing enzymes and electron carriers (ferredoxin), outcompete lactic acid bacteria when iron becomes available. The inability of *Clostridia* to outcompete lactic acid bacteria can, thus, be explained by their differential iron requirement which enables pathways with different n_{ATP}, and not necessarily by pursuing a rate/yield strategy (see Chap. 5).

3.5.4 Lowering n_{ATP} to Access Substrates at Low Concentrations

The lowest substrate concentration (threshold), that can sustain microbial growth, is determined kinetically, when the negative ΔG is large, or thermodynamically, when the negative ΔG is small.

Large negative ΔG pathways. As indicated above (Sect. 3.5.1), some aerobic microbes decrease pathway n_{ATP} by up to 50% when O_2 concentration becomes very low. This decrease in n_{ATP} occurs despite the concentration-dependent decrease of ΔG of respiration by only less than 10%. For example, respiration of glucose with 6 O_2 to 6 CO_2 is exergonic by ~−3000 kJ/mol at $pO_2 = 0.2$ bar (20% O_2) in the gas phase and $pCO_2 = 4.1 \times 10^{-4}$ bar (0.041%) and is ~−2800 kJ/mol at $pO_2 = 0.2 \times 10^{-6}$ bar and $pCO_2 = 4 \times 10^{-4}$ bar. At such low pO_2, *E. coli* replaces cytochrome *bo oxidase* by a cytochrome *bd oxidase* and NDH-1 by NDH-2, which do not translocate any proton although sufficient energy is available (see Fig. 3.4). This partial uncoupling of the respiratory chain from energy conservation correlates with the finding that *E. coli* utilizes O_2 at low concentration at a higher rate than in the coupled pathway. Thus, decreasing n_{ATP} and lowering thermodynamic efficiency makes aerobic growth of *E. coli* at low O_2 concentrations **kinetically** competitive. This metabolic strategy enables *E. coli*, which is a facultative aerobic bacterium, to metabolize O_2 at a concentration that is too low to be used for, e.g., the obligate aerobic *Pseudomonas*, which has a higher n_{ATP} than *E. coli*. In both cases, the lowest O_2 concentration possible at which either microbe can grow, the threshold O_2 concentration, is determined **kinetically** and not thermodynamically.

Small negative ΔG pathways. When the negative pathway ΔG is small, as for some strictly anaerobic microbes, adjusting metabolism and its rate to low substrate concentrations can be quite different because changes in substrate/product concentration have a large effect on the free energy change relative to that under standard conditions. For example, methanogenesis from 4 H_2 and CO_2 ($\Delta G^0 = -131$ kJ/mol) is associated with a free energy change of only about -136 kJ/mol when H_2 concentration in the gas phase is 80% (0.8 bar), CO_2 concentration 20% and methane concentration is 1%. At H_2 partial pressures of 10^{-4} bar, as in many methanogenic sediments, the free energy change is near $-$40 kJ/mol. Depending on the type of methanogen, n_{ATP} is either 0.5 or 1 (see Chap. 10). Notably and in contrast to the above *E. coli* example, n_{ATP} of the pathways remains constant in each microbe at different H_2 concentration. Because n_{ATP} does not change, lowering the substrate concentration from 0.8 bar to 10^{-4} bar increases the thermodynamic efficiency to near 100% and, by that, decreases the rate of the catabolism to near zero. Thus, the H_2 threshold concentration at which the methanogens cease to reduce CO_2 with H_2 to methane is determined **thermodynamically**. A microbe with a relatively low n_{ATP} methanogenesis pathway is adapted to low H_2 concentrations as they prevail in their natural habitats.

The theoretical H_2 threshold can be calculated from ΔG of the catabolic reaction, the ΔG of ATP synthesis, and from n_{ATP}, as exemplified for methanogenesis from H_2 and CO_2 (Fig. 3.8):

$$4H_2 + CO_2 + nADP + nPi \rightarrow CH_4 + nATP + 3H_2O$$

When $n_{ATP} = 0.5$, thermodynamic equilibrium at 25 °C ($\Delta G = 0$ kJ/mol) is reached at a pH_2 of approximately 2 Pa; when $n_{ATP} = 1$, equilibrium is reached at 30 Pa H_2 (calculated assuming that the partial pressure of CO_2 is equal to the partial pressure of methane and $\Delta G_{ADP \rightarrow ATP} = +50$ kJ/mol).

Figure 3.8 shows how ΔG' varies for different n_{ATP} as a function of pH_2. Indeed, H_2 oxidation by *Methanothermobacter marburgensis* with a pathway $n_{ATP} = 0.5$ ceases at pH_2 of 2–10 Pa, whereas in *Methanosarcina barkeri* with a pathway $n_{ATP} = 1$, H_2 oxidation ceases at a

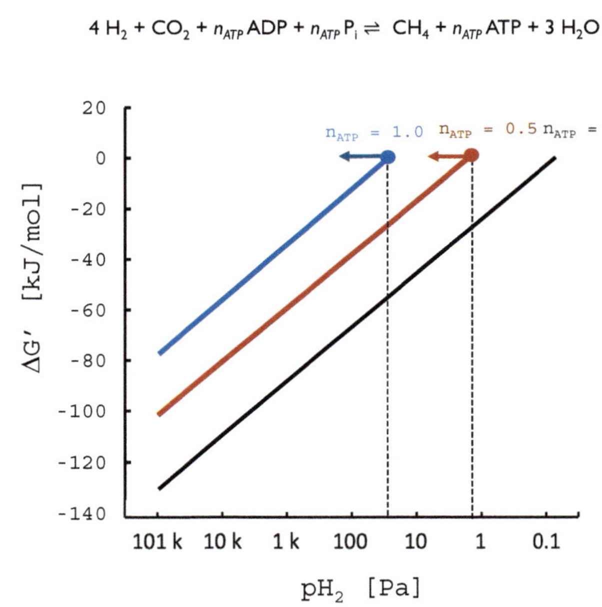

Fig. 3.8 Available energy ($-\Delta G'$) for hydrogenotrophic methanogenesis as a function of H_2 concentration in methanogens with different n_{ATP} of the methanogenic pathway. The common overall reaction is: $4H_2 + CO_2 \rightarrow CH_4 + 2H_2O$: *Methanothermobacter marburgensis* ($n_{ATP} = 0.5$, orange color) and *Methanosarcina barkeri* ($n_{ATP} = 1.0$, blue color). In the above $\Delta G'$ calculations, only pH_2 is varied. Dot (•) indicates the pH_2 when the catabolic reaction is 100% efficient; only at concentrations above those (to the left, see arrow (←)), when the thermodynamic efficiency is <100%, there is a net flux of CH_4, and with that, growth possible. The differences in n_{ATP} result in different theoretical pH_2 thresholds for ATP synthesis in each type: energy conservation in *M. barkeri* ceases at a higher pH_2 (~30 Pa) than in *M. marburgensis* at ~2 Pa. Experimentally determined H_2 thresholds are somewhat higher. See text for details

concentration that is at least tenfold higher (Fig. 3.8, see also Chap. 10). *M. barkeri*, using a high n_{ATP} pathway, will reach "death by equilibrium" at a threshold of ~30 Pa, while *M. marburgensis* with a $n_{ATP} = 0.5$ and a threshold of ~2 Pa will outcompete *M. barkeri* at low pH_2.

Generalizing for low ΔG microorganisms, hydrogenotrophic microbes with the lower H_2 threshold will outcompete at low H_2 concentration those with a higher H_2 threshold, because the rate of H_2 oxidation decreases with decreasing in situ H_2 concentrations to zero at the threshold H_2 concentration. As indicated above, the latter is determined by the relationship between n_{ATP}, ΔG of ATP synthesis, and ΔG of the H_2-consuming reaction which is dependent on the redox potential of the electron acceptor (CO_2 in the case of hydrogenotrophic methanogenesis, Fig. 3.8). For example in marine sediments, H_2 consumption by sulfate-reducing microbes with a H_2 threshold of < 0.1 Pa can lower the in situ H_2 concentration to a concentration, at which hydrogenotrophic methanogens with a H_2 threshold >1 Pa can no longer grow (Fig. 3.9). Such thermodynamic substrate concentration threshold can explain why methanogenesis from H_2 and CO_2 does not occur in marine sediments, in which the sulfate concentration is high. They can also explain why in the intestinal tract of ruminants, in which H_2 partial pressure seldom exceeds 10 Pa, acetogenesis from H_2 and CO_2 does not occur. In situ H_2 concentration in anaerobic sediments correlates with the redox potential of the predominant electron acceptor used by the H_2-consuming microbes (Fig. 3.9): the more positive the redox potential of the electron acceptor, the lower the H_2 concentration. Thus, H_2 consumption controls the redox potential of the $2H^+/H_2$ couple in sediments via its in situ concentration (see also Chaps. 9 and 10).

Besides for H_2, a threshold concentration for acetate is also ecologically important. For example, methanogenesis from acetate ($\Delta G^{0\prime} = -35$ kJ/mol) is only exergonic when acetate concentration is higher than 0.1 μM (at an assumed CO_2 concentration in the gas phase of 10% and a methane concentration of 90%, see Chap. 10). There are two types of acetoclastic methanogens, those with an acetate threshold in the mM range

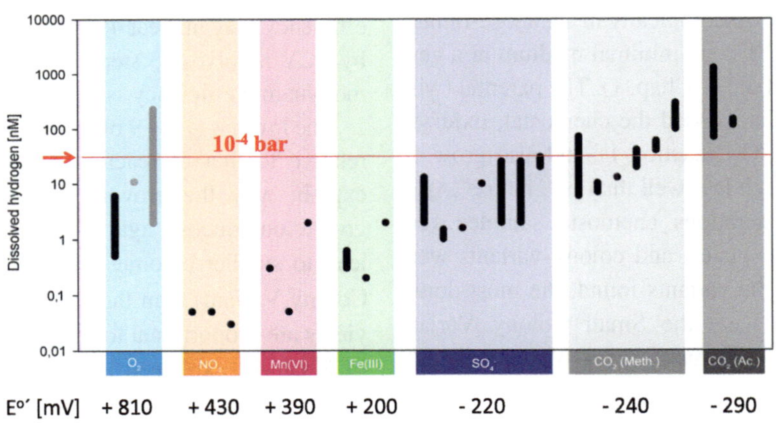

Fig. 3.9 Correlation of experimentally determined H_2 concentrations in sediments and the redox potential of the predominant electron acceptors. On the abscissa, the redox potentials of the predominant terminal electron acceptor present in the sediments are given. In sediments using H_2 as electron donor, microbes reduce nitrate (NO_3^-) to ammonia, Mn(IV) to Mn(II), SO_4^{2-} to H_2S, CO_2(Meth) to methane, and CO_2(Ac) to acetate. At 25 °C, 1 mM H_2 corresponds to a H_2 partial pressure (pH_2) of 1.28 bar (1.28 × 10^5Pa). At 2 °C, the temperature in many sediments, 1 mM H_2 corresponds to a H_2 partial pressure (pH_2) of 1.075 bar (1.075 10^5Pa). Courtesy Dr. A-C Girnth with data from Li et al. (2012) Geochimica et Cosmochimica Acta 77, 186–201, with permission

and those with an acetate threshold concentration in the μM range. Acetoclastic methanogens with the higher threshold have a higher n_{ATP} of 0.66 and activate acetate involving *acetate kinase* and *phosphotransacetylase* (1 ATP/acetate), and those with the lower threshold have a lower n_{ATP} of 0.33 and activate acetate involving an *acetyl-CoA synthetase* (2 ATP/acetate) (see Chaps. 8 and 10). Therefore, acetoclastic methanogens with a low n_{ATP} pathway and lower acetate threshold outcompete those with a high n_{ATP} pathway and high acetate concentrations, although their maximal growth rate is much lower than that of methanogens with the higher n_{ATP}. On the other hand, sulfate reducers oxidizing acetate to 2 CO_2 have an acetate threshold concentration significantly lower than 0.1 μM explaining why in marine sediments methanogenesis from acetate does not occur.

In all of the above examples, the substrate threshold is determined thermodynamically, and not kinetically.

3.5.5 Thermodynamic View of the Julian Adams Experiment

E. coli was grown aerobically in a glucose-limited chemostat at 30°C on minimal medium at a generation time of 3.5 h (Chap. 1). The parental (wild type) strain, that seeded the chemostat, oxidized glucose to 6 CO_2 (reaction 1), and also grow on acetate, although less well than on glucose. After nearly 800 generations, chemostat samples were plated on agar plates, and colony variants were observed. Of the variants found, the most dominant variants were the Small Colony Variant (SCV) and the Large Colony Variant (LCV). SCV metabolized glucose and adapted to utilizing glucose at concentration lower than wild type which prevailed in the chemostat. However, glucose was only metabolized incompletely to acetate. LCV grew much better using acetate than using glucose. These traits were inheritable.

1. $$\text{Glucose} + 6O_2 \rightarrow 6CO_2 + 6H_2O \qquad \Delta G^0 = -2870\,\text{kJ/mol};$$
   n_{ATP} = 24; steps = 25; thermodynamic efficiency = 42%
2. $$\text{Glucose} + 2O_2 \rightarrow 2\ \text{acetate}^- + 2H^+ + 2CO_2 + 2H_2O \qquad \Delta G^{0\prime} = -1148\,\text{kJ/mol};$$
   n_{ATP} = 9; steps = 18; thermodynamic efficiency = 41%
3. $$\text{Acetate}^- + H^+ + 2O_2 \rightarrow 2CO_2 + 2H_2O \qquad \Delta G^{0\prime} = -837\,\text{kJ/mol};$$
   n_{ATP} = 5.5; steps = 13; thermodynamic efficiency = 34%

ΔG^0 of the three catabolic reactions was calculated from the free energy changes of formation: From what is now known, reaction 1 is catalyzed by *E. coli* wild type involving 25 steps involving glucose transport (PEP transferase system), glycolysis, citric acid cycle, respiratory chain, and membrane-associated *ATP synthetase*; n_{ATP} is 24. Considering physiological concentrations of substrates and products, the thermodynamic efficiency of ATP synthesis of this catabolism is near 40%. Reaction 2 catalyzed by SCV with incomplete glucose oxidation involves 18 steps, n_{ATP} is 9, and thermodynamic efficiency is again near 40%. Reaction 3 catalyzed by LCV involves 13 steps, n_{ATP} is 5.5, and thermodynamic efficiency is near 35%.

The lower $n_{ATP} = 9$ of the pathway of the SCV relative to that of reaction 1 ($n_{ATP} = 24$) can explain why the growth yield of SCV, when grown on glucose agar plates, was lower and lead to smaller colonies (thus, the name: Small Colony Variant) than those of wild type; growth yields are proportional to n_{ATP}. The arising of the LCV can be explained by the new acetate niche that developed as a consequence of the adaptive evolution and metabolic change of the SCV, thereby, enabling LCV to escape competition with wild type and SCV for glucose. This outcome of two new dominant strains showed that the overall pathway of aerobic glucose oxidation to CO_2 of wild type had speciated into two non-competing strains, SCV and LCV, which resulted in a simple metabolic food chain. This thermodynamic approach, however, cannot explain the driver for speciation and the metabolic patterns and thermodynamic efficiencies of the evolved *E. coli* variants in the glucose-limited chemostat.

Further Reading

Alberty RA (1998) Calculation of standard transformed Gibbs energies and standard transformed enthalpies of biochemical reactants. Arch Biochem Biophys 353(1): 116–130

Amend JP, Schock EL (2001) Energetics of overall metabolic reactions of thermophilic and hyperphilic Archaea and bacteria. FEMS Microbiol Rev 25(2): 175–243

Lehninger AL (1982) Principles of biochemistry. Worth Publishers

Thauer RK, Jungermann K, Decker K (1977) Energy conservation in chemotrophic anaerobic bacteria. Bacteriol Rev 41(1):100–180

Catabolism and Its Coupling to Anabolism

4

Contents

4.1 Introduction

Thermodynamics of catabolism was described in the previous chapter and showed that the amount of ATP formed (n_{ATP}) in a catabolic pathway can be estimated from the free energy associated with the catabolic pathway divided by approximately 80 kJ/mol, which is the free energy change required for the synthesis of 1 ATP from ADP and phosphate under the irreversible cellular conditions. This relationship, however, is only valid when the growth rate is limited by catabolism and for microbes that are selected for maximal growth yield rather than for maximal growth rate.

This chapter provides a closer look on how catabolic pathways are generally structured and how catabolic and anabolic pathways are coupled. Fig. 4.1 considers that most catabolisms are redox processes and that catabolism and anabolism are not only coupled via a flux of ATP, as shown in Fig. 3.1, but also via common intermediates and electrons.

A. M. Spormann, *Principles of Microbial Metabolism and Metabolic Ecology*,
https://doi.org/10.1007/978-3-031-28218-8_4

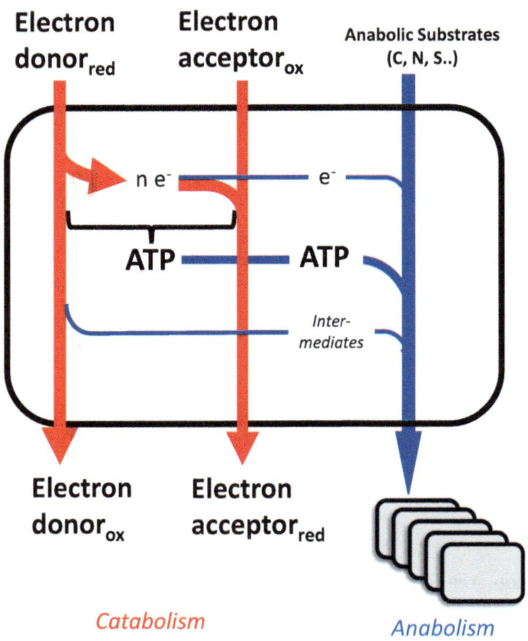

Fig. 4.1 Metabolic connections between catabolism (red) and anabolism (blue) in a respiring chemoorganotrophic microbe. Catabolism consists of oxidative and reductive reactions, where electrons, often in conjunction with H^+, are transferred between both pathways for ATP synthesis (see Chap. 6). ATP, electrons, and intermediates are exchanged between oxidative and reductive pathways based on the chemical nature of the organic substrate (e.g., oxidation state of carbon). Anabolism includes (1) the synthesis of the 20 amino acids, the four nucleotides (for rRNA, tRNA, mRNA, sRNA), of deoxynucleotides (for DNA), and long chain fatty acids for phospholipids, as well as cofactors and cell wall components, (2) transport of anabolic substrates, and (3) macromolecular syntheses. See text for details

4.2 Most Energy Metabolisms Are Redox Processes

Most catabolisms are based on redox reactions. There are only a few examples where catabolism is not a redox process, one being the *arginine deiminase* pathway, in which arginine is hydrolyzed via citrulline to ornithine, CO_2 and NH_3 (Chap. 6). Another exception is malate fermentation to lactate (Chap. 9). Catabolic processes can be deconvoluted into **an electron-donating, oxidative pathway** and an **electron-accepting, reductive pathway** (Fig. 4.1). These two "redox half reactions" are coupled by intracellular electron transfer. The electrons are not "free" electrons but bound to specific cellular electron carriers (see Chap. 7). The oxidative and a reductive half-reactions are each a conjugated redox couple with a redox potential: E_{Donor} being the redox potential of the e-donor$_{ox}$/e-donor$_{red}$ couple and E_{Acceptor} the redox potential of the e-acceptor$_{ox}$/e-acceptor$_{red}$ couple. The difference between the redox potentials, $\Delta E = E_{\text{Acceptor}} - E_{\text{Donor}}$, must be positive for the overall redox process to be exergonic (Fig. 4.1, see Chap. 3). For example in glucose respiration (Glucose +6 $O_2 \rightleftarrows 6\ CO_2 + 6\ H_2O$), catabolism can be thought of consisting of an **oxidative pathway** (Glucose $\rightarrow$ 6 CO_2 + + $24e^-$ + $24H^+$) and a **reductive pathway** ($6O_2^-$ + $24e^-$ + $24H^+$ $\rightarrow$ $12H_2O$). E^0 of the O_2/2 H_2O (electron acceptor) couple is +0.82 V and E^0 of the 6 CO_2/glucose (electron donor) couple is −0.42 V. Thus, $\Delta E^0 = E^{o}_{\text{Acceptor}} - E^{o}_{\text{Donor}} = -1.24$ V; at $n = 24$, the free energy change of this catabolism is $\Delta G^0 = -nF\Delta E^0 = -2870$ kJ/mol.

Glucose oxidation can be coupled to reduction of other electron acceptors, e.g., to nitrate, and O_2 reduction can be coupled to the oxidation of other catabolic electron donors in combinations as long as ΔE is positive.

Figure 4.2 shows a graphical summary of common, microbiologically-relevant, redox half-reactions and the associated redox potentials under standard conditions. The figure illustrates that many different combinations of oxidations of organic and inorganic compounds with reductions of inorganic or organic compounds are thermodynamically possible. For example, oxidation of organic matter (<CH_2O>) with NO_3^- is exergonic, and so is oxidation of H_2S with O_2, and H_2 oxidation with fumarate. On the other hand, Fe^{2+} oxidation with sulfate, or H_2S oxidation with CO_2 as electron acceptor, respectively, are endergonic reactions, and, thus, will not combine into catabolism. This figure also reveals **how uniquely modular microbial catabolism can be.** With a few interesting modifications, many half-reaction pathways can function in microbial metabolism either in the

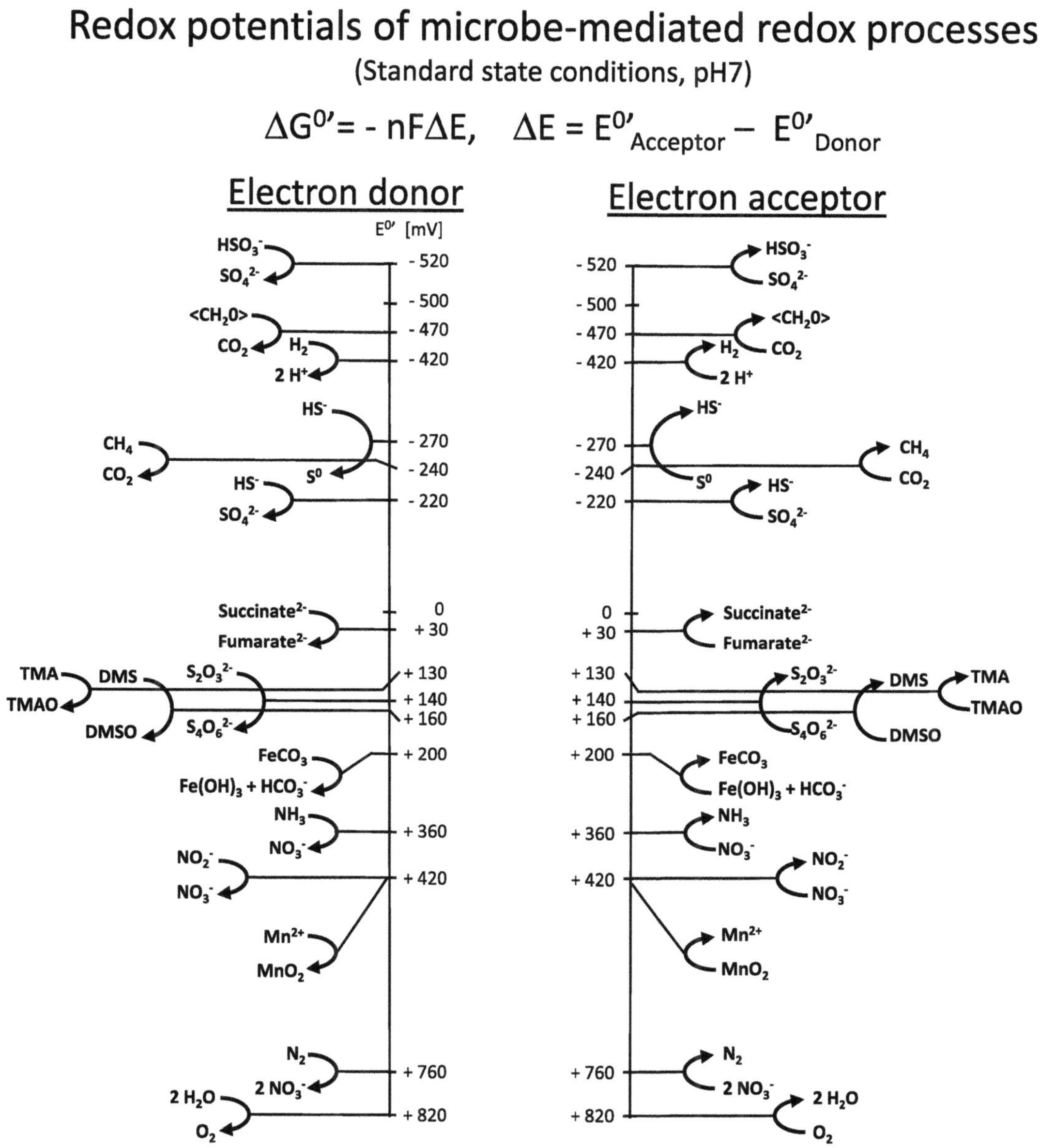

Fig. 4.2 Redox potentials of microbially-mediated, catabolic redox reactions grouped in electron-donating and electron-accepting half reactions according to their redox potential. The electron-donating half reaction is written in the oxidizing direction as they occur if the reduced compound is electron donor; the electron-accepting half reactions as they occur if the oxidized compound is electron acceptor. $E^{0'}$ is for soluble compounds at 1 M concentration in H_2O, pH 7, and for gases at 1 bar (10^5 Pa) pressure. S^o, F^{3+}, and Mn^{4+} are considered to be insoluble at pH 7. According to the Nernst equation, if the redox potential of the electron donor is more negative (i.e., higher up in the tower's left side) than that of the electron acceptor, then the reaction is exergonic (an arrow connecting the left and right side would point down). The number of electrons transferred in conjunction with the difference in redox potentials enables is required to calculate the ΔG^0 of the reaction via the Nernst equation. The abbreviation <CH_2O > indicates biomass carbon with an oxidation state of the average carbon (0) approximated by formaldehyde. The parenthesis indicates that it is a conceptual rather than an actual molecule. See text for details

oxidative or reductive direction depending on overall catabolic redox process (Chap. 8). For example, H^+ reduction to H_2 can be an electron-accepting pathway in syntrophic microbes, whereas H_2 oxidation can be an oxidative pathway in methanogens, sulfate reducers, or aerobic Knallgas bacteria (Chaps. 9 and 10). These couplings based on microbially-mediated redox reactions are of ecological and geochemical significance as cycles of chemical elements (e.g., H, O, S, C) on a global scale. For example, the carbon cycle is microbially coupled to the hydrogen, the sulfur, to the nitrogen, and the oxygen cycles; and the sulfur cycle to the oxygen cycle. Thus, Fig. 4.2 allows for a rapid assessment of which reactions are exergonic and, therefore, which may occur in microbial ecosystems and in individual microbes.

Notably, the values listed in Fig. 4.2 are $E^{0\prime}$ rather than in situ E values. For example, $E^\prime$ of the $2H^+/H_2$ couple can deviate in marine sediments by more than 200 mV from $E^{0\prime}$, due to the very low in situ H_2 concentrations. In case of the S^0/HS^- couple, $E^\prime$ in marine sediments can deviate from $E^{0\prime}$ by more than 200 mV, in which the HS^- concentrations are μM rather than M. The same is valid for $E^{0\prime}$of the Fe^{3+}/Fe^{2+} couple and the Mn^{4+}/Mn^{2+} couple considering that Fe^{3+} forms insoluble hydrates and Mn^{4+} an insoluble oxide. These four examples have in common that the concentration of the oxidized form, H^+, S^0, Fe^{3+}, and Mn^{4+} remains essentially constant, leading, therefore, to large changes in redox potential, when the concentration of the reduced forms decrease (see Chap. 3). In Fig. 4.2, the in situ redox potentials of the 2 H^+/H_2, S^0/H_2S, Fe^{3+}/Fe^{2+}, and Mn^{4+}/Mn^{2+} couples can, thus, be substantially more positive, which makes them considerably better electron acceptors than predicted from their position in the left column.

4.3 Classification of Microbial Metabolism

Based on whether a microbe uses light energy or chemical energy in its catabolism to drive ATP synthesis, the microorganism is a phototroph or a chemotroph. If a microbe derives at least 50% of its cell carbon in its anabolism from CO_2, it is defined as an autotroph, otherwise it is a heterotroph. Prokaryotes, which use N_2 as nitrogen source, are referred to as diazotrophs. Diazotrophy is widespread amongst prokaryotes but not found in eukaryotes.

Among all living chemotrophs, only prokaryotes can use an unprecedented diversity of organic and inorganic electron donors and electron acceptors in their catabolism. If both donor and acceptor are inorganic, a microbe is a lithotroph; if electrons are derived from an organic compound, a microorganism is an organotroph independent of whether the electron acceptor is organic or inorganic. If the electron acceptor is O_2, then the microbe is an aerobe, if not, then it is an anaerobe. If the electron acceptor, such as O_2, is acquired from a microbe's external environment, the process is referred to as respiration; if the electron acceptor, such as pyruvate or acetyl-CoA, is derived internally from oxidative catabolism of the organic electron donor or is H^+, the process is called fermentation. Lithotrophy and anaerobic respiration, with the exception of fumarate respiration, are found only in prokaryotes.

Catabolism of phototrophs (Sect. 4.4.4) is a redox process, similar to that of chemotrophs; however, the electron flow is cyclic and driven by light energy rather than by chemical energy. In case of phototrophs with anoxygenic photosynthesis, the electron donor for anabolic reduction reactions can be inorganic (e.g., H_2S or Fe^{2+}) or organic (e.g., isopropanol). In phototrophs with oxygenic photosynthesis, the electron donor is H_2O.

Organotrophs Versus Heterotroph

There is a often inconsistent use of nomenclature to describe the use of organic compounds as electron donor in the catabolic redox process of chemotrophs. Based on the above definitions, "organotroph" describes a microbe that uses an organic compound as catabolic electron donor,

(continued)

whereas heterotroph indicates merely that more than 50% of a cell's anabolic carbon is derived from organic compounds. The confusion is based on the fact that most organotrophs are also heterotrophs with only a few exceptions. For example, bacteria growing on formate (organic) as electron donor are organoautotrophs. The importance of a clear distinction is not only semantics, because catabolic fluxes can be 2–10 higher than anabolic fluxes, and the flux of catabolic substrates and end products is chemically much more homogeneous. Moreover, a catabolic proteome is under different selection than an anabolic proteome (see below).

4.4 Major Catabolic Groups of Microbes

Catabolism and anabolism are intricately connected via fluxes of ATP, electrons, and (organic) intermediates between both metabolic networks (see Fig. 4.1). How these fluxes are connected defines some of the distinctive features of a microbe's physiology and metabolic ecology. In general, four catabolic types of microbes are distinguished:

4.4.1 Respiring Prokaryotes

Most respiring prokaryotes use in their energy metabolism an **external** electron acceptor such as O_2, nitrate, nitrite, Fe^{3+}, Mn^{2+}, dimethysulfoxide, trimethylamine-N-oxide, fumarate, S^0, or H^+. The standard redox potentials range from +810 mV to −420 mV (Table A.1 in the Appendix). A general scheme of a respiring microbe is given in Fig. 4.1.

4.4.2 Fermenting Prokaryotes

Fermentative catabolism is defined here as the electron acceptor being **intracellular** and being a metabolite derived from the organic electron donor, or being a proton (Fig. 4.3a). This is indicated by the yellow circle in Fig. 4.3a. Well-known examples of fermenting microbes are lactic acid bacteria (pyruvate from glucose oxidation

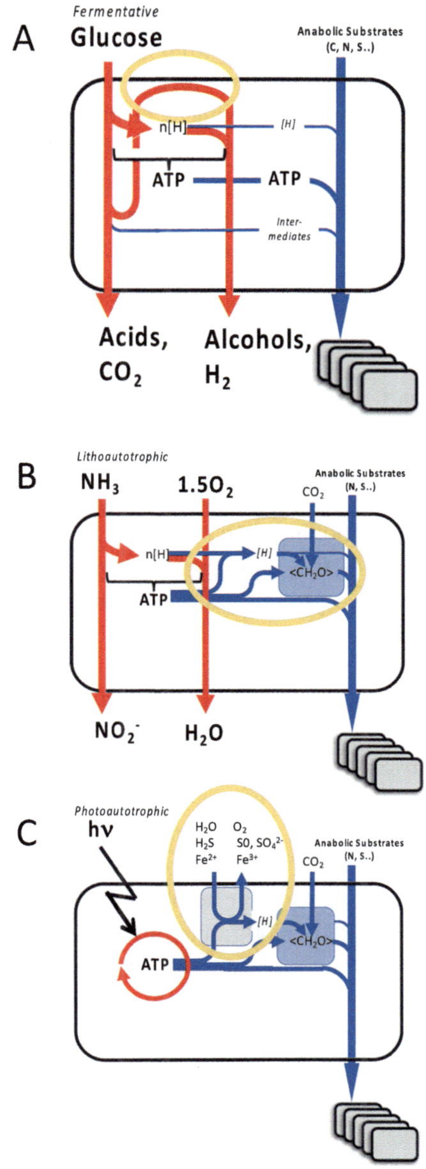

Fig. 4.3 Coupling between catabolism and anabolism in fermenting (**a**), lithoautotrophic (**b**), and phototrophic (**c**) microbes. In fermentation, the catabolic electron acceptor is internal and derived from the primary organic substrate (electron donor). Yellow circles highlight the distinctive metabolic features of each physiological group. See text for details

becoming the electron acceptor) or saccharolytic Clostridia (acetyl-CoA from glucose oxidation and H^+ becoming the electron acceptor). The pathways and underlying metabolic steps are extensively discussed in Chaps. 7–9. There are fermentations, such as of pairs of amino acids via "Stickland Fermentation," where a second and external organic substrate is metabolized into an electron acceptor (S_2) (see Chap. 9), which is an exception to the above definition.

The term "fermentation" is often used inconsistently, based on the scientific field, history, and context of discussion. The founding Latin verb "fervere" means "surging" or "boiling" and pictorially reflects the escape of CO_2 during yeast conversion of fruit juice or malt grain to wine and beer, respectively. As the understanding of this overall process became more refined in light of metabolic and physiological principles, "fermentation" has been used subsequently to describe catabolic transformations of organic compounds in the absence of air, or more specifically, in the absence of O_2. However even in today's biotech industry, the aerobic operation of a 50,000 or 100,000 L microbial "fermenter" is still referred to as "fermentation" although pure O_2 is pumped into the vessel for driving aerobic respiratory catabolism!

A second inconsistent use of "fermentation" is that it is used in conjunction with either the substrate being fermented or the product being formed. For example, "glucose fermentation" is used to indicate that glucose is the substrate, while "lactate fermentation" is often used to indicate that lactate is the end product. Lactate, however, can also be a substrate of a fermentation to propionate. Thus a priori, the term "lactate fermentation" leaves it ambiguous as to whether the meaning is on the substrate or end product of a fermentation process. To eliminate such ambiguity and to provide a consistent nomenclature throughout this book, we use "fermentation" in conjunction with the **organic catabolic electron donor substrate**, and **not** to indicate the end product formed. For example, conversion of 1 glucose to 2 lactate$^-$ in the absence of an inorganic electron acceptor is referred to as **glucose fermentation** or more specifically, **lactate-forming glucose fermentation** (glucose can also be fermented to butyrate, to acetate, or to ethanol).

On a few occasions, some terms frequently found in pertinent literature will be used although they are inconsistent with this definition here. For example, the intent of the description of "homofermentative and heterofermentative lactic acid fermentation" is to indicate that lactate is the exclusive end product in homofermentative lactic acid-forming fermentation but that products in addition to lactate, such as ethanol and CO_2, are formed in heterofermentative lactic acid-forming fermentation. Being technically correct would make the description awkwardly complex.

4.4.3 Chemolithoautotrophic Prokaryotes

The capacity for lithotrophy (Fig. 4.3b), i.e., **the use of inorganic compounds both as electron donor and electron acceptor**, is unique to prokaryotes! The combination of possible electron donor and acceptors is determined by thermodynamics, i.e., the Nernst equation and the difference in redox potential between electron acceptor and electron donor (Chap. 3, see Fig. 4.2). Although some lithotrophs can be aerobes, chemolithotrophic microbes have several intrinsic energetic and metabolic challenges. Because their inorganic catabolic substrate is NOT a carbon compound, most chemolithotrophic prokaryotes are also autotrophs where cell carbon is assimilated via CO_2 fixation.

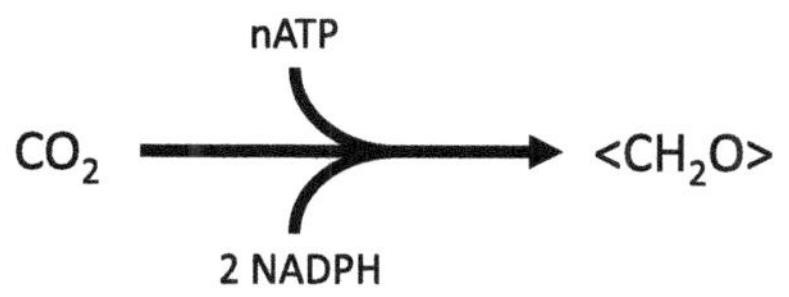

Fig. 4.4 Metabolic requirements for CO_2 fixation (assimilation) in autotrophy. Electrons, typically in form of NADPH or reduced ferredoxin, are required to reduce CO_2 to the oxidation level of cell carbon. Energy in form of ATP is required to convert the C1 compound CO_2 into assimilable carbon. See text for details

Autotrophic CO_2 fixation requires two metabolic processes (Fig. 4.4):

1. A reductive anabolic CO_2 fixation pathway to reduce the CO_2 carbon from the oxidation level of +4 to ~zero for assimilation into cell mass (<CH_2O>). This requires a net flux of electrons, typically via NADPH but also via reduced ferredoxin, from catabolism to anabolism, and
2. ATP as energy source for assimilating CO_2 into simple intermediates, such as acetate (acetyl-CoA), pyruvate, or glyceraldehyde-3-phosphate. Such ATP requirement is met by lithotrophic catabolism. Between 1 2/3 and 3 ATP are consumed per CO_2 assimilated, depending on the type of CO_2 fixation pathway involved.

These metabolic costs of (1) & (2) are in addition to the biosynthetic costs of monomer syntheses and polymerizations and explain the Y_{ATP} of <10 g cell(dw)/molATP in autotrophs (see below).

There is a second energy requirement in lithotrophic microbes, which can be more costly than the CO_2 fixation pathway: the cost of **activating electrons** from the inorganic compound to the redox potential of NADPH and ferredoxin. The standard state redox potential of $NADP^+$/NADPH is −320 mV. As many reduced inorganic compounds have redox potentials more positive than −320 mV (Fig. 4.2), metabolic energy is required to "activate" these electrons to the redox potential of −320 mV for $NADP^+$ reduction. Figure 4.5 illustrates this challenge: the more positive the redox potential of the inorganic environmental electron donor is, the more costly is the reduction of $NADP^+$ to NADPH. For example, for a nitrite-oxidizing chemolithoautotrophic microbe, such as *Nitrospira* or *Nitrobacter* sp., the equivalent of 5.7 mol ATP per mol CO_2 reduced to <CH_2O> (+ 285.6 kJ/4e$^-$,

Fig. 4.5 The cost of autotrophy. Energy (ΔG) required for activation of electrons derived from an inorganic electron donor for reducing $NADP^+$ as a function of the redox potential of the donor. Calculations are based on the free energy required for producing 2 NADPH from 4e$^-$. While the formation of NADPH from H_2 as electron donor is exergonic, the more positive the redox potential of the donor is, the more endergonic is the reaction. See text for details

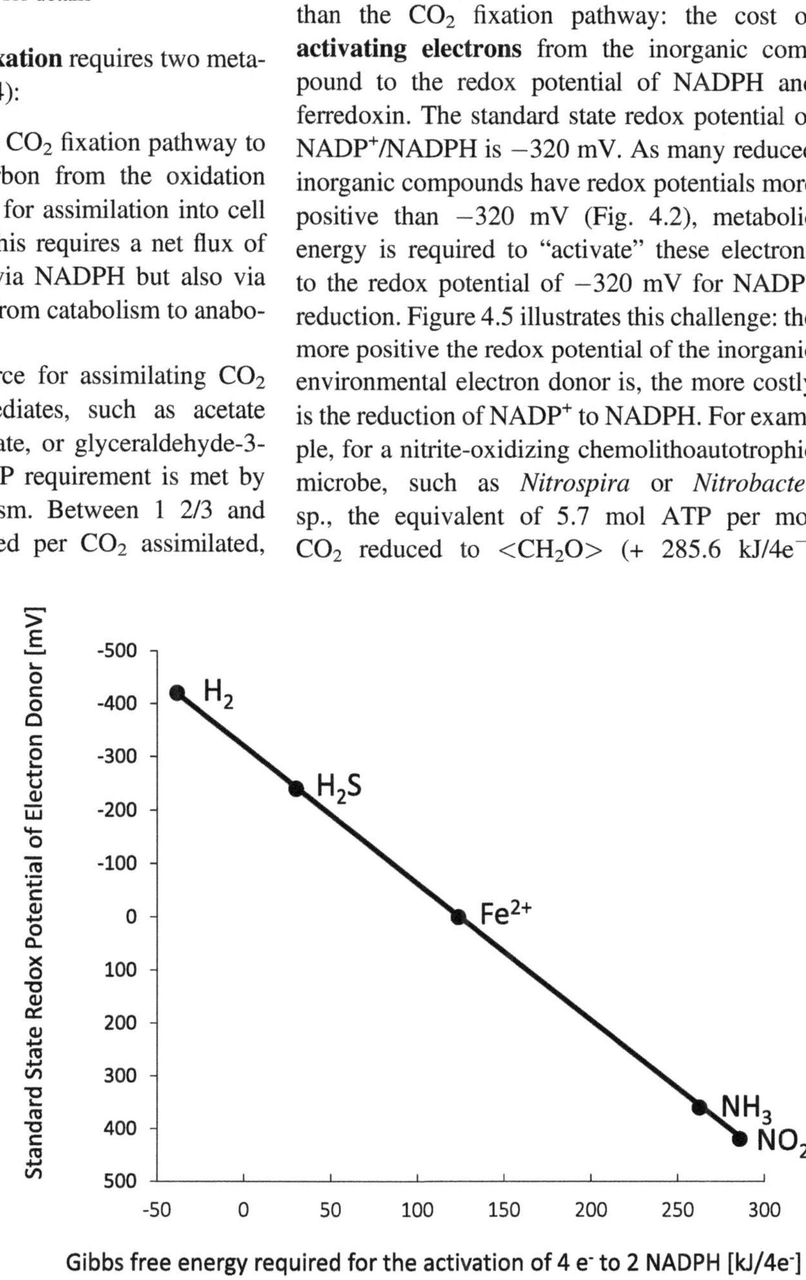

phosphorylation potential of ATP: + 50 kJ/mol ATP) are required **just for electron activation** and are in addition to the 1–3 ATP consumed in the CO_2 fixation pathway! This indicates that when only −74 kJ/mol nitrite (≈1 ATP) are conserved in the catabolic reaction of aerobic nitrite oxidation, the microbe needs to oxidize approximately 5 mol nitrite to nitrate to obtain just the energy for activating $4e^-$. These costs are in addition to 2 mol nitrite oxidized for obtaining the four electrons for reduction of 1 mol CO_2, because these electrons are NOT available for energy conservation! Therefore and as indicated in Fig. 4.5, significant metabolic energy (in form of ATP, $\Delta\mu H^+$) is required for the activation of electrons for CO_2 fixation in some chemolithoautotrophic microbes, in addition to reactions in the CO_2-fixing pathway.

4.4.4 Phototrophic Prokaryotes

Phototrophic microorganisms use light energy for ATP synthesis. Their energy metabolism is based on **cyclic photophosphorylation**, which does not involve a net flow of electrons in their catabolism (Fig. 4.4c, see Chap. 6). Consequently, no electrons can be re-routed to biosynthesis, including CO_2 fixation, unless they are replenished. Therefore, electrons for CO_2 fixation are acquired by phototrophs via pathways that oxidize anabolically reduced environmental compounds, such as H_2O, H_2S, S^0, or Fe^{2+}. As a consequence and stoichiometric to the amount of CO_2 fixed, the oxidized form of the environmental electron source, in the above example O_2, S^0, SO_4^{2-}, or Fe^{3+}, is released from cells to the environment. The latter compounds, in turn, can provide new metabolic opportunities for other microbes as electron acceptor, such as O_2, SO_4^{2-}, or Fe(III). This is the case, for example, in microbial mats in salt marshes (or the Winogradsky column in the lab), or in numerous aqueous environments leading, over geological times, e.g., to the rise of molecular oxygen in the atmosphere.

4.5 Major Anabolic Metabolism

In contrast to catabolic pathways, there is a relative paucity in diversity of anabolic pathways in the prokaryotic world. Pathways for amino acid and nucleotide biosynthesis are rather conserved across phylogenetically different prokaryotic groups, including Archaea, although alternative pathways exist. In a first approximation, the amount of energy required for monomer synthesis is comparable between pathways among different microbes. Monomer synthesis per se is only moderately costly (i.e., ATP-consuming). For example, the synthesis of the key anabolic metabolite acetate or pyruvate (both are intermediates in many synthetic pathways of monomers) from CO_2 and H_2 is exergonic by about −100 kJ/mol, suggesting that a priori no metabolic energy is required for monomer synthesis. ATP-dependence in many anabolic pathways is often to drive reactions (1) kinetically to offset demand on proteome space (see below) or (2) to mediate an irreversible, committed flux of carbon toward new biomass.

Protein synthesis is the energetically most costly process in a growing cell, amounting to about 75% of the cellular ATP consumption. In particular, the process of **translation** for polymerizing amino acid monomers into peptides is thermodynamically demanding and at the same time highly inefficient. The energy content of a peptide bond is ~−16 kJ/mol peptide bond. Yet, for the synthesis of one peptide bond involving ribosomes and aminoacylated-tRNA, four ATP equivalents (2 ATP equivalents for activation to aminoacyl-tRNA and 2 GTP for elongation) are required, amounting to a total of −200 kJ of cellular energy per mol peptide bond formed. In addition, six additional ATP equivalents are required to transcribe one codon (i.e., equivalent to one amino acid) into mRNA, although a single mRNA is "catalytic" in the sense that it participates in hundreds to thousands of translation reactions. Thus, only ~8% of the energy consumed in translation is retained in the actual peptide bond, and overall 92% of the energy consumed is released as heat (see Fig. 3.1)!

Because of the molecular architecture of the translation machinery, most of the heat is released in regulatory steps and is associated with the large decline in entropy by condensation of amino acids into specific peptide sequences.

4.6 Coupling of Catabolism and Anabolism

When growth rate is limited by the catabolic substrate, catabolism and anabolism are kinetically and thermodynamically fully coupled, because of the fixed stoichiometry between catabolic substrate utilization and the amount of newly formed biomass. Because the pathways and machineries for both monomer and macromolecular syntheses are generally conserved, the specific amount of ATP consumed in these pathways is also similar. Overall yield of how much biomass can be formed per mol ATP, $\mathbf{Y_{ATP}}$, is relatively constant across Bacteria and Archaea, when fully coupled. The average empirical value of Y_{ATP} is 10 g ((g cell dry weight)/mol ATP) and refers to microbes growing in a medium with an organic carbon source, such as carbohydrates, amino acids, acetate, or succinate. Thus, a catabolic reaction with $n_{ATP} = 2$ provides the energy to produce 20 g cells (dried mass) per mol catabolic substrate. Autotrophic microbes synthesize precursors for monomers from CO_2, and for the reasons listed above, their Y_{ATP} is lower, ~4–6 g/mol ATP to account for the energy required for electron activation and CO_2 fixation. On the other hand, some microbial pathogens, such as *Rickettsia* or *Chlamydia* that grow in eukaryotes as intracellular parasites, have a Y_{ATP} of ~15 g (dw)/mol ATP because most of the monomeric precursors do not have to be synthesized and are parasitized from the host (see also Chap. 3).

4.6.1 Coupling Based on the Flux of ATP, Electrons, and Small Intermediates

Flux of ATP: There is a flux of ATP from catabolic to anabolic pathways where most of anabolic ATP consumption is in translation (see above). It is generally assumed that during energy-limited growth, catabolism and anabolism are coupled, meaning that all ATP that is conserved via catabolic metabolism at a specific n_{ATP} is consumed in diverse anabolic pathways and processes (see also in Chap. 5 Fig. 5.1). On the other hand, a limitation by an anabolic substrate, e.g., N or P, can lead to catabolism in the absence of ATP consumption via biosynthesis. Under those conditions, there is an **"uncoupling"** between the flux of a substrate in a catabolic pathway and biosynthesis.

In coupled cells, catabolic and anabolic processes are kinetically linked through the ATP and ADP pool. Observed initially in electron transport phosphorylation in mitochondria, the rate of electron flux through the respiratory chain is high at low ADP concentration and low at high ATP concentration. In such "respiratory control," O_2 is only consumed when ATP is low. When respiration is mechanistically uncoupled from ATP synthesis in the laboratory, e.g., via addition of protonophores, the rate of O_2 consumption is increasing. A similar tight coupling between the rates of catabolism and ATP consumption is found in methanogens, such as *Methanosarcina barkeri*, when forming methane from methanol and H_2.

Transfer of electrons: A second important connection between catabolism and anabolism is the transfer of electrons for biosynthesis, in particular for adjusting the oxidation state of environment-derived C, N, or S for incorporation into new biomass. For example, if nitrate or N_2 is the only N source, net anabolic reductions are required to reduce N from an oxidation state of +5, or 0, respectively, to −3, which is the oxidation state of N in amino and nucleic acids.

Quantitatively important is the flux of electrons for CO_2 fixation, because half of a cell's dry mass is carbon. If the carbon source in the growth medium is at an oxidation state other than ~0, then electrons are diverted in chemotrophic prokaryotes either from or into catabolism, respectively, to change the oxidation of carbon to ~0 for assimilation into biomass. Catabolism, therefore, serves as a source or sink of anabolic electrons, depending on whether the environmental carbon is more oxidized, e.g., as in oxalate or formate, or more reduced, e.g., as in alkanes or aromatic hydrocarbons.

In general, the electron carrier in **anabolic reactions** is **NADPH** or **ferredoxin** (see Chap. 6); however, there are exceptions. Formation of NADPH can occur through $NADP^+$-specific reactions in oxidative catabolic pathways, for example in:

- *glucose-6-phosphate dehydrogenase* and *6-phosphogluconate dehydrogenase* in the oxidative pentose phosphate pathway,
- *isocitrate dehydrogenase* of the citric acid cycle,
- *glucose dehydrogenase* in the Entner–Doudoroff pathway
- non-phosphorylating *glyceraldehyde-3-phosphate dehydrogenase* in Embden–Meyerhof pathway

Furthermore, NADPH is also formed by energy-dependent *transhydrogenases*, by *ferredoxin:NADP$^+$ oxidoreductases*, by electron-bifurcating *NADH-dependent ferredoxin:NADP$^+$ oxidoreductase* and *NADP$^+$-reducing hydrogenases*. While the standard redox potential of the NAD^+/NADH and $NADP^+$/NADPH is −320 mV, the in vivo concentration resulting in redox potentials of −280 and −380 mV are sufficiently different (Table A.2 in the Appendix; see below).

Reduced ferredoxin, when formed anabolically, is formed in *electron-bifurcating hydrogenase* reactions or via reverse electron transport via Rnf (see Chap. 6).

Catabolic intermediates: In most chemoorganotrophic microorganisms, metabolic intermediates of catabolic pathways, such as acetyl-CoA, pyruvate, oxaloacetate, phosphoenol pyruvate, 2-oxo-glutarate, or succinyl-CoA are also precursors for syntheses of monomeric building blocks, such as fatty acids, nucleotides, amino acids, tetrapyrroles, or other cofactors.

It is important to distinguish in these microbes the source and fraction of carbon that is assimilated into cell biomass relative to the fraction of carbon used as catabolic electron donor.

4.6.2 Coupling Based on the Amount of Energy Conserved in Catabolism

As described above, catabolism and anabolism in growing cells are stoichiometrically coupled by the **ΔG of the catabolic reaction**. In chemoorganotrophic microbes, the organic substrate is both the catabolic substrate and the carbon source for anabolic biosynthesis. How does the ΔG of the catabolic reaction determine the fractional flow of carbon to either cell mass or catabolic end product? Consider the example of catabolic glucose utilization:

$$C_6H_{12}O_6 + 6O_2 \rightarrow 6CO_2 + 6H_2O \quad \Delta G^{0\prime} = -2872\ \mathrm{kJ/mol}$$

$$C_6H_{12}O_6 \rightarrow 2C_3H_5O_3^{\ -} + 2H^+ \quad \Delta G^{0\prime} = -198\ \mathrm{kJ/mol}$$

With an efficiency of ~60%, synthesis of 1 mol ATP under cellular irreversible pathway condition requires about −80 kJ/mol ATP (Chap. 3). During aerobic oxidation of glucose, the energy of ~−3000 kJ/mol glucose released in vivo enables synthesis of no more than 36 ATP, which is sufficient to produce 360 g cells (Y_{ATP} = 10 g (dw)/mol ATP). On the other hand, fermentation of glucose to 2 lactate releases ~−200 kJ/mol glucose, which is sufficient to synthesize 20 g cells per mol glucose. Consequently, in order to produce 360 g cells, microbes performing lactic acid-forming fermentation need to catabolically consume 18 times the molar amount of glucose relative to aerobic cells, in order to obtain the same amount of energy (ATP) to synthesize 360 g cells. Conversely, 80–90% of the carbon of glucose available to a lactic acid bacterium is converted to lactate, whereas in glucose respiration about 70% of glucose carbon is incorporated into biomass (Fig. 4.6).

This calculation illustrates that it is the ΔG of the catabolic reaction which determines in

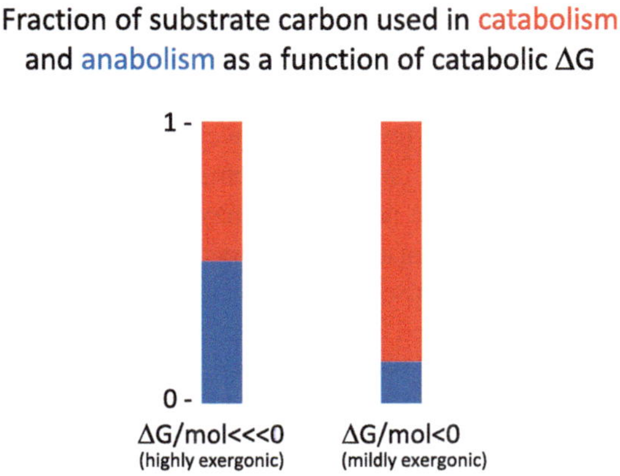

Fig. 4.6 Fraction of organic substrate used by a chemoorganotrophic microorganism in catabolism or anabolism as a function of catabolic ΔG where the organic compound is both a catabolic substrate and the carbon source for biosynthesis. See text for details

coupled microbes the ratio of end product-to-biomass produced from the organic substrate. One direct consequence of this relationship is that industrial fermentation processes with the objective to produce excreted end products, such as ethanol, succinate, or lactate use microbes with a small negative catabolic ΔG per mol substrate (Fig. 4.6). On the other hand, to produce significant amounts of single cell protein from glucose or methane, for example, a catabolic reaction with a large negative ΔG/reaction is desirable.

4.6.3 Coupling Based on the Overall Metabolic State

As discussed in Chap. 3, the amount of energy required for ATP synthesis in a cell is determined by $\Delta G = \Delta G^0 + 5.7 \lg \frac{[\text{ATP}]}{[\text{ADP}][P_i]}$. Thus, this **phosphorylation potential** is a useful indicator of the amount of energy available to synthesize ATP invivo as it integrates the interconnected concentration of ATP, ADP, and P_i. The ratio of $\frac{[\text{ATP}]}{[\text{ADP}][P_i]}$ can vary over 6 orders of magnitude depending on whether a cell is starving or growing (see Chap. 3).

The overall energetic state of a cell is assessed by the **adenylate charge**. Intracellular ATP and ADP concentrations are relative to the entire adenine nucleotide pool which includes AMP as well as the GTP pool, and, thereby, reflect anabolic energy use. Some anabolic reactions consume ATP and form either ADP or AMP, such as in amino acid activation (*aminoacyl t-RNA synthase*). Moreover, *myokinase*, which is present in most cells, catalyzes the conversion of AMP + ATP ⇌ 2 ADP, thereby tying AMP into the ATP/ADP pool. Steady-state concentrations of ATP, ADP, and AMP are enzymatically "connected" and in thermodynamic equilibrium. These concentrations are a good proxy for the **overall catabolic and anabolic fluxes**; in other words, for how well a cell kinetically coordinates and couples catabolism and anabolism (Fig. 4.1). Because many anabolic and catabolic enzymes catalyzing ATP- or ADP-dependent reactions are allosterically regulated by these adenine nucleotides, it has been useful to express their relationship as the **energy** or **adenylate charge** (see also Sect. 3.4.2), as defined by

$$\text{Energy charge} = \frac{[\text{ATP}] + 0.5[\text{ADP}]}{[\text{AMP}] + [\text{ADP}] + [\text{ATP}]}.$$

In growing cells, the adenylate energy charge is generally above 0.9 (Fig. 4.7). A low energy charge, when the adenylate pool is primarily in the AMP form, is indicative of a de-energized cell

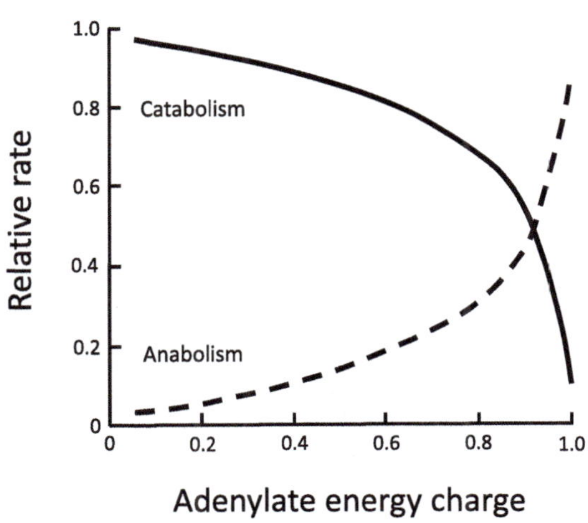

Fig. 4.7 The adenylate energy charge = $\frac{[\text{ATP}]+0.5[\text{ADP}]}{[\text{AMP}]+[\text{ADP}]+[\text{ATP}]}$, regulating activities of catabolic and anabolic enzymes involved in ATP formation and ATP utilization. See text for details

and of insufficient "trapping" of the ATP pool, because cells turn over ATP very rapidly, up to 2000 times per hour (see legend to Fig. 5.1). (The turnover of NADH (→NAD + →NADH) is almost at the same rate, although it is lower in aerobes but higher in many anaerobes). In fact, a value of ~0.5 in *E. coli* and other prokaryotes correlates well with a lack of viability. As active metabolism commences, more of the adenylate is in the ATP form and energy charge increases. In viable cells, the energy charge is relatively constant and close to 0.85, which helps to understand why a cell uses the adenine nucleotide pools for allosteric regulations to kinetically coordinate catabolic and anabolic enzyme activities. The close-to-invariant nature of ~0.85 in growing cells indicates again tight kinetic coupling of catabolism and anabolism (see also Chap. 3).

For obtaining meaningful metabolic data, experiments should be conducted only at an adenylate energy charge of ≥0.8.

4.6.4 Coupling Based on Proteome Allocation

When grown aerobically oxidizing glucose to CO_2 at moderate growth rate, *E. coli* uses the Embden–Meyerhof pathway and the TCA cycle to oxidize glucose to CO_2, and the respiratory chain to oxidize NADH (more in Chaps. 6–8). Most ATP is conserved by respiration. Under these conditions, the microbe's total catabolic proteome is composed of a "fermentative" proteome (Emden–Meyerhof pathway and fermentative enzymes) and a "respiratory" proteome (TCA cycle and respiratory enzymes) (Fig. 4.8). When grown anaerobically by fermenting glucose, *E. coli* also uses the Embden–Meyerhof pathway as well as enzymes of mixed acids-forming fermentation (see Chap. 9).

When *E. coli* grows fast, there is a demand for higher biosynthetic capacity, which is kinetically achieved by higher enzyme concentration (increasing $[E_T]$ in $v = k_{cat}\ [E_T]$, see Chap. 5) for monomer biosynthesis and a higher ribosome content. This shift increases the demand for the anabolic proteome in the finite cellular volume. Faster growth (higher growth rate) requires also an accompanying higher flux of ATP, J_{ATP}, to satisfy the increased demand of energy flux for biosynthesis and protein translation (Fig. 4.8). Since in *E. coli* growth rate is proportional to the anabolic proteome, the catabolic proteome needs to be smaller and mediate a higher J_{ATP} at higher growth rate. Theory predicts a faster growth can be traded off by a less efficient use of the energy resource (Chap. 5). Based on its distinctive glucose metabolism, *E. coli* reduces the size of its total catabolic proteome by reducing the size of the respiratory proteome coupled to a smaller increase in fermentative proteome

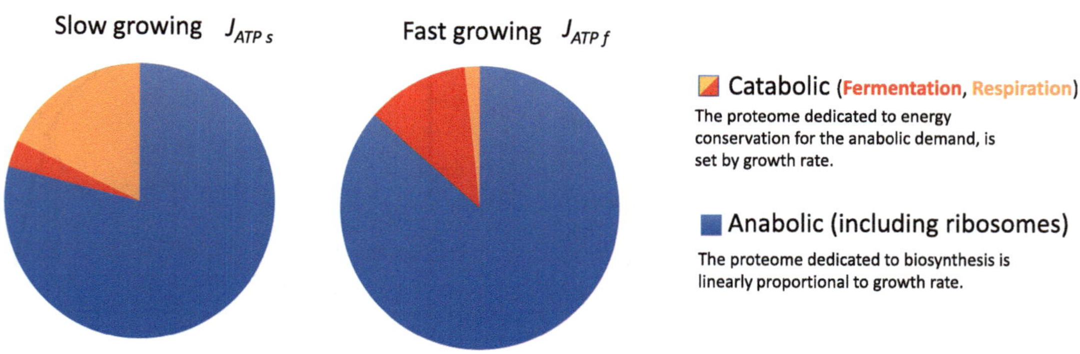

Fig. 4.8 Schematic representation of allocation of the catabolic and anabolic proteome in slow- and fast-growing *E. coli*. Ratios of both proteomes are not to scale. At faster growth rates, the relative fraction of the fermentative proteome increases relative to the respiratory proteome resulting in a higher J_{ATP}, but lower molar growth yield. About 10–20% of the total cell protein is catabolic proteome also in other microorganisms. Modified after Basan et al. (2015). See text for details

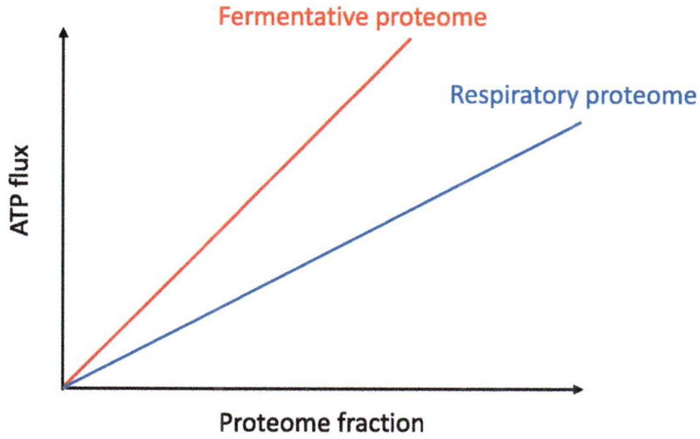

Fig. 4.9 Partitioning of proteome fractions into flux components in *E. coli*. For the same ATP flux, a fermentative proteome can be about 50% the size of a respiratory proteome in *E. coli*. Modified after Basan et al. (2015)

(Fig. 4.8). In E. *coli*, for the same fraction of proteome, J_{ATP} of the fermentative proteome is about twice as large compared to a complete oxidizing respiratory proteome (Fig. 4.9). In other words, *E. coli* can produce fermentatively a higher flux of ATP by a smaller, predominantly fermentative, proteome than by a fully glucose-oxidizing fermentative plus respiratory proteome. This is achieved at the trade-off that no or little respiration is involved and that the available resource glucose is only used inefficiently, and acetate is excreted.

A fractionation of an overall metabolic pathway in two shorter additive pathways became genetically fixed in the SCV (glucose → 2 acetate) and LCV (acetate → $2CO_2$) in the Julian Adams experiment. When *E. coli* grows slower and contains a small anabolic proteome, the catabolic proteome includes respiration to enable a more efficient use of all available energy (glucose plus oxygen) (Fig. 4.8).

It is important to note that while *E. coli* and many other fast-growing microbes of a similar ecological strategy and metabolism use such reallocation of cellular resources in response to changes in substrate flux and growth rate, other microbes pursue different strategies. When growth rate is limited by catabolism, the methanogenic archaea *Methanococcus maripaludis*, for example, contains an invariant catabolic and anabolic proteome, including a fixed set of ribosomes, and metabolism is not altered, which is mechanistically impossible in hydrogenotrophic methanogens. Thus, while cellular resource allocation in *E. coli* provides an interesting view into one strategy of microbial cellular resource allocation, this is by no means a universal law.

Further Reading

Basan M et al (2015) Overflow metabolism in *Escherichia coli* results from efficient proteome allocation. Nature 528:99–103

Hui S, Silverman JM, Chen SS, Erickson DW, Basan M, Wang J, Hwa T, Williamson JR (2015) Quantitative proteomic analysis reveals a simple strategy of global resource allocation in bacteria. Mol Syst Biol 11:e784

Li GW, Burkhardt D, Gross C, Weissman JS (2014) Quantifying absolute protein synthesis rates reveals principles underlying allocation of cellular resources. Cell 157:624–635

Microbial Kinetics

5

Contents

5.1 Introduction

Describing catabolic or growth processes as "fast" or "slow" is relative and contextual to the fluxes of energy a microbe is exposed to in its environment. As the specific activity of a catabolic pathway for a single substrate in two different microbes can vary greatly, comparing microbes and the architecture of their catabolic pathways needs to be seen in the context of fluxes in the specific microbial environment.

Catabolic and anabolic processes in a growing cell are stoichiometrically linked fluxes (J) (see Chap. 4, Fig. 5.1). For the discussion and mathematical descriptions of this chapter, an ideal, unstructured, planktonic environment is considered, where the flux J_S of the catabolic substrate is limiting growth and ecological success. In other

A. M. Spormann, *Principles of Microbial Metabolism and Metabolic Ecology*,
https://doi.org/10.1007/978-3-031-28218-8_5

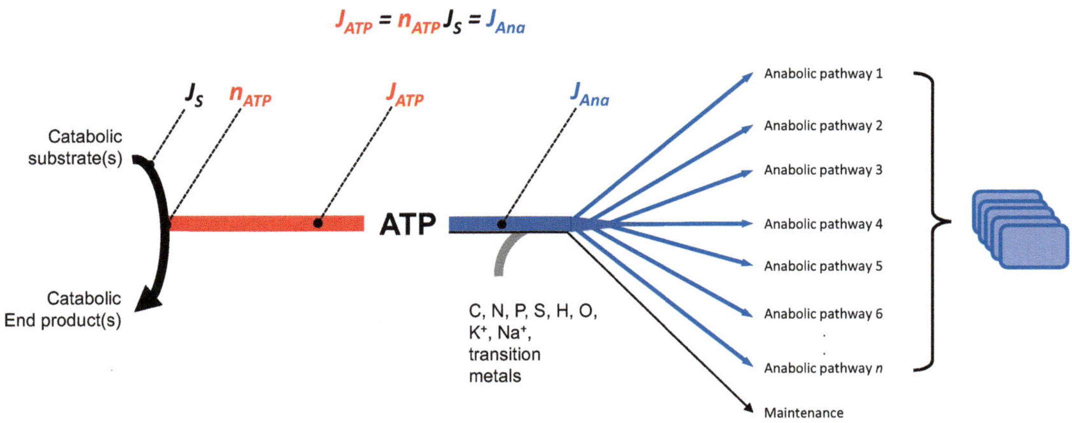

Fig. 5.1 Coupled catabolic and anabolic fluxes in growing, energy-limited cells. In living cells, the rate of ATP synthesis (J_{ATP}) in catabolism and the rate of ATP consumption in anabolism (J_{Ana}) are tightly coupled. Energy charge = ([ATP] + 0.5 [ADP])/([ATP] + [ADP] + [AMP]) (see Chap. 3). Assuming a doubling time of 1 h, for example, an intracellular concentration of ATP at ~10 mM and considering that 100 mmol ATP are required for the synthesis of 1 g cells (dried mass) from glucose, the ATP pool turns over 2000 times in 1 h. In growing microbes with a doubling time of 1 h, close to 10% of the ATP generated in catabolism are consumed for non-biosynthetic processes. Since there is typically only one catabolic pathway for the synthesis of ATP compared to hundreds of anabolic pathways, which consume ATP, catabolic enzymes are present in much higher concentration and specific activities than anabolic enzymes. Only about 10–20% of the cytoplasmic proteome space is allocated to catabolic enzymes in fast-growing cells, and this appears to be independent of whether cells grow using one or substrate combinations at high concentrations or on several different substrates at low concentrations (mixotrophic growth)

words, metabolism and growth rate μ are proportional to J_S; $\mu \propto J_S$. In most planktonic environments, the rate of ATP synthesis, J_{ATP} ($J_{ATP} = n_{ATP} J_S$), determines the growth rate.

The terms "rate" and "flux" are used interchangeably in this chapter with the understanding that flux is equivalent to rate, normalized to an area, in our case that of the cell surface. Using "flux" to characterize a biological process emphasizes the link of a microbe's physiology to its environment because all cellular activities result from compounds take up by or excreted from a cell through its cell surface area (see Chap. 2). This chapter first introduces key aspects of growth kinetics and physiology followed by the link of kinetics and thermodynamics in shaping metabolic pathways and enzyme kinetics.

Some specific examples mentioned in this chapter are introduced more thoroughly in later chapters of this book. The intent of giving these examples here, though, is to emphasize the importance and show early the pervasive implications of rates and fluxes on metabolism and ecophysiology. Thus, examples that appear too specific and seem to lack sufficient background information might need to be re-read after the proper introduction in later chapters.

5.2 Microbial Growth: In Brief

Microbial growth in nature is mostly energy-limited, where an increase in J_{ATP} is followed by a corresponding increase in the rate of ATP consumption in anabolism: $J_{Ana} = J_S\, n_{ATP}$, where n_{ATP} is the number of mol ATP formed per mol catabolic substrate (Fig. 5.1). If fluxes of ATP formation and of ATP consumption in anabolism are imbalanced, e.g., when there is a kinetic bottleneck in one of the many catabolic pathways, then either the catabolic pathway is down regulated and/or switched to a pathway with a smaller n_{ATP} to generate less ATP. Also, the rate of ATP consumption can be increased by coupling to ATP hydrolysis and/or by ATP-dependent production of secondary metabolites. The latter two possibilities are used mainly in aerobic microorganisms because, for thermodynamic reasons described in Chap. 3, aerobes relative to anaerobes generate much more ATP/mol catabolic substrate, and, therefore, growth of aerobes is more frequently anabolism-limited.

In many cases, the catabolic substrate is also used as C and/or N source for biosynthesis. For example, many microbes can grow using glucose both in catabolism and anabolism. The apparent substrate utilization kinetics for glucose in catabolism and anabolism, however, are different. At increasing [S], the rate of anabolism is frequently more rapidly saturated than the rate of catabolism, indicating that at higher [S], growth is kinetically not limited by energy supply. In nature, a higher maximal anabolic rate is generally not selected for.

In most natural habitats, multiple catabolic substrates are present at low concentration, and microbes can metabolize them at the same time (mixotrophy). For example *E. coli* does not metabolize glucose and lactose simultaneously at high concentrations; however, both are metabolized simultaneously at low concentrations. For the sake of simplicity, kinetics of mixotrophic substrate utilization and growth are not discussed here; they are not significantly different relative to a single or pair of substrate(s), although somewhat less straightforward.

Under energy-limiting growth conditions, growth (increase in cell mass, or cell number per unit time) is proportional to J_{ATP}:

$\mu \propto J_{ATP}$ (Fig. 5.1).

Under these conditions, growth is generally exponential. Truly exponential growth is also referred to as balanced growth to indicate that under such conditions all cell components increase at the same rate and, thus, biomass has exactly the same composition at any point in exponential growth. Only when cell growth is balanced, reproducible experimental results can be expected. When the rate of anabolism limits growth rate, then growth is generally linear:

$$(C_2 - C_1)/t = \text{constant}.$$

During **linear growth,** the cellular composition changes between time points, and a specific growth rate or a doubling time cannot be determined in a meaningful way as they change constantly in the linear growth.

5.2.1 Specific Growth Rate μ and Doubling Time t_d

As outlined above, in balanced microbial growth, increase in biomass by cell division is proportional to the increase of all cellular components. The rate of biomass formation when catabolism is limiting is proportional to the dividing biomass. In ideal microbial growth, therefore, increase in biomass is described by:

$$\frac{dC}{dt} = \mu C,$$

where C is the actively dividing biomass or dividing cell population, and μ is the proportionality constant, **the specific growth rate** (Fig. 5.2). The **doubling time** is related to the specific growth rate by $t_d = \frac{\ln 2}{\mu} = \frac{0.693}{\mu}$.

The growth equation can also be derived from the cell number increase from N_1 to $N_2 = N_1\, 2^n$ after n divisions and that cells double with a constant rate $\frac{n}{t} = \upsilon$, where ν is the cell doubling rate, and $\Delta N = 2 \times 10^{\nu\, \Delta t}$. When $N_2 = 2\, N_1$, then $2 = 2 \times 10^{\nu g}$ or $1 = 10^{\nu g}$, where g is the generation time. Solving for g yields $g = 1/\nu$. t_d and g ideally have the same value. Therefore, $\nu = \mu/0.693$. The generation time g is, however, often somewhat smaller than t_d because cultures generally contain a certain percentage of inactive, including dead, cells that do not multiply.

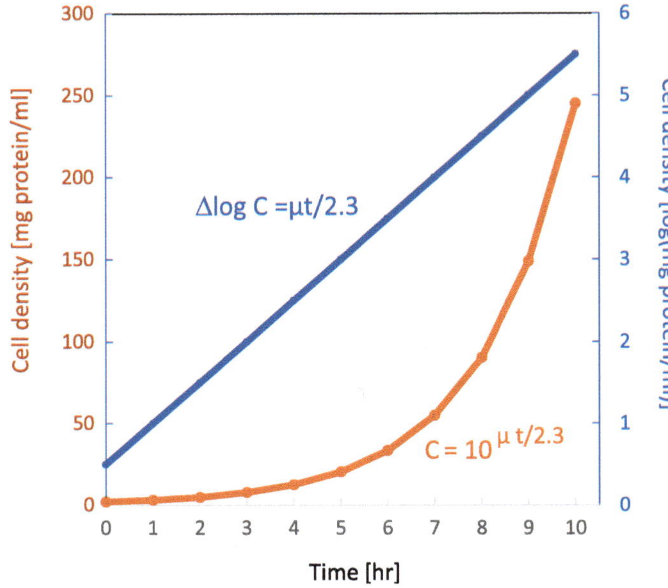

Fig. 5.2 Growth of a microorganism in batch culture with a doubling time $t_d = 1.4$ h and a specific growth rate $\mu = 0.5$ h^{-1}. *C*, cells; *t*, time

Growth of microorganisms can be measured either by counting the cell number per volume or by measuring a parameter proportional to the cell number, such as light scattering or cell mass in mg/volume, or by molecular methods, such as qPCR. Densities of 10^7 cells/ml or less can hardly be detected visually as turbidity or by the most sensitive photometers. At a wavelength of 600 nm and a light pathway length of 1 cm, 10^7 bacterial cells per ml yield an Optical Density (OD_{600}) of ~0.001; 10^{10} cells/ml at an OD_{600} is ~1, which corresponds to, depending on the microbe, ca. 2 mg cells (wet mass) per ml (corresponding to 0.4 mg dried mass per ml). Conversely, maximally 10^{12} to 10^{13} microbial cells can be packed into a volume of 1 mL of cell paste or microbial colony. Counting cells below a concentration of 10^7 cells per ml requires that prior to measurement in the photometer cells are concentrated or that cells are plated on nutrient rich agar, if cells can readily grow at high plating efficiency. A bacterial suspension has to be diluted such that per agar plate less than 100 colonies are formed.

Bacterial suspensions with more than 10^7cells/ml can also be counted by microscopy in special counting chambers. Measuring cell numbers turbidometrically or by counting yields the total cell number (viable and dead cells), whereas counting after growth on agar plates yields the viable cell number. The plating method is only useful when plating efficiency is high in invariant of growth rate. Calibrated qPCR or Fluorescence In Situ Hybridization (FISH) approaches allow for quantifying cells at low density.

5.2.2 Dependence of μ on the Substrate Concentration: Rate and K_S Specialists

The specific growth rate μ generally increases with increasing concentration of the catabolic substrate and saturates at relative high substrate concentrations (Fig. 5.3, see below). The dependence is empirically described by the Monod equation, which resembles the Michaelis–Menten kinetics of enzymes:

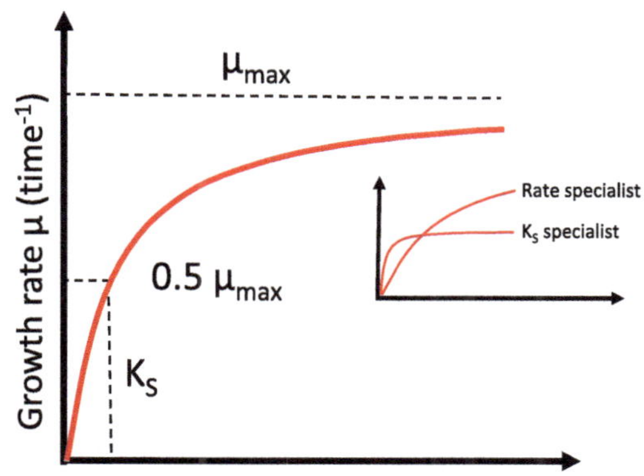

Fig. 5.3 Dependence of specific growth rate on catabolic substrate concentration as determined in a substrate-limited chemostat (see below). K_S is the substrate concentration at which the maximal growth rate is 50%. The inset shows that growth curves of a rate specialist with a higher μ_{max} and a higher K_S relative to a K_S specialist with a lower μ_{max} and a lower K_S. At high environmental [*S*] (right of the crossover concentration), the rate specialist prevails while at low [*S*] (left of the crossover concentration), the K_S specialist wins the race. See text for details

$$\mu = \frac{\mu_{max}[S]}{K_S + [S]}$$

where μ_{max} is the maximal specific growth rate, K_S is the apparent affinity constant of a cell to the catabolic substrate S and indicates numerically that substrate concentration where the specific growth rate is half maximal. The maximal rate μ_{max} can easily be determined in batch cultures because the substrate concentration is generally very high relative to K_S. On the other hand, growth at substrate concentration near K_S and, thus, where the actual growth rate is near $\frac{1}{2}\mu_{max}$ can only be studied in chemostat cultures (see below). K_S values are often similar to the K_M values of the transporter proteins involved in substrate uptake, indicating that the rate of catabolism is transport-limited. K_S values are useful to know in applications such as wastewater treatment because they indicate how low the concentration of a contaminant can be reduced via biodegradation in a reasonable period of time.

Empirically, there are two types of microbes: those with a very low *Ks* (in the nM range), which makes these microbes (K_S specialists) specialized to grow at very low catabolic substrate concentration, and those with a high μ_{max}, which makes them (rate specialists) specialized to grow very rapidly at high substrate concentrations. K_S specialists, also often referred to as oligotrophs, generally grow very slowly and to low cell densities. They dominate in environments, where the energy flux is relatively low and constant. Rate specialists, also referred to as copiotrophs, dominate in environments, where the energy flux is high and fluctuating. In the inset of Fig. 5.3 it is shown that at substrate concentrations below the concentration cross-over, K_S specialists can outcompete rate specialists.

5.2.3 Fast and Slow-Growing Microbes

Doubling times of microbes vary widely between minutes (e.g., *Vibrio natriegens*, $t_d < 10$ min) and estimated 50–100 years of microbes in deep-sea sediments. K_S values vary between mM and nM concentrations. From all what is currently known, the architecture of catabolic and anabolic pathways and processes in such microbes growing at different rates are qualitatively similar. Thus, there are large microbes that grow relatively slowly and small ones that grow relatively rapidly. On the other hand, cell size of *E. coli* depends on the growth rate: when growing slowly, the cell size decreases relative to when growing fast. To understand why microbes have such different growth rates and K_S, the conditions in their natural habitat provide a possible answer. In the following discussion, growth rates will be discussed for the same overall catabolic reaction, unless otherwise indicated.

Fast-growing microbes. Environments with high catabolic substrate concentration and a constant and high energy flux select for the fastest growing microbe(s), and microbes with the highest μ_{max} will dominate (and are rate specialists) (see Chap. 2). In environments with large fluctuations of energy fluxes, microbes of low *Ks* are present as they have a growth advantage at low substrate concentration unused by the rate specialists. Depending on the type of fluctuations, both a high μ_{max} and a low K_S microbes can be found in the same environment. Generally, fast-growing microbes can be cultured readily in the laboratory because they grow in batch culture in a reasonable period of time to relatively high cell density.

Slow-growing microbes: Environments with typically low catabolic substrate concentration and low energy fluxes select for microbes with the lowest K_S, which will dominate in these environments and are K_S specialists (see Chap. 2). The maximum growth rate in such low K_S microbes, μ_{max}, will also be low because of the Haldane relationship between K_{eq}, K_M, and v_{max} (see Sect. 5.4.2). A very low μ_{max} is the reason why most oligotrophs are very difficult to grow in the laboratory. For most microbes in the open ocean or the deep biosphere, energy flux is low, and most microbes from these environments have not been cultures yet.

K_S strategists of low ΔG catabolic reactions. Because $\mu_{max} \propto J_{ATP}$, the maximal growth rate μ_{max} also depends on the magnitude of free energy change associated with the catabolic reaction, $\Delta G_{S \to P}$. J_{ATP} is related to $\Delta G_{S \to P}$ of a catabolic reaction's via n_{ATP}:

$$J_{ATP} = \frac{dATP}{dt} = n_{ATP} \frac{dS}{dt}$$

As shown in Chap. 3, if $\Delta G^0{}_{S \to P}$ is large, then even at low substrate concentration sufficient energy (n_{ATP}) is available for ATP synthesis at a reasonable efficiency. For example, microaerophilic microbes with a very low K_S for both O_2 and glucose still grow more rapidly than syntrophic anaerobes fermenting butyrate with a low Ks for butyrate. This is because the free energy change associated with glucose oxidation to 6 CO_2 at pM concentrations of both glucose and O_2 is near −2500 kJ/mol, whereas the free energy change of syntrophic butyrate fermentation to 2 acetate and 2 H_2 is only −14 kJ/mol at mM concentrations. In both cases, flux of energy J_{ATP} through the habitat determines the growth rate but this flux is higher when O_2 is the electron acceptor than when protons are reduced.

How do slow-growing microbes with a small catabolic pathway ΔG escape "death in equilibrium" when competing with other K_S specialists. Consider the example of syntrophic butyrate disproportionation to acetate and H_2. A steady-state butyrate concentration can only be lowered to a concentration at which the thermodynamic efficiency of ATP synthesis is close to 100%; at a 100% efficiency, the rate is zero. As indicated in Chap. 3, anabolism is highly irreversible, and more than 90% of the free energy change released during ATP consumption dissipates as heat. This energy dissipation drives catabolism by lowering ATP concentration, when catabolism operates near thermodynamic equilibrium. Therefore, it is ultimately the consumption of ATP in anabolism that allows cells to conserve energy in catabolism near thermodynamic equilibrium when the net rate of catabolic substrate consumption is very low. Under such low $\Delta G_{S \to P}$ conditions, it is clearly the rate of anabolism rather than that of the catabolism that determines the growth rate.

At pM concentration of catabolic substrates, enzymes involved in substrate transport into the cells or in intracellular activation of the substrates are present at concentrations that are higher than the catabolic substrate concentration. If the substrate concentration is lower than the enzyme concentrations, then Michaelis–Menten kinetics no longer applies (Fig. 5.3); the dependence of the rate on substrate concentration, in reality, does not extrapolate to zero rate at zero concentration but to zero rate and at some pM concentration. In the pM range, the rate increases much less with the substrate concentration than at K_S concentrations. Therefore, the growth rate of oligotrophs at very low substrate concentration is lower than predictable from the K_S. This minimum substrate concentrations for an observable rate of growth or metabolism, S_{min}, is important in predicting and controlling bioremediation and wastewater treatment.

Predicting Growth Rates from Genomic Data
Predicting from genomic data whether a microbe will pursue a K_S or μ_{max} strategy makes use of the codon usage bias. The genetic code for encoding a specific amino acid is degenerate by a variation in the third nucleotide position of a codon: several codons encode the same amino acid and differ only in the nucleotide of the third position. The frequency of how all codons for a specific amino acid are used by a cell, however, is unequal across different genes, resulting in a bias for using certain codons over others in some genes. For a codon to be translated, and, thus, a protein to be synthesized, a particular amino acid-charged tRNA needs to be present that is specific for the third nucleotide in the codon. Thus, differences in codon usage are most pronounced when charged tRNAs become limiting, i.e., during fast growth. The codon usage bias is particularly noticeably in highly expressed genes, such as those encoding ribosomal RNA, tRNA, and catabolic proteins in fast-growing microbes. This correlation between codon usage bias and growth rate has been used to predict from genomic data whether a microbe might be fast or slow growing.

5.2.4 Dependence of Growth Yield on Growth Rate: The Costs of Maintenance

As indicated in Fig. 5.1, a fraction of the ATP generated in catabolism in growing microbes that is not used for the synthesis of new cells is consumed for maintenance. What is maintenance energy used for? ATP is consumed among others to maintain an electrochemical potential across the cytoplasmic membrane, to maintain osmotic pressure, to drive motility, for proofreading of DNA, and for turnover of RNA, enzymes, and lipids. Such non-growth-related energy is consumed whether or not cells grow and varies with the particular condition a cell is in. Non-growth-related energy is difficult to quantify. Estimates come from energy flux- and cell mass measurements in soils and sediments.

In rapidly growing cells this fraction of energy used for maintenance is relatively low, however, the fraction increases with decreasing growth rate. As discussed previously, Y_{ATP} of rapidly growing cells is ~10 g/mol ATP (Y_{ATP}) when the cell carbon is derived from sugars or amino acids, and ~5 g/mol ATP when the cell carbon is mainly derived from CO_2. These yields decrease considerably when growth rate decreases, e.g., in a chemostat, by lowering the dilution rate (see below).

A quantitative relation is given by the Pirt equation that is derived from the Monod equation and $\frac{dY_S}{dt} = \mu Y_S - aY_S$, where a is the specific maintenance rate (1/time):

$$\frac{1}{Y_S} = \frac{m}{\mu} + \frac{1}{Y_{\max}}$$

A plot of $\frac{1}{Y}$ versus $\frac{1}{\mu}$ yields a straight line with the intercept on the ordinate yielding $\frac{1}{Y_{\max}}$ and with the slope m, the maintenance coefficient (mol substrate/g cells per time). Y is the cell yield in g cells (dried mass) per mol substrate, and $m = \frac{a}{Y_{\max}}$, where a is the specific maintenance rate (mol substrate consumed for maintenance per time). Considering that $Y = n_{ATP}\ Y_{ATP}$, the rate of ATP consumed for maintenance can be calculated to ~0.4 mmol ATP $g^{-1}\ h^{-1}$ for aerobically growing *E. coli*, ~6.4 mmol ATP $g^{-1}\ h^{-1}$ for aerobically growing *Bacillus subtilis*, and ~12 mmol ATP $g^{-1}\ h^{-1}$ for anaerobically growing *Zymomonas mobilis* at their respective optimum growth temperature. Assuming a cell doubling time of 1 h and considering that about 1 mol ATP is required for the synthesis of 10 g cells (dried mass) from glucose, the total ATP consumption rate for anabolism in these bacteria is estimated to be 100 mmol ATP $g^{-1}\ h^{-1}$. Thus, between 0.4% (*E. coli*) and 12% (*Z. mobilis*) of the ATP generated in catabolism are used for maintenance under maximal growth conditions.

ATP maintenance coefficient of *E. coli* was shown to increase almost sevenfold to 2.5 mmol ATP $g^{-1}\ h^{-1}$ when cells grow at 40 °C rather than the optimum temperature at 37 °C, and the coefficient is only 0.25 mmol ATP $g^{-1}\ h^{-1}$ at 25 °C.

Salt and other environmental stresses increase the maintenance coefficient. Thus, the absolute amount of ATP per cell and time consumed for maintenance in growing cells varies considerably between microbes and growth conditions, however, is generally higher by a factor of 10–1000 than in nongrowing cells. ATP maintenance coefficient is even lower in resting cells such as spores.

5.2.5 Growth in Batch Culture

In batch culture, such as in a test tube during pure culture isolations (Chap. 2), catabolic substrate is typically present at high concentration relative to the K_S of the enzymes involved. Substrate concentration is chosen high to allow the build-up of sufficient biomass to monitor growth either gravimetrically or turbidometrically and to be used in experiments. For the following discussion it is assumed that the catabolic substrate concentration determines the growth rate and that anabolism is not rate-limiting.

Catabolism in batch culture (Chap. 2) is frequently limited kinetically by transport of the catabolic substrate into cells or by activation of the substrate inside. The reason for this is that the high concentrations chosen in the laboratory do not occur in nature. In natural environments, many substrates are derived from polymers as monomers or oligomers, such as sugars or amino acids, before taken up and catabolized. The rate of extracellular hydrolysis of polymers, such as cellulose, starch, or proteins, is generally slow relative to the catabolic consumption of the derived monomers. As a result, steady-state concentration of the monomers in nature is mostly very low. Microbes simply have not been selected for metabolism at high batch concentration, and, therefore, growth at high substrate concentration in the laboratory is still limited by transport or activation.

Batch cultures are semi-open systems, in which gaseous substrates can enter and gaseous products escape, whereas nongaseous substrates decrease in concentration while metabolic end products as well as newly formed cells accumulate in the vessel during growth (Chap. 2). Thus, microbes growing aerobically (e.g., *E. coli*), air (20% O_2) needs to be continuously bubbled through the culture to provide sufficient O_2 to cells because media when equilibrated with air contain less than 0.2 mM O_2. At a cell density of 10^8 cells/ml, the capacity of *E. coli* for O_2 consumption is much higher than the rate of diffusion of O_2 into and within the aqueous phase. Therefore, cultures have to be aerated and stirred vigorously to increase convection and total bubble surface. As soon as aeration stops or is slowed down, growing culture will become anoxic within less than a minute, and anaerobic metabolism is induced if possible. When catabolic substrates, such as nitrite or formaldehyde, are inhibitory at higher concentration, these nongaseous substrates are fed continuously into the fermenter, often referred to as fed-batch reactor.

A semi-log plot of the increase in cell number versus time and a plot of the specific growth rate versus time, as might be observed in an idealized batch system, are shown in shown Fig. 5.4. Five phases are defined, a lag phase, a log- and a late log phase, a stationary phase, and a death phase.

Lag phase: Batch cultures are often inoculated with cells that have been in stationary phase or in some other growth medium. Directly after inoculation, cells acclimate to the new environment by new gene expression or re-routing of metabolic fluxes if necessary. The length of a lag phase is often determined by the time period it takes to coordinate cellular metabolism to fit the niche of the fresh culture. More than often, the exact microbe's physiology and reason(s) for the duration of the lag phase are unknown. Length of lag phase depends also on (1) the type of substrates that cells had previously catabolized, (2) how long cells were exposed to accumulating end products, and (3) whether cells were subjected to starvation and the length of that.

Log Phase: Once acclimated, cell growth is ideally exponential, which is reflected as a straight line in the semi-log plot of Fig. 5.4 A. In exponential phase, growth rate is maximal (μ_{max}), because catabolic substrate concentration is above K_S by order(s) of magnitude. In log

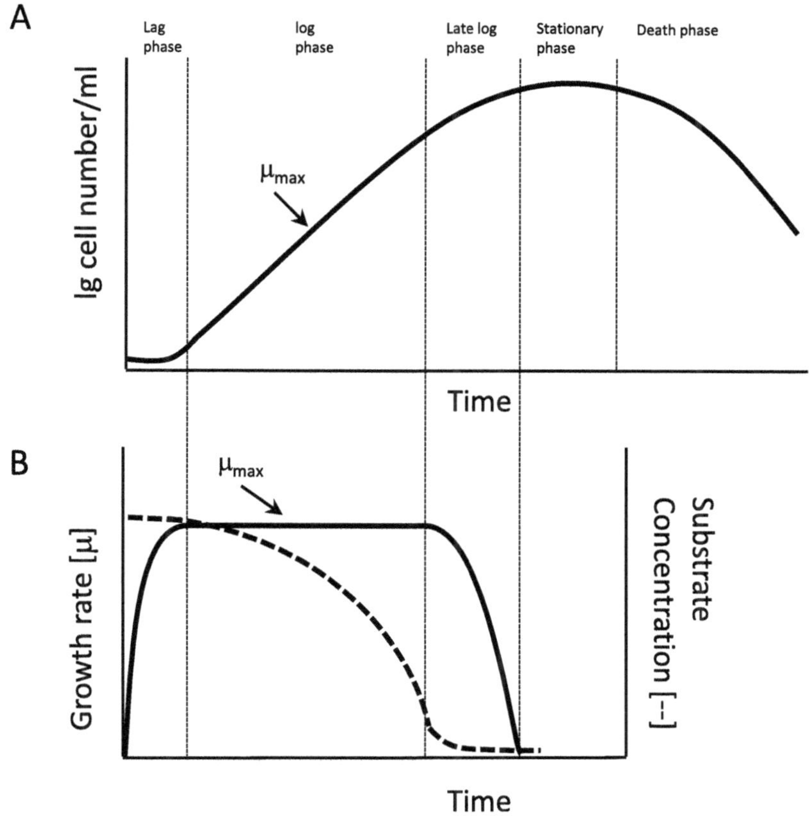

Fig. 5.4 Idealized growth of a microbe in a batch culture. (**a**) Semi-log plot of the cell number versus time and (**b**) specific growth rate (solid line) and substrate concentration (dashed line) versus time. Note the five growth phases. The growth rate is maximal in the log phase, and substrate concentration decreases exponentially during log phase

phase, substrate concentration decreases exponentially. As concentration of resources decreases and product concentrations increase, growth rate decreases continuously while cells pass through a **late log phase** into a **stationary phase**, where there is no net growth ($\mu = 0$). The cessation of growth can be due to either depletion of the catabolic substrate or end product toxicity, including change in pH. For example, during wine making, sugar fermentation to ethanol does not stop because of a depletion of sugar (plenty of sugar left) but because the limit of ethanol tolerance by the *Saccharomyces* strain was reached (a 14% (v/v) Cabernet Sauvignon contains 2.4 M ethanol!).

Stationary phase: While there is no net growth of cells in stationary phase, there still can be growth using "alternate" substrates if present or metabolites excreted previously during exponential growth, depending on (1) the type of growth medium and metabolism (e.g., metabolism of excreted acetate by *E. coli*), (2) end products formed, and (3) cell lysis or other aspects enabling subpopulations to grow. While growth of such populations in stationary phase is measurable once it is clear what to look for, the amount of growth is relatively small considering the density of the growing subpopulation and available resources. Thus, there can be metabolism in stationary phase that is invisible if focusing only on cell density. Cell shape and cell size as well as storage compounds (endogenous metabolism) can change dynamically as well. The length of stationary phase, as defined by a constant cell density, can vary greatly as well as viability of cells. In the following **death phase**, cell lysis occurs to different extent in a microbe-specific manner.

Batch cultures after inoculation with 2–5% of a preculture typically grow for 3–4 generations. Longer serial batch cultivation can select for microbes or variants with higher μ_{max} and an accelerated rate to restart growth after stationary phase.

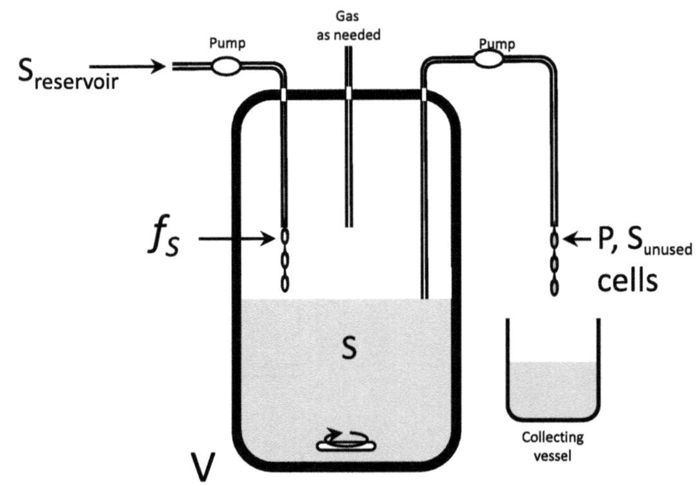

Fig. 5.5 Microbial growth in a chemostat. See text for details

5.2.6 Growth in a Chemostat

In contrast to a batch system, culturing of microbes at low substrate concentrations near K_S is only possible in a **chemostat,** which is an open system with a constant flux of nutrients (resources) into the reactor vessel and a constant efflux of cells, unused nutrients, and metabolic end products out (Fig. 5.5).

Substrate is introduced from a reservoir, containing high concentration [$S_{reservoir}$] at a constant flow rate into the reactor volume *V*, where it is diluted and rapidly consumed by the microbes, so the steady-state concentration of *S* in the vessel is low, typically at or below K_S. Reactor volume in the vessel is constant. A chemostat is in steady state when flow rate, volume, and cell density are constant, and only under those conditions, the mathematical description below applies. A chemostat approximates a "microbial environment" as previously defined (Chap. 2, Fig. 2.1) consisting of a volume around a microbe and a flux of resources, except that the chemostat is not on a µm scale.

In a chemostat, growth rate can be precisely controlled experimentally and maintained constant. This is in contrast to batch culturing, where the experimentalist has no control over the inherently changing growth rate. Decrease of cell mass or cell number (*C*) in a chemostat is described by $-\frac{dC}{dt}=DC$, where *D* is the dilution rate. Increase in *C* via growth is $\frac{dC}{dt}=\mu C$. In steady state, the rate of biomass formation via growth is equal to the loss of biomass by dilution:

$$-\mu C=DC.$$

Thus, the dilution rate, which can be experimentally controlled via a pump and the fixed volume reactor vessel, set the in situ growth rate as

$$D=\mu_{\max}\frac{S}{K_S+S}$$

where *S* is the substrate concentration. It follows that substrate concentration *S* in the chemostat depends on the kinetic parameters of the microbe (K_S and $\mu_{\max}$) but also on the dilution rate:

$$S=K_S\frac{D}{\mu_{\max}-D}.$$

In steady state, substrate concentration *S* in the vessel is constant, so that the rate of total substrate (nutrient) addition from the reservoir, S_r, is equal to the rate of cell loss plus the rate of removal of unused substrate:

$$DS_r=\frac{dS}{dt}+DS.$$

Substituting *dS* in the above equation, based on the relationship of the substrate to the growth yield $\left(\frac{dC}{dt}=\mu C,\ \frac{dS}{dC}=\frac{1}{Y},\ \frac{dS}{dt}=\frac{\mu*X}{Y}\right)$, it follows that $DS_r=\frac{\mu C}{Y}+DS$, and in steady state:

$$C = Y(S_r - S)$$

($S_r - S$) is the amount of catabolic substrate used for growth (energy and carbon if S is organic). In other words, growth rate in a chemostat is determined by the dilution rate, and cell mass formed (productivity) is determined by growth yield and amount of substrate consumed. Thus, the chemostat, when in equilibrium, enables to measure how well cells couple resource utilization to cell growth.

Chemostats are operated typically at dilution/growth rates of less of μ_{max} (e.g., <50% μ_{max}), because a dilution rate exceeding μ_{max} leads to washout. Such washout, however, can be important in environments such as the rumen of ruminants, a sewage treatment plant, or the mammalian gut. For example, if via overdemand or increased intake of water in a wastewater treatment plant, e.g., after rain, or under conditions of diarrhea in the human gut, the dilution rate exceeds μ_{max}, the microbial community will be washed out. This removal of microbes can lead to opening up of niches and metabolic opportunities which can be occupied by different microbes, including facultative pathogens, that have not been part of the previous community. In addition, newly developing niches select for microbes with fast growth to occupy the new niche rapidly. The resulting shifted community may have a different overall phenotype.

Notably, a chemostat can also serve as an experimental platform to investigate growth under anabolic limitation. For example, growth in a chemostat can be limited by phosphate or a N-source. Under these conditions, catabolic substrate concentration in the chemostat is high and the rate of ATP synthesis exceeds that of ATP consumption by anabolism. Under such conditions, either n_{ATP} of the catabolic pathway is decreased and/or the ATP maintenance coefficient is increased, e.g. via ATP hydrolysis catalyzed by *ATPases*.

In a **turbidostat**, the dilution rate is regulated by the cell concentration in the outflow such that when cell concentration decreases, pumping decreases and vice versa. Cell concentration in the outflow is continuously followed turbidometrically, and the photometer regulates the pump.

5.3 Flux of Energy and Microbial Fitness

In Chap. 3.5, three examples showed that the growth rate of microbes negatively correlates with thermodynamic efficiency of ATP synthesis: the higher the efficiency is, the lower is the growth rate and vice versa. Thermodynamics and kinetics are linked via the equilibrium constant: $\Delta G^0 = -RT\ln K_{eq}$ and $K_{eq} = k_{+1}/k_{-1}$, where k_{+1} and k_{-1} are rate constants of the forward- and back reactions, respectively. The negative correlation between growth rate and thermodynamic efficiency of ATP synthesis, thus, has a theoretical basis which is discussed below.

Microbial conversion of $S \rightarrow P$ in a catabolic pathway and coupled formation of ATP have an associated rate, or flux, which is described by $J_S = \frac{dS}{dt}$ and by $J_{ATP} = \frac{dATP}{dt}$. J_S and J_{ATP} are related via $J_{ATP} = n_{ATP} J_S$ (Fig. 5.6). Under energy-limiting growth conditions, growth yield is proportional to the flux of ATP formation (J_{ATP}) and to n_{ATP} and J_S; $\mu_{max} \propto J_{ATP}$. When the rate of ATP production increases, the rate of biomass formation increases proportionally and vice versa. In energy-limited planktonic environments, those microbes with the highest J_{ATP} should be ecologically the most successful (i.e., fittest). That should be the microbes that have the largest possible n_{ATP} (largest possible thermodynamic efficiency $= \frac{n_{ATP}\Delta G'_{ATP}}{\Delta G_{S \rightarrow P}} \times 100$) and the enzymatically highest rate of substrate consumption.

The Julian Adams experiment (Chap. 1) revealed, however, that it is not as simple: in the chemostat environment and at a medium dilution rate, the evolved *E. coli* variant that incompletely oxidized glucose to 2 acetate and 2 CO_2 ($n_{ATP} = 9$) was ecologically more successful than wild type, which oxidized glucose completely to 6 CO_2 ($n_{ATP} = 24$) (Sect. 3.5.4). Apparently, J_{ATP} and, thus, growth rate are dependent also on parameters other than n_{ATP} and J_S. It will be

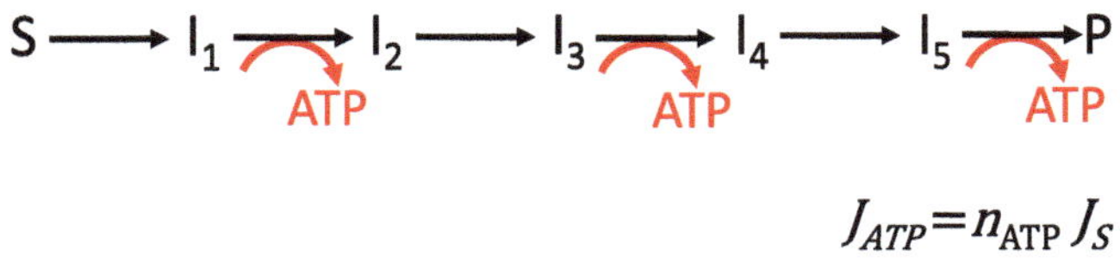

Fig. 5.6 Kinetic coupling of substrate to product conversion with ATP formation in a simple catabolic pathway. J_S is the rate or flux of substrate conversion to the product. J_{ATP} is the rate of ATP synthesis and n_{ATP} the number of ATP synthesized per substrate during substrate conversion to product(s). In this scheme, $n_{ATP} = 3$ (3 mol ATP/mol S). I_1 to I_5 are intermediates. See text for details

shown that there is a relationship between the growth rate and the thermodynamic efficiency of ATP synthesis (Sect. 5.3.1) and between the growth rate and the length of the catabolic pathway, which is the number of reactions and enzymes involved (Sect. 5.3.2).

5.3.1 Growth Rate and Thermodynamic Efficiency of ATP Synthesis: The Rate: Yield Trade-Offs

The discussion of energetics of catabolic pathways (Chap. 3) showed that for an overall energy-conserving pathway $S \rightarrow P$, the total available free energy is only partially used for ATP synthesis, while the rest dissipates as heat:

$$\begin{aligned}\Delta G_{S\rightarrow P} &= n_{ATP}\,\Delta G_{ATP\rightarrow ADP} + \text{Heat}\\ &= n_{ATP}\,\Delta G_{ATP\rightarrow ADP} + \Delta G_{unused}.\end{aligned}$$

Consequently, release of heat (Fig. 3.1) is necessary for a pathway to proceed forward at a finite rate.

The relationship between the heat released and the rate of a pathway can also be derived based on the energetics and flux through a catabolic pathway. Consider the previously-introduced relationship:

$$\Delta G_{S\rightarrow P} = n_{ATP}\Delta G_{ATP} + \Delta G_{unused}, \text{ and therefore}$$
$$\Delta G_{unused} = \Delta G_{S\rightarrow P} - n_{ATP}\Delta G_{ATP}. \tag{5.1}$$

When in thermodynamic equilibrium, the reaction S + ADP + $Pi \rightarrow P$ + ATP is completely reversible and $\Delta G_{unused} = 0$ as all energy from the transition of S→P is conserved as ATP. Under those conditions, n_{ATP} is maximal, $n_{ATP} = n_{max}$, and

$$n_{max} = \frac{\Delta G_{S\rightarrow P}}{\Delta G_{ATP}}. \tag{5.2}$$

When in equilibrium, there is also no net flux of S to P and, therefore, J_S and J_{ATP} are zero. However, **if n_{ATP} is less than n_{max}, then ΔG_{heat} is released ($\Delta G_{unused} < 0$)**, and there will be a net flux of S and of ATP. In other words, the thermodynamic efficiency of ATP synthesis needs to be less than 100% in order for a pathway to have a net forward flux. Therefore, J_S is proportional to the release of ΔG_{unused}:

$$J_S \propto \Delta G_{unused}. \tag{5.3}$$

And ΔG_{unused} (=heat) provides a driving force for a forward flux through a pathway.

Substituting ΔG_{unused} in (5.3) with ΔG_{unused} derived from (5.1) and (5.2), and combining the constant value of ΔG_{ATP} to a proportionality constant k_1, it follows that:

$$\begin{aligned}J_S &= k_1(n_{max} - n_{ATP}) \text{ and } J_{ATP}\\ &= k_1\, n_{ATP}\,(n_{max} - n_{ATP}).\end{aligned}$$

This relationship indicates that the smaller n_{ATP}, the higher the flux J_S or J_{ATP}, and vice versa. n_{ATP} reflects the thermodynamic efficiency $\frac{n_{ATP}\Delta G'_{ATP}}{\Delta G_{S\rightarrow P}} \times 100$. This inverse correlation is displayed in Fig. 5.7.

Hence, for fundamental thermodynamic reasons, there is always a trade-off between rate (moles of ATP per unit of time) of ATP

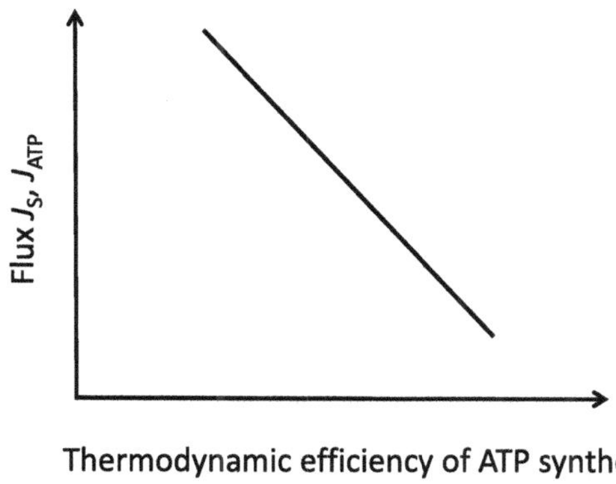

Fig. 5.7 Inverse relationship between rate of ATP synthesis (J_{ATP}) and the thermodynamic efficiency of ATP synthesis $\left(\frac{n_{ATP}\Delta G'_{ATP}}{\Delta G_{S \to P}} \times 100\right)$ of a given catabolic pathway at high-energy flux. The lower the thermodynamic efficiency, the higher the flux of S→P and ATP formation, but at the cost of a lower growth yield = n_{ATP} Y_{ATP}. Thus, there is a high rate–low growth yield trade-off. Mathematically, this relationship is not linear, and this display emphasizes this relationship only qualitatively. See text for details

production and growth yield ($n_{ATP}Y_{ATP}$, [g cells per mol catabolic substrate]) in chemotrophic microbes. The trade-off is energy flux dependent. A decrease in ATP production, however, comes at a cost in ATP yield: because of a decrease in n_{ATP}, less usable energy is retrieved from the substrate.

High rate–low growth yield trade-off. Metabolism of glucose provides several real-world examples for rate-yield trade-offs: Glucose fermentation to ethanol by yeast proceeds as $n_{ATP} = 2$ and a thermodynamic efficiency = 44%, and by *Zymomonas mobilis* at $n_{ATP} = 1$ with thermodynamic efficiency = 22%. Glucose respiration at high O_2 concentration by *Pseudomonas* sp. proceeds at $n_{ATP} = 35$ and a thermodynamic efficiency near 60%, and by *E. coli* at $n_{ATP} = 24$ and a thermodynamic efficiency near 40%. The lowest thermodynamic efficiency observed up to today is, thus, near an efficiency of 20%. Under energy-limited growth conditions, *Z. mobilis* outcompetes yeast, and *E. coli* outcompetes *Pseudomonas* sp. when growing in batch on glucose at concentrations well above K_S.

Control of n_{ATP} by pathway architecture. Chap. 3 (Fig. 3.6) showed conceptually how multiple pathways exit in different microbes for the same overall catabolic reaction, yet with different n_{ATP}. In addition, individual microbes can control whether to employ a high or low n_{ATP} pathway by differential expression of genes encoding for pathway enzymes with different energetic coupling. For example, diverse H^+-translocating enzymes in a respiratory chain differ in the stoichiometry of H^+ translocation (e.g., Fig. 3.7). As already pointed out, ecophysiological consequences of operating pathways with low n_{ATP} and, thus, with a low thermodynamic efficiency are two-fold: at high catabolic substrate concentration relative to K_S, low n_{ATP} pathways enable a microbe to **kinetically outcompete** those microbes that have a higher n_{ATP}; at very low substrate concentrations relative to K_S, the relationship inverses; those microbes with a larger n_{ATP} **thermodynamically outcompete** those with a lower n_{ATP} and, thus, those with a smaller growth yield.

5.3.2 Growth Rate and Length of the Catabolic Pathway

Compare *E. coli* wild type and the SCV growing aerobically and energy-limited on glucose in a chemostat at medium dilution rate (50% of D_{max}). Wild type is forming 6 CO_2 at $n_{ATP} = 24$ in 25 steps while the LCV is forming 2 acetate and 2 CO_2 at $n_{ATP} = 9$ in 18 steps. A change in n_{ATP} is here achieved by reducing the number of pathway steps in a pathway which results in a shorter pathway. Although thermodynamic efficiency of ATP synthesis in both pathways is almost identical (near 40%), SCV with a low n_{ATP} in this chemostat environment is more successful than the high n_{ATP} wild type.

Figure 5.8 shows how, in an ideal case, J_{ATP} and, thus, growth rate depends on the pathway length *m*. The model assumes that (1) n_{ATP} increases linearly with increasing m and (2) the coupling sites, in which ATP is formed, are equally distributed in the pathway. Whereas the flux J_S continuously decreases with increasing m

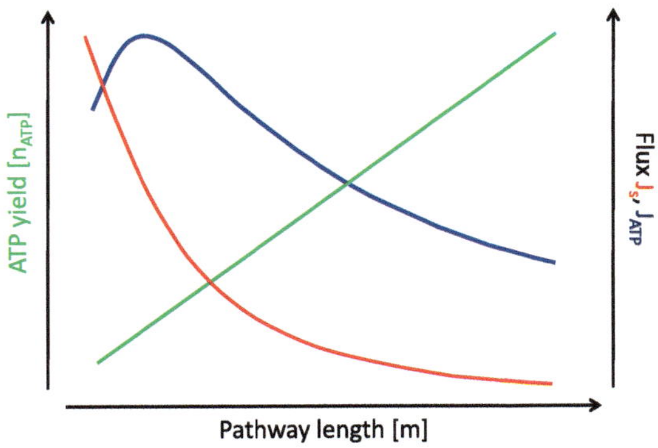

Fig. 5.8 Dependence of n_{ATP}, substrate flux J_S, and ATP flux J_{ATP} on the number of pathway steps m (i.e., length of a pathway). Note that in growing cells, the flow of metabolites through a pathway will be in a quasi-steady state because growth is a slower process than catabolic substrate turnover. The substrate flux J_S monotonically decreases according to $J_S = k[S]\ E_{max}X_{max}/X_{max} + [S]\ m^2$, where k is the rate constant for all enzymes and $[S]$ is the catabolic substrate concentration. The total concentration of all enzymes is restricted to E_{max} and the total concentration of all intermediates is restricted to X_{max}. The ATP flux J_{ATP} is the product of substrate flux and n_{ATP}, $J_{ATP} = n_{ATP}\ J_S$. If n_{ATP} is simply proportional to pathway length, the optimal length is given by $m = \sqrt{(X_{max}/[S])}$. In other words, the optimal length is shorter when the substrate concentration and the metabolic costs of intermediates (which restricts X_{max} to lower values) is high. To show the shape of these functions, all parameters were set to unity apart from $[S] = 1/4$, therefore, $m_{opt} = 2$. (modified from Costa et al. 2004)

and n_{ATP} increases linearly with m, predicted flux J_{ATP} follows a maximum function of m. The model assumes that at a given pathway length there is a maximal total concentration of catabolic enzymes (E_{max}) and of intermediates (X_{max}) and that these concentrations are cumulative and remain constant when pathway length increases. This model predicts two important points:

1. in catabolic pathways with many steps and a high n_{ATP}, shortening of the pathway results in an increase in growth rate; and
2. shortening a pathway too much can result in a decrease in J_{ATP} and with that in growth rate. The existence of an optimal length of a catabolic pathway for a given n_{ATP} (in the model of Fig. 5.8 when $n_{ATP} = 2$) implies that there is natural selection on a pathway architecture with respect to a trade-off of high J_{ATP} versus high growth rate. A predictable chemostat environment at relatively high dilution rate, therefore, can select for shorter catabolic pathways, if mechanistically possible.

The impact of pathway length on the rate of ATP synthesis is on three levels:

Catabolic proteome allocation. Since the volume of a cell's cytoplasm is finite, the total protein content is also finite (Chap. 2) and with that also the size of a catabolic proteome. The fraction of the proteome that is allocated to the catabolic proteome of a growing cell is in the order of 10–20%. If this fraction accommodates enzymes of a short pathway, the relative concentration of each pathway enzyme can be high ($[E_T]$), resulting in an overall capacity for high flux (high $V_{max} = [E_T]*k_{cat}$) of J_S and J_{ATP} through the pathway. On the other hand, if the catabolic proteome is shared by many different catabolic enzymes of a long pathway, the relative concentration of each pathway enzyme is low and, consequently, the rate of each pathway step is low as well. Indeed, in *E. coli* the same proteome fraction of a fermentative proteome mediates a ~2-fold higher J_{ATP} than the longer respiratory proteome (Fig. 4.9).

Another solution to accommodate a higher fraction of catabolic enzymes for a long pathway could be by decreasing the pool of anabolic enzymes. However, a reduction in anabolic proteome reduces the overall rates for biosynthesis, and, with that, the growth rate.

Leakage of intermediates. The higher the number of enzymatic steps in a pathway, the higher is the cumulative concentration of all catabolic intermediates and, thus, the osmotic pressure. A high cumulative concentration results in unspecific enzyme inhibition and in a greater loss of metabolites by unspecific leakage from a cell, thereby, lowering the effective intracellular intermediate concentration. Low concentration of each metabolic intermediate slows down the rate of the catalysis (because of the Michaelis–Menten kinetics, see below), which reduces the rate of S→P transformation and a microbe's competition for S. Loss of intermediates of cyclic pathways, such as the citric acid cycle, is particularly consequential as such loss will reduce the concentrations of the acceptor metabolite or even halt transformations through that pathway (see Chap. 8). In addition, an increase in osmolarity based on a higher concentration of cumulative intermediates comes at a cost of requiring metabolic energy for osmotic work instead of using this energy for biosynthesis.

Allocation of translational capacity. For all enzymes of a long catabolic pathway to be expressed, a higher fraction of ribosomal capacity needs to be allocated to this catabolic proteome, which at the same time compromises ribosomal capacity for synthesis of biosynthetic enzymes (anabolic proteome) and decreases the rate of producing viable offsprings (growth rate).

5.3.3 Competition Favors Growth Rate Over Growth Yield

When is it beneficial for a microbe to be fast but inefficient, and when does it become advantageous to be slow but grow with a high growth yield? The outcome of the Julian Adams experiment showed that there are environmental conditions that favor one over the other. Can these findings be related to substrate flux and competition in microbial natural environments?

The overall amount of energy resources secured as ATP over a given period of time determines ecological success, because in a fully-coupled microbe, all ATP is used for biosynthesis to form viable offsprings. Fluxes J_S and J_{ATP} are not indicative of success per se unless we know how long a specific flux occurs. Thus, the total productivity $P_{(n_{ATP})}$ is the product of the flux and time *t*:

$$P_{(nATP)} = tJ_{ATP(nATP)}$$

where $J_{ATP(nATP)}$ is the flux of ATP formation at a given n_{ATP}. The corresponding total substrate utilized

$$S = tJ_{S(nATP)}.$$

Resolving the latter equation for *t* and including this in the previous equation, considering that $J_{ATP}/J_S = n_{ATP}$, it follows that

$$P_{(nATP)} = S\frac{J_{ATP(nATP)}}{J_{S(nATP)}} = S \cdot n_{ATP}$$

This relationship indicates that productivity depends only on n_{ATP} and the substrate concentration *S* in the **absence of competition**.

When there is **competition,** some flux of the substrate *S* is diverted to the competitor *C* (J_C). Consequently, less substrate is available for the primary microbe and flux J_S is reduced by J_C. $P(n_{ATP})$ then becomes:

$$P_{(n_{ATP})} = S\frac{J_{ATP(nATP)}}{J_{S(nATP)} + J_c}.$$

Substituting *Js* with $k_1(n_{max} - n_{ATP})$ and J_{ATP} with $k_1 n_{ATP}(n_{max} - n_{ATP})$ (see above), it follows that

$$P_{(n_{ATP})} = S\frac{k_1 n_{ATP}(n_{max} - n_{ATP})}{k_1(n_{max} - n_{ATP}) + J_c}.$$

When competition is strong, J_C is large and numerically dominates the denominator, and $k_1(n_{max} - n_{ATP})$ in the denominator can be neglected. Consequently, a high productivity is achieved only when the enumerator is large, and this is the case when n_{ATP} is small. This means that in the presence of significant competition, it is more advantageous for a microbe to have a low n_{ATP}. And low n_{ATP} enables a high rate. Thus, **competition favors microbes to express a rate over a yield phenotype and to grow faster**.

This principle can explain, for example, why *E. coli* with a $n_{ATP} = 24$ of a pathway for glucose utilization can outcompete in batch culture a more efficient *Pseudomonas aeruginosa* with an $n_{ATP} = 35$ for the same overall catabolic reaction. Another example is the outcome of the Julian Adams evolution experiment (Chap. 1), which selected for higher rate rather than for higher n_{ATP} variants.

5.3.4 Energy Flux-Driven Speciation

Speciation means that multiple genotypes arise from a single ancestral clone which can coexist over evolutionary time. Two examples are discussed here where the driving force for speciation is the adaptation of energy metabolism to local flux conditions, one in an environment of constant *high energy flux* and one in an environment where *energy flux is relatively low* and repeatedly varies with time.

5.3.4.1 Speciation in Constant, High Energy Flux Environments

As indicated in Sect. 5.2.1 and 5.2.2, there are long, slow catabolic pathways with high n_{ATP} and short, fast catabolic pathways with low n_{ATP}. In high-energy flux environments and with competition, natural selection favors those planktonic microbes with the shorter pathway.

In the laboratory: The Julian Adams experiment describes speciation observed under laboratory conditions. When *E. coli* grows using glucose in a stable, predictable environment (constant J_S), there is a significant fitness gain associated with shorten its catabolic pathway length; the associated decrease in n_{ATP} is offset by a higher flux, $J_{ATP,}$ through the proteome of the shorter pathway. However, this evolutionary outcome is probably only observed at medium- and high glucose concentration. At very low glucose concentrations as they prevail in a chemostat at very low dilution rates, different *E. coli* variants would be expected to be selected for, e.g.,with the higher n_{ATP}, and a SCV may not persist. The latter conditions resemble those of biofilms, where selection mainly on growth yield rather than on growth rate. The main observation from the Julian Adams experiment is that under certain environmental conditions, a long pathway can segregate and speciate into two complementary, short pathways, and such speciation is ecologically beneficial. In other words, there are evolutionary trajectories to explain metabolic speciation in microbes and the arising of metabolic networks and food chains as they are observed in nature, e.g., in denitrification or glucose fermentation under methanogenic conditions.

In nature:

1. Nitrification was considered for a long time to require two species, one oxidizing ammonia to nitrite and the other oxidizing nitrite to nitrate (Table 5.1). *Nitrobacter* and *Nitrosomonas sp.* have been isolated from high flux denitrifying environments. If such fragmentation of an overall denitrifying pathway is the result of speciation events that are driven by high substrate flux and competition, it should be possible to isolate from a low flux denitrifying environment, a microbe that contains an entire, complete denitrification pathway. Indeed, such Comammox bacterium was recently isolated.
2. Another example of metabolic speciation is hexose fermentation to three acetate (Table 5.1). Most acetogenic bacteria ferment carbohydrates to 3 acetate and at a $n_{ATP} = 4.3$ when growing in batch culture, i. e. at high hexose concentrations and high energy flux (See Chap. 10). In high flux acetogenic environments such as the intestinal tract of termites or marsupials that ferment cellulose, hexoses are also fermented to 3 acetate involving two different bacteria: one that ferments sugars to 2 acetate, 2 CO_2 and 4 H_2 ($n_{ATP} = 4$) and a second fermenting 4 H_2 and 2 CO_2 to acetic acid ($n_{ATP} = 0.3$). It appears that varying energy fluxes in the intestinal tract determine whether the un-fragmented- or the fragmented pathway prevails.
3. There has been an interesting, long lasting discussion on whether a methanogen exists that ferments glucose to 3 CO_2 and 3 methane in a single species (Table 5.1). All presently known methanogens are niche-speciated to C1 and C2 compounds (Chap. 10). At mesophilic temperatures the overall fermentation of glucose involves three different microbes, (i) bacteria that ferment glucose to 2 acetate, 2 CO_2 and 4 H_2, (ii) hydrogenotrophic methanogenic archaea that ferment H_2 and CO_2 to methane, and (iii) acetoclastic methanogenic archaea that ferment acetate to CO_2 and methane. Methanogenesis from carbohydrates or amino

Table 5.1 Examples of pathway speciation in natural environments

Catabolic reaction	$\Delta G^{0\prime}$	Microorganism
1) Nitrification		
$NH_3 + 1.5\ O_2 \rightarrow NO_2^- + H^+ + H_2O$	−275 kJ/mol	*Nitrosomonas* sp.
$NO_2^- + 0.5\ O_2 \rightarrow NO_3^-$	−74 kJ/mol	*Nitrobacteria* sp.
$NH_3 + 2\ O_2 \rightarrow NO_3^- + H^+ + H_2O$	−349 kJ/mol	Comammox bacteria
2) Hexose fermentation to three acetate		
Glucose → 3 acetate− + 3 H^+	−311 kJ/mol	Most acetogenic bacteria
Glucose + 2 H_2O → 2 acetate− + 2 H^+ + 2 CO_2 + 4 H_2	−216 kJ/mol	Many anaerobic procaryotes
$4H_2 + 2\ CO_2 \rightarrow$ acetate$^-$ + H^+ + 2 H_2O	−95 kJ/mol	Many acetogenic bacteria
3) Glucose fermentation to methane		
Glucose + 2 H_2O → 2 acetate− + 2 H^+ + 2 CO_2 + 4 H_2	−216 kJ/mol	Many anaerobic procaryotes
Acetate$^-$ + H^+ → CO_2 + CH_4	−35 kJ/mol	Acetoclastic methanogens
$4\ H_2 + CO_2 \rightarrow CH_4 + 2\ H_2O$	−131 kJ/mol	Hydrogenotrophic methanogens
Glucose → $3CO_2$ + 3 CH_4	−418 kJ/mol	Microbe not yet found

acids has been looked-for but has not been found but this does not exclude that such archaea will be discovered in the future.

5.3.4.2 Speciation in Fluctuating Environments

In environments fluctuating between high and low energy fluxes, speciation is driven by successions. For example, in anoxic freshwater sediments, plant material, such as celluloses, hemicelluloses, pectins and starch, is successively converted to CO_2 and methane (Fig. 5.9). This succession occurs every time new plant material reaches a sediment.

In a first phase (Fig. 5.9), kinetically more readily fermentable sugars from hemicelluloses pectins and starch are fermented to acetate, propionate, butyrate, ethanol, CO_2 and H_2; the latter being consumed by hydrogenotrophic methanogenic archaea. In this phase, the rate of H_2-formation determines the steady H_2 partial pressure ($pH_2 >> 10^{-4}$ bar). At $pH_2 >> 10^{-4}$ bar, propionate-, butyrate- and ethanol fermentations to acetate, CO_2, and H_2 are insufficiently exergonic to allow growth (Chap. 9). Growth on these intermediates is only possible at $pH_2 \leq 10^{-4}$ bar. Thus, these short chain fatty acids and alcohols accumulate in the first phase.

Cellulose is fermented much slower than hemicellulose or pectin. In a second phase (Fig. 5.9), cellulose is converted to CO_2 and CH_4 via the reactions and microbes indicated above (see Chaps. 9 and 10). Because cellulose hydrolysis is rate-limiting, the rate of H_2 consumption by methanogens determines steady state H_2 partial pressure ($pH_2 \leq 10^{-4}$ bar). At this very low H_2 partial pressures, glucose is mainly fermented to 2 acetate, 2 CO_2, and 4 H_2. And only under those conditions, butyrate, propionate and ethanol are fermented to 2 acetate, CO_2 and 2 H_2 by obligately syntrophic bacteria.

The shift in glucose fermentation as a function of pH_2 has been studied in the laboratory with several bacteria (Table 5.2). Different fermentation pathways exist at different pH_2 although with the same or nearly the same n_{ATP}.

The above discussion and examples show that there is a correlation between speciation of metabolic pathways, the metabolic fluxes, and thermodynamics involved. Theory provides a clear link between both and is predictive to explain speciation in nature and in the laboratory. This provides the basis for understanding microbial food chains and metabolic interactions (Chaps. 9 and 10).

5.4 Enzyme Kinetics: In Brief

Growth and metabolic flux are mediated by enzymes. Enzymes do not change the equilibrium constant of the reaction they catalyze. Kinetic

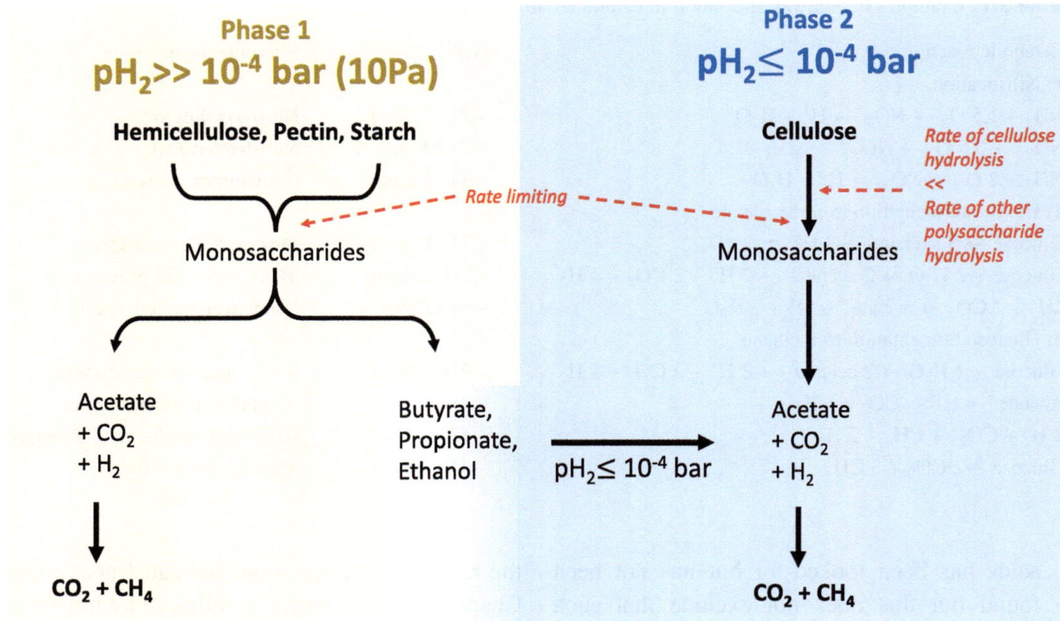

Fig. 5.9 Fermentation in two phases of polysaccharides from plant residues to CO_2 and CH_4 in freshwater sediments. (**left**) In the first phase 1, the kinetically easily fermentable hemicelluloses pectins and starch are fermented first. Therefore, the steady-state H_2 concentration is much higher than when slow-fermenting cellulose is the substrate. At the relatively high pH_2, butyrate, propionate, and ethanol cannot be further converted to acetate and H_2 for thermodynamic reasons. (**right**) In the second phase 2, the much slower cellulose fermentation dominates the degradation kinetics, and pH_2 decreases to 10^{-4} bar. Only at this low pH_2, the conversion of butyrate, propionate, and ethanol to methanogenic substrates is sufficiently exergonic to sustain growth of the respective obligate syntrophic bacteria

Table 5.2 Glucose fermentations and ATP gain as a function of pH_2

Reaction	$\Delta G^{0\prime}$	ATP gain
Glucose → 1.3 Acetate + 0.7 Ethanol + 2 CO_2 + 2.6 H_2	−222 kJ/mol	3.3 ATP at $pH_2 >> 10^{-4}$ bar
Glucose → 0.6 Acetate + 0.7 Butyrate + 2 CO_2 + 2.6 H_2	−260 kJ/mol	3.3 ATP at $pH_2 >> 10^{-4}$ bar
Glucose → 2 Acetate + 2 CO_2 + 4H_2	−300 kJ/mol	4 ATP at $pH_2 \leq 10^{-4}$ bar

features of a catabolic enzyme are the result of natural selection acting within the constraints of an enzyme's evolvability, cellular physiology, and flux patterns of a microbe's resource. While the fundamentals of enzyme kinetics are introduced in general biochemistry textbooks, the focus here is on those aspects that link enzyme kinetics to ecological fitness.

K_M **and** V_{max}**:** Consider a single enzyme, catalyzing conversion of a substrate S to a product P in a single reaction rather than in a reaction pathway, where conversion of S to P involves an entire pathway with multiple enzymatic steps. In the laboratory, substrate concentrations are in excess of enzyme concentration ($[S] >> [E_T]$). Enzyme activities are measured by starting the reaction with substrate or enzyme and then following the rate (v) of substrate consumption ($-\Delta S/t$) or product formation ($\Delta P/t$). The initial rate ($v_{t=0}$) of an enzyme-catalyzed reaction follows Michaelis–Menten kinetics:

$$v_0 = \frac{v_{max}[S]}{K_M + [S]} = k_{cat}[E_T]\frac{[S]}{K_M + [S]}$$

where v is the rate of the reaction, $[S]$ the substrate concentration, K_M the Michaelis–Menten constant, and E_T the total enzyme concentration. The course of the saturation kinetics is identical to the Monod growth kinetic (see below). Michaelis and Menten assumed that the enzyme reacts reversibly with the substrate to form an enzyme-substrate complex, ES, that in turn is converted to an enzyme-product complex, EP, from which the enzyme and the product are released in an irreversible, rate-limiting step (i.e., at $t = 0$, when the product concentration is zero, there is no reverse reaction). K_M was first interpreted to be the dissociation constant k_{-1}/k_{+1} (Michaelis and Menten) but was later shown to be a more complex constant $(k_{-1} + k_{+2})/k_{+1}$ (Briggs–Haldane). The rate constant k_{+2} is also referred to as k_{cat}

$$E + S \underset{k_{-1}}{\overset{k_{+1}}{\rightleftarrows}} ES \rightleftarrows EP \overset{k_{+2}}{\rightarrow} E + P$$

K_M is *not* the equilibrium constant K_{eq} of the enzyme-substrate complex reaction. A high K_M indicates a low affinity between the enzyme and substrate or product and vice versa. K_M is equal to the substrate concentration at which the reaction rate is half maximal (1/2 v_{max}). At a concentration of 10× K_M, the rate of the reaction is 90% of V_{max}. At low $[S]$ (when $[S] << K_M$), enzyme kinetics simplifies to

$$v = \frac{k_{cat}}{K_M}[E_T][S],$$

where $\frac{k_{cat}}{K_M}$ is the catalytic efficiency. Enzymes are often compared based on $\frac{k_{cat}}{K_M}$ values.

In microbial cells in their natural environment, the kinetics of enzymes can be quite different from what was determined in the laboratory for several reasons. First of all, product concentration is never zero in vivo, thus, some reverse reaction occurs always. Second, concentration of enzymes, especially of some catabolic enzymes, can be in the range of the concentrations of substrates and products, with the consequence that the amount of substrate bound to the enzyme cannot be neglected. An excess of free substrate relative to an enzyme's concentration, however, is a prerequisite for solving the equation relating reaction rate to substrate concentration of enzyme-catalyzed reactions.

Inside a cell, substrate concentration is generally in the order of the "apparent" K_M of the enzymes. The apparent K_M is mostly higher than K_M because inside a cell products compete with the substrates for binding to the enzyme. An apparent K_M, however, can be determined in the lab by measuring the rate dependence of the enzyme on the substrate concentration in the presence of different product concentrations.

There are at least two levels where natural selection acts on catabolic enzymes and the fluxes that they control: on k_{cat} and on the enzyme concentration [E]. $\mathbf{k_{cat}}$ reflects mechanistic constraints, such as the arrangement of catalytically-active amino acid residues at the active site, the positioning of the active site in the protein scaffold, as well as an enzyme's hydrophilicity/hydrophobicity, to name a few. There are only an estimated 2000 different natural protein folds known, and an even smaller number of reaction mechanisms; so selection of enzymes with a better k_{cat} is under significant constraints, in particular, as selection in natural environments is mostly weak and acts on microbes close to their fitness peak (see Chap. 2). The other relevant parameter under natural selection is the cellular enzyme concentration at the level of gene expression.

K_M of catabolic enzymes is also under natural selection, because K_M determines use of resources within the constraints of the prevailing concentration and because of the trade-off between k_{cat} and K_M. Rate response to fluctuations of $[S]$ is greatest in the quasi-linear range of K_M. This is a "sweet spot" of regulation as the linear response at low $[S]$ results only in rate changes at low numerical value. At high $[S]$ ($>5 \times K_M$), the rate of the enzyme-catalyzed reaction is relatively invariant of $[S]$. While a given K_M of an enzyme may be sufficient for fitness of the microbe in one environment, in another environment with the same resource, however, at a different concentration may select for enzyme variants with different K_M. For example, when *E. coli* grows aerobically on acetate in a chemostat at low dilution rates, variants with a lower apparent K_M for acetate are selected for (mediated by expression of *acetyl-CoA synthetase*) resulting in a more efficient recruitment of resources at low concentration. On the other hand, in fluctuating environments selection for a similar low K_M may not act sufficiently long for an increased number of such variants to emerge (too few generations).

Depending on a pathway context, some enzymes function physiologically the forward or reverse reaction. For example, *fumarase* catalyzes hydration of fumarate to malate in the oxidative TCA cycle (see below) while in the reverse pathway (reductive TCA cycle for CO_2 fixation), *fumarase* catalyzes the abstraction of water from malate. In the former pathway, *fumarase* catalyzes fumarate hydration to malate with a higher k_{cat} and K_M than malate dehydration to fumarate, while in the latter pathway, the kinetic parameters of the corresponding fumarase are reverse (see below Haldane equation).

Catabolic enzymes generally have turnover rates (mol S per mol enzyme and per second) that are at least 30-fold higher than anabolic enzymes. The reason for this is that the flux through a catabolic pathway has to provide **all** ATP that is consumed in **hundreds of diverse** anabolic reactions operating simultaneously to coordinately produce a new cell (Fig. 5.1). Thus, the metabolic fluxes through ATP-consuming anabolic pathways is generally more than 100 times lower (in the case of prosthetic group biosynthesis more than 1000 times lower) than through catabolic pathways, and there is, therefore, no positive selection on anabolic enzyme variants to increase further their k_{cat}.

5.4.1 Growth Rate and Thermodynamically-Based Kinetic Bottlenecks

In many catabolic pathways, one or two reactions are involved that are endergonic under standard conditions. This creates a thermodynamics-based kinetic bottleneck, because the relationships $\Delta G^0 = -RT\ln K_{eq}$ and $K_{eq} = \frac{k_{+1}}{k_{-1}}$ indicate that the more positive ΔG^0 of a reaction is, the lower the rate will be relative to that of the reverse reaction. There are, in principle, two ways for microbes to overcome such a thermodynamics-based kinetic bottleneck: (1) to increase the catalytic efficiency or concentration of the enzyme or (2) to couple the endergonic reaction to exergonic ATP hydrolysis. This will be discussed in the following for the *β-ketothiolase* reaction, for the *butyryl-CoA dehydrogenase* reaction, and for the *methyl-coenzyme M reductase* reaction.

The *β-ketothiolase* reaction 2 acetyl-CoA $\rightleftharpoons$ acetoacetyl-CoA + CoA; $\Delta G^0 = +25$ kJ/mol is involved in the synthesis of butyrate in butyric acid-forming bacteria (Chap. 9). This endergonic reaction (under standard conditions) is the kinetic bottleneck in this catabolism. The reaction is in equilibrium when acetyl-CoA concentration is near 1 mM and acetoacetyl-CoA and CoA concentrations are near 10 μM. At these concentrations, $\Delta G = 0$ kJ/mol and the rate is zero. Lower product concentrations are constrained by the thermodynamics and kinetics of the reactions consuming acetoacetyl-CoA and CoA; a high acetyl-CoA concentration is constrained mainly by the total cellular pool of HS-CoA (= free HS-CoA + bound in acyl-CoA thioesters).

At given non-equilibrium substrate- and product concentrations, the rate can be increased by increasing *β-ketothiolase* concentration or its catalytic efficiency (k_{cat}/K_M). Indeed, intracellular concentrations of HS-CoA, acetyl-CoA, as well as of *β-ketothiolase* are higher in rapidly growing, butyric acid-forming bacteria relative to slow-growing cells. In case of *Clostridium kluyveri*, which ferments ethanol and acetate to butyrate and where butyric acid formation solely drives ADP phosphorylation, *β-ketothiolase* accounts about 10% of the soluble proteins. However, there is a limit to the fraction that *β-ketothiolase* can occupy in the cellular proteome space. A high fraction comes at a fitness cost, because occupying a larger fraction in the finite cellular proteome space necessitates concentrations of other proteins to be reduced, which reduces the kinetics of those other, e.g., biosynthetic, pathways and with that growth rate.

The rate of *β-ketothiolase* reaction could be much further increased if it would be possible to couple this endergonic reaction with exergonic hydrolysis of ATP resulting in an overall exergonic reaction ($\Delta G^0 = -7$ kJ/mol). This possibility is, as far as known, not observed in catabolism of known butyric acid-forming bacteria. However, it is present in the pathway of anabolic, long-chain fatty acid synthesis from acetyl-CoA, which involves three of the four following reactions:

$$\text{Acetyl} - \text{CoA} + \text{CO}_2 + \text{ATP} \rightleftharpoons \text{Malonyl} - \text{CoA} + \text{ADP} + \text{Pi}$$

$$\text{Acetyl} - \text{CoA} + \text{ACP (Acyl} - \text{Carrier Protein)} \rightleftharpoons \text{Acetyl} - \text{ACP} + \text{CoA} \quad \Delta G^0 = 0\ \text{kJ/mol}$$

$$\text{Malonyl} - \text{CoA} + \text{Acetyl} - \text{ACP (Acyl} - \text{Carrier Protein)} \rightleftharpoons \text{Acetoacetyl} - \text{ACP} + \text{CoA} + \text{CO}_2$$

$$\text{Acetoacetyl} - \text{ACP} + \text{CoA} \rightleftharpoons \text{Acetoacetyl} - \text{CoA} + \text{ACP} \quad \Delta G^0 = 0\ \text{kJ/mol}$$

Indeed, it has been possible to engineer ATP-dependent acetoacetyl-CoA forming reaction sequence into butyric acid-forming fermenters by metabolic engineering and, thereby, to increase the rate of butyric acid formation from glucose. However, the trade-off is a lower n_{ATP} (2 instead of 3) and a longer pathway leading to a lower growth rate and yield. This, conversely, can explain why the ATP-dependent pathway is not selected for in catabolic pathway in nature.

The *butyryl-CoA dehydrogenase* reaction,

$$\text{Butyryl-CoA} + \text{MK} \rightleftharpoons \text{Crotonyl-CoA} + \text{MK}_{red} \quad \Delta G^{0\prime} = +12\ \text{kJ/mol}$$

is involved in anaerobic butyric acid oxidation by syntrophic microbes to 2 acetate and 2 H_2 ($\Delta G^{0\prime\prime}$ = +48 kJ/mol) (Chap. 9). Oxidation of butyryl-CoA to crotonyl-CoA with protons via reduced menaquinone (MK_{red}) is the thermodynamic bottleneck ($\Delta G^{0\prime}$ = + 80 kJ/mol). The ΔG of this reaction becomes exergonic only if H_2 concentration is ~10^{-6} bar and the ratio of crotonyl-CoA–/butyryl-CoA concentration is at 1/100. However, syntrophic microbes grow in anoxic environments where the H_2 partial pressure is near 10^{-4} bar. To overcome this thermodynamic bottleneck, energy input of at least −34 kJ/mol is required, which is mediated by "reversed electron transport," where the two electrons from butyryl-CoA ($E^{0\prime}$ = −10 mV) are "pushed" by proton motive force from MK_{red} ($E^{0\prime}$ = −70 mV) thermodynamically uphill to reduce 2 protons (E'of the $2H^+/H_2$ couple = −300 mV), consuming the energy equivalent of 80% of an ATP (80% of −55 kJ/mol = −44 kJ/mol). Anaerobic growth on butyrate and nitrate, however, is rapid due to the fact that butyryl-CoA oxidation to crotonyl-CoA with nitrate as electron acceptor is exergonic under standard conditions.

The *methyl-coenzyme M reductase* reaction,

$$\text{Methane} + \text{CoM-S-S-CoB} \rightleftharpoons \text{Methyl-S-CoM} + \text{HS-CoB} \quad \Delta G^{0\prime} = +30\ \text{kJ/mol}$$

is involved in the anaerobic oxidation of methane by methanotrophic Archaea. This endergonic reaction (under standard conditions) is the kinetic bottleneck in this catabolism. The reaction becomes exergonic at mM concentration of CoM-S-S-CoB and μM concentrations of methyl-S-CoM and HS-CoB, when pCH_4 = 1 bar. There appears to be no mechanism to couple this reaction to an exergonic reaction. Therefore, anaerobic methane oxidation involving *methyl-coenzyme M reductase* (MCR) is intrinsically very slow, explaining why the concentration of MCR in the methanotrophs is very high (>10%). Rapid oxidation of methane to CO_2 can only be achieved, if the first step in methane oxidation is exergonic under standard conditions, as in the *methane monooxygenase* (MMO) reaction ($CH_4 + O_2 + NADH + H^+ \rightleftharpoons CH_3OH + H_2O + NAD^+$; $\Delta G^{0\prime} = -340$ kJ/mol). MMO is involved in methane oxidation to methanol in aerobic bacteria and in some anaerobic bacteria (Chap. 10) oxidizing methane with nitrate that is reduced to N_2 via NO, 2 mol of which react to O_2 and N_2.

5.4.2 The Haldane Relationship

The Haldane relationship connects kinetics and free energy change of a reaction at the enzyme level. This is explained here for the *β-ketothiolase-* and *methyl-coenzyme M reductase* catalyzed reactions that have been discussed above. The Haldane equation is first introduced with the more simple example of the *fumarase* reaction.

The rate of a chemical reaction ($S \rightarrow P$), in which a substrate is reversibly converted to a product is given by $dS/dt = k_{+1}\ [S]$ (forward direction) and $dP/dt = k_{-1}\ [P]$ (reverse direction). At equilibrium $dS/dt = dP/dt$ and $k_{+1}\ [S]_{eq} = k_{-1}\ [P]_{eq}$. Rearrangement yields $[P]_{eq}/[S]_{eq} = k_{+1}/k_{-1} = K_{eq}$, where k_{+1} and k_{-1} are rate constants and K_{eq} is the equilibrium constant. In case of the enzyme-catalyzed reaction ($S \rightarrow P$) k_{+1} and k_{-1} are substituted by the catalytic efficiency in the forward reaction $\left(\frac{k_{cat\ f}}{K_{M\ f}}\right)$ and the catalytic efficiency of the reverse reaction $\left(\frac{k_{cat\ r}}{K_{M\ r}}\right)$ yielding

$$K_{eq} = \frac{k_{cat\ f} K_{Mr}}{k_{cat\ r} K_{Mf}}$$

which is referred to as the Haldane relationship. **This relationship shows that k_{cat} and K_M of an enzyme are constrained by the reaction thermodynamics** ($\Delta G^0 = -RT\ln K_{eq}$). In case of two or more substrates/products, as in most enzymatic reactions, and as will be shown below, the Haldane equation is more complex and depends also on the catalytic mechanism. The Haldane equation works only with "real" K_M, not with apparent K_M.

Fumarase catalyzes the reversible dehydration of malate to fumarate ($\Delta G^{0\prime} = -0.9$ kJ/mol; $K_{eq} = 1.4$). The catalytic efficiency in both directions is almost the same. In catabolism of aerobic microbes, this enzyme catalyzes the hydration of fumarate in the citric acid cycle, and in the anabolism of some autotrophic anaerobes the enzyme catalyzes the dehydration of malate in the reversed citric acid cycle. Catabolic *fumarase* catalyzes fumarate hydration with a higher k_{cat} and K_M than malate dehydration. Anabolic *fumarase* is the opposite: k_{cat} and K_M in direction of malate dehydration are higher than those in direction of fumarate hydration. Thus, there is a clear functional (kinetic) bias in both enzyme variants catalyzing the same reaction with the same ΔG. Based on the Haldane equation, it is intrinsically impossible to change the catalytic efficiency k_{cat}/K_M in one direction, e.g., by solely decreasing K_M, without increasing catalytic efficiency in the other direction.

β-Ketothiolase catalyzes the formation of acetoacetyl-CoA and CoA from 2 acetyl-CoA ($\Delta G_0{}' = +25$ kJ/mol; Keq $\approx$ 0.0001), as introduced above. The two substrates—two products enzyme has a catalytic ping pong mechanism. The corresponding Haldane relationship is

$$K_{eq} = \frac{\frac{k_{cat^2\ f}}{K_A\ K_B}}{\frac{k_{cat^2\ r}}{K_C\ K_D}}$$

where K_A and K_B are the two K_M for the two acetyl-CoA and K_C and K_D are the K_M for acetoacetyl-CoA and CoA, respectively. Assuming the four K_M to be identical,

$$\frac{k_{cat^2\ f}}{k_{cat^2\ r}} = 10^{-4}$$

Because of the catalytic mechanism, it is the ratio of the squared maximal rates that is 10^{-4}. If $k_{cat}f = 1$, then $k_{cat}r = 100$. This ratio of 1/100 allows the enzyme to catalyze acetoacetyl-CoA formation at the rate necessary to function in catabolic butyric acid formation. If $k_{cat}f/k_{cat}r$ were 10^{-4}, as in case of a ternary complex mechanism, this would not be possible.

Methyl-Coenzyme M reductase catalyzes oxidation of methane with heterodisulfide CoM-S-S-CoB to methyl-coenzyme M and HS-CoB ($\Delta G^{0\prime}$ = + 30 kJ/mol; Keq $\approx 10^{-5}$) (Chap. 10). The two substrates—two products enzyme has a sequential catalytic mechanism. The corresponding Haldane relationship is

$$K_{eq} = \frac{\frac{k_{cat\ f}}{K_A\ K_B}}{\frac{k_{cat\ r}}{K_C\ K_D}}$$

where K_A is the K_M for methane, K_B for the heterodisulfide, K_C for methyl-coenzyme M, and K_D that for HS-CoB. Assuming the four K_M to be identical,

$$\frac{k_{cat\ f}}{k_{cat\ r}} = 10^{-5}$$

This can explain why anaerobic methane oxidation involving *methyl-coenzyme M reductase* is such a slow process relative to methanogenesis, in which this enzyme catalyzes methane formation. The observed rates of anaerobic methane oxidation in marine sediments with high concentrations of methane are higher than predicted from $k_{cat}f/k_{cat}r = 10^{-5}$. This can be explained by the finding that the K_M for methane of the enzyme and concentration of intracellular enzyme is very high as predicted from the Haldane relationship.

5.5 A Kinetic View on the Julian Adam Experiment

As an experimental proof of an ecological concept, the Julian Adams experiment showed that metabolic speciation and segregation of pathways into distinct microorganisms can occur and that evolved strains can stably co-exist because a metabolic food web enabled microbes to escape competition.

The outcome of this adaptive laboratory evolution experiment was contingent on the specific organismal and environmental parameters chosen, but showed nevertheless that metabolic speciation can be explained by the interplay between thermodynamics and kinetics, the flux of energy, and the physical structure of the environment. In Chap. 3.5.5 a thermodynamic view on the Julian Adam experiment was given.

5.5.1 *E. coli* Glucose Metabolism and Strain Characteristics

The ancestral *E. coli* strain (JA122, derived from MG1622) oxidized glucose to 6 CO_2 ($n_{ATP} = 24$). The thermodynamic efficiency of ATP synthesis is about 40% (Sect. 3.5.4). A particularity of aerobic glucose metabolism by *E. coli* is that at high substrate flux, such as in a batch culture, glucose is used only incompletely, which results in excretion of acetate (acetate overflow metabolism). Acetate is taken up again and metabolized further once glucose is depleted. Since selection in batch conditions is for a high rate phenotype, incomplete glucose oxidation is favored in *E. coli* because by switching from a respiratory to a predominantly fermentative metabolism, a higher cellular flux of ATP per proteome can be achieved by a smaller catabolic proteome (see Sects. 4.6 and 4.6.4). At the same time, the smaller catabolic proteome frees up space to accommodate an increased anabolic proteome for higher growth rate in batch. This trait is also beneficial at high glucose concentration and at high fluxes as in the chemostat. Chapters 7 and 8 provide a mechanistic explanation for why acetate is a good point for pathway speciation in glucose oxidation. Strain JA122 was a mutator strain that had a ~30-fold higher mutation rate than MG1622.

5.5.2 The Specific Chemostat Environment

The chemostat was operated at a dilution rate of 0.2 h^{-1}, which was little less than 50% of $D_{max} = 0.44\ h^{-1}$. Glucose concentration in the chemostat was, therefore, close to the K_S for glucose. At the selected dilution rate, *E. coli* wild type metabolized glucose at the transition point between respiratory and primarily fermentative metabolism since acetate was released from wild-type cells into the chemostat at low concentration. This "sweet spot" condition in conjunction with the idiosyncrasies of *E. coli*'s glucose metabolism set the stage for genetic fixation of the primarily fermentative metabolism in SCV, which was greatly enhanced by further adaptive mutations acquired in this mutator strain background.

Metabolism of SCV was primarily fermentive by oxidizing glucose to 2 acetate and 2 CO_2 ($n_{ATP} = 9$) despite O_2 being present. Describing glucose metabolism of the SCV as "fermentative" is not entirely accurate (see definition Sect. 4.4.2) as the actual redox balance is unclear, but used here to indicate that the major catabolic pathway is glycolysis, and neither TCA cycle or respiration is significantly involved (see Chaps. 7 and 8). Evolved SCV had accumulated mutations that enhanced its rate of glucose metabolism as well as resulted in a lower K_S for glucose for efficient use at low concentration. In monospecies chemostat environment at $D = 0.2\ h^{-1}$, both wild-type and SCV grow at the same rate, at $n_{ATP} = 24$ and 9, respectively, and at thermodynamic efficiencies around 40%. Thus, it is not a difference in growth rate or growth yield that explains the outcome of the competition and speciation in the chemostat. Rather, it is the difference in rate of glucose metabolism. Incomplete glucose oxidation effectively involves a shortened pathway of glucose metabolism which enables *E. coli* to operate the shorter fermentative pathway faster (see above Sect. 5.3). A faster but incomplete glucose metabolism enables SCV at the selected dilution rate to utilize glucose faster than the parental wild type and, thereby, to outcompete the ancestral strain for glucose. Although n_{ATP} in the shorter pathway of SCV is only 9 compared to 24 in the ancestor, the cellular flux of ATP per proteome space in SCV is twice as high, which offsets the low n_{ATP}. The lower molar growth yield of SCV is not visible under chemostat conditions because glucose is both constantly supplied and metabolized at a high rate. This is in contrast to the structured environment of an agar plate, where the variants were identified. The concomitant lower growth yield of SCV (=9/24 of wt) explains the small colony phenotype on diffusion-limited agar plates.

The arising and establishing in the chemostat of LCV was a consequence of the genetic fixation of the SCV phenotype and its outcompetition of wild type. LCV, which is genetically derived from the SCV, oxidized acetate to 2 CO_2 (Chap. 3.5.4), and at $n_{ATP} = 5.5$ and a thermodynamic efficiency = 34%.

LCV escaped competition with SCV and wild type by speciating on the new niche acetate, and by genetically fixing a phenotype that used acetate faster and at lower concentration, e.g., by constitutive expression of high affinity *acetate synthetase*. In fact, the apparent K_S for acetate was much lower in LCV than in wild type or SCV.

5.5.3 General Implications of the Julian Adams Experiment

The Julian Adams experiment showed that metabolic speciation can occur under certain conditions, here under constant, high flux conditions. The observed molecular features of the speciated pathways are contingent to the pathways (glycolysis, *pyruvate dehydrogenase* complex, TCA cycle, and respiratory chain) as well as the environment. The drivers for speciation are based on thermodynamics (growth yield, n_{ATP}), kinetics (fluxes, rate-yield trade-off), and ecophysiological (proteome allocation) traits. Because *E. coli* did speciate in this experiment, this indicates that *E. coli* wild type is not well-adapted to this particular chemostat environment (its natural environment in the cecum is much more fluctuating). Notably, the outcome of this laboratory experiment certainly would be different using different dilution rates, different catabolic substrates, or different microbes, and different combinations of the above. A low dilution rate glucose-limited, oxic chemostat system is predicted to select for variants with increased growth yield rather than faster glucose consumption. Also, structured environments with constant but low flux are predicted to select for efficient (high n_{ATP}) rather than fast (low n_{ATP}) pathways.

It can be argued that the specific conditions of the Julian Adams experiment in terms of glucose concentration and dilution rate were favorable to the observed outcome. Many microbes and their metabolism described in the following chapters appear to have speciated that can be understood based on the material discussed (Chaps. 1–6) so far, and Chaps. 9 and 10 will provide examples.

It is important to note that also other metabolic features, such as absence of special cofactors, can also explain specific metabolic features, however, rate-yield trade-off is very common and has its imprint in pathway architecture. The intriguing observations of the Julian Adams experiment outline a path for understanding metabolic speciation as found in extant environments.

Further Reading

Basan et al (2015) Overflow metabolism in *Escherichia coli* results from efficient proteome allocation. Nature 528:99–104

Costa E, Pérez J, Kreft JU (2006) Why is metabolic labour divided in nitrification? Trends Microbiol 14:213–219

Karlin S, Mrazek J (2000) Predicted highly expressed genes of diverse prokaryotic genomes. J Bacteriol 182(18):5238–5250

Novak et al (2006) Experimental test for an evolutionary tradeoff between growth rate and yield in *E. coli*. Am Nat 168(2):242–251

Pfeiffer, Bonhoeffer (2002) Evolutionary consequences of tradeoffs between yield and rate and ATP production. Z Phys Chem 216:51–63

Wortel et al (2016) Evolutionary pressure on microbial strategies in the chemostat. Sci Rep 6:29503. https://doi.org/10.1038/srep29503

Mechanisms of Microbial Energy Conservation

6

Contents

6.1 Introduction

The previous chapters outlined the physical-chemical and biological context within which microbial metabolism, in particular catabolism, functions. Specifically, diffusion, molecular crowding of the cytoplasm, proteome allocation, thermodynamics, kinetics, and the rate-yield trade-off shape the molecular architecture of microbial metabolism. This chapter and the following examine the limited molecular rules and mechanistic details of energy conservation, and how they are expressed within the above-described constraints. The recurrent use of these mechanisms in different combinations gives rise to the diversity of microbial metabolism.

Microbial catabolism is the conservation of environmentally available energy in form of ATP. At its core, energy metabolism, and with that Microbial Life is redox chemistry where in chemoorganotrophic microbes ATP synthesis is coupled to oxidative and/or reductive catabolic

A. M. Spormann, *Principles of Microbial Metabolism and Metabolic Ecology*,
https://doi.org/10.1007/978-3-031-28218-8_6

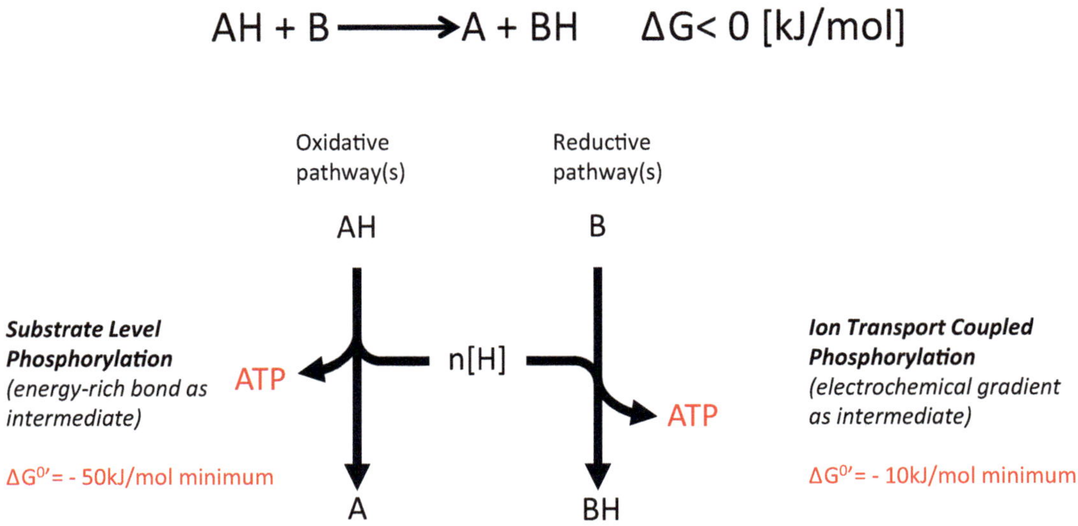

Fig. 6.1 Microbial energy conservation in chemotrophic prokaryotes. Energy conservation can occur in the oxidative or in the reductive pathway of a cell's catabolism. AH →A + [H] (oxidative half reaction); B + [H] → BH (reductive half reaction). Oxidative and reductive pathways are stoichiometrically coupled by the transfer of electrons (shown here as reducing equivalent [H] = $e^- + H^+$). The most common catabolic [H] carriers are ferredoxins, flavodoxins, NAD^+, Factor F_{420}, ubi/menaquinone, and cytochromes. Minimum free energies for each energy conservation are indicated

pathways (Fig. 6.1). In phototrophic microbes, light energy initiates (1) in chlorophyll-based energy conservation a cyclic redox process for ATP formation, or (2) in bacteriorhodopsin-based energy conservation a redox-independent isomerization chemistry leading to ATP synthesis. Despite the diverse chemical substrates and the difference in chemical and light energy, there are only two mechanisms known for ATP formation: (i) **Substrate Level Phosphorylation (SLP)**, or (ii) **Ion Translocation Coupled Phosphorylation (TCP)**. The molecular principles of both modes of energy conservation are illustrated here primarily for chemotrophic microbes; energy conservation from light energy is discussed as well.

Substrate Level Phosphorylation is a cytoplasmic process and occurs in chemoorganotrophic microbes in some oxidative pathways, while **Ion Translocation Coupled Phosphorylation** is a cytoplasmic membrane-bound process and occurs in reductive pathways (Fig. 6.1, see also Fig. 9.1). Most biochemistry textbooks introduce TCP as "electron transport phosphorylation" or as "respiration." The two latter processes are specific variations of TCP, and the more ubiquitous TCP is introduced as the general principle.

The amount of energy that can be conserved via SLP or TCP differs (Fig. 6.1). For SLP, the minimum energy necessary under reversible, cellular conditions is that of the phosphorylation potential of one ATP, typically around +50 kJ/mol (see below, Sect. 6.4.3), while that required for TCP is less than one ATP and equivalent to the **electrochemical proton (or sodium) potential** across the cytoplasmic membrane, typically −15 to −20 kJ/mol. This difference has implications for which mechanism of energy conservation is involved in microbes growing with small amounts of energy; the mechanisms will be discussed in this chapter.

Some familiar examples of pathways involved in the **oxidation** of organic compounds are the Embden–Meyerhof pathway (glycolysis), the Entner–Doudoroff pathway, β-oxidation, and the citric acid cycle. Examples of **reductive** pathways include aerobic electron transport chains, the sulfate reduction pathway, the denitrification/ammonification pathways, reductive fermentative pathways, the Wood–Ljungdahl pathway, and CO_2 reduction to methane. The oxidative and reductive pathways resemble oxidative and reductive half reactions (Fig. 6.1) that function in an overall exergonic catabolic pathway based on the redox potentials of the half reactions (see Table A.1 in the Appendix). ATP synthesis via both modes is linked to an electron flux via cellular electron carriers between the oxidative and reductive pathway. Therefore, this chapter discusses first general features of cellular electron carriers as they are key for both modes of energy conservation, followed by the mechanism of TCP and SLP.

6.2 Electron Carriers

Cellular electron carriers connect oxidative and reductive pathways by mediating the transfer of a hydride ($H^- = 2\,e^- + H^+$), of two single electrons ($2\ e^-$), or of a single electron ($1\ e^-$). While the general concept of "reducing equivalents", [H], was occasionally used in the previous chapters to describe general electron transfer in metabolism, this and the following chapters differentiate between the specific electron carriers involved as their differences are key for understanding the idiosyncratic features of a pathway.

Based on the mode of electron transfer, three major types are distinguished:

1. The one-electron carrier ferredoxins, flavodoxins, and cytochromes,
2. The one- AND two-electron carrier flavins, quinones, and methanophenazine, and
3. The two-electron carrying pyridine nucleotides (NAD^+, $NADP^+$) and Factor F_{420}.

These electron carriers differ in their mechanism of electron transfer and redox potentials.

6.2.1 One-Electron Carriers

6.2.1.1 Ferredoxins

(oxidized form is abbreviated as "Fd") are low molecular weight (6–12 kDa) proteinaceous electron carriers containing one or two redox-active **[2Fe-2S], [4Fe-4S], or [3Fe-4S] clusters**, which are coordinated in the protein by cysteine-derived S bridges (Fig. 6.2). Oxidized [Fe-S] clusters give purified ferredoxins a characteristic brownish color ($\varepsilon_{390} \approx 30.6$ (mM protein)$^{-1}$ × cm^{-1}, depending on the number and type of [Fe-S] cluster), although they are typically distinguished via their g-value by NMR spectroscopy. Ferredoxins in strictly anaerobic chemotrophic microbes typically contain two surface-exposed [4Fe-4S] clusters (Fig. 6.2b), where each [4Fe-4S] cluster can be reduced reversibly by one electron. Each electron of a cluster is transferred individually and not simultaneously as in $NAD(P)^+$ or F_{420} (see below). For the following discussion, it is assumed that ferredoxin-dependent *oxidoreductases* in 2-electron transfer reactions reduce or oxidize each [Fe-S] cluster ($Fd^{2\times1-} = Fd^{2-}$) as was shown for *[Fe-Fe] hydrogenase* of *Clostridium pasteurianum*. This is different from plant *ferredoxin:NADP⁺ oxidoreductase*, where two ferredoxin proteins each undergoing each a 1-electron transfer.

A single microbial genome may encode multiple different ferredoxins including polyferredoxins, where the latter consist of multiple ferredoxin domains and with more than two [4Fe-4S] clusters (e.g., 14 [4Fe-4S] clusters in the dimer of polyferredoxin of *Methanobacterium wolfei*, Fig. 6.2c). For electron relay across multiple [4Fe-4S] clusters in polyferredoxins or Fe-S cluster-containing *oxidoreductases*, the required distance between clusters is 10–14 Å (Fig. 6.2c). Ferredoxins and ferredoxin-dependent *oxidoreductases* are sensitive toward molecular oxygen, because O_2 can oxidatively destruct the surface-exposed [Fe-S] clusters and form toxic O_2^-. Therefore, ferredoxins are either restricted to strictly anaerobic microbes or contain Fe-S clusters that are deeply buried in the protein in aerobic microbes. The surface exposure of the Fe-S cluster explains also why many ferredoxins can substitute for each other in vitro, while they are specific to cognate *oxidoreductases* in vivo.

The rate of diffusion of ferredoxins in the cytoplasm is only about 1/10 relative to NAD(P)$^+$ or F_{420}; moreover, they form transient

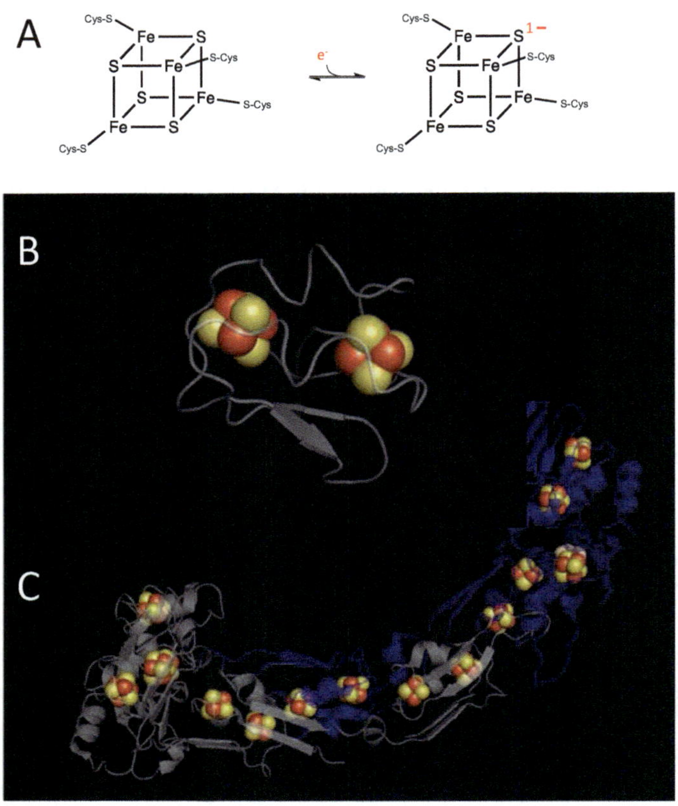

Fig. 6.2 [4Fe-4S] cluster ferredoxins. Four Fe and four S atoms are coordinated as cluster by cysteinyl groups in ferredoxin proteins. (**a**) A cubane [4Fe-4S] cluster can be reversibly reduced by 1 electron. (**b**) Structure of the 55 amino acid, two [4Fe-4S] cluster-containing ferredoxin from *Clostridium pasteurianum.* Most catabolic ferredoxins in strictly anaerobic microbes contain two [4Fe-4S] clusters. The cluster is indicated with brown Fe and yellow S atoms, peptide scaffold is in gray. Each [4Fe-4S] can take up or release a single electron independent of each other. (**c**) The polyferredoxin dimer FwdF from *Methanobacterium wolfei* of *formylmethanofuran dehydrogenase*. Protein scaffold of a monomer is indicated in gray and dark blue color. The 14 [4Fe-4S] clusters align to form an electron relay chain through the polyferredoxin dimer in the enzyme. See text for details

protein–protein complexes with cognate *oxidoreductases*. As there is only one ferredoxin molecule in such complex, there is no meaningful ratio of the concentration of Fd and Fd^{2-} that would determine its redox potential as defined by $E' = E^{o\prime} + \frac{RT}{nF} \ln \frac{[C_{Fd}]}{[C_{Fd(2-)}]}$. This is different from cytoplasmic pools of $NAD(P)^+$, F_{420}, or membrane-diffusible quinones (see below). Instead, the redox potential of a ferredoxin is determined by (1) the protein scaffold of the [4Fe-4S] cluster as encoded by a specific amino acid sequence, (2) conformational changes upon binding to a cognate *oxidoreductase*, as well as whether (3) the electron flux through the oxidative and the reductive pathways is electron donor- or acceptor-limited. Therefore and in contrast to $NAD(P)^+$, F_{420}, or quinones, redox potentials of ferredoxins are fine-tuned and range between −300 to −500 mV. For an electron transfer from ferredoxin to $NAD(P)^+$ or F_{420}, flavins (see below) are required for switching a two times one-electron transfer into a simultaneous transfer of 2 e^- in a hydride.

Ferredoxins function as catabolic electron carrier with *hydrogenases*, *formate dehydrogenases*, *pyruvate:ferredoxin oxidoreductases*, in energy-conserving Rnf and Ech, in electron-bifurcating enzymes, and in *ferredoxin:NADP⁺ oxidoreductases*, among others and are found in, but not limited to, strictly anaerobic fermenting microbes, methanogens, or sulfate-reducing microbes. Ferredoxins also function as low redox potential "electrical wires" to directly couple pathway-related *oxidoreductases* into supercomplexes, which increases the electron transfer rate within a pathway such as in the *formylmethanofuran dehydrogenase-heterodisulfide reductase* complex in hydrogenotrophic methanogens via a 22 [Fe-S] cluster relay (Chap. 10)!

6.2.1.2 Flavodoxins

(Fld) are also proteinaceous (14–24 kDa) one-electron carriers in strictly anaerobic microbes containing a non-covalently-bound, redox-active **flavin mononucleotide (FMN)** as cofactor (Figs. 6.3). Although flavins can carry both one and two electrons, the function relevant for flavodoxins is that of an exclusive one-electron carrier. The mode of action of a flavin as a dedicated one-electron carrier is discussed below after introduction of the chemistry of flavins. Of the three redox transitions of flavin possible, the semiquinone/hydroquinone transition is relevant for the single electron transfer function of flavodoxin.

Flavodoxins, if present, are expressed in microbes when in Fe-limited environments and can substitute for ferredoxins and *vice versa*. This is because the redox-active cofactors ([Fe-S] clusters and FMN) are located at the surface of each protein (Fig. 6.3). However, cognate *oxidoreductases* have distinct specificities and affinities to either. There are short-chain and long-chain flavodoxins, which are due to on insertion of up to 38 amino acids in the long chain, leading to lower redox potentials of long-chain flavodoxins.

6.2.1.3 Cytochromes

Cytochromes are proteinaceous (~104 amino acids) one-electron carriers containing a **heme** as redox-active cofactor (Fig. 6.4a). A heme is a porphyrin (side chain substituted, cyclic tetrapyrrole) containing Fe as the central atom that is coordinated by four imidazole nitrogen atoms and by axial amino acid side chains (mostly histidine, methionine) of the protein (Fig. 6.4a, b). The heme iron undergoes reversibly reduction/oxidation between Fe^{3+} and Fe^{2+}. The redox potential of cytochrome hemes varies between −500 to +400 mV, which is a much larger range than that of Fe-S clusters or flavins. There are several different types of heme cofactor that differ in their structure, which, in conjunction with a cytochrome's amino acid sequence, defines the physico/(electro)chemical properties. The most common cytochromes in microbial electron transport are *b*- and *c*-type cytochromes that differ also

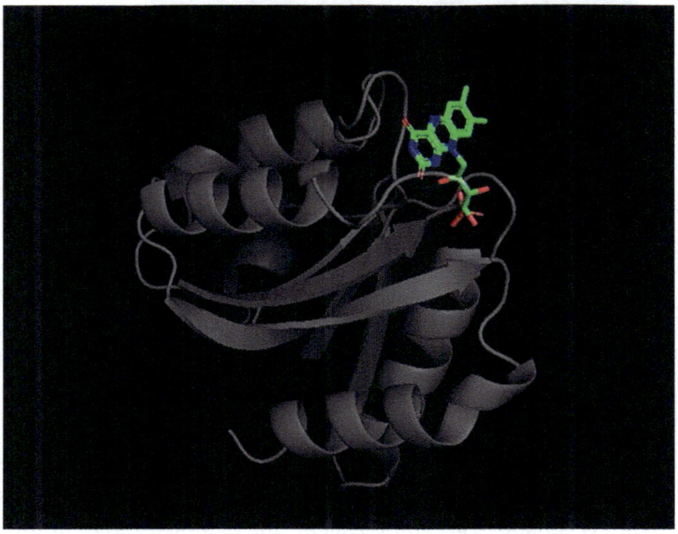

Fig. 6.3 Flavodoxin from *Clostridium beijerinckii*. The FMN cofactor is exposed to the surface of the 136 amino acid flavodoxin

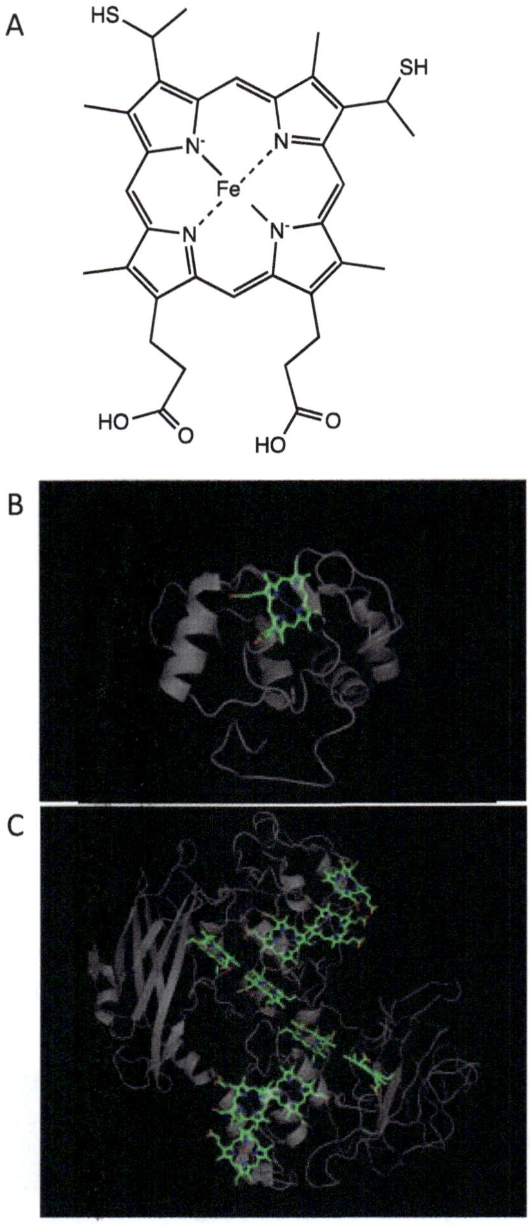

Fig. 6.4 Microbial *c*-type cytochromes. (**a**) Structure of heme *c*. Heme *c* is covalently attached to the protein backbone by a thioether bond. (**b**) Cytochrome c_{550} from *Paracoccus denitrificans*. The *c*-type heme is at the surface of the protein. (**c**) MtrF from *Shewanella oneidensis* MR1. Cytochrome-based electron relay through 10 hemes of the dimeric cytochrome *c*. See text for details

in the attachment of the heme group to the protein: the heme group in a *c*-type cytochrome is bound covalently to the protein, whereas a *b*-type heme is bound via non-covalent interactions. Cytochromes are typically identified by an idiosyncratic UV-VIS absorption in the long wavelength range.

b-Type cytochromes are present in membrane-bound *oxidoreductases* of energy-conserving respiratory chains and are either embedded in the membrane or face the cytoplasmic side. The redox potential of different *b*-type cytochromes of *E. coli* varies between −50 mV and +260 mV and is less variable than between *c*-type cytochromes.

c-Type cytochromes are on the outside of the cytoplasmic membrane and can range in redox potentials between −500 mV and +100 mV. They accept electrons from *quinol oxidases* where a hydride from a hydroquinone is switched to single electrons in cytochromes in a respiratory chain; *c*-type cytochromes are prosthetic groups and mediate electron transport between redox enzymes of a respiratory chain. They are involved in reduction and oxidation of many S, N, and Fe compounds as catabolic electron acceptor or electron donor, such as in denitrification, nitrate ammonification, sulfate reduction, H_2S oxidation, and anaerobic ammonium oxidation. While most microbial *c*-type cytochromes contain one or two heme per protein, multiheme *c*-type cytochromes are involved in extracellular and interspecies electron transfer from the outer surface of the cytoplasmic membrane via the periplasm and through the outer membrane as porin-cytochrome complex in Gram-negative microbes (Fig. 6.4c). Such electron relay networks function in *Shewanella* and *Geobacter* species, for example, to electrically connect insoluble Fe/Mn-oxide minerals or anodic surfaces to the cellular quinone pool. A *c*-type cytochrome of the fumarate-, Fe^{3+}-, and anode-reducing δ-proteobacterium *Desulfuromonas soudanensis* is predicted to contain 69 hemes or 58 in the thermophilic Fe^{3+}-reducing Gram-positive *Thermincola ferriacetica*. Multiheme *c*-type cytochromes are also proposed in direct electron transfer of anaerobic methane-oxidizing Methanosarcinales (ANME-1 and ANME-2) with sulfate-reducing partner microbes (Chap. 9, Fig. 9.31). Multiheme *c*-type cytochromes are present in other Bacteria and Archaea, such as Thermoproteales (*Desulfurococcales* sp., *Pyrobaculum* sp.) and the Fe^{3+}-reducing *Ferroglobus placidus* of the Archaeglobales. The typical distance between hemes for efficient intramolecular electron transfer is 4–8 Å which is shorter than the 10–14 Å distance between [4Fe-4S] clusters in ferredoxins or in Fe-S cluster-containing oxidoreductases. A difference in redox potential between hemes of multiheme *c*-type cytochromes of different cytochromes in respiratory chains provides a thermodynamic directionality of electron flux.

6.2.2 One- AND Two-Electron Carriers

Flavins and **quinones** are both one- AND two-electron carriers and can switch between two 1-electron and one 2-electron transfers (Fig. 6.5). Biologically active flavins are either flavin mononucleotide (FMN, or riboflavin-5′--phosphate) or flavin adenine dinucleotide (FAD). Flavins are not free inside a cell but are protein-bound, where their redox potentials are strongly determined by the protein environment.

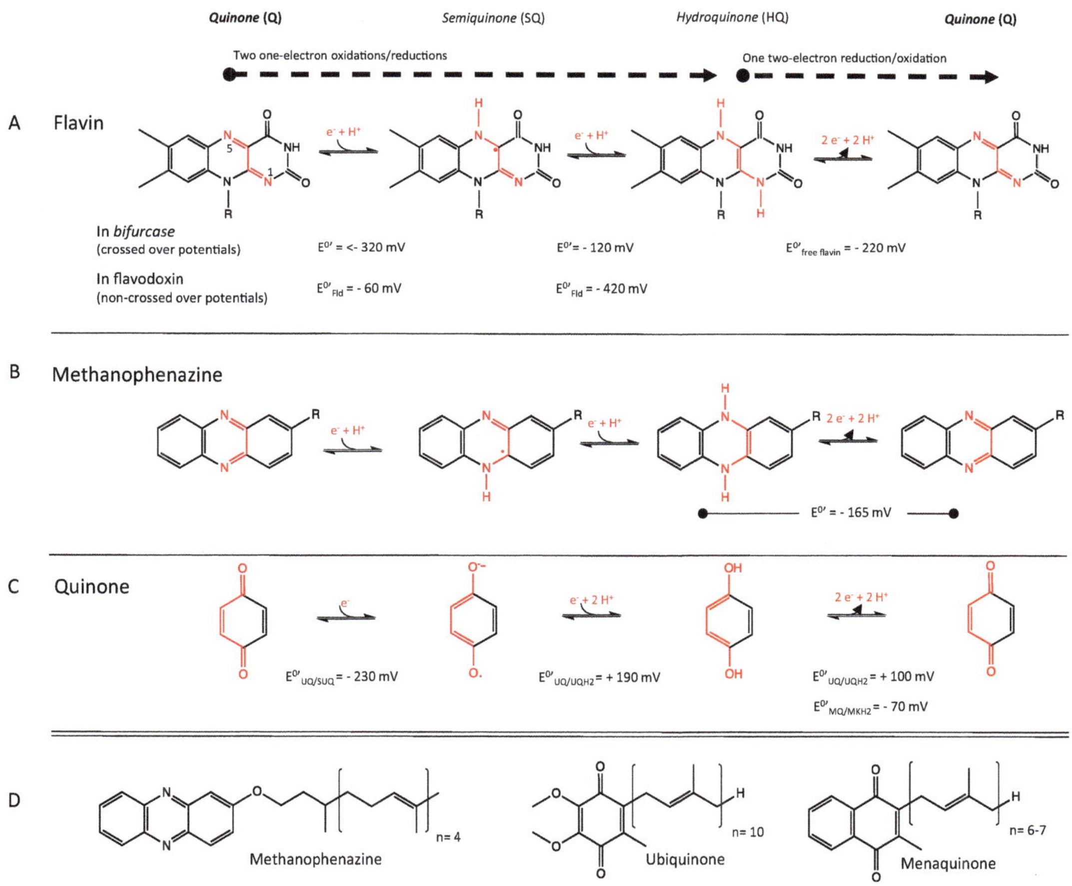

Fig. 6.5 One- AND two-electron carriers in microbial catabolism. Only structures relevant for their redox chemistry are shown. (**a**) Reversible reduction and oxidation of flavin from quinone to hydroquinone can proceed either by a two 1-electron or a one 2-electron transfer. Left part of **a**: Stepwise 1-electron reductions of a quinone via the semiquinone radical anion to hydroquinone (quinol). Right part of **a**: Single 2-electron oxidation of hydroquinone to quinone. (**b**) Structure and reduction of methanophenazine functioning as a quinone analog in electron transport phosphorylation in methanogens with cytochromes. (**c**) Reversible reduction and oxidation of a quinone/hydroquinone. (**d**) Structures of methanophenazine, ubiquinone, and menaquinone. Number of the membrane-anchoring isoprene units is indicated as n. Cyanobacterial plastoquinone, not shown here, involved in cyclic electron flow in cyanobacteria is at a redox potential of $E^{0\prime} = + 80$ mV. See text for details

FMN is the one-electron carrier prosthetic group in flavodoxin (see above) that functions in low redox potential electron transfers. Both FMN and FAD can function as hydride carrier in various *oxidoreductases* including in *bifurcases* to reversibly split a hydride into two single-electrons, each at a different redox potential (see below and Fig. 6.5).

The major classes of quinones (Fig. 6.5d) are the (1) ubiquinones (benzoquinone) and (2) menaquinones (naphthoquinone), functioning as electron carrier in aerobic and in anaerobic respiratory chains, respectively, and (3) plastoquinone (benzoquinone, not shown here) in photosystem II. Methanophenazine is a functional quinone analog in some methanogens (Fig. 6.5b, d; Chap. 10). Covalent ring modifications of ubiquinone and menaquinone as well as their binding affect their redox potential. For example, the binding of free 1-methylmenaquinone to a cognate membrane enzyme changes the redox potential from −124 mV to −250 mV, which is sufficiently low to couple electron transfer from formate to polysulfide ($E^{0\prime}$ = −270 mV) reduction via menaquinone. Quinones are anchored to the cytoplasmic membrane by their isoprene side chain of variable length (up to 10 in ubiquinone, 6–7 in menaquinone, and 4 in methanophenazine) where they are freely diffusible (Fig. 6.5).

The free isoalloxazine ring of flavin and the quinone ring have similar reduction/oxidation properties for 1 ↔ 2 electron switching (Fig. 6.5a, c), which are discussed below in more detail for flavins. Flavins/quinones have three oxidation states: the quinone state (*Q*, = 0), the semiquinone state (SQ, = −1), and the hydroquinone state (HQ, = −2) (Fig. 6.5a). Flavins/quinones function as (1) two 1-electron, (2) one 2-electron carrier, and (3) a 1 ↔ 2 electron switch. This property allows to bifurcate or "split" a hydride (H^-) into two 1-electron transfers, where each single electron in the respective transfer is at a different redox potential (crossed potentials). Reduction of a free FMN or FAD from the Q to the SQ state has a redox potential that is lower by at least 200 mV (i.e., $E^{0\prime}$ −300 to −420 mV) than the following one-electron reduction of SQ to HQ, i.e., $E^{0\prime}$ −60 mV to −120 mV. The redox potential of free flavin of the two-electron transition Q/HQ is −200 mV.

6.2.2.1 Flavins in Flavodoxin

In flavodoxins, the redox potentials of Q/SQ and of SQ/HQ couple are non-crossed over, meaning that the more negative potential of the flavin SQ/HQ (e.g., −420 mV in *Acidaminococcus fermentans*) makes the SQ more stable for a 1-electron reduction to HQ, relative to the Q/SQ couple at –60 mV (Fig. 6.5c). This is different from free FMN/FAD and flavins functioning in electron bifurcation (see below). The different potentials are the result of specific amino acid side chain interactions with the redox-active N1 and N5 of the isoalloxazine ring, which collectively stabilize the SQ and HQ and destabilize the quinone state. Also, the two-electron Q → HQ transition of flavodoxin-bound FMN is thermodynamically much more difficult (+ 60 to +180 mV) than in free FMN. Single electrons from two flavodoxin HQ can thermodynamically reduce 2 H^+ to H_2.

Besides the function of quinones in energy conservation by the cytochrome *bc*$_1$ complex, the switching between 1-electron and 2-electron transfers is important for coupling diverse membrane-bound oxidation reactions: Oxidation of NADH, alcohols, and some aldehydes is coupled to diverse reduction reactions of inorganic electron acceptors and of cytochromes (Fig. 6.6). For example, quinones accept two single electrons in succinate oxidation to fumarate. Conversely, in anaerobic fumarate respiration, single electron reductions of fumarate to succinate by flavin-containing *fumarate reductase* involves a one 2-electron ↔ two 1-electron switch via a quinol. Figure 6.6 provides an overview of the modularity of quinone-based respiratory electron flux coupling via ubiquinone or menaquinone in *E. coli*. For catabolic pathways, the specific combinations follow thermodynamic constraints (not considered in this figure).

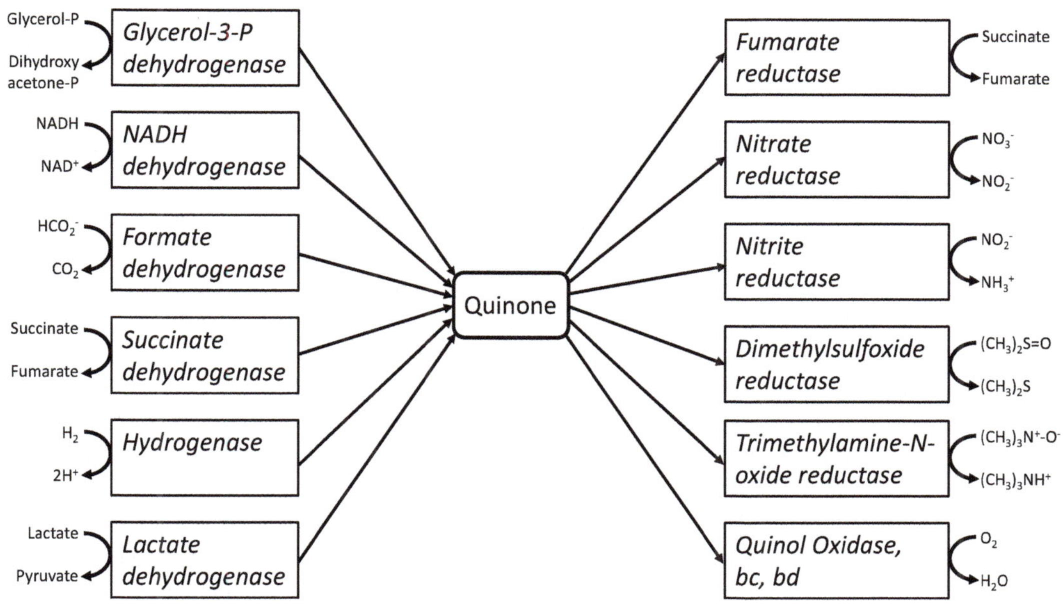

Fig. 6.6 Central role of quinones in *E. coli* respirations and coupling of diverse membrane-bound oxidative reactions (left panel) to reductions (right panel) of diverse electron acceptors via electron transport phosphorylation. Only those combinations are possible in catabolism where the difference in redox potential between electron acceptor and electron donor is positive, i.e., $\Delta G = -nF\Delta E$. See text for details

6.2.2.2 Methanophenazine

Methanophenazine is a quinone analog and functions in methanogens with cytochromes in energy-conserving electron transport with HdrDE *heterodisulfide reductase* (Figs. 6.5d, 10.8b, Chap. 10). The standard redox potential of methanophenazine/methanophenazineH_2 is −165 mV.

6.2.3 Two-Electron Carriers

6.2.3.1 NAD+ and NADP+

The pyridine nucleotides **NAD+** and **NADP+** are two-electron (hydride) carriers (Fig. 6.7). When 2 e− + 2 H+ are abstracted from a substrate in a NAD(P)+-dependent oxidation reaction, 2 e− + H+ (hydride, H−) are transferred to the nicotinamide ring (Fig. 6.7); the second proton is released to the solvent water. The difference between NAD+ and NADP+ is a phosphorylation at the 2′ OH group of the adenosine ribose, which does not affect electron transfer properties and redox potential, but determines binding specificity (K_M) to *oxidoreductases*. Thus, the standard

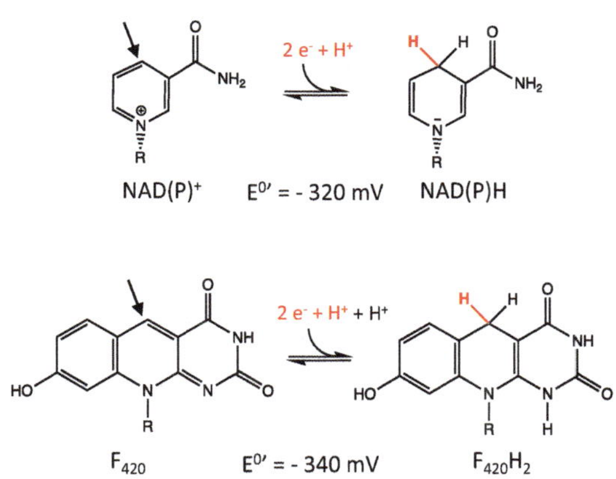

Fig. 6.7 Two-electron carriers involved in chemotrophic prokaryotes. Only structures of NAD(P)+ and F_{420} relevant to their redox chemistry are shown. (**top**) The pyridine nucleotides NAD+ and NADP+ and their respective reduced forms. The arrow indicates the carbon of the nicotinamide ring which is reversibly reduced by a hydride, i.e., 2 e− + 1 H+. The second proton from substrate oxidation is released as H+ to water. NADP+/NADPH carries a phosphoryl group at the 2′ OH group of the adenosine ribose, which does not affect electron transfer properties and redox potential, which is $E^{0\prime} = -$320 mV for both pyridine nucleotides. (**bottom**) The deazaflavin Factor F_{420} and its reduced form. Arrow indicates the carbon that is reversibly reduced by a hydride; the second proton is added to the conjugated N. See text for details

redox potential for both electron carriers is $E^{0\prime} = -320$ mV. In contrast to ferredoxins, there is only one in situ redox potential for each pyridine nucleotide pool which is depending on whether cellular electron flux is donor- or acceptor-limited. The resulting different ratios of $NAD(P)^+/NAD(P)H$ change the redox potential (E') modestly in the order of ±30 mV per order of magnitude, due to the competitive binding of the oxidized and reduced forms to their cognate oxidoreductases. In an aerobically growing, glucose-catabolizing *E. coli* cell, the in situ concentration of NAD^+ is in the mM range while that of NADH is <0.1 mM, which corresponds to an in situ redox potential of $E' \approx -280$ mV. The intracellular ratio of $NADP^+/NADPH$ (~ 2 μM and 100 μM, respectively) is inverse to that of $NAD^+/NADH$ and corresponds to an in situ redox potential of $E' \approx -380$ mV. Thus, the in situ redox potential varies between those maximal values and $E^{0\prime}$ (−320 mV), depending on whether the rate of NADH oxidation and $NADP^+$ reduction is limiting. Precise measurements of the in situ ratio of both pyridine nucleotides, and, thus, of their in situ redox potential is challenging, because of their rapid turnover. Measurements are reliable when consistent with an energy charge of >0.9 (see Sect. 3.4.2).

$NAD^+/NADH$ is the electron carrier in many oxidative pathways of organotrophic microbes, in particular in oxidations of alcohols and aldehydes, in reductions of enoates/enoyl-CoA (fumarate, crotonyl-CoA), in C1 metabolism in methylene-tetrahydrofolate reduction and oxidation, as well as a co-substrate in some *monooxygenases*. $NADP^+/NADPH$ is the electron carrier mostly in anabolic pathways; however, NADPH is electron donor in some *formate dehydrogenases* in acetogens (Fig. 10.14). The flux of electrons between catabolic or anabolic pathways is tightly controlled by *transhydrogenases*.

6.2.3.2 Factor F_{420}

Factor F_{420} is a 5-deazaflavin derivative and was first discovered in hydrogenotrophic methanogens (Fig. 6.7 Bottom). This electron carrier has since been found in phylogenetically diverse microbes including Euryarchaeota (Methanogens, Halobacteria, and Archaeoglobi), Chloroflexi (*Thermomicrobium roseum*), Actinobacteria (*Mycobacterium* sp., *Streptomyces* sp., *Rhodococcus* sp.), and some Proteobacteria (*Oligotropha carboxidovorans*). F_{420} has a catabolic function only in methanogens and methanotrophic archaea. Although structurally similar to flavin, the isoalloxazine ring of F_{420} contains a carbon instead of a nitrogen atom at the position equivalent to N5 in flavins (see Fig. 6.5), which prevents a single electron transfer. This restricts F_{420} to function as a 2-electron (hydride) carrier (Fig. 6.7 Bottom). Its standard redox potential is at $E^{0\prime} = -340$ mV more negative than that of $NAD(P)^+/NAD(P)H$ which explains why this electron carrier is involved in tetrahydromethanopterin-dependent and not in tetrahydrofolate-dependent C1 metabolism (Chap. 10). In methanogens, F_{420} is electron carrier in F_{420}-dependent *hydrogenases* and *formate dehydrogenases* in H_2 and formate oxidation, respectively, and in oxidation and reduction of methylene-tetrahydromethanopterin. In other microbes, including *Mycobacteria* and *Rhodococcus* sp., F_{420} is involved in enoate reductions, alcohol oxidations/ketoreductions, imine reductions, and nitro reductions.

6.3 Energy Conservation by Substrate Level Phosphorylation (SLP)

ATP is formed either by Substrate Level Phosphorylation (SLP) as it occurs in some oxidative pathways, or by Ion Translocation Coupled Phosphorylation (TCP) via a chemiosmotic mechanism as it occurs in reductive pathways (Fig. 6.1). Oxidative and reductive pathways are stoichiometrically coupled by the transfer of electrons via the above-discussed cellular electron carriers.

Not all oxidative or reductive catabolic pathways mediate energy conservation by SLP or TCP. In some microbes, energy is conserved exclusively either by SLP or by TCP; in others, both SLP and TCP work in tandem, with different fractions of ATP formed by each (see also Fig. 9.1). In some microbes, the amount of energy conserved (n_{ATP}) for a given overall catabolic reaction is variable and biologically controlled, e.g., by expressing different enzymes with different "coupling" properties, resulting in altered energy conservation by SLP or TCP, or by both. These traits of controlling of n_{ATP} are ecologically consequential and result in rate-yield trade-offs (e.g., short and inefficient but fast *vs* efficient but slow pathways).

6.3.1 Substrate Level Phosphorylation via Organic Compounds

Although in the microbial world energy conservation by TCP is more widely found, for the ease of understanding Substrate Level Phosphorylation is introduced here first.

Energy conservation by SLP occurs on organic substrates and intermediates that are converted to energy-rich intermediates. Energy-rich intermediates are typically anhydrides, thioesters, enol ester, or amides (Fig. 6.8), where their free energy of hydrolysis is sufficiently negative to drive ATP synthesis by at least −32 kJ/mol under standard conditions. Thus, if an oxidative pathway can mechanistically produce any of these intermediates, the energy of their hydrolysis can be recovered and linked to ATP synthesis. One notable exception to SLP functioning in the oxidative pathway is energy conservation by *glycine reductase* (see Sect. 9.5.3.1). Here, *glycine reductase* functions in the reductive pathway of amino acid-fermenting microbes, where the reduction of the C2 of glycine results in the formation of energy-rich compound acetyl-phosphate (see below).

The paradigm example for energy conservation via SLP is in carbohydrate degradation in the oxidation of the **carbonyl group of** glyceraldehyde-3-phosphate (GAP) **to a carboxyl group** (Fig. 6.9). Carbonyl group oxidation can result in an activated carboxyl group (mixed anhydride) such as in 1,3-bisphosphoglycerate, which is coupled via *phosphoglycerate kinase* to stoichiometric formation of ATP and 2[H] release as shown in Fig. 6.9. This reaction will be discussed in greater detail in Chap. 7. This mode of energy conservation is so central to carbohydrate degradation pathways that it is an intrinsic and main **energy-conserving "module"** found in diverse pathways such as the Embden–Meyerhof, Entner–Doudoroff, and *Phosphoketolase* pathway (Chap. 8). Similar as in TCP, the *GAP dehydrogenase* module can have different n_{ATP}, in this case 0 or 1, where these differences have implications for the microbe's fitness and ecology (see Chap. 7).

Energy conservation via SLP is restricted to those organic compounds that can be converted to

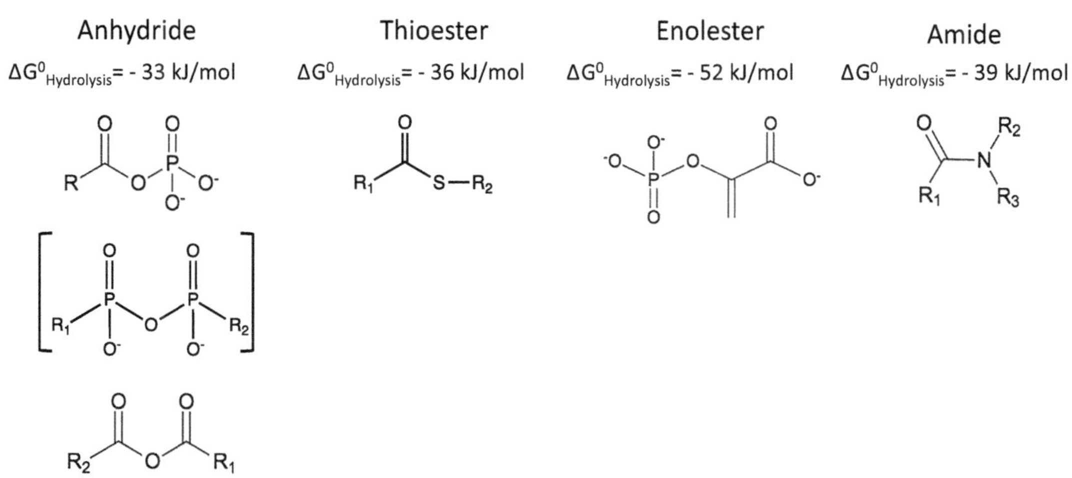

Fig. 6.8 Energy-rich bonds and intermediates in microbiological systems. See text for details

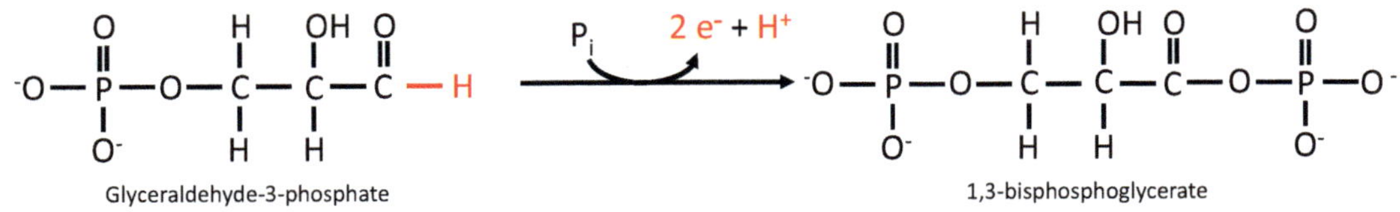

Fig. 6.9 The *glyceraldehyde-3-phosphate dehydrogenase* reaction of forming an energy-rich bond of 1,3-bisphosphoglycerate. For details, see Chap. 7

the energy-rich intermediates described (Fig. 6.8). Thus, SLP cannot function in energy conservation using H_2, formate, or carbon monoxide as electron donor, for example.

6.3.2 Substrate Level Phosphorylation via Inorganic Compounds

While fermentation is defined as an organic, electron donor-derived intermediate functioning as electron acceptor in a reductive pathway (Fig. 4.3), substrate level phosphorylation has also been used synonymously with the term "fermentation." A key feature of substrate level phosphorylation is the formation of a mixed anhydride as the result of an oxidation reaction. There is an example in inorganic sulfur catabolism that resembles the principle of oxidation of a "carbonyl to a carboxyl group." A mixed acid anhydride is formed in inorganic thiosulfate "fermentation" during oxidation of sulfite to sulfate in the adenosylphosphosulfate pathway (Fig. 6.10). During fermentation of thiosulfate, which is a disproportionation to sulfate and H_2S, sulfite is oxidized to sulfate with the intermediate formation of adenosylphosphosulfate, which carries an anhydride bond between a phosphoric acid of ADP and sulfuric acid. The energy of this bond is conserved by an *ATP sulfurylase*, which converts adenosylphosphosulfate plus P_i into ATP plus sulfate. The pair of electrons released during sulfite oxidation to adenosylphosphosulfate is activated by reverse electron transport (see below) with $\Delta\mu H^+$ and used to reduce thiosulfate to H_2S and sulfite. This pathway also fulfills another characteristic of fermentation: that both the electron donor AND electron acceptor are derived from the same substrate molecule.

6.3.3 *Arginine deiminase* Pathway

The *arginine deiminase* pathway is an example of formation of an energy-rich bond in the absence of a net redox reaction. Arginine is converted to ornithine and carbamoyl phosphate by *arginine deiminase* and *ornithine carbamoyl transferase* (Fig. 6.11). This non-redox pathway leads to the formation of energy-rich compound carbamoyl phosphate. Carbamoyl phosphate is the phosphoryl donor for ADP in *carbamate kinase*, forming ATP, CO_2, and NH_3. Arginine is imported into a microbial cell via an *arginine:ornithine*

Fig. 6.10 Proposed thiosulfate disproportionation ("fermentation") in *Desulfovibrio sulfodismutans*, *Desulfocapsa thiozymogenes*. Oxidation of sulfite to sulfate is coupled with formation of ADP from AMP in the *ATP sulfurylase* reaction. See text for details

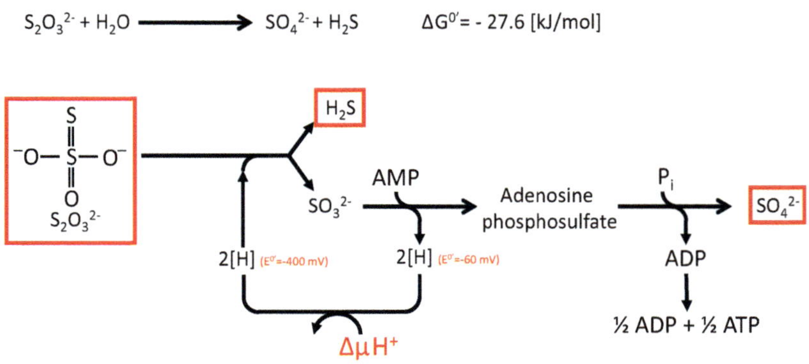

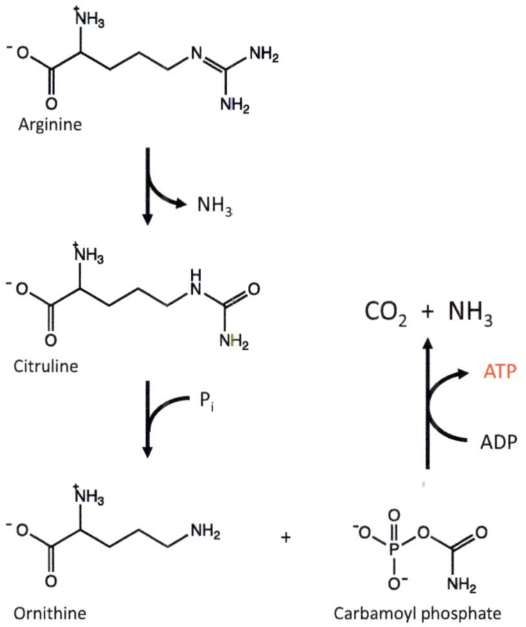

Fig. 6.11 *Arginine deiminase* pathway for energy conservation in a catabolic pathway without a redox reaction. Arginine is taken up into a cell by an *arginine:ornithine antiporter* and deaminated by *arginine deiminase* to citrulline. Citrulline is converted to ornithine and carbamoyl phosphate by a phosphate-dependent *ornithine carbamoyl transferase*. Carbamoyl phosphate is an energy-rich compound (see Fig. 6.8), and the activated phosphoryl group is transferred to ADP by *carbamate kinase*, releasing CO_2 and NH_3. See text for details

antiporter. This *arginine deiminase* pathway is found in some medically- and environmentally-relevant bacteria, such as *Streptococci*, *Mycoplasma*, and *Pseudomonas* species, and in catabolic ATP formation.

6.4 Energy Conservation via Ion Translocation-Coupled Phosphorylation

6.4.1 Proton or Sodium Ion Translocation-Coupled Phosphorylation (TCP)

By far the most common and most diversity-including mode of microbial energy conservation is Proton/Sodium Ion **Translocation Coupled Phosphorylation** (TCP), which occurs at the cytoplasmic membrane**.** TCP couples the translocation of H^+ or Na^+ with the synthesis of ATP from ADP + P_i, driven by an electrochemical membrane potential across the cytoplasmic membrane. For simplicity, only bioenergetics using H^+ as the coupling ion is discussed here in detail with the understanding that the principles also apply to Na^+. Important differences are indicated.

The cytoplasmic membrane is impermeable for ions and carries a proton motive force (Δp) (or Na^+ motive force, Δp_{Na+}, when Na^+ is the coupling ion) of −180 to −200 mV (Fig. 6.12). The cytoplasmic membrane is about 100–1000-fold less permeable for Na^+ than for H^+. Charged molecules at physiological concentrations can only pass the membrane via ion-specific transporters which is highly controlled. While the permeability for protons increases with temperature, this is not the case for Na^+, suggesting that hyperthermophiles may rely more on a sodium ion rather than a H^+ energetics. Moreover, the ratio of internal to external Na^+ concentration can be varied more easily, in contrast to $\Delta\mu H^+$, because protons are always present at physiological pH. Because of the lower membrane permeability, a Na^+ energetics is advantageous over a H^+-dependent energetics.

How much energy is released when a proton translocates from the outside to the inside in the electrochemical gradient? The total force driving energy conservation is the **proton motive force** (Δp), composed of an electrical and chemical component (Fig. 6.12a), is given by

$$\Delta p = \Delta\psi - \frac{2.3RT}{F}\Delta pH = \Delta\psi - (59.1\text{mV})\Delta pH$$

where Δp is the overall driving force, ΔΨ the electrical gradient (difference between the electrical potential Ψ inside vs Ψ outside), and ΔpH the chemical gradient (pH difference between inside and outside) (Fig. 6.12b). In most microbes, the cytoplasmic membrane is negatively charged inside and positively charged outside (Fig. 6.12a). ATP synthesis in TCP by *ATPases* makes use of Δp. The translocation of a H^+ from the outside to the inside is driven by an electric and by a chemical gradient (Fig. 6.12b). The

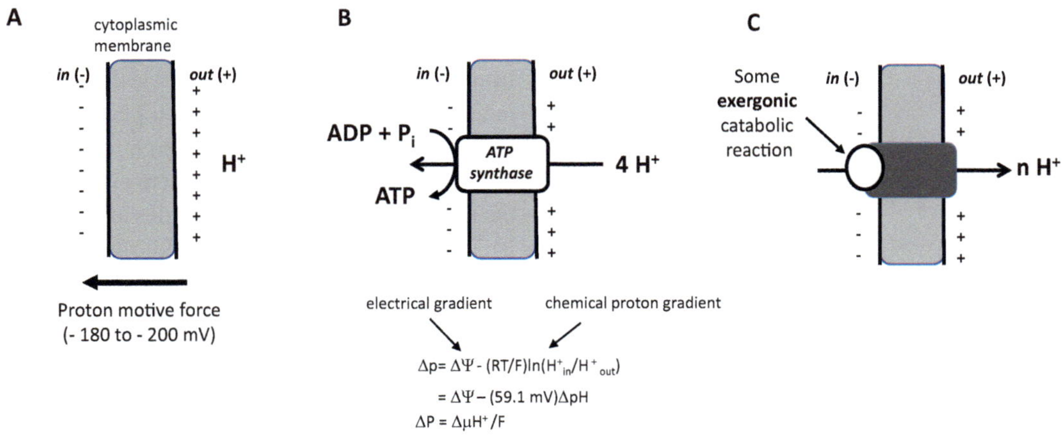

Fig. 6.12 Proton Translocation-Coupled Phosphorylation in microbial energy conservation. (**a**) The phospholipid bilayer of the cytoplasmic membrane as capacitor for keeping electrical charges separated and maintaining a homogeneous electric field. (**b**) Energy conservation by Proton Translocation Coupled Phosphorylation in mesophilic microorganisms. Protons are translocated in the electric field via F_oF_1 *ATP synthase*, driven by Δp, the proton motive force, which is composed of the voltage across the cytoplasmic membrane (in mV) and ΔpH, the difference in pH between inside and outside. The H^+/ATP stoichiometry of different *ATPases* can vary as described in the text and is assumed here to be 4 as in *E. coli*. Energy conservation is reversible as ATP hydrolysis can be converted into an electrochemical proton potential. (**c**) Converting chemical energy into a $\Delta\mu H^+$. The energy of an exergonic catabolic reaction is used to "pump" one or more H^+ against Δp (proton motive force) from the inside to the outside. Thus, the energy is "stored" in form of ΔμH until the energy is used and transduced by the *ATPase* (see **b**). See text for details

electric force acts on the proton's positive charge in the electric field in the direction of the negative charge. The chemical force results from a difference in proton concentration between inside and outside ($\Delta pH = pH_{in} - pH_{out}$). Considering that the bulk pH inside and outside of a cell (with exception of acido-or alkalophilic prokaryotes and in some fermentations, see below) is in the same concentration range, this component of ΔpH can be neglected in a first approximation. This implies that most of the force that is acting in proton translocation for energy conservation is determined by the electric component, or membrane potential: $\Delta p \approx \Delta\psi$.

$\Delta\mu H^+ = \Delta G_{H+}$ and is the energy that is released when a H^+ translocates from the outside to the inside along the electrochemical gradient. The upper limit of Δp of –200 mV in microbial membranes is most likely due to an increase in proton permeability of the cytoplasmic membrane when Δp increases, thereby, effectively dissipating $\Delta\mu H^+$. $\Delta\mu H^+$ ranges between −20 kJ/mol H^+ to −10 kJ/mol H^+ as a function of the membrane potential between −200 mV to ~ −100 mV.

How much energy is required to move a proton from the inside to the outside in the electrochemical gradient? Exergonic reactions in catabolic pathways are needed to provide the energy to "pump" a proton against Δp from the inside to the outside (Fig. 6.12c), and a minimum energy of −10 to −20 kJ/reaction is required, depending on $\Delta\mu H^+$. The stoichiometry of how many of such "electrogenic" protons are translocated per reaction depends on the amount of energy released as well as on the coupling mechanisms involved. The latter is subject to biological control, and examples (see also Chap. 3) show that this n_{ATP}, i.e., how much of the available energy is used for translocating a proton against Δp from the inside to the outside, varies from fully-coupled to completely-uncoupled!

Bioenergetics at Low Environmental pH

Neutrophilic microbes have a proton motive force Δp across the cytoplasmic membrane of around −180 to −200 mV, where the most contributing component is the difference in electrical potential Δψ (inside negative, outside positive). For acidophilic microbes with an inside pH of ~6.8 and an outside of, e.g., pH2 as in some hot sulfur springs, the pH difference between inside and outside is 4.8 orders of magnitude and equal to a $\Delta p_{\mathrm{pH}} = -\ 290$ mV. At such pH gradient the cytoplasmic membrane becomes leaky for protons resulting in a gradual acidification of the interior. One solution that acidophilic microbes employ to reduce this Δp is by overlaying the proton gradient with an inverted membrane potential. By establishing a Δψ of +100 mV (negative outside and positive inside), the overall Δp is reduced by $\Delta p = +100$ mV −290 mV = −190 mV, which is a workable bioenergetic for proton pumping and prevents proton leakage. Such inverted membrane potential can be built up by selectively taking up Na^+ or K^+ cations. Notably, *E. coli*, when surviving at pH values around 4, also maintain an inverted membrane potential.

6.4.2 Energy Conservation via F_oF_1 *ATP Synthase*

At the core of microbial energy conservation by TCP is the activity of the *F_OF_1 ATP synthase* (Fig. 6.13). This enzyme is also referred to as *ATP synthetase*, because the enzyme catalyzes the hydrolysis of ATP in the reverse reaction. *F_OF_1 ATP synthase* is an integral cytoplasmic membrane protein, facing the inside of the cell, and catalyzes the reversible phosphorylation of ADP to ATP. Mechanistically, this interconversion is facilitated in *F_OF_1 ATPase* by two coupled rotary motors, F_O and F_1, where the F_O motor (primarily the *c* ring in Fig. 6.13) induces a torque through a central stalk (ε, γ) that leads to ATP synthesis in the static F_1. (The subscript "o" of F_O is derived from "oligomycin-sensitive," not "zero"). The F_1 motor is always composed of three dimeric subunits, collectively catalyzing the formation of three ATP from 3 ADP and 3 P_i per 360° rotation. The number of *c* subunits in the F_O part of the rotor is variable between 8 and 15 (8 in mitochondria, 15 in cyanobacteria). The torque of the crank shaft rotating the F_1 motor is powered by a proton flux through the F_O motor (Fig. 6.9), where each *c* subunit binds one H^+. Typically, a genome of a prokaryote encodes only one *F_OF_1 ATP synthase.*

The stoichiometry for *ATPases* in many microbes discussed in this book, including those in Chaps. 9 and 10, have not been experimentally determined. Thus, assumptions were made based on experimentally characterized *ATPases* such as a stoichiometry of 4 H^+/ATP for *E. coli* catalyzing catabolic reactions distant from equilibrium (ΔG < −50 kJ/mol), while for syntrophic microbes that operate close to the thermodynamic equilibrium ($\Delta G = -15$ to −50 kJ/mol), a stoichiometry of 5 H^+/ATP was assumed.

The specificity of microbial *ATPases* for either H^+ or Na^+ is determined by the amino acid sequence of the *c* ring subunit. The *c*-ring subunit in the Na^+-dependent *ATPases*, such as of *Propiogenium modestum*, *Ilyobacter tartaricus*, and *Acetobacterium woodii*, has sequence motifs for binding Na^+ that are distinct from H^+-dependent *c*-ring subunits.

6.4.3 The Triangular Relationship in TCP

Phosphorylation potential—ΔμH^+—number of *c*-ring subunits in *ATPase*

Since the basis of ATP synthesis via TCP is an electric field-driven proton flux through *F_oF_1 ATP synthase*, this raises the question: how much energy is necessary for synthesis of 1 ATP; how many protons are required to be translocated per ATP formed; and how is this mechanistically coupled in F_oF_1 *ATP synthase* to ATP formation?

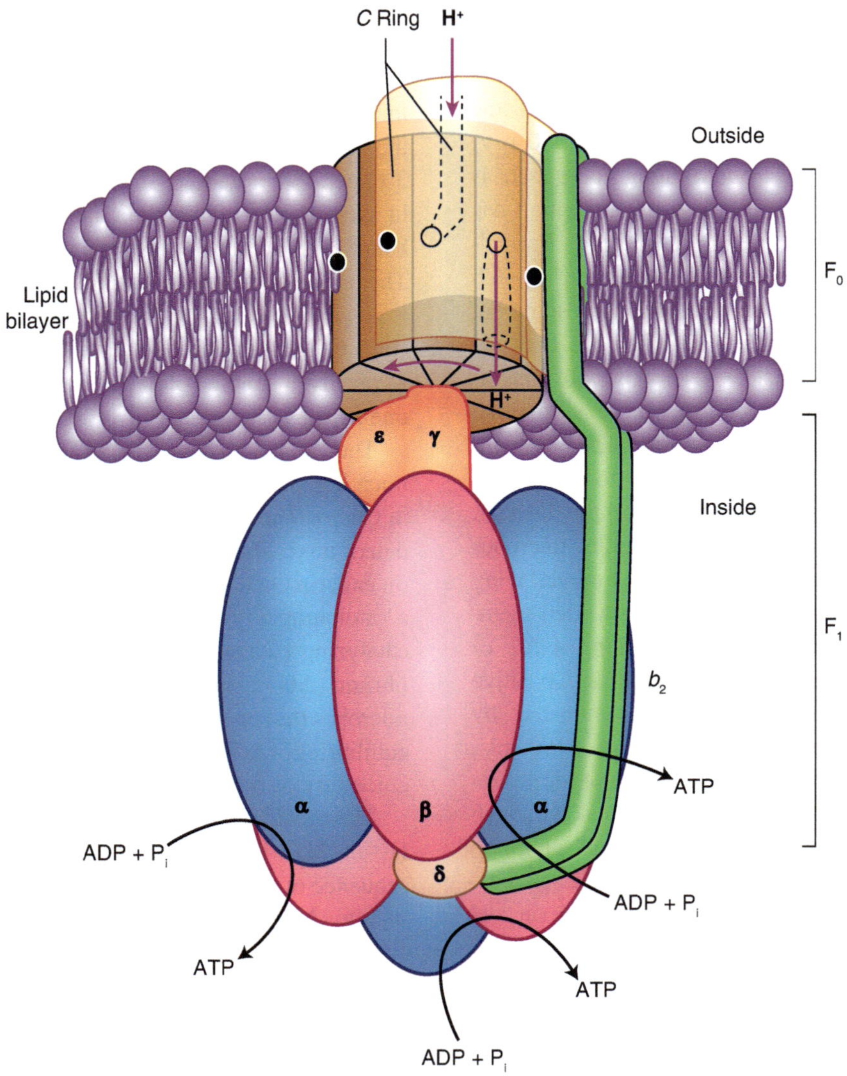

Fig. 6.13 Structure of the bacterial H^+ F_OF_1 ATP synthase from Bacillus subtilis (from Guo and Rubinstein 2019). See text for details

The in vivo **phosphorylation potential, ΔG_P,** as defined by the energy released in ATP hydrolysis ($ATP + H_2O \rightleftharpoons ADP + P_i$), depends on the intracellular ATP and ADP concentration, which is related to the physiological and growth state of a microbe. At 10 mM ATP, 1 mM ADP and P_i for a growing *E. coli* cell, ΔG_P is ~ − 55 kJ/mol (Chap. 3). In slow-growing microbes and with catabolic ΔG close to equilibrium as in syntrophs (e.g., $\Delta G = -15$ kJ/mol, Chap. 9), cellular ATP concentrations are lower so that ΔG_P can be as low as −45 kJ/mol. In other words, the amount of energy required for in vivo ATP synthesis can vary between +45 to +55 kJ/mol depending on the available energy and physiological state. Cellular stress where ATP hydrolysis is required to mitigate that stress, e.g., salt stress and pumping ions out of the cell, results also in lower ATP concentration.

Since *F_OF_1 ATPase* transduces the electrochemical energy of $\Delta\mu H^+$ into the chemical energy of ATP, the magnitude of $\Delta\mu H^+$ matters for this coupling and ranges between −20 to − 10 kJ/mol. Neglecting a ΔpH across the membrane, this is equivalent to a membrane potential Ψ of ~ −200 mV to −100 mV (considering

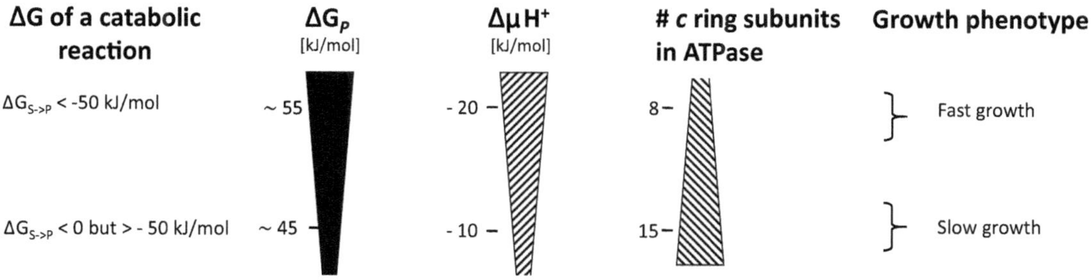

Fig. 6.14 Simplified relationship between phosphorylation potential of ADP ΔG_P, electrochemical H^+ potential $\Delta\mu H^+$, number of *c*-ring subunits in ATPase, and growth rate as a function of free energy in catabolism. A range in phosphorylation potential (see Sect. 3.4) was assumed here. See text for details

experimental errors). This means that by controlling Ψ, the minimum amount of energy that can be harvested by TCP via translocating a single H^+ is −20 to −10 kJ/mol. Thus theoretically, with a ΔG_P of −55 kJ/mol and a $\Delta\mu H^+$ of −20 kJ/mol, a minimum of 3 H^+ per ATP is required. On the other hand, with a ΔG_P of −45 kJ/mol and a $\Delta\mu H^+$ of −10 kJ/mol, a minimum of ~5 H^+/ATP is required.

Proton flux through the F_o motor of the F_oF_1 *ATP synthase* is through the *c* ring, and each *c* ring subunit carries one proton (Fig. 6.13). Since the number of *c* subunits varies between 10 and 15 in prokaryotes, between 3.3 and 5 protons are translocated per mol ATP formed in different prokaryotes. Unless otherwise indicated, the stoichiometry of the *E. coli ATPase* of 4 H^+/ATP is assumed to operate in the microbes discussed in the following chapters.

This intrinsic co-variation of ΔG_P, $\Delta\mu H^+$, and number of ATPase *c*-ring subunits as a function of catabolic energy available ($\Delta G_{S \rightarrow P}$) and the growth kinetic consequences is qualitatively shown in Fig. 6.14.

6.4.4 Mechanisms of Transducing Energy from Catabolic Pathway Reactions into an Electrochemical Gradient

There are four known mechanisms to generate an electrochemical gradient via ion translocation; three involve an electron transport and the fourth one is independent of a redox reaction:

1. Linear electron transport
2. Fumarate respiration
3. Photophosphorylation
4. Translocation coupled phosphorylation coupled to non-redox reactions

6.4.4.1 Linear Electron Transport in Chemotrophic Microbes

Electron transport leading via TCP to ATP synthesis is often introduced by using the aerobic respiratory chain as example and, thereby, implying that this as the "most common mechanism." Such perspective neglects evolutionary and mechanistic insights from the diversity of electron transport phosphorylation. The introduction below takes a more evolution-guided perspective.

Two modes of catabolic electron transport phosphorylation are distinguished based on the characteristic electron carriers involved: (1) **ferredoxin-dependent**, and (2) **quinone-dependent** electron transport phosphorylation (ETP) (Fig. 6.15). Both have in common that membrane-associated redox reactions generate a $\Delta\mu H^+$ or $\Delta\mu Na^+$. There is great diversity, however, in how both modes accomplish this task. Ferredoxin-dependent ETP is involved in prokaryotes where the standard redox potential of the **environmental** electron acceptor is more negative than ~ −220 mV, such as in catabolic CO_2 and H^+ reduction pathways. Quinone-dependent ETP is found in microbes with electron

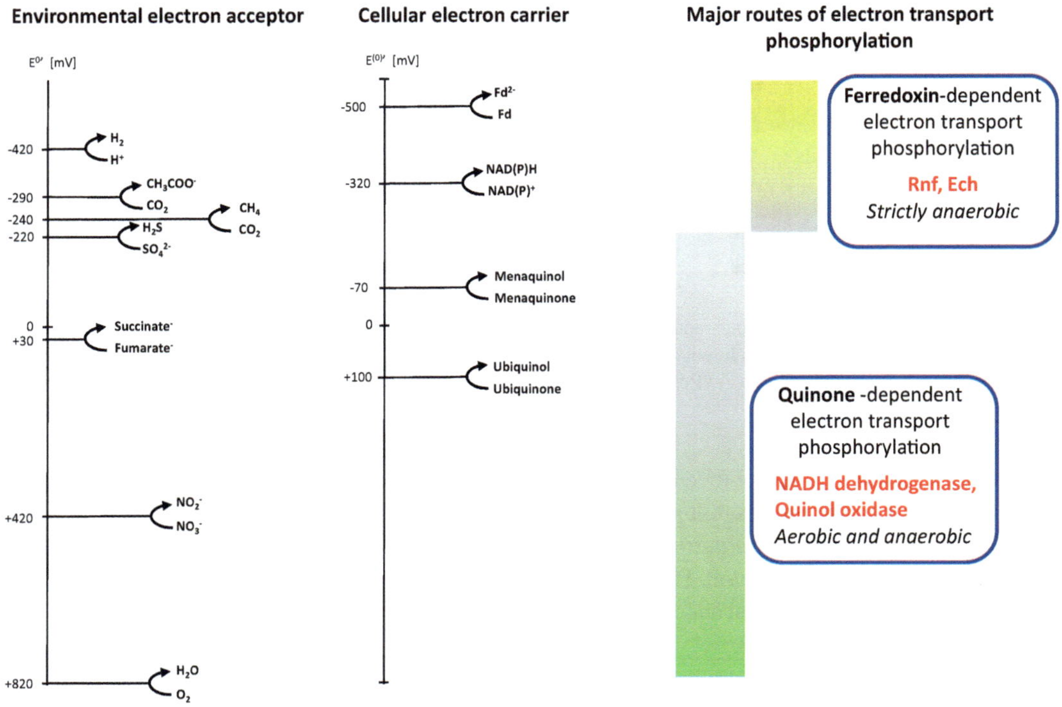

Fig. 6.15 Major microbial electron transport phosphorylation as a function of the redox potentials of electron acceptors and cellular electron carriers. Ferredoxin-dependent ETP is primarily found in strictly anaerobic, fermenting, CO_2-reducing, and probably also in sulfate-reducing microbes. Quinone-containing ETP is found in those microbes where quinones are involved in H^+ or Na^+ pumping rather than just being electron transfer component. Gradient yellow and green areas indicate coarsely the range of redox potentials covered by Rnf and Ech, and by quinone/cytochrome-based electron transport phosphorylation, respectively. See text for details

acceptors at potentials more positive than ~−220 mV, such as in catabolic SO_4^{2-}, S^0 Fe^{3+}, NO_2^-, NO_3^-, or O_2 reductions (Fig. 6.15). In the latter group, quinones are involved in proton-pumping rather than serving exclusively in electron transfer (see below).

As Fig. 6.15 shows, the involvement of quinones correlates with electron acceptors, whose redox potentials are more positive than that of CO_2/CH_4, CO_2/CH_3COO^- and H^+/H_2. Since SO_4^{2-}, S^0 Fe^{3+}, NO_2^-, NO_3^-, or even O_2 was formed on Earth only after anoxygenic and oxygenic photosynthesis had evolved for retrieving electrons for CO_2 fixation from H_2S, Fe^{2+}, and H_2O (Fig. 4.3c), quinone-based ETP evolved after ferredoxin-based ETP, which is considered as the more ancient and foundational mechanism for electron transport phosphorylation.

6.4.4.1.1 Ferredoxin-Dependent Electron Transport Phosphorylation

In strictly anaerobic prokaryotes that metabolize H_2, formate, or CO as electron donors with electron acceptors that have a standard redox potential of more negative than ~−220 mV, **ferredoxin (Fd)** is the intracellular electron carrier. Reduced ferredoxin (Fd^{2-}) is entry point of electrons into a **H^+- or Na^+-translocating electron transport chain consisting of Rnf** or **Ech,** respectively. The physiological electron acceptor in the Rnf and Ech reaction is NAD^+ or H^+, respectively (Fig. 6.12). Rnf stands for *Rhodobacter* nitrogen fixation (where this mechanism was first discovered in the reverse direction), and Ech for energy-converting hydrogenase. The electron transport chain is short and includes only a single heteromeric, membrane-bound enzyme complex

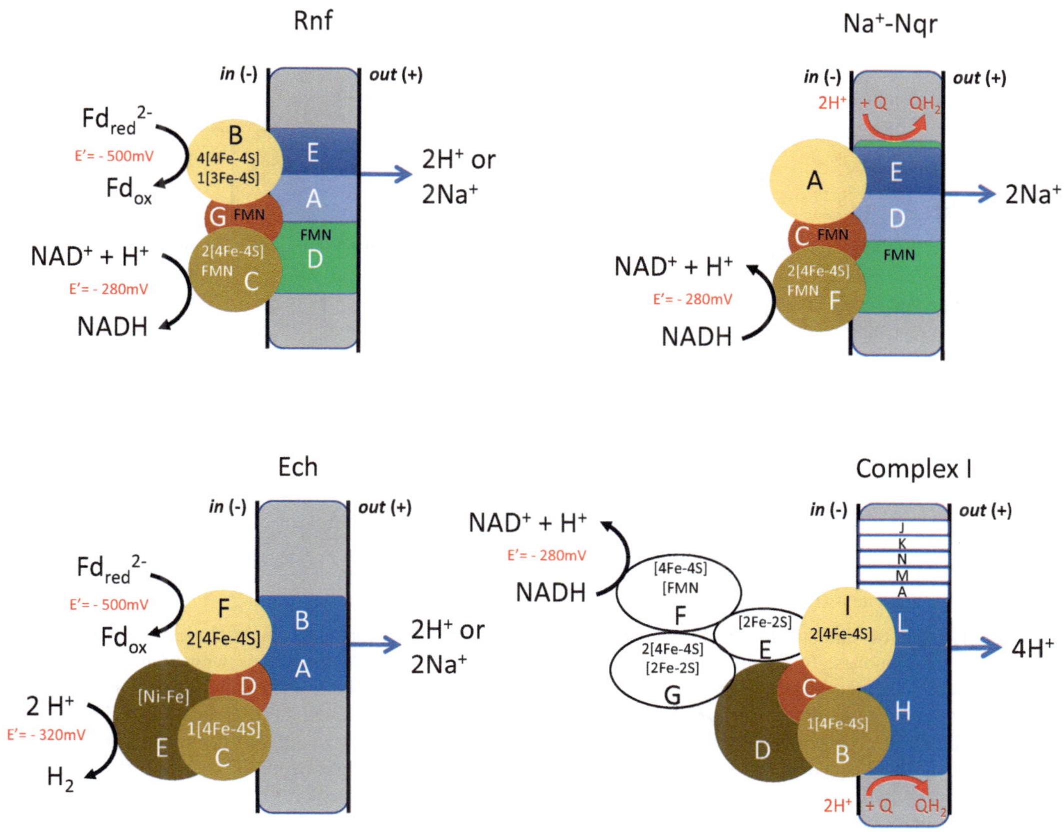

Fig. 6.16 Energy converting Rnf and Ech complexes and their relationship to evolutionarily more recent, respiratory *NADH dehydrogenases*. (RnfA–G) from *Clostridium tetanomorphum*; (Na^+NqrA-F) Na^+-translocating *NADH-quinone oxidoreductase* from *Vibrio cholerae*; (EchA–F) energy converting *[NiFe]-hydrogenase* from *Rhodospirillum rubrum*; Complex I from *E. coli* (Modified after Buckel and Thauer 2013). The electrochemical H^+ or Na^+ gradients generated in these reactions are converted to ATP by an *ATP synthase*. See text for details

(Fig. 6.16). The Rnf and Ech complexes consist of several protein components and Fe-S clusters as well as a flavin (FMN), as in the case of Rnf. Both enzymes face the cytoplasmic side of the cytoplasmic membrane. In Rnf, reduced ferredoxin is oxidized and electrons are transferred to NAD^+, forming NADH, or to H^+ to form H_2 in Ech. In each case, 2 H^+ or 2 Na^+ is translocated across the cytoplasmic membrane, respectively, per two electrons transferred.

There is a striking similarity in the molecular composition, subunit homology, and architecture of the Rnf and Ech complexes with the evolutionarily more recent Complex I and *NADH:ubiquinone oxidoreductase* (Nar) of the respiratory chain in aerobic microbes, such as in *E. coli* and *Vibrio cholerae*. However, in Rnf **NAD^+** is the cellular **electron acceptor** while in modern Complex I *NADH dehydrogenases*, **NADH** is the **electron donor** (Fig. 6.16)! Frequently, Rnf or Ech functions each in concert with oxygen-sensitive, flavin-based electron-bifurcating enzymes (*bifurcases*) to conserve energy from small redox potential differences at low redox potentials (see below, and Chaps. 9 and 10).

6.4.4.1.2 Quinone-Based Electron Transport Phosphorylation

Quinones couple diverse oxidative reactions in electron transport chains involving terminal electron acceptors with redox potentials more positive than −100 mV. Menaquinone ($E^{0\prime} = -70$ mV)

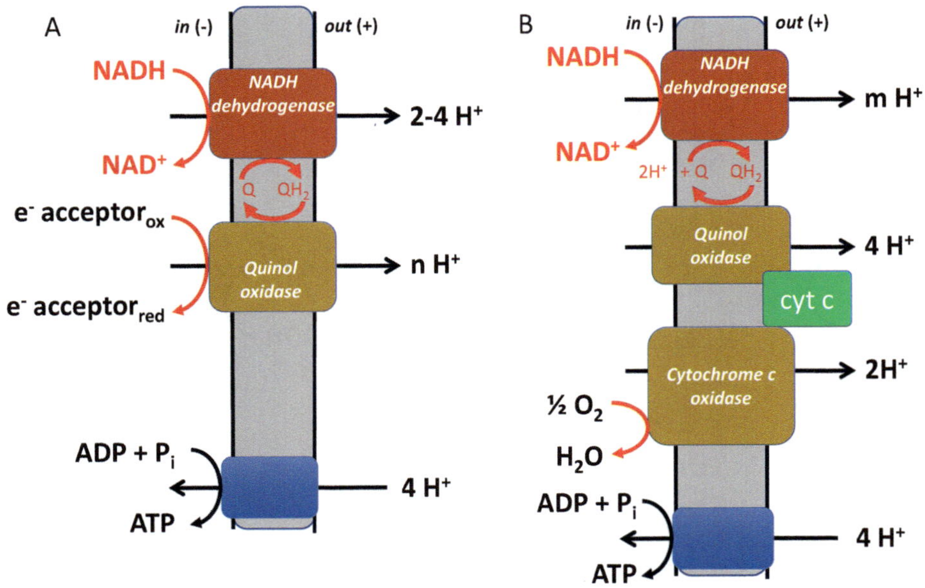

Fig. 6.17 Generic (**a**) and specific (**b**) electron transport chain involving quinones. In the more general electron transport chain (**a**), membrane-integrated quinones are involved in electron transport from *NADH dehydrogenase* to a terminal electron acceptor. During reduction and oxidation of quinones and quinols, protons may be translocated depending on the electron donor and acceptor used by the microorganisms as well as the microbe's ecological strategy. (**b**) shows the function of quinones in an aerobic electron transport chain involving cytochrome *bc*1 as cellular electron acceptor for *quinol oxidase.* In this case, the bifurcating activity of quinones, i.e., the capacity to convert a single, 2-electron reduction step into two 1-electron oxidations with high and low redox potentials, is key for translocating 2 additional H^+ across the cytoplasmic membrane (see below Sect. 6.5.1.1). In (**b**), electrons from the *quinol oxidase* are transferred to a cytochrome *c*, and oxidation of cytochrome *c* by the terminal electron acceptor O_2 by *cytochrome c oxidase* results in another pumping of 2 H^+ across the membrane. Q quinone; H_2 quinol (hydroquinone). Stoichiometries of H^+ translocation, although indicated with a defined values in this figure, are variable depending on the enzyme involved. See text for details

or ubiquinone ($E^{0\prime}$ = + 90 mV) are the membrane-integrated electron carriers that couple electron transfer to a terminal electron acceptor with H^+ translocation (Fig. 6.17).

In some anaerobic respiratory processes, menaquinones that are methylated at the aromatic ring, such as 5- or 8-methyl-menaquinones, are involved. Depending on the methylation state, methyl-menaquinones can have a more negative standard redox potential. Methyl-menaquinones are associated, for example, with *polysulfide reductase* in *Wolinella succinogenes* as part of the polysulfide respiratory chain. In that system, the standard redox potential of the free and the *polysulfide reductase*-bound methyl-menaquinone is −90 mV and −230 mV, respectively. Functionally related to quinones is methanophenazine ($E^{0\prime}$ = − 170 mV) (Fig. 6.5) that operates as quinone equivalent in the respiratory chain of methanogens with cytochromes, such as *Methanosarcina* species. As discussed in Chap. 10, when catabolizing H_2 and CO_2, *Methanosarcina sp.* operate a membrane-bound electron transport chain, coupling H_2 oxidation via methanophenazine with the reduction of a heterodisulfide as cellular electron acceptor.

When quinones are part of a respiratory chain, the redox potentials of the electron donor and the electron acceptor must be more negative and positive, respectively, than that of the quinone. In quinone-containing electron transport phosphorylation, the predominant proton-translocating coupling sites are *NADH dehydrogenases* and/or *quinol oxidases* Fig. 6.17). Typically, in Gram-negative bacteria, ubiquinones function under oxic conditions, and menaquinones under anoxic conditions. Gram-positive bacteria growing aerobically use only menaquinone. A "generic" quinone-based electron transport chain is depicted in Fig. 6.17a. The stoichiometry of H^+ translocated to the oxidation reaction(s) is variable, depending on the enzymes involved. The entry of electrons in the electron transport chain is typically at the level of NADH via membrane-bound *NADH dehydrogenases*. These enzymes transfer electrons to a quinone that is anchored in the cytoplasmic membrane because of the quinone's long hydrophobic isoprene-based side chain. The quinone reduction step is often coupled to H^+ translocation with different stoichiometry of $H^+/2e^-$ for different *NADH dehydrogenases*, (e.g., Complex I, Na^+-Nqr, NDH-1, NDH-2, see below). For quinol oxidation, the type of *quinol oxidase* involved also determines the number of H^+ translocated per quinol oxidized. Several types of membrane-anchored *quinol oxidases* contain cytochromes (such as cytochrome *bc*1, Fig. 6.17b) and can have different degrees of energy conservation.

Some microbes, such as *Pseudomonas* sp. or *Paracoccus* sp., contain a cytochrome coupling site, where cytochrome *c* is oxidized by *cytochrome c oxidase* and electrogenic protons are translocated: the electrons are transferred from reduced cytochrome *c* to the terminal electron acceptor in intracellular or periplasmic reductase reactions (complex IV, if O_2 is terminal electron acceptor, Fig. 6.17).

6.4.4.1.3 Plasticity in Energetic Coupling of Respiratory Processes

There are variations in energy coupling of respiratory chains resulting in differences in n_{ATP} for the same overall reaction. Examples were introduced in Chap. 3 (see Fig. 3.7). NADH oxidation by *NADH dehydrogenase* coupled to ubiquinone reduction is exergonic by −61 kJ/mol NADH. This amount of energy is thermodynamically sufficient to translocate 3–4 protons across the cytoplasmic membrane (depending on $\Delta\mu H^+$). There are isoforms of *NADH dehydrogenase* that couple proton translocation in variable stoichiometry: NDH-1 in Complex I of *Paracoccus denitrificans* translocates four electrogenic H^+ per ubiquinone reduced; NDH-1 in *E. coli* translocates 2 H^+, and NDH-2 is not pumping any proton at all! As expected, the turnover rate of NDH-2 is 3–5 times faster than of NDH-1. Nqr *NADH dehydrogenase* is Na^+-pumping (2 Na^+/ubiquinol) and present in many marine and pathogenic microbes, such as *Vibrio* species including *V. cholerae*, *Haemophilus influenza*, *Yersinia pestis*, *Shewanella putrefaciens*, and *Pseudomonas aeruginosa*. These three *NADH dehydrogenase* families are of different evolutionary origin. Some prokaryotes carry *NADH oxidases* or electron-transferring flavoproteins which oxidize NADH with O_2 directly and without a respiratory chain and dissipate all redox energy ($n_{ATP} = 0$)! Some lactic acid bacteria have an *NADH oxidase* activity, which allows for the formation of acetate, a weaker acid (p*Ka* = 4.6) than lactate (pKa = 3.7) (see Chap. 9).

A similar variable coupling (n_{ATP}) is found for quinol oxidation in some microorganisms. Enteric bacteria, such as *E. coli*, *Salmonella enterica*, or *Vibrio cholerae* contain two evolutionarily distinct classes of *quinol oxidases*. Cytochrome *bo oxidase*, which is related to cytochrome c *oxidase*, translocates 2 H^+ when oxidizing ubiquinol with O_2 at higher pO_2. Under oxygen-limiting conditions such as in the human gut or when pathogenic, which is often electron acceptor (O_2)-limited, a cytochrome bd_1 *oxidase* that translocates 1 H+/e− in O2 reduction, is used for the same reaction. Other microbes containing several *quinol oxidases* with variable stoichiometries are *Klebsiella pneumoniae*, *Mycobacterium tuberculosis*, or *Shigella flexneri*. In the chemoorganotrophic *Shewanella oneidensis* when growing under O_2-limiting conditions or Fe(III)-reducing conditions, no energy is conserved in quinol oxidation when using catabolic electron acceptors with redox potentials substantially more positive than quinones such as Fe^{3+} ($E^{\circ\prime} \sim + 100$ mV), NO_3^- ($E^{\circ\prime} = + 430$ mV), DMSO ($E^{\circ\prime} = + 160$ mV), or TMAO ($E^{\circ\prime} = + 130$ mV).

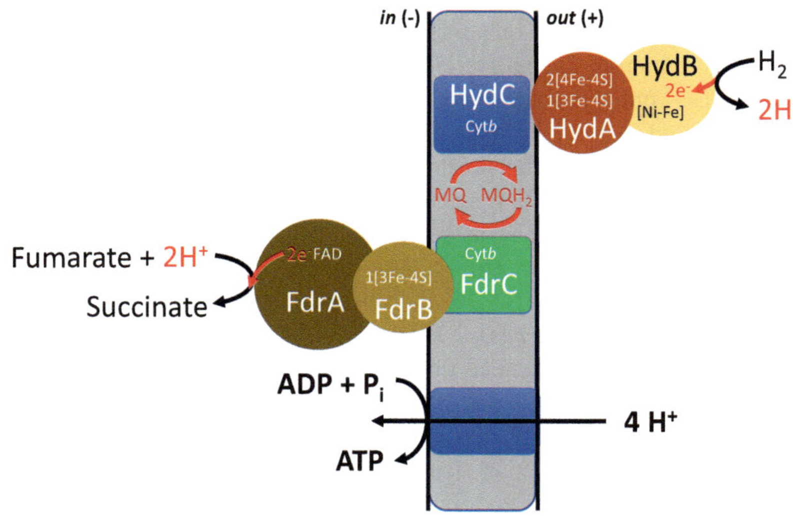

Fig. 6.18 Electron transport phosphorylation in H_2 oxidation with fumarate by *Wolinella succinogenes*. See text for details

6.4.4.2 Fumarate Respiration

An interesting variation of electron transport phosphorylation involving no proton pumping in quinol oxidation is in fumarate respiration. Prerequisite is that oxidation of the electron donor occurs outside the cytoplasm, opposite of the site of the reduction reaction. One example is H_2 oxidation with fumarate by *Wolinella succinogenes*. A *hydrogenase* is associated with the outside of the cytoplasmic membrane and oxidizes H_2 to 2 H^+ and $2e^-$ (Fig. 6.18). Electrons are then transferred via menaquinone to *fumarate reductase* at the inside. Rather than pumping protons, charge separation occurs by separating electrons from H_2 via *hydrogenase* on the outside of the membrane, leaving 2 H^+ with their positive charge on the outside. Only electrons are transferred via electron transport chain components to the inside where terminal electron acceptor reductase, *fumarate reductase*, is located. *Fumarate reductase* consumes protons at the inside, thereby contributing equally as the outside *hydrogenase* to generation of a proton motive force. A similar electron transport in the absence of proton pumping has also been implicated in H_2 metabolism in the sulfate-reducing bacteria *Desulfovibrio vulgaris* and in *Dehalobacter restrictus*.

6.4.4.3 Photophosphorylation

In photophosphorylation, ATP is synthesized from ADP and P_i using light as energy source. The physical energy of light is given by $E = hcN_A\lambda^{-1}$, where λ is the wavelength, h the Planck constant ($6.6 \times 10^{-34} Js^{-1}$), c the speed of light ($3 \times 10^8 ms^{-1}$), and N_A the Avogadro constant (6.022×10^{23} particles per mol). For example, one mol photons at 400 nm wavelength contains ~−300 kJ/mol photon, while in the infrared at 1000 nm, the energy is only −120 kJ/mol photon.

> The term **photosynthesis** is often used synonymously with **photophosphorylation**. Photosynthesis, however, implies two metabolic processes that are not necessarily linked: energy conservation by photophosphorylation ("photo-") AND autotrophic CO_2 fixation for biomass synthesis ("-synthesis"). Autotrophic CO_2 fixation, as present in many phototrophic as well as chemolithoautotrophic microbes, requires electrons for CO_2 reduction that are derived from a reduced, environmental compound, such as H_2S, S^0, Fe^{2+}, or H_2O. The cellular reductant for CO_2 fixation is NAD(P)H ($E^{0\prime} = -320$ mV) with a

(continued)

redox potential that is more negative than these compounds ($E^{0\prime} > -220$ mV). Thus, coupling oxidation of H_2S, S^0, Fe^{2+}, or H_2O with reduction of cellular $NADP^+$ requires metabolic energy in form of $\Delta\mu H^+$. This energy can be generated by TCP, either via respiratory electron transport (as in chemolithoautotrophic bacteria) or by cyclic photophosphorylation (as in phototrophic prokaryotes). For example, cyanobacteria and the chemolithoautotrophic aerobic H_2-oxidizing *Cupriavidus necator* are both autotrophs using the Calvin cycle for CO_2 fixation; however, their energy metabolism is different. Photophosphorylation, in fact, can be used in photoheterotrophic microbes in the absence of CO_2 fixation when organic carbon rather than CO_2 is assimilated into biomass. In a strict sense, these microbes would not be classified as "photosynthetic" (because of the absence of electron activation for $NAD(P)^+$ reduction and CO_2 fixation) but as "phototrophic" because they use photophosphorylation. It is, therefore, useful to distinguish photosynthesis and photophosphorylation. Nevertheless, in the literature and to be consistent with it in the following text, some microbes are described to perform "photosynthesis" while deriving energy from light even in the absence of an electron activation and CO_2 fixation.

Two types of **photophosphorylation** are distinguished based on the difference in molecular mechanism: **chlorophyll-based photophosphorylation** and **bacteriorhodopsin-based photophosphorylation**. Chlorophyll-based photophosphorylation is described first and below.

6.4.4.3.1 Chlorophyll-Based Photophosphorylation

There are two types of chlorophyll-based photosynthesis in prokaryotes: Anoxygenic photosynthesis in Purple and in Green Bacteria, and oxygenic (O_2-producing) photosynthesis in cyanobacteria (Fig. 6.19). Anoxygenic photosynthesis evolved first about 3.2–3.5 billion years ago, while oxygenic photosynthesis evolved by combination of molecular components of anoxygenic photophosphorylation (see below) some 2.4 billion years ago. Anoxygenic photosynthesis in Purple and Green Bacteria can be further differentiated based on whether H_2S is the electron source for CO_2 fixation (Purple Sulfur Bacteria and Green Sulfur Bacteria, forming transiently intra- or extracellular deposition of the H_2S oxidation product S^0) or whether no electrons for CO_2 fixation are required and reduced carbon is simply assimilated (Purple non-Sulfur Bacteria and Filamentous Green Bacteria) (Fig. 6.19). These distinctions are not rigorous as, for example, some Purple non-Sulfur Bacteria can still use H_2S as electron donor.

Chlorophylls are the light absorbing pigments as the "special pair" in the photosynthetic reaction center (RC) and in the light harvesting complex. Chlorophylls are derived from porphyrins and contain Mg^{2+} as the central coordinated metal atom (Fig. 6.20a). (Bacterio)chlorophylls are more reduced than porphyrins and are defined by the presence of a fifth ring in addition to the four pyrrole-like rings. (Bacterio)pheophytins are involved in electron relay from the excited (bacterio)chlorophyll in the reaction center. (Bacterio)pheophytins are also based on a porphyrin structure; however, they lack the central metal atom. A phytol side chain (Fig. 6.20) anchors chlorophylls and pheophytins into the membrane-integral photosynthetic reaction center.

The porphyrin system, including the side chains, contains an extended conjugated double bond system, where the extent of conjugation determines the wavelength of light that is absorbed and with that the energy: the more extended the double bond system is, the longer the wavelength that can be absorbed. Thus, modifying the (bacterio)chlorophyll side chains as well as the degree of saturation of the pyrroles enables phototrophic microbes to adapt to different niches, i.e., wavelengths, thereby, escaping

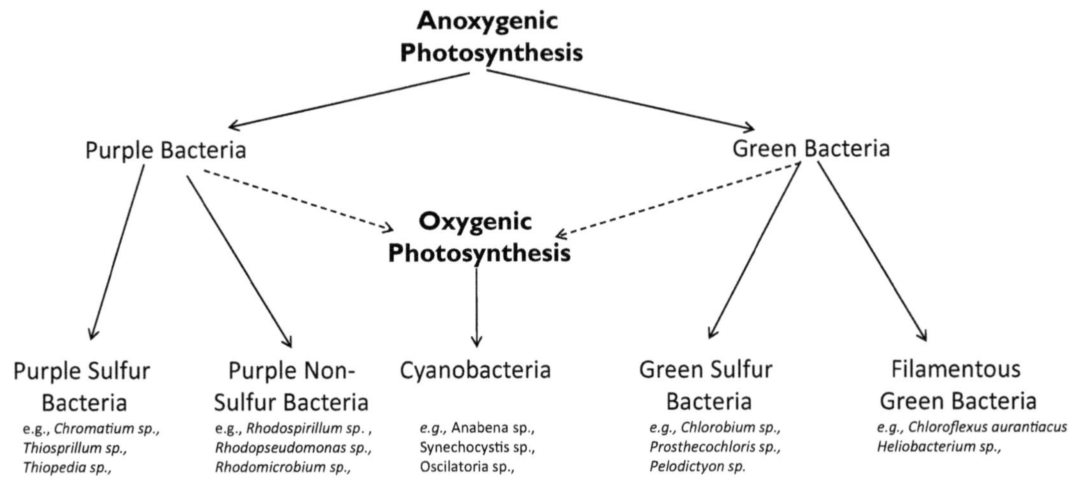

Fig. 6.19 Chlorophyll-based photosynthesis in Prokaryotes: Anoxygenic photophosphorylation in Purple and in Green Bacteria, and oxygenic photosynthesis in cyanobacteria. Anoxygenic photosynthetic microbes can be further distinguished based on whether H_2S is used as electron source for CO_2 fixation or whether organic carbon is assimilated. In contrast to anoxygenic photophosphorylation, two reaction centers are involved in oxygenic photosynthesis, where it is assumed that the water splitting PSII is derived from photophosphorylation of purple bacteria, and the energy conserving PSI is derived from that of Green Bacteria. See text for details

competition for photons of the same wavelength. Bacteriochlorophyll and chlorophyll differ in a -CHO side chain (Fig. 6.20). Bacteriochlorophylls *a* and *b* are found in Purple Bacteria, occupying the longer wavelength spectrum of visible light, while bacteriochlorophyll c, d, e, and g enable Green Bacteria to utilize shorter wavelengths. Cyanobacteria, algae, and plants

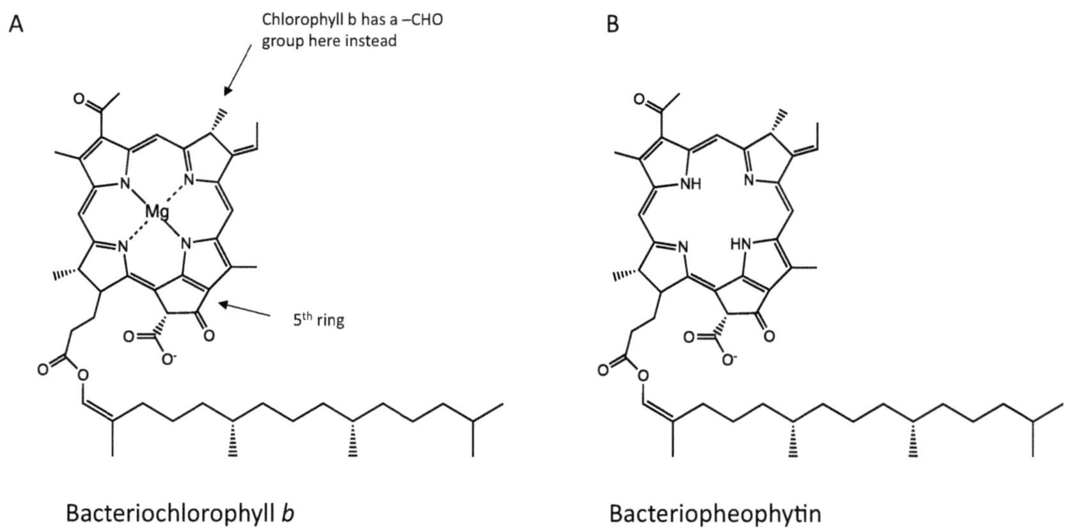

Fig. 6.20 Structure of porphyrin-derived cofactors in chlorophyll-based photophosphorylation. (**a**) Bacteriochlorphyll *b* (Bchl-*b*) differs from chlorophyll a in the -CHO side chain. Other Bchls have different side chain substitutes (see Table 6.1). (**b**) Bacteriopheophytin is structurally similar as BCh, except it does not contain a central metal atom, two protonated pyrrole nitrogen atoms, and the redox potential is more positive by 200–300 mV than that of the excited Bchl-*b** (see Fig. 6.21). See text for details

Table 6.1 Photoactive pigments, absorption range, and phototrophic microbes

Photoactive pigment	Absorption range [nm]	Phototrophic microbes
Bacteriophlorophyll a, b	800–1400	Purple bacteria
Bacteriophlorophyll c, d, e	720–755	Green sulfur bacteria
Bacteriophlorophyll c_S	720	Filamentous green bacteria
Bacteriophlorophyll g	670/788	Heliobacterium
Chlorophyll a	680	Cyanobacteria, higher plants, algae
Chlorophyll b	660	Prochlorophytes, higher plants
Chlorophyll d	700–730	Cyanobacteria
Chlorophyll c	447–452	Eukaryotic algae, dinoflagellates, diatoms

use wavelengths in between both (see Table 6.1) while chlorophylls are found in cyanobacteria and plants.

Energy conservation by chlorophyll-based photophosphorylation is a variation of the above discussed respiratory electron transport phosphorylation via the cytochrome bc_1 complex with quinones and cytochromes as core electron transporting and energy conserving components (see Sect. 6.4.4.1.2). However, the fundamental difference is **that the flux of electrons is cyclic when driven by light energy, and linear when driven by a net electron flux in chemotrophic microbes.** Photophosphorylation is a membrane-integral process and consists of a Reaction Center (RC) and a short electron transport chain (Fig. 6.21). The RC in Purple non-Sulfur Bacteria such as of *Rhodopseudomonas* sp. consists of a heterodimeric core of three protein, M, L, H, and a *c*-type cytochrome, four bacteriochlorophyll (BChl), two bacteriopheophytin (BPhe), two quinones, a carotenoid, and Fe^{2+} (pheophytin-quinone type RC). Reaction centers follow phylogenetic relationships (Fig. 6.21a). Chlorophyll-based photophosphorylation has not been found in Archaea.

The architecture of the reaction center has a two-fold symmetry with 2 Bchl forming a “special pair” to which the electronic excitation by light is transferred to from antenna pigments (not shown here) (Fig. 6.21a). There is variability in reaction center chlorophylls; some Purple Bacteria contain BChl b (*Rhodopseudomonas viridis*) while others contain BChl a (*Rhodobacter sphaeroides*). The excited BChl of the special pair has a redox potential (here Bchl b of Purple Bacteria) of more negative than $E^{0\prime} \sim -700$ mV relative compared to the unexcited state (+ 475 mV) (see also Fig. 6.21a). This change in redox potential initiates the primary charge separation of an electron from a BChl (= Bchl-b^+ + e^-). The electron is transferred via a monomeric bacteriochlorophyll (Bchl (B_A)), bacteriopheophytin (BPhe), and one ubiquinone (Q_A) to ubiquinone Q_B (Fig. 6.21a). Two sequential electron transfers result in complete reduction of ubiquinone Q_B to ubiquinol. Only one branch is active in electron relay.

Ubiquinol Q_B is oxidized by a quinone from the membrane quinone pool (red *Q* in Fig. 6.21b), while two protons are taken up from the cytoplasmic side. Oxidation of the pool quinol by *cytochrome bc1 oxidase* releases 2 H^+ at the periplasmic side. When the two electrons are transferred to the *c*-type cytochrome, another 2 H^+ are translocated (see Fig. 6.21). The reduced *c*-type cytochrome reduces the “parental” oxidized $Bchl^+$, and the cyclic process of photophosphorylation is closed.

The RC complex in Green Bacteria is different from Purple Bacteria in that their RC is homodimeric, and does not contain bacteriopheophytin and quinone (Fig. 6.22ab). Electron transport from BChl to the membrane-diffusible quinone is mediated by Fe-S clusters in the RC complex (Fe-S-type RC). Energy conservation proceeds also via a cyt bc_1 complex.

Cyanobacteria (and plants) contain two different photosynthetic reaction centers, PSI and PSII. In order to access H_2O as electron donor for $NADP^+$ reduction (and subsequent CO_2 fixation), a redox difference of $\Delta E = 1140$ mV needs to be

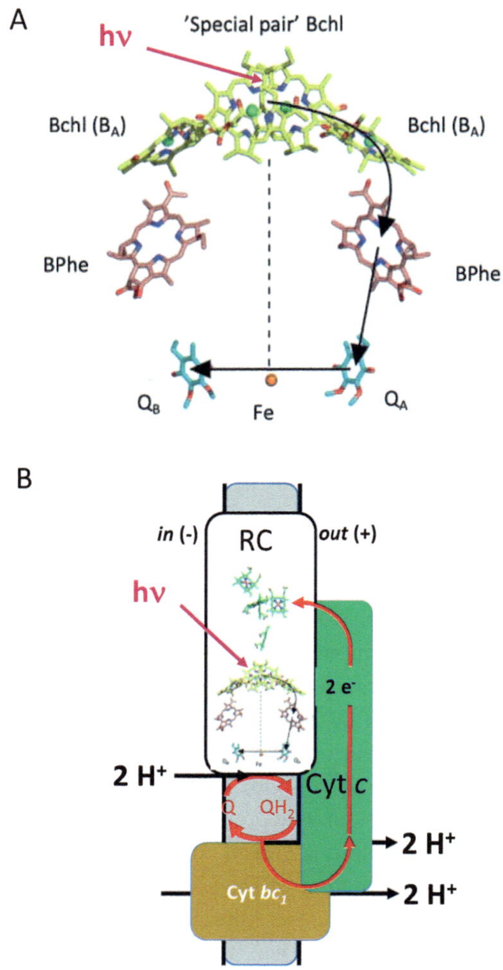

Fig. 6.21 Energy conservation by photophosphorylation in the Purple Bacterium *Rhodobacter sphaeroides*. (**a**) Two-fold symmetry of cofactor organization in the reaction center. Only cofactors relevant for light absorption and electron relay are shown. Note, that only the right branch is active in electron relay (arrow). Modified after Lin et al. PNAS (2009) 106 (21) 8537-42. (**b**) Simplified representation of cyclic electron flow and energy conservation by photophosphorylation in *Rb. sphaeroides*. Antenna pigments were omitted for simplicity. See text for details

overcome, which requires mechanistically two photons, each exciting chlorophyll in the two different RCs, to drive this reaction irreversibly (Fig. 6.22). Photosystem I functions in cyclic electron flow for energy conservation via generating a $\Delta\mu H^+$, while PS II is the water-splitting RC that works in conjunction with PSI to generate a linear electron flow and reduction of $NADP^+$. Water oxidation by PSII involves a sequential 4 photon reaction that removes oxidatively 4 e^- sequentially from 2 H_2O. This linear electron flow release O_2, 4 H^+ in the lumen, and 4 e^- to reduce $NADP^+$. Thus, PSI is required for both cyclic and linear electron flow. Note that in cyanobacteria NADPH is formed, while Purple and Green Bacteria produce NADH. PSII is of the pheophytin-quinone-type while PSI is of the Fe-S-type, indicating that oxygenic photosynthesis resulted from combination of the two distinct RCs of anoxygenic photosynthesis (Fig. 6.22).

There is also a modularity in how electron activation for NAD^+ reduction is coupled to photophosphorylation in anoxygenic phototrophs that fix CO_2 (Fig. 6.22a, b). In Purple Sulfur Bacteria, H_2S is oxidized and NAD^+ is reduced by a reverse electron transport, where photophosphorylation generates the $\Delta\mu H^+$ to drive endergonic reduction of NAD^+ with H_2S (Fig. 6.22a). In Green Sulfur Bacteria, H_2S oxidation and coupled NAD^+ reduction are mechanistically part of the electron flow in photophosphorylation. Here, menaquinone initiates H_2S oxidation and transfers the electrons to a Cyt bc_1 complex (Fig. 6.22b). Light absorption by bacteriochlorophyll and subsequent linear electron flow to NAD^+ results in an oxidized cyt bc_1 complex at a redox potential that is sufficiently positive to oxidize H_2S to S^0. Thus, this mechanistic link in Green Sulfur Bacteria effectively converts cyclic photophosphorylation of energy conservation to linear photophosphorylation for energy-dependent H_2S oxidation and NAD^+ reduction.

6.4.4.3.2 Bacteriorhodopsin

Haloarchaea, such as *Halobacterium salinarum* and other archaea, use a bacteriorhodopsin-based mechanism for energy conservation from light that is auxiliary to their chemoorganotrophic energy metabolism. The chromophore all-*trans*-retinal is bound to a membrane protein as bacteriorhodopsin and functions as a light-dependent H^+ pump to generate a $\Delta\mu H^+$ (Fig. 6.23). In contrast to chlorophyll-based photophosphorylation, this transduction of light energy into chemical energy is independent of any redox reaction.

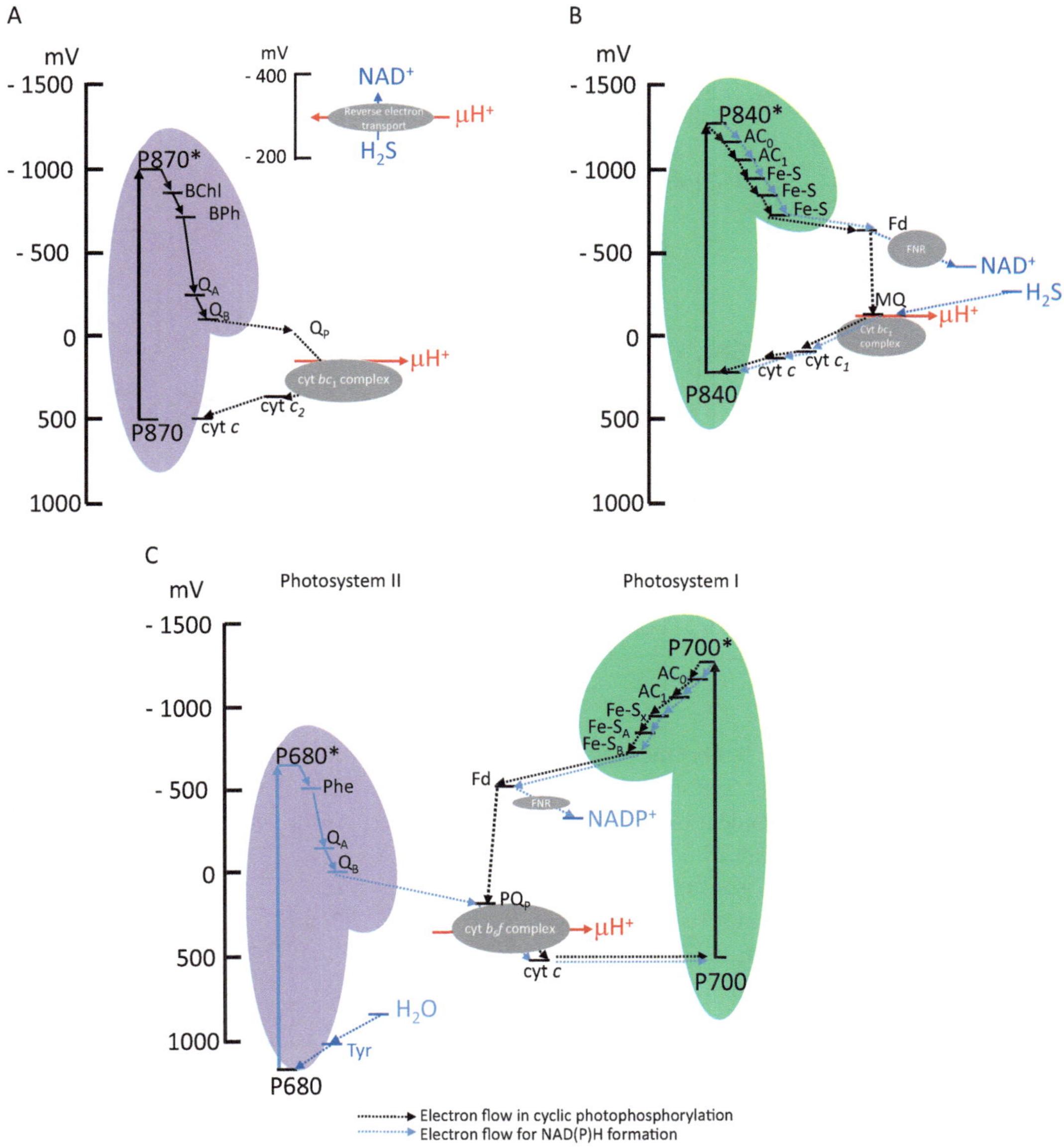

Fig. 6.22 Comparison of the cofactor components and their thermodynamics in the chlorophyll-based photosynthetic reaction centers (RC). Components of the photosynthetic reaction center are included in the respective shaded areas. (**a**) Pheophytin-quinone-type RC (purple background) in Purple Bacteria, (**b**) FeS-type RC (green background) in Green Bacteria, and (**c**) Cyanobacteria containing a pheophytin-quinone-type in PSII and a Fe-S-type in PSI. Black solid arrows indicate energy-conserving electron flow in RC, whereas black dotted arrows indicate electron flow outside of RC. Blue solid arrows indicate electron flow for electron activation for NADP$^+$ reduction, whereas blue dotted arrows indicate corresponding anabolic electron flow outside of RC. Abbreviations: P870, P840, P680, P700, special pair chlorophyll with absorption maximum at 870 nm, 840 nm, 680 nm, and 700 nm; BChl, monomeric RC bacteriochlorophyll; BPh, bacteriopheophytin; Q$_A$ RC complex quinone A; Q$_B$, RC complex quinone B; Q$_p$, quinone of membrane quinone pool; cyt$_c$, *c*-type cytochrome; ACo, AC1; Fe-S, Fe-S cluster; Fd, ferredoxin; MQ, menaquinone; Phe, pheophytin. Modified after Blankenship (2010)

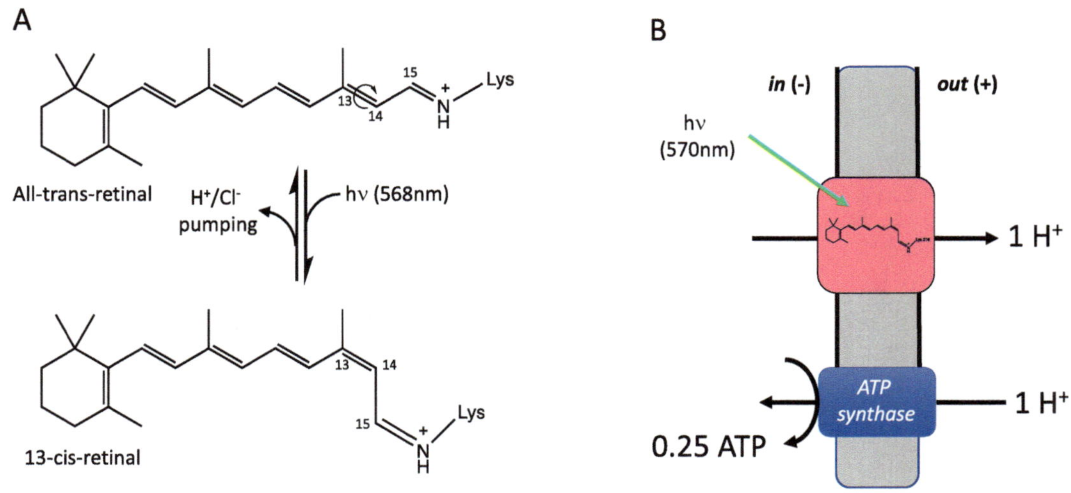

Fig. 6.23 Proposed conservation of light energy via bacteriorhodopsin. (**a**) Bacteriorhodopsin, bound to a lysine residue in the protein, is a light-driven proton pump absorbing light between 500 and 650 nm with a maximum at 568 nm. (**b**) The active chromophore is all-*trans*-retinal, which isomerizes after light absorption to 13-*cis*-retinal. Exergonic re-isomerization to all-trans-retinal is coupled to the translocation of a proton. Interestingly, in the rhodopsin-dependent vision process of animals, 11-*cis*-retinal is the active form of retinal. There upon absorption of a photon, the excited 11-*cis*-retinal isomerizes to the all-*trans*-retinal, sending off the signal of light perception

Bacteriorhodopsin absorbs light in the range of 500–650 nm (max absorption 568 nm). All-*trans*-retinal undergoes a light-dependent isomerization reaction to 13-*cis*-retinal (Fig. 6.23a). The energy released when 13-*cis*-retinal isomerizes back to the all-trans-retinal configuration is coupled to translocation of a proton from the inside to the outside, thereby generating a $\Delta\mu H^+$, which is then converted to ATP via an F_oF_1 *ATP synthase* (Fig. 6.23b).

6.4.4.4 TCP Coupled to Non-redox Reactions

There are exceptions to the general statement that "all catabolic pathways involve redox reactions." They all have in common that a $\Delta\mu H^+$ or $\Delta\mu Na^+$ is generated typically in short specialized pathways.

6.4.4.4.1 Decarboxylation of β-Keto Acids

β-keto acids (Fig. 6.24) can undergo "spontaneous" (i.e., exergonic) decarboxylation. In such reactions, for example in the *oxaloacetate decarboxylase*, the exergonic decarboxylation ($\Delta G^{o\prime}$ = – 23.8 J/mol) is coupled to the endergonic translocation of a sodium ion (Na^+) in the electric field of the membrane potential. The generated $\Delta\mu Na^+$ can be used either by a

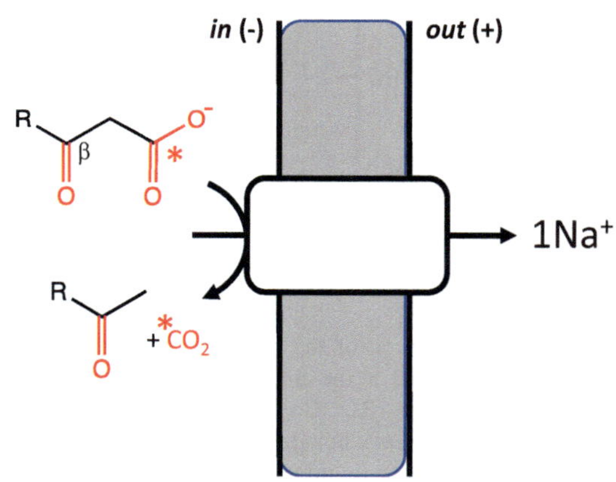

Fig. 6.24 Proposed energy conservation by decarboxylation of a β-keto-acid, a non-redox reaction. Organic acids where the carboxylate is in β position to a functional carbonyl group, i.e., keto group or thioester, can be enzymatically decarboxylated. In membrane-associated, biotin-containing enzymes, the free energy of the decarboxylation reaction of ~ −20 kJ/mol CO_2 is conserved by translocating a Na^+ in $\Delta\mu Na^+$ across the cytoplasmic membrane. R is typically an acyl carrier protein (ACP) with a modified coenzyme A. See text for details

Na^+-dependent *ATP synthase* or be converted via a Na^+/H^+ antiporter into a $\Delta\mu H^+$, which drives ATP synthesis via the more common H^+-dependent F_oF_1 *ATP synthase.*

Examples for energy conservation coupled to decarboxylation of a β-keto acid are in microbes fermenting dicarboxylic acids:

1. Oxaloacetate fermentation in *Klebsiella pneumoniae* and *Salmonella typhimurium*

$$\text{Oxaloacetate}^- + H^+ \rightarrow \text{Pyruvate}^- + CO_2$$
$$\Delta G^{o\prime} = -24\ \text{kJ/mol}$$

Decarboxylating β-keto acid:

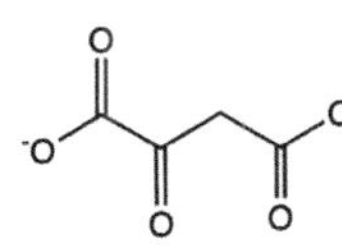

Oxaloacetate^{2-}

2. Malonate fermentation in *Malonomonas rubra*

$$\text{Malonate}^{2-} + H^+ \rightarrow \text{Acetate}^- + CO_2$$
$$\Delta G^{o\prime} = -17\ \text{kJ/mol}$$

Decarboxylating β -keto acid:

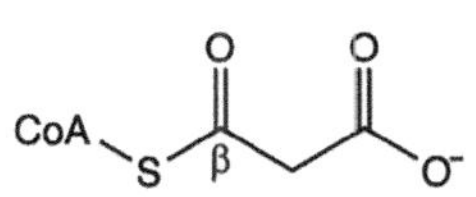

Malonyl-CoA

3. Glutamate fermentation *Acidaminococcus fermentans, Fusobacterium nucleatum, Peptostreptococcus asaccharolyticus, Clostridium symbiosum, Syntrophus gentianae*:

$$5\,\text{Glutamate}^- + 6\,H_2O + 2H^+ \rightarrow 6\,\text{Acetate}^- + 5\,CO_2 + 2\,\text{Butyrate}^- + NH_4^+$$
$$\Delta G^{o\prime} = -64\,\text{kJ/mol glutamate}$$

Decarboxylating β -keto acid equivalent:

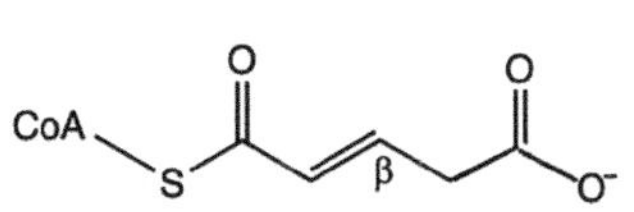

Glutaconyl-CoA

4. Succinate fermentation in *Propiogenium modestum* or *Veillonella parvula*:

$$\text{Succinate}^{2-} + H^+ \rightarrow \text{Propionate}^- + CO_2$$
$$\Delta G^{o\prime} = -25\ \text{kJ/mol}$$

Decarboxylating β -keto acid:

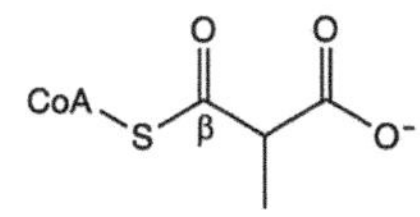

Methylmalonyl-CoA

6.4.4.4.2 *Methyltransferase* in Methanogenic Archaea

Another example of a catabolic, Na^+-translocating non-redox reaction is in methane formation in methanogenic archaea, catalyzed by the coenzyme B_{12}-containing *methyltransferase* (MtrA-H) (Fig. 6.25). This membrane-bound enzyme catalyzes the transfer of the methyl group from CH_3-tetrahydromethanopterine to coenzyme M:

$$CH_3 - H_4MPT + HS - CoM \rightarrow CH_3 - CoM + H_4MPT\ \Delta G^{o\prime} = -30\ \text{kJ/mol}$$

6.4.4.4.3 TCP Coupled to Metabolite Efflux

Some bacteria conserve energy via a $\Delta\mu H^+$ that is generated by efflux of a negatively charged end product. When lactic acid bacteria ferment glucose into two lactic acids (see Chap. 9), the environmental lactate concentration at the beginning of a batch fermentation or in situ in the

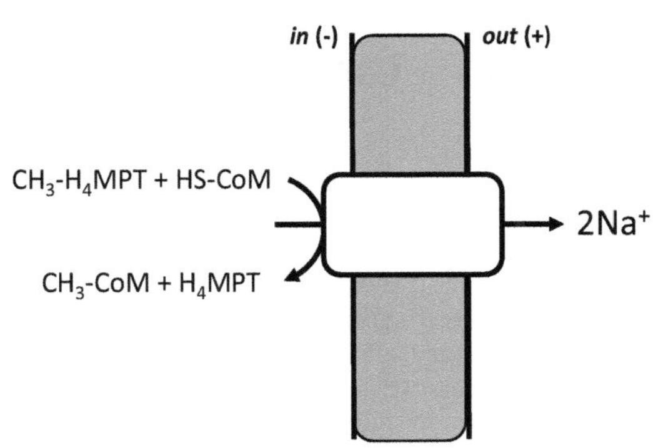

Fig. 6.25 Proposed energy conservation in the non-redox methyl transferase (Mtr) in methanogenic archaea. See text for details

environment is typically low. However, intracellular lactate concentration is high because the dissociated form of lactic acid cannot diffuse through the cytoplasmic membrane (Fig. 6.26). Using a specific exporter enzyme, lactate is exported in symport with 2 protons: One of which is neutralizing the negative charge of the lactate anion; the second proton is electrogenic and driven by the outward-directed gradient in lactate concentration. This translocated H^+ can drive the phosphorylation of 0.25 ATP (assuming C ring composed of 12 subunits). Via this mechanism lactic acid bacteria come closer to using all available free ΔG (high n_{ATP}) in the catabolic reaction (Glucose $\rightarrow$ 2 $Lactate^-$ + $2H^+$) at the beginning of the fermentation when extracellular lactate is low. As the fermentation proceeds in batch culture, extracellular lactate increases and reduces the concentration gradient and, thus, the available.

Such export-coupled TCP will not operate in fermenting microbes that produce as significant end products weaker acids such as butyrate, propionate, or acetate (see Chap. 9). The pKa of these acids is ~4.7, resulting in a more significant fraction being protonated, that can diffuse at a significant rate through the membrane. The permeability coefficients are higher for the protonated forms, 10^{-2} [cm s^{-1}] for acetic acid *vs* 10^{-7} [cm s^{-1}] for lactic acid. Thus, a cell cannot maintain a significant gradient of intracellularly-produced short-chain fatty acids. When accumulating in the extracellular environment at circumneutral pH and at concentrations higher than 1–10 mM, these weaker acids diffuse back at a relatively higher amount through the cytoplasmic membrane into a cell and acidify the cytoplasm. Thus, these end products effectively act as metabolic uncouplers at higher concentration, which leads to dissipation of $\Delta\mu H^+$. This would not occur if lactic acid were formed. This is one of the reasons why acetate-, propionate-, or butyrate-forming microbes either cannot produce a high titer of these acids (because they run into inhibition) or they need to allocate a significant fraction of their cellular energy toward a (futile) export.

Another example for ion translocation coupled phosphorylation in the absence of a net redox reaction is oxalate fermentation to formate by *Oxalobacter formigenes* (Fig. 6.27). The metabolic activity of this and metabolically similar microbes has been implied in oxalate degradation in the human body and in prevention of kidney stone formation (calcium oxalate can be a substantial fraction of the calcium mineral in kidney stones). Oxalate is taken up by an *oxalate:formate antiporter* (Fig. 6.27). Intracellular oxalate is

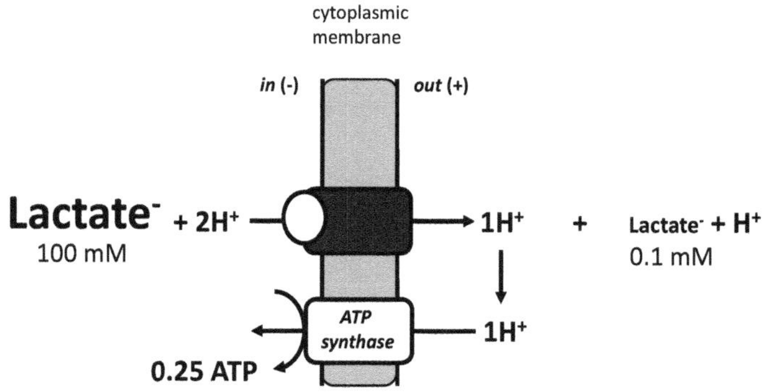

Fig. 6.26 Proposed translocation coupled phosphorylation (TCP) associated with lactate export. Membrane permeability of lactate ($P = 10^{-7}$ cm s^{-1}) is very low, indicating that lactate formed inside lactic acid bacteria does not exit a cell at a significant rate by diffusion through the membrane. Since the pKa of lactic acid is 3.8, at an intracellular pH of 7 less than 0.1% is protonated and lactate accumulates inside. The resulting concentration difference between inside and outside is the driving force to co-translocate a second proton that is used for energy conservation. In the scheme, it is assumed that in lactic acid bacteria four electrogenic protons are required for synthesis of ATP from ADP and inorganic phosphate. See text for details

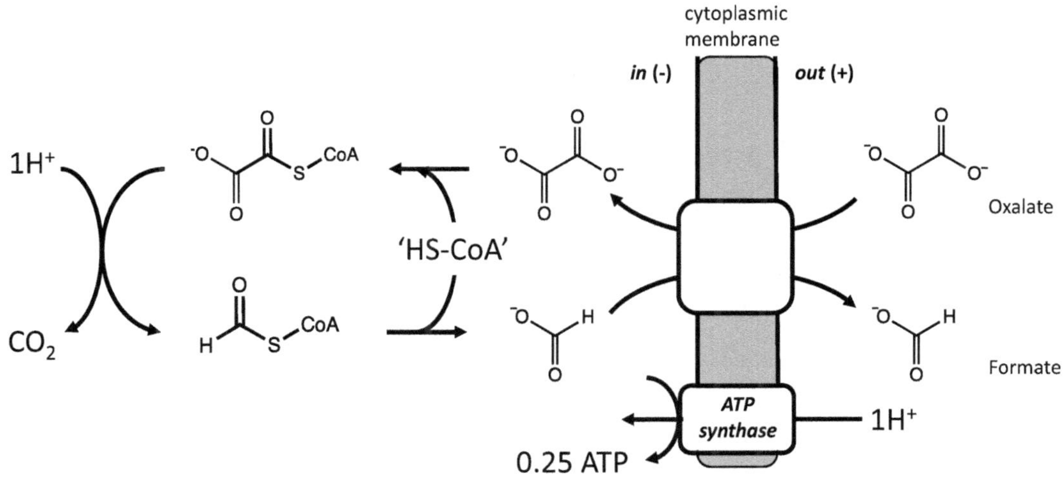

Fig. 6.27 Proposed energy conservation in oxalate fermentation to formate in *Oxalobacter formigenes*. See text for details

enzymatically activated by a *CoA transferase* to oxalyl-CoA and decarboxylated in a non-redox reaction to formyl-CoA, which results in consumption of an intracellular H^+. Formate is exported by the electrogenic *oxalate:formate antiporter*. Thus, these two metabolic processes effectively function as a proton pump, generating a $\Delta\mu H^+$, where intracellular oxalate consumption and formate accumulation drive the electrogenic antiporter. The energy of the translocated proton is conserved as 0.25 ATP by *F_OF_1 ATP synthase*.

6.5 Energy Interconversions

There are important metabolic "modules" that enable the **interconversion** of diverse redox energies so that the above-discussed mechanisms for energy conservation can operate under wider metabolic scenarios. The overall "purpose" of catabolic pathways is the conservation of available energy as ATP. As shown above and in the following chapters, some catabolic pathways contain steps for activation of a substrate or intermediate to thermodynamically/kinetically enhance a metabolic flux (see Sect. 5.4.1). One such activating step is the ATP-dependent transfer of a phosphoryl group or thiol from coenzyme A to an intermediate/substrate (see Chaps. 5, 9, and 10). A second "activation" step, introduced below, is the activation of electrons. All interconversion modules discussed in the following mediate an energy-dependent change of redox potential of electrons.

The two examples discussed are:

- Electron bifurcation/confurcation
- Reverse electron transport

6.5.1 Electron Bifurcation/Confurcation

In quinone-based transport-coupled phosphorylation, a difference in redox potential is conserved when coupled to an electron flow from NADH to quinone via, e.g., *NADH dehydrogenases* (see Fig. 6.17). NADH is formed in numerous oxidation reactions of diverse organic compounds, and such reactions are described in the following chapters. However, many microbes do not contain quinones, or, if they do, quinones may not be involved in quinone-based-TCP. These microbes typically use a reduced ferredoxin-based TCP for energy conservation. Thus, an electron flux needs to connect the oxidative metabolic reactions (NADH) with the prevailing energy-conserving reactions of the Fd^{2-}-depending Rnf or Ech modules. For this, NADH electrons at −280 mV in situ need to be activated to the redox potential of Fd/Fd^{2-} (~ −500 mV) (Fig. 6.28). This is

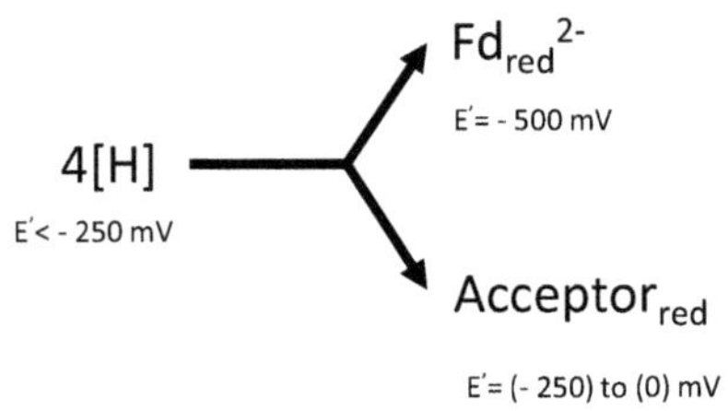

Fig. 6.28 General principle of flavin-based electron bifurcation. Only electron bifurcation using ferredoxin as low potential acceptor is discussed below. See text for details

particularly important in fermentations of many *Clostridia*, acetogens, methanogens, and sulfate reducers. Flavin-based **electron bifurcation** provides a mechanism to activate low redox potential electrons, e.g., NADH, NADPH, H_2, or formate, and couples them to ferredoxin reduction. In its reverse, electron confurcation provides a mechanism to drive thermodynamically difficult oxidations.

Flavins and Quinones in Electron Bifurcation
There are two types of electron bifurcation known, a flavin-based and a quinone-based electron bifurcation. Flavin-based electron bifurcation is the evolutionary more ancient process and is relevant for Rnf- and Ech-based energy conversion as discussed above. Quinone-based electron bifurcation is relevant as a mechanism in generating a $\Delta\mu H^+$, and, therefore, involved in energy conservation not conversion. Quinone-based electron bifurcation is discussed here only in a comparative context.

6.5.1.1 Flavin-Based Electron Bifurcation/Confurcation

Flavin-based electron bifurcation is a cytoplasmic, flavoenzyme-based mechanism and **couples an exergonic redox reaction to an endergonic, electron-activating reaction** resulting in the formation of reduced ferredoxin or of reduced flavodoxin (Fig. 6.28). Reduced ferredoxin can then be used in conjunction with Rnf or Ech for direct energy conservation, or in Fd^{2-}-dependent catabolic reactions such as the *formylmethanofuran dehydrogenase* reaction in methanogens, or carbon monoxide formation in acetogens (Chap. 10).

The biochemical principle of bifurcation is based on the properties of flavins (here FAD), as both a 2-electron AND 1-electron carrier (Fig. 6.5). This allows to bifurcate or "split" the electron pair of a hydride (H^-) into two 1-electron transfers, where each single electron in the respective transfer is at a different, crossed-over redox potential. A well-studied example is the electron-transferring flavoprotein (Etf) *electron bifurcating butyryl-CoA dehydrogenase* complex (Fig. 6.29). The intracellular redox potentials of the conjugated redox couples of NAD^+/NADH, Fd_{red}/Fd_{ox}, and butyryl-CoA/crotonyl-CoA are −280 mV, −500 mV and −10 mV, respectively. This enzyme complex performs two identical bifurcations by oxidizing 2 NADH, one NADH in each bifurcation cycle, for reduction of one Fd to Fd^{2-} and one crotonyl-CoA to butyryl-CoA (Fig. 6.29).

In the first step of each bifurcation, transfer of a hydrid from NADH to flavin (red) produces hydroquinone (HQ; FADH, red). Oxidation of the HQ to the SQ generates a high potential single electron (e.g., − 60 mV of free flavin) in Etf that is transfer to a second flavin (magenta) in Etf, which relays the single electron to a flavin (black) located in *butyryl-CoA dehydrogenase* and stored there, awaiting a second single electron from the second bifurcation. The redox potential of the short-lived SQ anion radical ($[FADH^\bullet]$, red) remaining at the flavin in Etf is sufficiently negative (<−500 mV) to reduce one [4Fe-4S] cluster of a ferredoxin; thereby, regenerating that oxidized flavin for the second bifurcation. The second bifurcation of the Etf hydroquinone (FADH, red) releases the second electrons for (1) reduction of $FAD^\bullet$ to $FADH^-$ (black), which reduces crotonyl-CoA to butyryl-CoA and (2) reduction of the second [4Fe-4S] cluster in Fd^{1-} to form Fd^{2-}, respectively. When electron-bifurcating *butyryl-CoA dehydrogenase* functions in conjunction with Rnf, the energy of one NADH oxidation with crotonyl-CoA is conserved as $\Delta\mu Na^+/H^+$, a metabolic task that could not be achieved by quinone-based TCP (Fig. 6.30)!

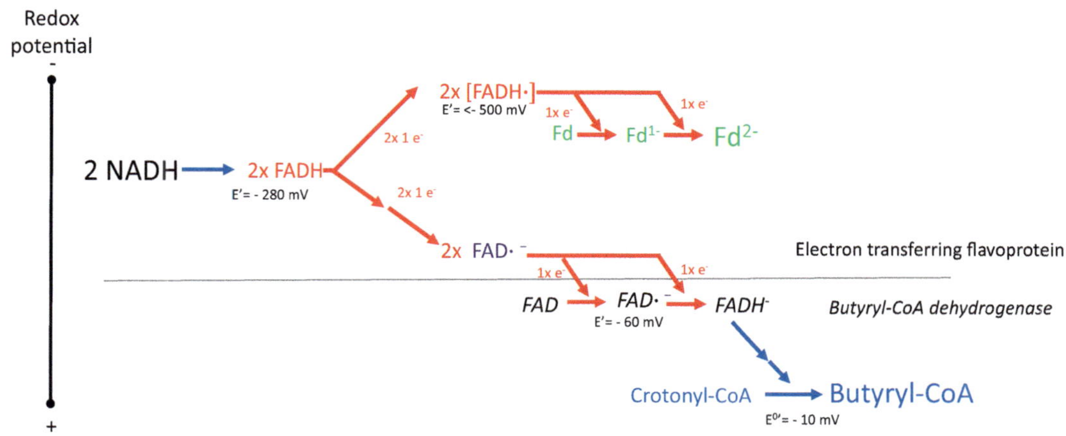

Fig. 6.29 Proposed flavin-based electron bifurcation of *butyryl-CoA dehydrogenase* in complex with electron-transferring flavoprotein (Etf), ordered by redox potential of electron carriers involved. Reactions involving single electron transfer are drawn in red; reactions involving two electron (hydride) transfer are drawn in blue; green indicates low redox potential Fd; blue indicates high redox potential electron acceptor. For simplicity, not all flavins in *butyryl-CoA dehydrogenase* are shown and protons have not been accounted for. "2×" indicates that the same flavin goes through the cycle twice per 2 NADH; "1×" is a single electron transfer. The flavin shown in magenta is a second flavin in Etf relaying 2× $1e^-$ from the high potential (red) FADH to a flavin (black) in *butyryl-CoA dehydrogenase*. This latter flavin collects and stores 2 single electrons for the reduction of crotonyl-CoA. FADH$^•$ (red) indicates the highly reactive, low redox potential semiquinone anion radical, reducing an [FeS] cluster in ferredoxin. Redox potentials indicated for flavins are for free flavin. Adapted and modified from Buckel and Thauer (2013). See text for details

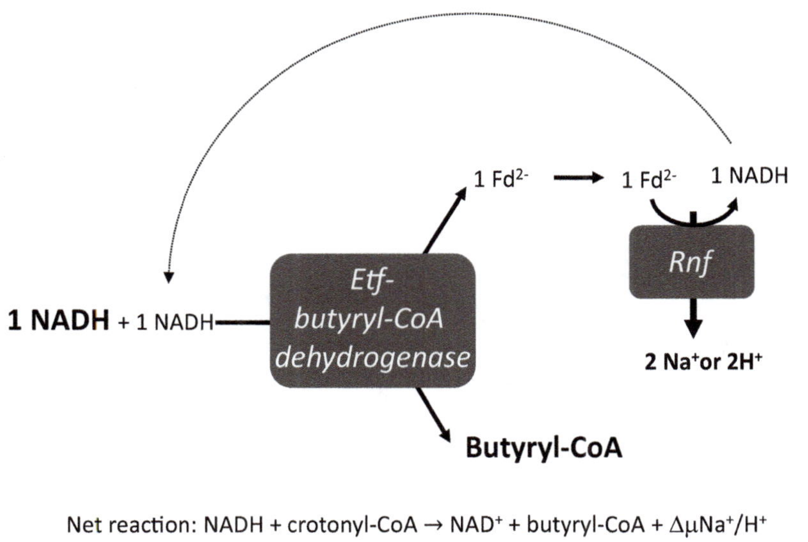

Fig. 6.30 Proposed electron bifurcating *butyryl-CoA dehydrogenase* in conjunction with Rnf. Net oxidation of 1 NADH with crotonyl-CoA as electron acceptor leads to conservation of the redox energy between NADH (−280 mV in situ) and crotonyl-CoA (−10 mV) ($\Delta G = -2\,F\,[(-0.1) - (-0.28)] = -34$ kJ/mol) in form of $\Delta\mu H^+$ or $\Delta\mu Na^+$. 1 NADH is cyclic in the process. Such energy conservation would be mechanistically impossible with a quinone-based TCP. Compounds in bold indicate the net flux of electrons and energy conservation: 1 NADH is oxidized with 1 crotonyl-CoA coupled to the formation of $\Delta\mu H^+/Na^+$

Quinone-Based Electron Bifurcation and Quinone-Dependent Reactions

While flavin-based bifurcation mediates an interconversion of redox energy, quinone-based bifurcation results in conservation of metabolic energy (Fig. 6.31). Ubi-and mena-quinone-based electron bifurcation follows essentially a similar chemistry as flavin-based electron bifurcation with the exception that

1. the redox potentials involved are more positive,
2. one of the two bifurcations is cyclic,
3. the net flux of electrons is only 2 compared to 4 in flavin-based bifurcation, and
4. quinone-based bifurcation results in generation of 4 electrogenic protons across the cytoplasmic membrane. In the first quinone-based bifurcation, two hydroquinones, Q_aH_2, and Q_bH_2 at +90 mV, are bifurcated into each one electron to reduce each one cyt c_1 at +250 mV and a cyt b at −60 mV via a semiquinone anion $Q_{a/b}^{\bullet -}$ as intermediate (Fig. 6.31). As only electrons are bifurcated, the protons from Q_aH_2, and Q_bH_2 oxidation are released to the outside of the cell as 4 electrogenic H^+. The two reduced cyt c_1 are oxidized by a downstream electron carrier of an electron transport chain such as cytochrome *c* or by a terminal electron acceptor. The two reduced cyt b_h are oxidized by the regenerated Q_a to from Q_aH_2 where two H^+ from the inside are taken up. Thus, hydroquinone-quinone Q_aH_2 formation is the cyclic part of this bifurcation. As a net result, this cyclic bifurcation in the cytochrome bc_1 complex mechanistically conserves the difference in redox energy between QH_2 (+ 90 mV) and cyt c_1 (+ 285 mV) by translocating 4 electrogenic H^+ that is coupled to a net flux of 2 electrons. Q_bH_2 is formed by reduction of Q_b by a physiological electron donor such as NADH.

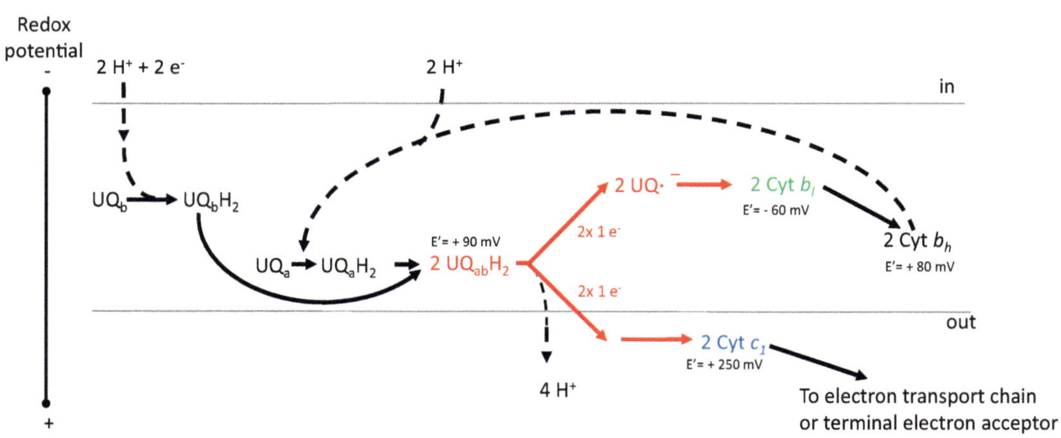

Fig. 6.31 Quinone-based electron bifurcation in the mitochondrial cytochrome bc_1 complex. Riske Fe-S center accepting transiently high redox potential electrons from HQ is not shown. For clarity, the two quinones are arbitrarily label Q_a and Q_b. Quinone-based electron bifurcation is cyclic; only two electrons are net bifurcated. Two ubihydroquinones are bifurcated and 4 H^+ are released to the outside. Adapted and modified from Buckel and Thauer (2013). See text for details

Quinones in Electron Transport *vs* Quinones in Electron Bifurcation

It is important to distinguish the function of quinones in electron transport (Figs. 6.1 and 6.18) and in bifurcation of the cyt *bc*1 complex (Fig. 6.31). Firstly, quinone molecules in electron transfer between a *NADH dehydrogenase* and a *quinol oxidase* are in a freely diffusible pool in the cytoplasmic membrane. Secondly, cyt *bc*1 quinones are stoichiometric to and integrated in the enzyme as opposed to being part of a free membrane quinone pool. Thirdly, cyt *bc*1 complex is so far the only known example of quinones mediating electron bifurcation, and only in context to translocating H^+ across the cytoplasmic membrane. The redox potentials of hydroquinone and semiquinone of flavin are −280 mV and −120 mV, and of quinone in cyt $_{bc}$1 are +90 mV and +190 mV.

Examples of cellular electron acceptors driving the endergonic reduction of Fd include reduction of crotonyl-CoA in butyrate-forming *C. kluyveri*, of caffeyl-CoA in *Acetobacterium woodii*, of methylene-tetrahydrofolate in *Moorella thermoacetica*, as well as of the heterodisulfide (CoB-S-S-CoM) in hydrogenotrophic methanogens (Fig. 6.32). Also, NAD^+ can serve as acceptor for a *transhydrogenase* coupling NADPH oxidation to Fd reduction. Therefore, electron bifurcation allows for converting small differences in redox potential, i.e., small amounts of available redox energy, from reducing equivalents that are at redox potentials of less than 0 mV (Fig. 6.30). Further examples for electron bifurcation as a key element of a microbe's catabolism are shown in Fig. 6.32.

Many but not all *bifurcases* are physiologically reversible and can function as *confurcases*. Electron confurcation, which is the reverse of the above-described flavin-based electron bifurcation, is a metabolic step in strictly anaerobic microbes to drive thermodynamically unfavorable oxidations of organic intermediate by exergonic Fd^{2-} oxidation. For example and as described in more detail in Chap. 9, the primary physiological direction of *electron confurcating lactate dehydrogenase* is Fd^{2-}-driven lactate oxidation with NAD^+ as electron acceptor in *Acetobacterium woodii:*

$$\text{lactate} + Fd^{2-} \leftrightarrow 2NADH.$$

Electron bifurcating/confurcating [FeFe] hydrogenase (Hyd) in *Thermotoga maritima*, *Ruminococcus albus*, and *Acetobacterium woodii* catalyze the Fd^{2-}-driven oxidation of NADH to H_2:

$$NADH + Fd^{2-} \leftrightarrow 2\,H_2.$$

Also, *bifurcating transhydrogenases* function in both the con- and bi-furcating direction:

$$2\,NADPH + Fd^{2-} \leftrightarrow NADH.$$

On the other hand, Etf-*butyryl-CoA dehydrogenase,* probably for thermodynamic reasons, is physiologically irreversible to catalyze at appreciable rate butyryl-CoA oxidation with Fd^{2-} to 2 NADH; this might be the reason why butyrate-oxidizing syntrophic microbes use reverse electron transport phosphorylation and not *bifurcating butyryl-CoA dehydrogenase* to drive butyryl-CoA oxidation (see below and Chap. 9).

6.5.2 Reverse Electron Transport

An alternative mechanism for electron activation, typically acting at more positive redox potential and involving quinones, is **reverse electron transport** where the **endergonic, electron-activating reaction is driven by $\Delta\mu Na^+/H^+$** (Fig. 6.33). This membrane-associated reverse electron transfer can be thought of as literally the reverse of electron transport phosphorylation. For example, oxidation of succinate to fumarate ($E^{0\prime}$ = + 30 mV) or of butyryl-CoA to crotonyl-CoA ($E^{0\prime}$ = − 10 mV) coupled to reduction of H^+ to H_2 ($E^{0\prime}$ = − 420 mV) as in syntrophs is endergonic. And so is the anabolic reduction of

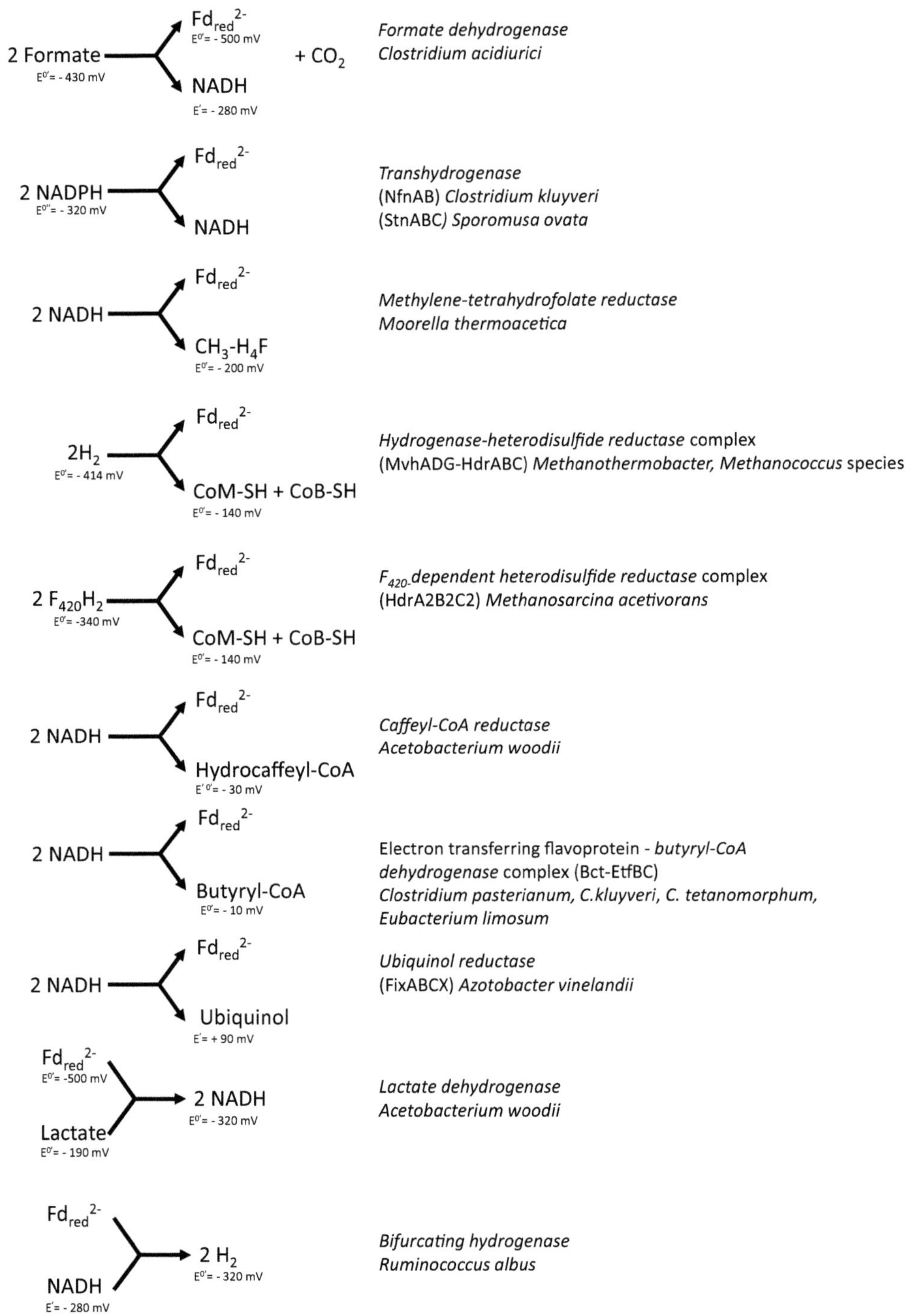

Fig. 6.32 Examples of electron bi- and con-furcation in strictly anaerobic prokaryotes, ordered by redox potential of the more positive electron acceptor. See text for details

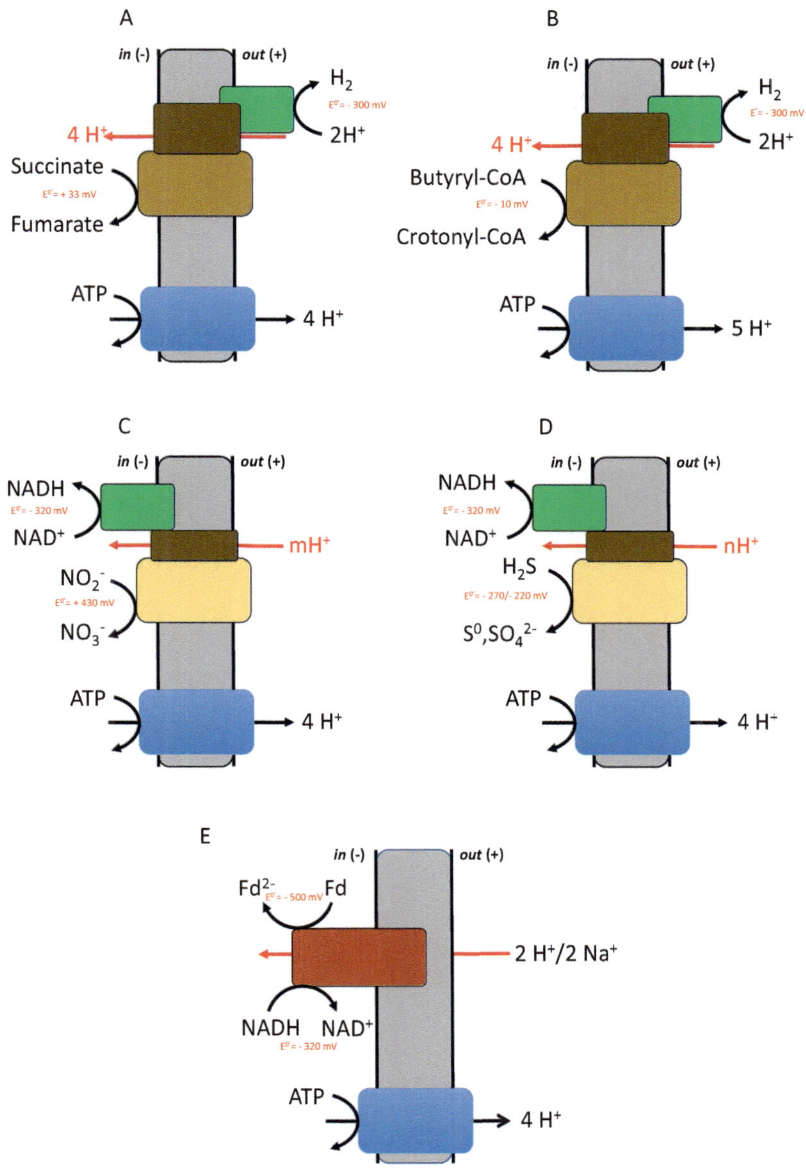

Fig. 6.33 Reverse Electron Transport as in succinate fermentation (**a**), syntrophic butyryl-CoA oxidation (syntrophic fatty acid oxidation) (**b**), and anabolic formation of NADH from NO_2^- (**c**), H_2S (**d**), or of Fd^{2-} from NADH by Rnf (E), respectively. See text for details

NAD^+ ($E^{0\prime}$ = − 320 mV) with electrons from NO_2^- or H_2S oxidation by chemolithotrophic NO_2^- and H_2S oxidizers, respectively, or in the anoxygenic phototrophs (Fig. 6.22).

Further Reading

Buckel W, Thauer RK (2013) Energy conservation via electron bifurcating ferredoxin reduction and proton/Na+ translocating ferredoxin oxidation. BBA 1827(2): 94–113

Buckel W, Thauer RK (2018) Flavin-based electron bifurcation, a new mechanism of biological energy coupling. Chem Rev 2018(118):3862–3886

Blankenship (2010) Early evolution of photosynthesis. Plant Physiol 154(2):434–438

Prototypic Reactions of Prokaryotic Carbon Catabolism

7

Contents

7.1 Introduction

The metabolic strategy of catabolic pathways is to convert a substrate into configuration(s) that mechanistically enable energy conservation via substrate-level phosphorylation and/or Ion Translocation-Coupled phosphorylation. Such conversions are mosaics of a few **prototypic reactions** acting in different combinations, thereby, giving rise to diverse pathway modules and pathways. "Prototypic" emphasizes the reaction, not the enzyme that is catalyzing the reaction or its substrate specificity. In fact, a single prototypic reaction can be catalyzed by multiple different enzymes with different reaction mechanisms and different substrate specificities as shown below.

What defines a **prototypic reaction** is the **reactivity of functional groups** (e.g., alcohol, aldehyde, ketone, hemiacetal, carboxylate) in an

The original version of the chapter has been revised. A correction to this chapter can be found at https://doi.org/10.1007/978-3-031-28218-8_11

A. M. Spormann, *Principles of Microbial Metabolism and Metabolic Ecology*,
https://doi.org/10.1007/978-3-031-28218-8_7

organic substrate molecule, and the relative spatial vicinity to other functional groups. Relevant vicinities of reactive groups are often limited to carbon atoms no more than three atoms apart. Microbial degradation of more complex organic compounds often proceeds via iterative use of a few ordered prototypic reactions (e.g., β-oxidation for fatty acid oxidation). The sequence of degradation steps is determined by the substrate specificity of enzymes involved in iteratively functioning pathways.

Adaptive evolution of substrate specificity of enzymes to accommodate new (more complex) substrates is frequent, while evolution of enzymes catalyzing novel prototypic reactions (mechanisms) is rare.

How to recognize relevant structural features in an organic compound and predict prototypic transformations.

A **prototypic enzymic transformation** is defined by structural features and reactivity of functional groups and the type of transformation acting on a substrate molecule. For example, the C_4 compound oxaloacetate contains four relevant structural features (Fig. 7.1) that allow five different prototypic transformations commonly found in microbial catabolism. Oxaloacetate is a dicarboxylic acid in central metabolism and an intermediate in numerous degradation pathways. In Fig. 7.1a, b, c, d, e, the relevant atoms and bonding electron pairs are colored in red to indicate those structural features that are necessary and sufficient for a particular prototypic reaction (provided that a substrate specificity of the catalyzing enzyme exists): **reduction (A), nucleophilic addition** (B), **decarboxylation of a 2-oxo-acid** (C), **oxidative decarboxylation** (D), and **carboxylate reduction** (E). Not all of these transformations have been observed to occur with oxaloacetate, yet they are possible and occur with other structurally related organic compounds in diverse pathways. The important structural determinants are mostly limited to one-to-three neighboring carbon atoms (ketone reduction requiring one carbon, *vs* decarboxylation of β-keto acid functions across three carbon) (Fig. 7.1). Viewing metabolism from the perspective of prototype reactions reveals simple organizing principle and the metabolic logic of catabolic pathways.

What is this set of prototypic enzymatic reactions?

This chapter will introduce the seven most common prototypic reactions (summarized in Table A.3 in the Appendix):

- alcohol oxidation (Sect. 7.2.1),
- hemiacetal oxidation (Sect. 7.2.2),
- carbonyl oxidation (Sect. 7.2.3),
- acyl-CoA oxidation (Sect. 7.2.4),
- aldol cleavage (Sect. 7.3.1),
- ketol cleavage (Sect. 7.3.2),
- hydrogen oxidation/formation (Sect. 7.2.5)

and six "auxiliary" reactions:

- mono−/di-oxygenation (Sect. 7.4.1),
- fumarate addition (Sect. 7.4.2),
- benzoyl-CoA reduction (Sect. 7.4.3),
- carbon–carbon rearrangements (Sect. 7.4.4),
- additions/eliminations reactions (Sect. 7.4.5),
- hydroxy/amino elimination (Sect. 7.4.6).

Prototypic enzymatic reactions combine into recurrent metabolic modules and diverse pathways that operate, with some modifications, in the forward or reverse direction depending on the thermodynamic and ecophysiological context (Chap. 8). For simplification and consistency, the reactions and enzyme names are introduced here primarily in the C–C cleavage and oxidative direction, with the understanding that many of these prototypic enzymes can function also in the reverse direction, for example in reductive pathways of fermenting microbes. Accordingly, enzyme names are introduced describing the direction for oxidative catabolism, although on occasion, historically more established enzyme names will be used. For example, when discussing fermentative conversion of glucose to lactate, *lactate dehydrogenase* will be used in lieu of "*pyruvate reductase*." Prototypic enzymes of

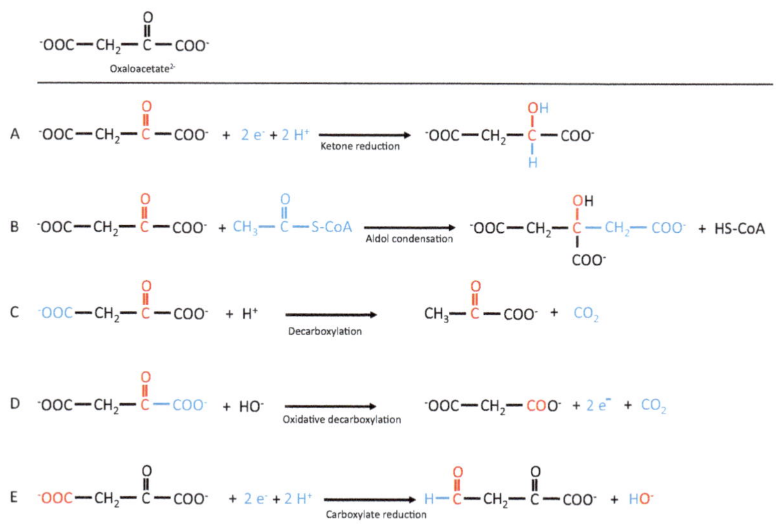

Fig. 7.1 Overview of structural features of organic metabolites required for prototypic transformations using oxaloacetate as an example. Indicated in red are those structural features that allow a specific transformation indicated. The reaction equations are not balanced to emphasize the structural features. For example, for reduction reactions of carbonyls (**a**) or carboxylates (**e**), only the carbonyl or carboxylate function is required. A functional carbonyl is also required for aldol condensation with the nucleophilic C2 carbon of acetyl-CoA (**b**). The structural requirement for decarboxylation (**c**) necessitates a three carbon-atom spanning configuration with a carboxylic group in β position to a functional carbonyl. The structural requirement for oxidative decarboxylation (**d**) necessitates a carbonyl in α position to a carboxyl group; thereby, spanning two vicinal carbon atoms. While neither oxidative decarboxylation nor carboxylate reductions of oxaloacetate have been reported in catabolism so far; analogous reactions are found in pathways of other organic molecules and are included here to illustrate the spectrum of metabolic transformations that an organic molecule can be subjected to

C1 metabolism are introduced separately in Chap. 10.

As introduced in Chaps. 3 and 5, a thermodynamically exergonic reaction results in a kinetically irreversible reaction, which provides a directionality of carbon and electron flux. While all pathway enzymes are mechanistically reversible, the rate in the reverse direction of a strongly exergonic pathway reaction, for example, is only a fraction of the rate of the forward reaction (see Haldane relationship, Sect. 5.4.2 for details). A low rate in the reverse direction is in most cases too slow to have a significant physiological impact in a microbe that grows at a finite rate. Such enzyme reactions are described here as (kinetically) "**irreversible**" or "**physiologically irreversible**." Physiological irreversibility in *oxidoreductase* reactions is often related to the difference in redox potential between the substrate and the cellular electron-carrying coenzyme involved. One metabolic strategy to control whether or not an *oxidoreductase* reaction, for example, is reversible or irreversible is to involve different electron-carrying coenzymes.

7.2 Prototypic Oxidation/Reduction Reactions

7.2.1 CH–OH Group Oxidations

Dehydrogenation of a **CH–OH group** is the abstraction of a hydride (H^-) from a primary or secondary alcohol (Fig. 7.2a). In the course of the reaction, the OH group is deprotonated, where the resulting electron pair of the alkoxide assists in the elimination of H^- as leaving group by forming a carbon-oxygen double bond (Fig. 7.2a). Based on this hydride abstraction, enzymatic CH–OH group oxidations occur only on primary and

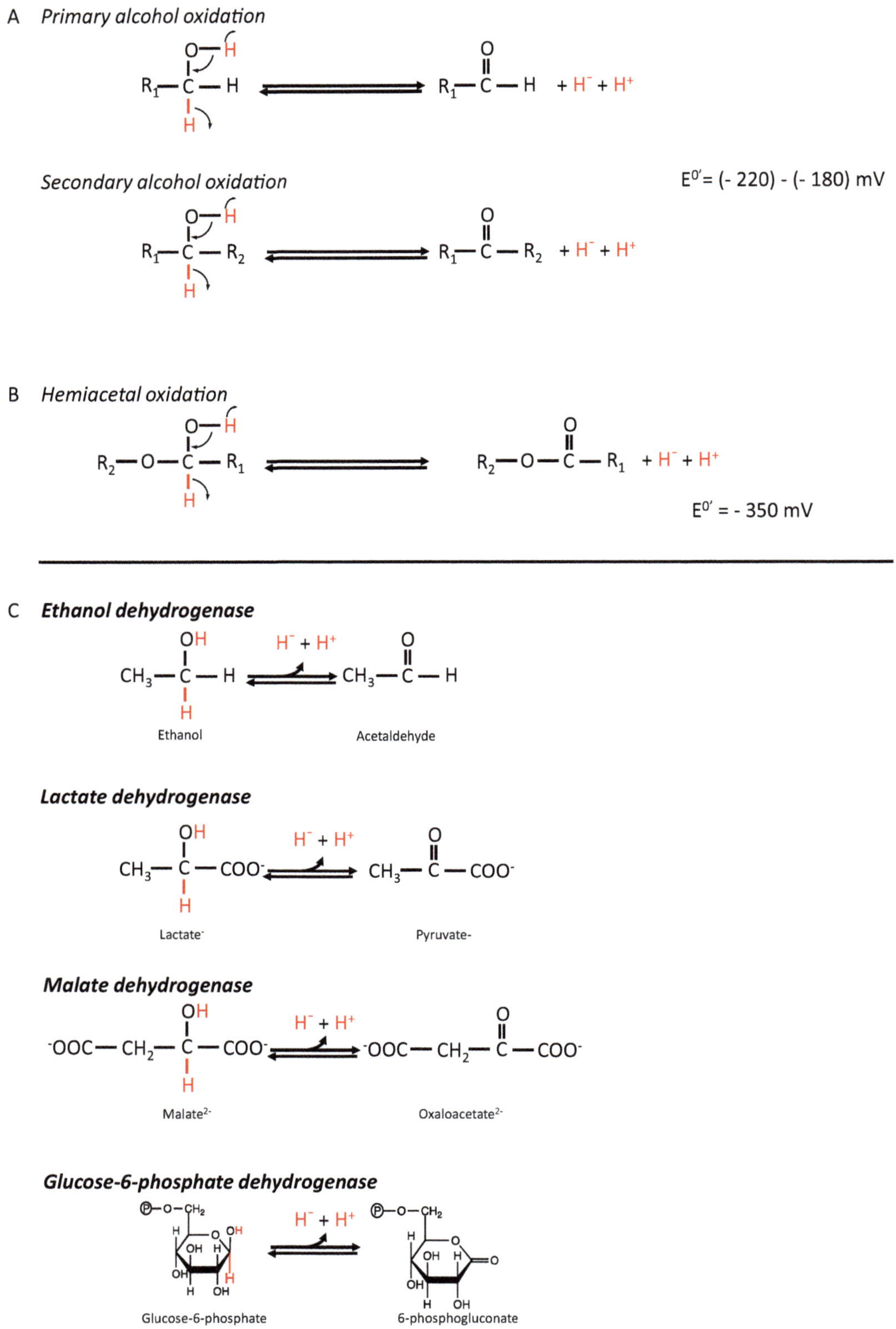

Fig. 7.2 CH-OH (alcohol) group and hemiacetal oxidation reactions. (**a**) Prototype of enzymatic CH-OH group oxidation; a hydride (H^-) is abstracted by *alcohol dehydrogenases*. The hydride is transferred to a cellular two-electron carrier. Depending on the R moiety, the standard redox potential of the aldehyde/alcohol redox couple is in the order of $E^{0\prime} = -220$ to -180 mV. (**b**) Hemiacetal oxidation proceeds via a similar hydride abstraction;

secondary, but not on tertiary, alcohols. The standard redox potential of the alcohol/aldehyde couple in diverse organic compounds is in the range of $E^{0\prime} = -220$ to -180 mV; the specific redox potential varies with the type of substituent(s) at the carbon being oxidized and the vicinal carbon (s). Figure 7.2c show frequent examples of prototypic alcohol *dehydrogenases*, such as *ethanol dehydrogenase*, forming acetaldehyde, or *lactate dehydrogenase* forming pyruvate. With NAD^+ as hydride electron acceptor, *ethanol dehydrogenases* are generally physiologically reversible, meaning that ethanol oxidation can occur at a rate that is physiologically relevant, whereas *lactate dehydrogenases* are not. The reason for this difference is that *ethanol dehydrogenase* forms an intracellular complex with *acetaldehyde dehydrogenase* so that the concentration of the free oxidation product, acetaldehyde, can be kept very low. In contrast, *lactate dehydrogenase* does not form a complex with many enzymes that use pyruvate as substrate with the consequence that the resulting equilibrium concentration of pyruvate would be too low for pyruvate to be metabolized further.

Lactate Dehydrogenase

Oxidation of lactate to pyruvate and the reverse, pyruvate reduction to lactate, are common reactions in numerous fermenting microbes; yet, they proceed by different enzymes. In the reductive direction, NADH is the typical electron-carrying coenzyme, driving thermodynamically the pyruvate reductive reaction because of the difference in redox potentials (NAD^+/NADH $E^{0\prime} = -320$ mV; pyruvate/lactate $E^{0\prime} = -190$ mV). For example, in lactic acid bacteria, D-or L-lactate is formed from pyruvate by an *NAD^+-dependent lactate dehydrogenase*. On the other hand, the direction of lactate oxidation with NAD^+ is endergonic and would be physiologically too slow or require significant proteome space for *lactate dehydrogenase*, which is why lactate oxidation in known prokaryotes does not proceed with NAD^+ as cellular electron acceptor. Instead, two alternate modes are known, depending on the environmental electron acceptor and the presence of an electron transport chain. In aerobic microbes that contain quinone-based respiratory chains, a FMN- or FAD-containing *lactate dehydrogenase* couples lactate oxidation with reduction of menaquinone ($E^{0\prime} = -70$ mV) or ubiquonine ($E^{0\prime} = +90$ mV) in an exergonic reaction as present in the anaerobic metabolism of propionate-forming *Propionibacterium freudenreichii* (Fig. 9.9). In the absence of quinone-based electron transport phosphorylation such as in the strictly anaerobic *Acetobacterium woodii*, a *confurcating lactate dehydrogenase* catalyzes lactate oxidation with a simultaneous, exergonic oxidation of Fd^{2-} with NAD^+ as electron acceptor in the same enzyme (Fig. 10.13).

7.2.2 Hemiacetal Oxidation

Enzymes mediating **hemiacetal dehydrogenation** catalyze a similar hydride transfer to a pyridine nucleotide, facilitated by an enzyme-stabilized alkoxide (Fig. 7.2b). Also here, the hydroxy group is deprotonated first, facilitating the leaving of a hydride under formation of

Fig. 7.2 (continued) however, the standard redox potential is lower ($E^{0\prime} = -350$ mV). (**c**) Examples of common alcohol and hemiacetal *dehydrogenases* that can operate in the forward (i.e., alcohol oxidation) or reverse (aldehyde reduction) direction depending on the metabolic pathway context and electron carrier employed: *ethanol dehydrogenase* in ethanol fermentation, *lactate dehydrogenase* in lactate fermentation, *malate dehydrogenase* in the TCA cycle, and *glucose-6-phosphate dehydrogenase* in glycolytic Embden–Meyerhof pathway. See text for details

carbon–oxygen double bond of a lactone. The standard redox potential of the hemiacetal/lactone redox couple of glucose-6-phosphate is $E^{0\prime} = -350$ mV, and NAD^+ or $NADP^+$ typically function as electron acceptors. Some of the catabolically important hemiacetal dehydrogenases are *glucose-6-phosphate dehydrogenase* in the extended heterofermentative lactic acid pathway as in *Leuconostoc mesenteroides* and other heterofermentative lactic acid bacteria (Fig. 9.6) and *2-hydroxy-4-carboxymuconate-6-semialdehyde dehydrogenase* in *Sphingomonades* and *Pseudomonas* species in aromatic compound degradation via the *meta* cleavage pathway.

The Fate of Substrate Protons in Enzymatic Reactions

Figure 7.2 introduces the first enzymatic oxidation reaction where a hydroxy group assists in oxidation by hydride removal. Key for this reaction is that the proton dissociates from the hydroxy group first. In many alcohol dehydrogenases this is achieved by binding of Zn^{2+} to one of the lone electron pairs of the OH group or by binding of the proton to a lone electron pair of an amino group at the active site. While following the fate of such abstracted H^+ is interesting for understanding a detailed reaction mechanism, for the purpose of this perspective, such details have been omitted for simplification with the understanding that H^+ abstraction in reaction mechanisms can be more complex.

7.2.3 Carbonyl Oxidation

Carbonyl oxidation comes in two variations: (1) oxidation of an **aldehyde** and (2) oxidation of a **ketone in α-position to a carboxylate** (2-oxo acid or α-keto acid) (Fig. 7.3).

Oxidation of an aldehyde to a carboxylate releases electrons at a standard redox potential of ~ $E^{0\prime} = -580$ mV (Fig. 7.4a). There are two types of enzymatic aldehyde oxidations: those associated with energy conservation by forming an energy-rich bond in the end product (Fig. 7.4b) and those that don't (Fig. 7.4c).

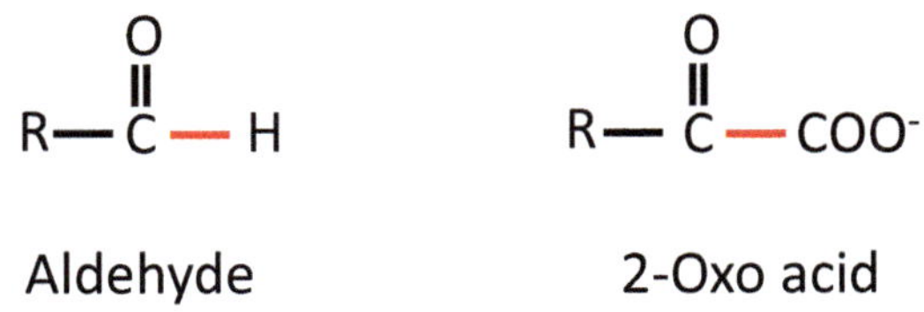

Fig. 7.3 General structure of an aldehyde and 2-oxo acid (a-keto acid). The bonding electron pair that are abstracted in the oxidation reaction is indicated in red. Note, ketones cannot be oxidized unless they are in α position to a carboxyl group and then only via oxidative decarboxylation

Aldehyde oxidation to a carboxylate occurs with either ferredoxin (Fd) or NAD^+ as cellular electron carrier by an ***aldehyde:ferrodoxin oxidoreductase*** or an ***NAD^+-dependent aldehyde dehydrogenase***, respectively. *NAD^+-dependent aldehyde dehydrogenase*, such as non-acetylating *acetaldehyde dehydrogenase* (Fig. 7.4a) is found in many catabolic pathways of aerobic, anaerobic, and facultative microbes, whereas *aldehyde:ferredoxin oxidoreductase* (AOR) is mostly but not exclusively found in anaerobic microorganisms. Because the redox potential of ferredoxin ($E' = -500$ mV) is close to that of the electrons being released from the aldehyde substrate ($E^{0\prime} = -580$ mV), AORs are reversible, while the exergonic reduction of NAD^+ ($E^{0\prime} = -320$ mV) with aldehyde-derived electrons renders this reaction physiologically irreversible (Fig. 7.4b). Reduced ferredoxin (Fd^{2-}) in strictly anaerobic prokaryotes can be used for ATP synthesis via a chemiosmotic coupling through Rnf or Ech (see Chaps. 6 and 9).

A frequently occurring aldehyde oxidation is the oxidation of glyceraldehyde-3-phosphate by *glyceraldehyde-3-phosphate dehydrogenase (GAP dehydrogenase)* (Fig. 7.5a, b, c). Currently, there are three different forms of catabolic *GAP dehydrogenases:* ***(1) NAD^+-dependent, phosphorylating GAP dehydrogenase, (2) ferredoxin-dependent, non-phosphorylating GAP dehydrogenase, and (3) $NADP^+$-dependent, non-phosphorylating GAP dehydrogenase*** (Fig. 7.5a, b, c). These three variations are possible because different electron-carrying

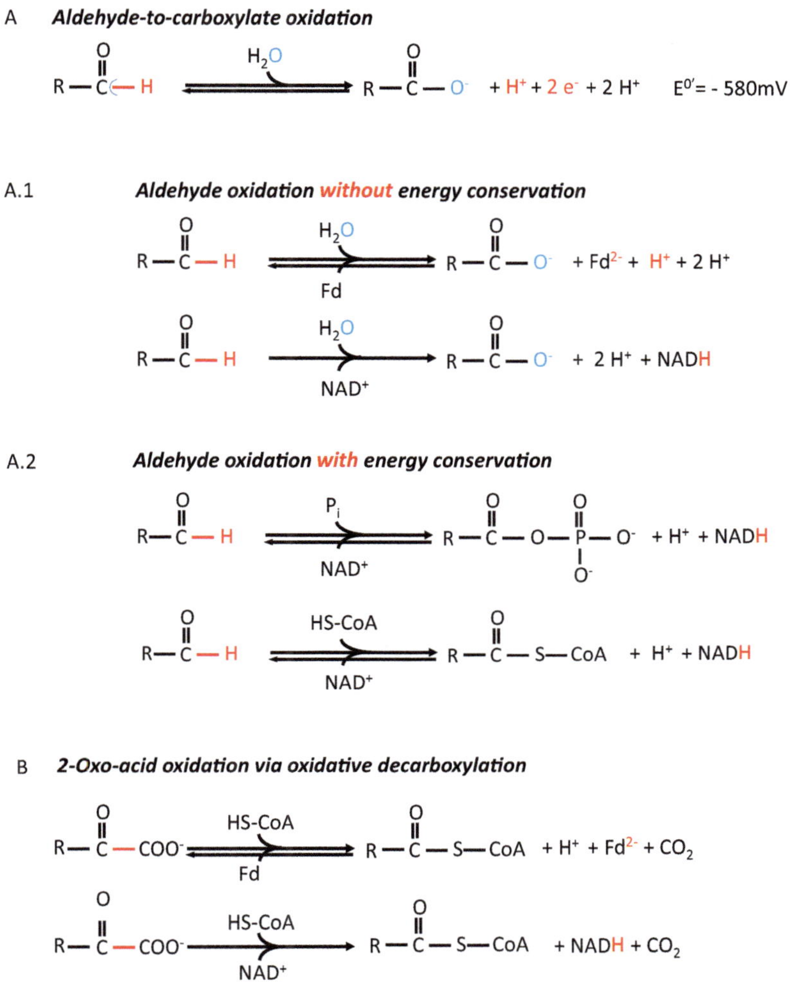

Fig. 7.4 Overview of prototypic aldehyde/2-oxo-acid oxidations. (**a**) General reaction of an *aldehyde dehydrogenase*. (A.1) Aldehyde oxidation without energy conservation. If ferredoxin is electron acceptor, the reaction is reversible; if NAD^+ is electron acceptor, the reaction is thermodynamically irreversible because the redox potential of $NADH/NAD^+$ at $E°' = -320$ mV is more positive than the aldehyde/carboxylate couple at ~ −580 mV. (A.2) Aldehyde oxidation coupled to energy conservation. Energy is conserved in form of a carboxylate-phosphate anhydride, such as in 1,3-bisphosphoglycerate mediated by *glyceraldehyde-3-phosphate dehydrogenase*, or a thioester with HS-CoA as mediated by *acetaldehyde dehydrogenase* in *E. coli* and *Pseudomonas* sp.. (**b**) 2-Oxo-acid oxidation (oxidative decarboxylation). Similar as in A.2, oxidation of a 2-oxo-acid, such as pyruvate or 2-oxo-glutarate, is reversible with ferredoxin as electron acceptor (*pyruvate:ferredoxin oxidoreductase, 2-oxo-glutarate:ferredoxin oxidoreductase*; the latter enzyme, when anabolic, is also called *pyruvate synthase, 2-oxo-glutarate synthase,* respectively), whereas oxidation with NAD^+ as hydride acceptor as in *pyruvate dehydrogenase* or *2-oxo-glutarate dehydrogenase*, are irreversible. See text for details

coenzymes are involved. As a consequence, the pathways involving these different variations result in different recovery of ATP (n_{ATP}): *NAD^+-dependent GAP dehydrogenase* recovers 1 ATP (in from of the energy-rich bond of 1,3-bisphosphoglycerate) and 1 NADH at ($E' = -270$ mV), while *ferredoxin-* and *$NADP^+$-dependent GAP dehydrogenase* recover no energy-rich bond in the end product, but electrons at a more negative redox potential (Fd^{2-}, $E' \sim -500$ mV; NADPH, $E' = -380$ mV), respectively. *$NADP^+$-dependent,*

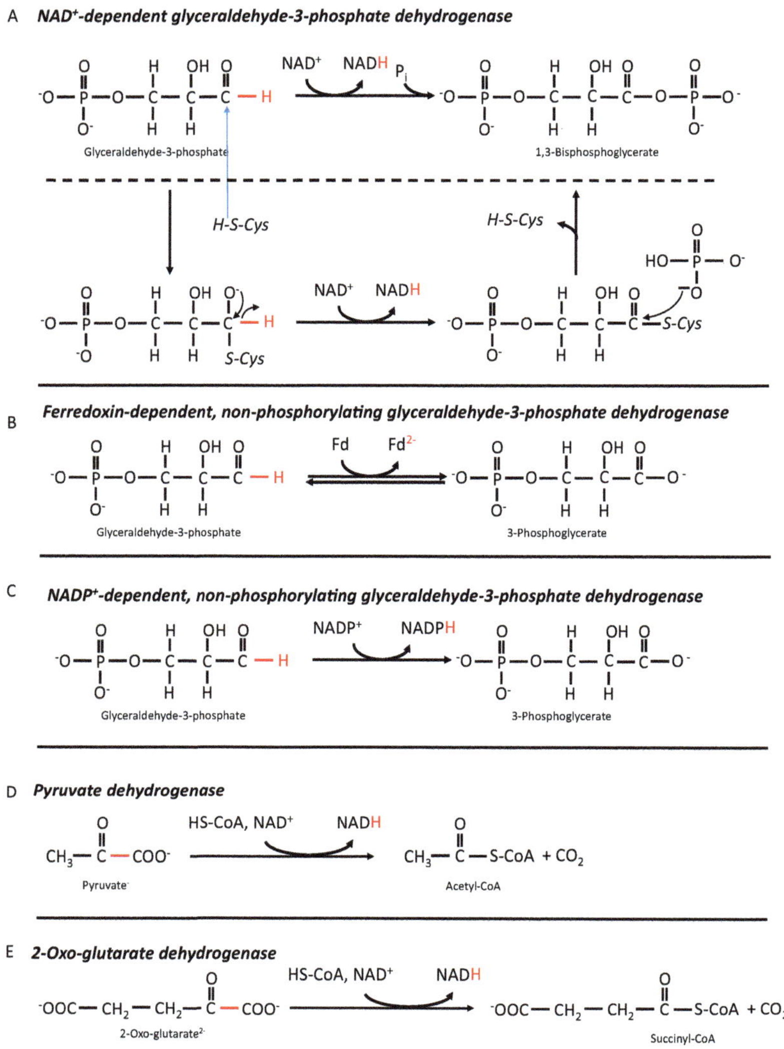

Fig. 7.5 Prokaryotic *glyceraldehyde-3-phosphate dehydrogenases* and 2-oxo-acid *dehydrogenases*. There are three types of *glyceraldehyde-3-phosphate dehydrogenases* (*GAP-DH*) (**a**–**c**): (**a**) *NAD⁺-dependent, phosphorylating GAP-DH* and the enzyme's reaction mechanism, (**b**) *ferredoxin-dependent, non-phosphorylating GAP-DH* and (**c**) *NADP⁺-dependent, non-phosphorylating GAP-DH*. They differ in whether or not energy is recovered in an anhydride bond (ATP equivalent) as well as in the electron-carrying coenzymes involved. (**d**) *pyruvate dehydrogenase* catalyzing oxidative decarboxylation of pyruvate and (**e**) *2-oxo-glutarate dehydrogenase* catalyzing oxidative decarboxylation of 2-oxo-glutarate. See text for details

non-phosphorylating GAP dehydrogenase (commonly referred to as GAPN) is irreversible. Here, the loss of ATP is traded-off against recovering low redox potential electrons (Chaps. 8 and 9).

Oxidation of glyceraldehyde-3-phosphate via the common *NAD⁺-dependent, phosphorylating GAP dehydrogenase* (GAP-DH) is shown in detail in Fig. 7.5a, and mechanistically coupled to form an energy-rich anhydride in 1,3-bisphosphoglycerate. The reaction mechanism of this *GAP dehydrogenase* is, to some extent, similar to that of an *alcohol dehydrogenase* because an enzyme thiolate (from an enzyme cysteine) makes a nucleophilic attack on the carbonyl carbon of glyceraldehyde-3-phosphate, converting this effectively into a CH–OH group, that is oxidized as described above (Sect. 7.2.1). The enzyme-bound thioester is phosphorylytically cleaved by an inorganic phosphate releasing 1,3-bisphosphoglycerate. Since the energy released by a *NAD^+-dependent, phosphorylating GAP-DH* is conserved in the reaction product, this reaction is reversible. The thioester of the energy-rich 1,3-bisphosphoglycerate is subsequently metabolized in the glycolytic pathway to phosphorylate ADP to ATP. This is one important mechanism for ATP formation via substrate-level phosphorylation (see Sect. 6.3.1).

In γ-proteobacteria, such as *E. coli*, *glyceraldehyde-3-phosphate dehydrogenase* is NAD^+-dependent and phosphorylating forming 1,3-bisphosphoglycerate plus NADH. In the hyperthermophilic Archaea *Pyrococcus furiosus*, *Pyrobaculum aerophilum*, and *Thermococcus spp.*, *glyceraldehyde-3-phosphate dehydrogenase* is ferredoxin-dependent and non-phosphorylating forming 3-phosphoglycerate. In some bacteria (e.g., *Streptococcus pyogenes*), *glyceraldehyde-3-phosphate dehydrogenase* is $NADP^+$-dependent and non-phosphorylating, also forming 3-phosphoglycerate. These differences in *glyceraldehyde-3-phosphate dehydrogenase* result in different pathway n_{ATP} and metabolic and eco-physiological consequences as discussed in Chaps. 8 and 9. A thermal stability issue of 1,3-bisphosphoglycerate is unlikely to be the reason for the inefficient energetic coupling by the non-phosphorylating enzymes in hyperthermophiles because the hyperthermophilic bacterium *Thermotoga maritima* uses a catabolic *NAD^+-dependent, phosphorylating GAP dehydrogenase*, indicating that a *glyceraldehyde-3-phosphate:ferredoxin oxidoreductase* is an Archaea-specific trait rather than a biochemical necessity. In microbes using a catabolic *ferredoxin-dependent, non-phosphorylating GAP dehydrogenase*, the energy may be only partially dissipated as heat since energy released during re-oxidation of the Fd^{2-} with NAD^+ via a membrane-associated Rnf is conserved by phosphorylation of 0.5 ADP (Fig. 6.16).

Thus, depending on the type of *GAP dehydrogenase* and cellular electron carrier involved, energy released in aldehyde oxidation can be **conserved**, **partially conserved,** or **completely dissipated**. These metabolic options enable a "yield" or a "rate" mode of growth (see Chaps. 4 and 6).

Oxidation of a 2-oxo acid, such as pyruvate or 2-oxo-glutarate, via oxidative decarboxylation is a third class of carbonyl-oxidizing enzymes: ***2-oxo acid:ferredoxin oxidoreductases*** or ***2-oxo acid dehydrogenases*** (Figs. 7.4d and 7.5e). In a formal sense, these enzymes catalyze the oxidation of a C–C-bond rather than of a C–H bond. Since these oxidations also involve a ketolytic carbon–carbon cleavage (see Sect. 7.3.3), *oxidoreductases* catalyzing oxidative decarboxylation contain thiamine pyrophosphate as coenzyme (see a more detailed discussion in Sect. 7.3.2, Figs. 7.12 and 7.13). Reactions catalyzed by *2-oxo acid:ferredoxin oxidoreductases* are reversible, whereas reactions catalyzed by *$NAD(P)^+$*-dependent *2-oxo acid dehydrogenases* are not. Because of the irreversibility, employing a *NAD^+-dependent 2-oxo acid dehydrogenase* (e.g., *2-oxo-glutarate dehydrogenase*) results in a metabolic directionality as in the oxidative TCA cycle, whereas Fd-dependent *2-oxo-glutarate:ferredoxin oxidoreductase*, for example, can operate both in the forward and reverse directions, e.g., for anabolic CO_2 fixation. For the same reason, a *pyruvate dehydrogenase* "commits" pyruvate carbon to acetyl-CoA-consuming pathways (e.g., oxidation via TCA cycle, or reduction in fatty acid biosynthesis). *2-Oxo acid:ferredoxin oxidoreductases* are important enzymes in energy conservation in carbohydrate-fermenting prokaryotes such as *Clostridium acetobutylicum* and *C. pasteurianum*, or amino acid-fermenting Bacteria such as in *Clostridium difficile*, *Clostridium sticklandii*, or the Archaea *Thermococcus litoralis* and *Pyrococcus furiosus*. *2-Oxo acid dehydrogenases* are found primarily in aerobic and facultative microbes including Propionibacteria.

The reverse of aldehyde oxidation is carboxylate reduction. Reduction of a carboxylate or an acyl-CoA to the corresponding aldehyde (alcohol) are key reactions in solventogenesis by some carbohydrate-fermenting Clostridia, such as *C. acetobutylicum* (See Fig. 9.17). These reactions are used in pathway sequences of carbohydrate fermentation when metabolism shifts from butyrate to butanol formation, typically induced by a low pH. For butyrate → butanol conversion, for example, there are two routes: One starting from butyryl-CoA using a NAD(P)$^+$-specific *aldehyde dehydrogenase* and an *alcohol dehydrogenase*, or a bifunctional *aldehyde/alcohol dehydrogenase* as in AdhE. In these reduction reactions, the energy-rich bond of the butyryl-CoA thioester is consumed and no longer available for energy conservation. A second route, as found in *Clostridium autoethanogenum*, starts with reduction of butyrate rather than butyryl-CoA to butyraldehyde using a Fd^{2-}-dependent AOR (Fig. 7.4d).

7.2.4 Acyl-CoA Oxidation

The fourth prototypic carbon oxidation reaction is the C–H oxidation in 3-position of an acyl-CoA mediated by *acyl-CoA dehydrogenases* (Fig. 7.6a). The best-studied reaction is the *butyryl-CoA dehydrogenase* reaction (Fig. 7.6b), oxidizing butyryl-CoA to crotonyl-CoA. The standard redox potential of electrons of butyryl-CoA/crotonyl-CoA at $E^{0\prime} = -10$ mV is significantly more positive than those in alcohol or aldehyde oxidations, and of NAD^+ reduction. Moreover, although the overall butyryl-CoA oxidation is the removal of a hydride, the actual mechanism of *butyryl-CoA dehydrogenase* involves a radical mechanism (Fig. 7.6b) via a flavin semiquinone. In the oxidative direction, two single electrons from the substrate are abstracted after deprotonation and transferred to the flavin of **electron transferring flavoprotein (Etf)** to form a flavin semiquinone and subsequently a hydroquinone (Fig. 7.6). Neither NAD^+ nor $NADP^+$ are physiological electron carriers in catabolic acyl-CoA oxidative pathways. The fate of the reduced flavin in Etf is contingent on whether or not the reaction proceeds in the metabolic context in the oxidative or reductive direction (Fig. 7.6c). In the direction of butyryl-CoA (acyl-CoA) oxidation, electrons from Etf are transferred to quinones (ubiquinone or menaquinone, which are also one and two electron carriers) in respiring and syntrophic microbes. In the direction of crotonyl-CoA (enoyl-CoA) reduction, there are two variations of this electron transfer: Reduction of 1 crotonyl-CoA with 1 NADH (i.e., *non-bifurcating butyryl-CoA dehydrogenase*), and reduction of 1 crotonyl-CoA with 2 NADH (i.e., *bifurcating butyryl-CoA dehydrogenase*) and concomitant formation of Fd^{2-} (Fig. 7.6c).

For many fermenting, butyrate-forming microbes, condensation of 2 acetyl-CoA and reduction to butyryl-CoA is a sink of 2 NADH from the cellular pool of electrons derived from carbohydrate oxidation (see Chap. 9). When using *electron bifurcating butyryl-CoA dehydrogenase* in the overall reductive pathway of 2 acetyl-CoA → butyryl-CoA^{2-} + Fd^{2-}, 3 NADH are consumed, and the difference in redox potential between NADH and crotonyl-CoA/butyryl-CoA is recovered as reduced ferredoxin, which can be oxidized to NAD^+ via Rnf recovering this energy; Chap. 6, Fig. 6.7.

Figure 7.13 also shows that a carbon substituent (e.g., CH_3-) at the 4- and 5-position, such as in 3-methyl butyryl-CoA (isovalerate fermentation) does not interfer mechanistically with the oxidation in 3-position, which makes *acyl-CoA dehydrogenases* quite versatile for degradation of linear and branched fatty acids (Fig. 7.6d).

Acyl-CoA dehydrogenases are often found in metabolic pathways in conjunction with *enoyl-CoA hydratases*, which add H_2O across a carbon–carbon double bond to form the corresponding 3-hydroxy acyl-CoA, (see auxiliary reactions in Sect. 7.4.5). An *acyl-CoA dehydrogenase* in conjunction with *enoyl-CoA hydratase,* an *alcohol dehydrogenase (β-hydroxyacyl-CoA dehydrogenase),* and a *β-ketothiolase* are the prototypic reactions forming the **β-oxidation module** (see Sect. 8.3.2).

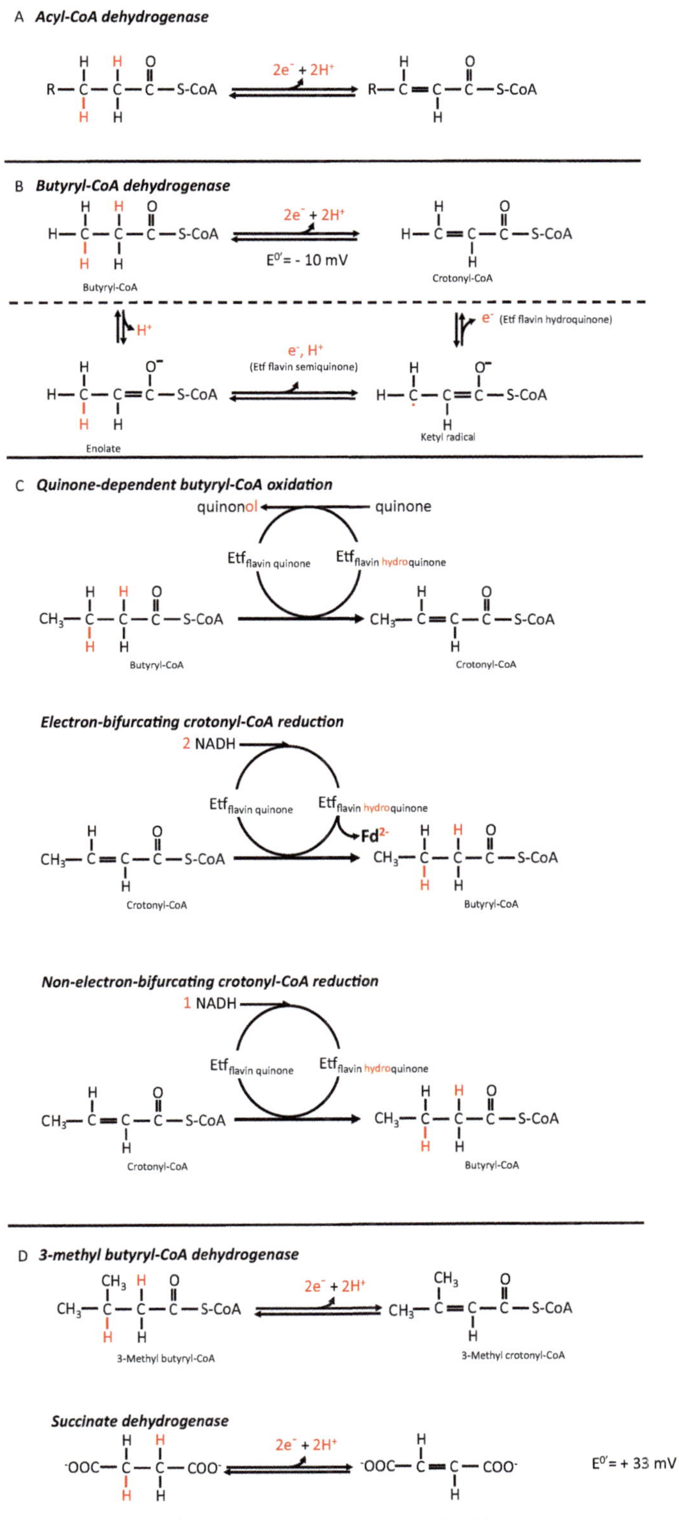

Fig. 7.6 *Acyl-CoA dehydrogenases*. (**a**) Prototype β-oxidation of an activated fatty acid. (**b**) Proposed reaction mechanism of *butyryl-CoA dehydrogenase*. Butyryl-CoA oxidation via *butyryl-CoA dehydrogenase* involves two 1-electron oxidations. In the oxidative direction, electrons from butyryl-CoA are transferred to an Etf, where the flavin cofactor is reduced by two 1 e- transfers via a flavin semiquinone as intermediate. (**c**) When operating in the butyryl-CoA oxidizing reaction, Etf transfers the hydroquinone electrons to a mena- or ubiquinone of a respiratory chain. In the reductive direction of crotonyl-CoA as substrate, 2 NADH are bifurcated by Etf in complex with *butyryl-CoA dehydrogenase* to reduce crotonyl-CoA to butyryl-CoA and to form Fd^2. In non-bifurcating *butyryl-CoA dehydrogenase*, the two hydroquinone electrons of Etf are transferred to crotonyl-CoA without Fd reduction and only 1 NADH is consumed. (**d**) Examples of other frequently occurring acyl-oxidation reactions: *3-methyl butyryl-CoA dehydrogenase* in isovalerate degradation, and *succinate dehydrogenase* in the TCA cycle. Note, that in *succinate dehydrogenase*, no CoA thioester is involved; instead the second carboxyl group assists in the hydride leaving

7.2.5 Hydrogenases

Although ***hydrogenases*** do not catalyze carbon oxidation/reduction reactions, they are introduced here as an important prototypic reaction. *Hydrogenases* emerged as one of the earliest enzymes in the metabolic evolution of Life and still play an essential role in extant, in particular anaerobic, microbial metabolism. In fact, metabolism of strictly anaerobic microbes and their ecophysiology can only be understood in the context of *hydrogenases*. Although the concentration of H_2 in anoxic environments is low ($pH_2 < 10$ Pa, <0.1 μM in aqueous solution), annually ~10^8 tons of H_2 are produced microbially, a pool that is turned over twice per year by *hydrogenases*. This reflects the enormous flux of H_2 electrons in catabolism of anaerobic microorganisms (primary, secondary fermenters, acetogens, and methanogens) and ecosystems as interspecies hydrogen transfer. Hydrogen is also found as intracellular hydrogen cycling in some fermenting microbes, connecting a cell's oxidative and reductive pathways. *Hydrogenases* and hydrogen metabolism are key to geochemical, environmental, ecological, and biotechnological processes.

Hydrogenases catalyze the reversible oxidation of H_2:

$$H_2 \rightleftharpoons 2e^- + 2H^+$$

All hydrogenases are metalloenzymes with a bioinorganic catalytic site. Currently, three types of *hydrogenases* are distinguished based on their metal cofactor content: *[FeFe]-hydrogenase*s, *[NiFe]-hydrogenases,* and a *[Fe-guanylylpyridinol] (Fe-GP) hydrogenase* (Fig. 7.7). All *hydrogenases* have in common that the symmetric H–H bond is activated by polarization by a low spin, non-redox-active Fe, complexed with at least one CO molecule (Fig. 7.7a). The resulting asymmetric electron distribution of the polarized H-H bond facilitates nucleophilic attack by Ni, Fe, or by the carbocation of methenyl-H_4MPT, respectively. These nucleophilic attacks result in subsequent H_2 oxidation either via two single-electron

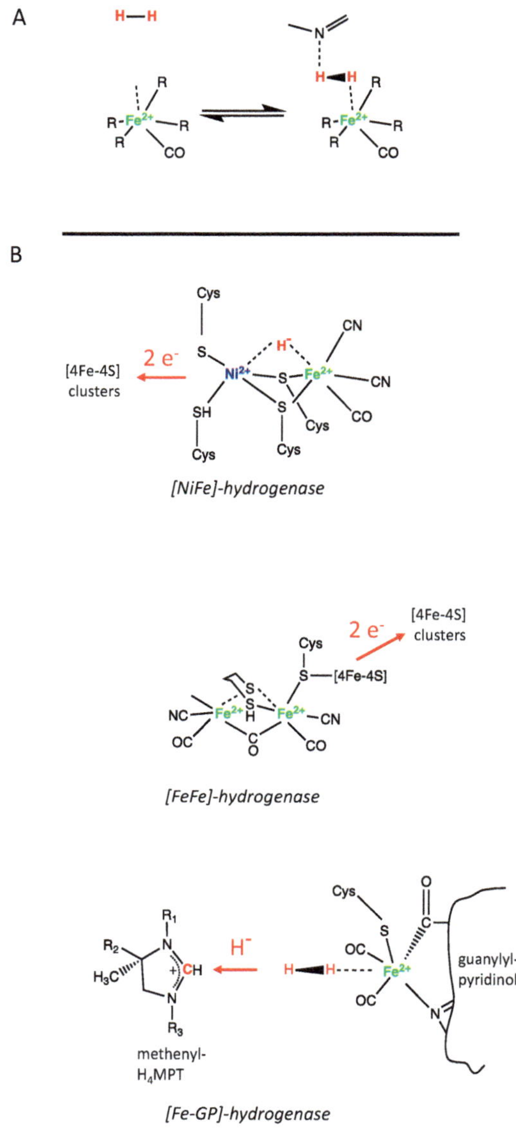

Fig. 7.7 The bioinorganic active sites of *hydrogenases*. All *hydrogenases* have in common a low spin, non-redox-active Fe, complexed with CO and sulfydryl groups. (**a**) A generic active site for activating H_2 by polarization of the H–H bond. (**b**) The three types of known *hydrogenases*: For *[NiFe]*-and *[FeFe]-hydrogenase,* the binuclear center is bridged by sulfur atoms and inorganic CO and CN ligands. Sulfhydryl groups of cysteine residues provide additional ligands. In *Fe-GP hydrogenase*, the guanylylpyridinol cofactor provides ligands as well as two CO and one cysteine. The polarized hydride attacks the carbocation of methylene-H_4MPT, while the *[NiFe]*- and *[FeFe]-hydrogenases* transfer two single electrons via the Ni or second Fe atom to the [4Fe-4S] clusters. See text for details

transfer, as in the *[NiFe]-* and *[FeFe]-hydrogenases*, or via hydride transfer to methenyl-H_4MPT in [Fe-GP] hydrogenase (Fig. 7.7b). The three *hydrogenases* are phylogenetically unrelated, and the metal center arose three times by convergent evolution.

Clostridial *[FeFe]-hydrogenases* are monomeric proteins with the dinuclear iron center, complexed by CN and CO, and three [4Fe-4S] and one [2Fe-2S] cluster (Fig. 7.7b). One [4Fe-4S] cluster bridges to the dinuclear Fe center, while the other two are involved in single electron relay to and from ferredoxin, which is the cellular electron acceptor/donor. *[FeFe]-hydrogenases* are also found in *Thermotoga maritima* and *Desulfovibrio* spp. but not in Archaea. They function primarily as H_2-releasing *hydrogenases* under fermentative conditions. Typically, their specific activity is higher (~10×) and they are more sensitive to O_2 than *[NiFe]-hydrogenases.*

[NiFe]-hydrogenases are heterodimeric proteins with a large subunit containing the dinuclear [NiFe] center and a small subunit containing three [4Fe-4S] clusters (the central one can be [3Fe-4S]) (Fig. 7.7b). Either ferredoxin or an [FeS] cluster of another redox enzyme is a cellular electron acceptor. Ni is redox active and involved in electron relay through the [4Fe-4S] clusters. *[NiFe]-hydrogenases* are uptake *hydrogenases* and are found in most Gram-negative bacteria, including *E. coli*, as well as in aerobic Knallgas bacteria such as in *Cupriavidus necator* (formerly *Ralstonia eutropha*); [NiFe] *hydrogenases* are present in Archaea.

[Fe-GP] hydrogenase (Hmd) is a homodimer and catalyzes the H_2-dependent reduction of methenyl-H_4MPT in methanogenic archaea in the methanogenesis pathway (Fig. 7.7b, Chap. 10). Consistent with the transfer of a hydride, the enzyme lacks [Fe-S] clusters. Next to methanogenic archaea where *[Fe-GP] hydrogenase* is essential under Ni-limiting growth conditions, Hmd of *Desulfurobacterium thermolithotrophicum* catalyzes the reduction of methylene-H_4F.

***Hydrogenases* in Complex with Other Enzymes**

By coupling the ferredoxin interface or subunit of *[FeFe]*-and *[NiFe]-hydrogenases* with other redox enzymes, *hydrogenases* can be modularly combined with other low redox potential catabolic *oxidoreductases*. For example, membrane-associated [NiFe]-containing *formate:hydrogen lyase* of *E. coli,* disproportionating formate to H_2 and CO_2 (see Sect. 9.2.3 and Figs. 9.3 and 9.7), is in complex with *pyruvate:formate lyase* in carbohydrate fermentations as a *pyruvate:formate lyase/formate:hydrogen lyase* complex to mitigate the acidity of accumulating formate (p*K*a = 3.7). The *hydrogenase* subunit of *formate:hydrogen lyase* is similar to Ech *hydrogenase.*

EchABC is a *hydrogenase*, where the binuclear [NiFe] subunit combines with a ferredoxin and a membrane-associated subunit (Fig. 6.16) to translocate 2 Na^+ or 2 H^+ across the cytoplasmic membrane coupling Fd^{2-} oxidation with 2 H^+ as electron acceptor. Ech plays a key role in energy conservation of strictly many anaerobic microbes (Chap. 9).

HydABC is an *electron-bifurcating/confurcating [FeFe] hydrogenase*, where the dinuclear iron center HydA subunit couples to a flavin-containing subunit with binding sites for Fd and NAD^+, respectively (Sect. 6.5.1). The physiological consequence of HydABC activity in *Thermotoga maritima*, *Ruminococcus albus*, and *Acetobacterium woodii* is to remove as hydrogen NADH electrons from oxidative fermentative pathways (Sect. 9.3.3, Fig. 9.14). The same enzyme, operating in the bifurcating direction, catalyzes H_2 bifurcation to Fd^{2-} and NADH in *A. woodii* during acetogenic growth on H_2/CO_2 via the Wood–Ljungdahl pathway (Chap. 10).

MvrAGD *hydrogenase* is a bifurcating *hydrogenase* of the MvhAGD-HdrABC complex in hydrogenotrophic methanogens (Chap. 10, Fig. 10.20). The binuclear [NiFe] subunit of MvrAGD couples via [Fe-S] clusters to a flavin-containing subunit to bifurcate H_2 into Fd^{2-} and reduced heterodisulfide (CoM-SH + HS-CoB, Fig. 10.10). In hydrogenotrophic methanogens without cytochromes, this *hydrogenase/*

heterosulfide reductase complex couples stoichiometrically the first (*Fd-dependent formylmethanofuran dehydrogenase*) step to the last (*methyl coenzyme M reductase*) in the methanogenesis pathway (see Chap. 10).

H_2-dependent CO_2 reductase in *A. woodii* is a *hydrogenase*, where in H_2 oxidation the electron relay from the binuclear [NiFe] center-containing subunit to the molybdopterin-containing *formate dehydrogenase* subunit proceeds via [4Fe-4S] clusters. This enzyme catalyzes the reversible reduction of CO_2 to formate, which is the first step of the acetogenic Wood–Ljungdahl pathway (see Fig. 10.11).

7.3 Prototypic C–C Cleavage Reactions

The second important set of prototypic reactions in catabolic degradation of organic compounds are **carbon–carbon bond-cleaving reactions**. The most common are the **aldol** and **ketol cleavage** reaction. Both types require both an **alcohol** and a **carbonyl** function, however, in different proximity relative to each other. Equally important but less frequent are de/carbonylation and fumarate addition reactions.

7.3.1 Aldol Cleavage/Condensation

The structural requirement for an organic molecule to undergo enzymatic aldol C–C cleavage is that of a hydroxy group in 3-position to a functional carbonyl (Fig. 7.8a). The carbon in 2-position is activated by the carbonyl (Fig. 7.8a).

For aldolytic cleavage, an electron-withdrawing group at the carbon in 3-position relative to a carbonyl is essential. The cleavage products is a new aldehyde where this aldehyde carbon is the former C3 carbon, and an enolate anion that becomes protonated (Fig. 7.8a). ***Aldolases*** catalyze this reversible reaction and the corresponding formation of a C–C bond by "head-to-tail" condensation of an aldehyde and an enolate.

In fatty acids, the carboxyl carbon does not have a carbocation character and, therefore, does not induce a partially positively charged 3-C carbon necessary for aldol cleavage. Therefore, forming a thioester with coenzyme A converts the unreactive carboxyl carbon of a fatty acid into a functional, "activated" carbonyl carbon (Fig. 7.8b). A fatty acid, when activated as acyl-CoA, contains a partially positively charged carboxy carbon and a partially negatively charged carbon in the 2-position. If the carbon in 3-position contains a OH group as substituent, the C–C bond between the 2 and 3 carbon can be cleaved aldolytically. If the carbon in 3-position contains a keto group as substituent, the C–C bond between the 2 and 3 carbon can be reversibly cleaved as well, when assisted by a nucleophile such as the thiolate of a free coenzyme A: acetoacetyl-CoA is thiolytically cleaved with HS-CoA by *β-ketothiolase* to 2 acetyl-CoA. In the reverse direction, *β-ketothiolase* condenses two acetyl-CoA to the C4 acetoacetyl-CoA. Making use of a CoA thioester to do carbonyl chemistry is a cell's version of avoiding cellular "toxic aldehydes" because thioesters are enzymatically formed and short-lived because of rapid subsequent metabolism.

The structural feature of a hydroxy group in 3-position to a carbonyl is present in many carbohydrates, when not in the hemiacetal form, such as in trioses, 1,6-bisphosphorylated hexoses, or 6-phosphorylated uronic acids. Thus, prototypic aldol cleavage/condensations are involved in many carbohydrate degradative pathways where *aldolases* are key enzymes, such as in Embden–Meyerhof pathway (***fructose-1,6-bisphosphate aldolase***) or the Entner–Doudoroff pathway (***2-keto-3-deoxy-6-phosphogluconate aldolase***), as well as in the (oxidative) pentose phosphate cycle (***transaldolase***) (Fig. 7.9).

In ***citrate synthase*** of the citric acid cycle (Fig. 7.9), the acetyl-CoA enolate anion condenses with the ketone of oxaloacetate to form citrate, releasing HS-CoA. The physiological consequence is that in fast-growing microbes employing ***citrate synthase***, acetyl-CoA is irreversibly committed into the oxidative citric acid cycle (i.e., into catabolism) and away from

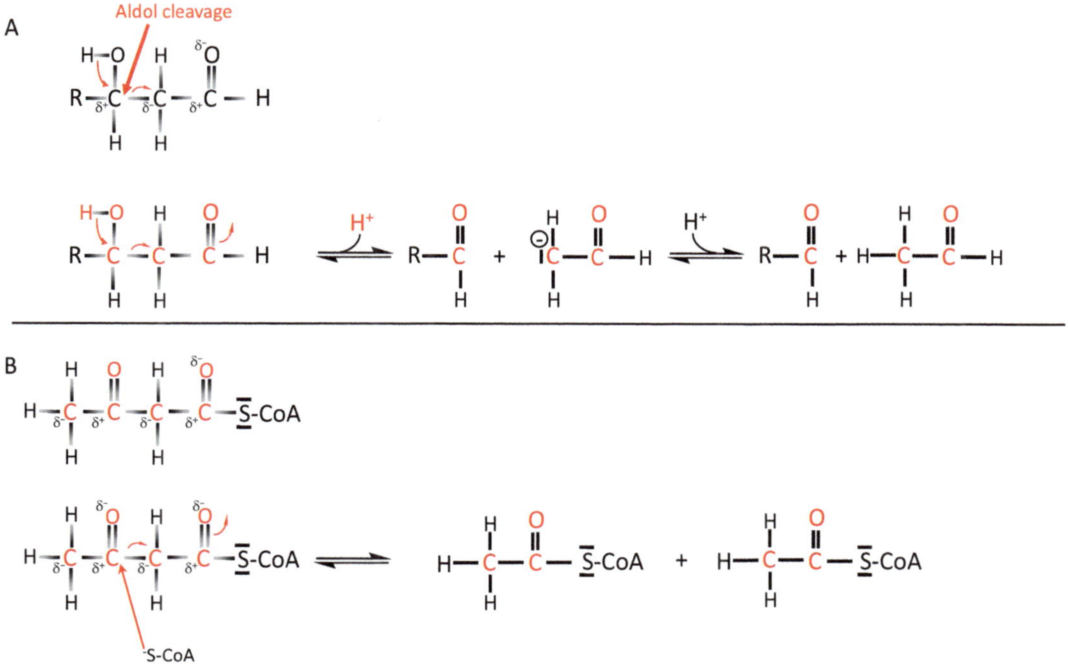

Fig. 7.8 Prototypic aldol cleavage/condensation. (**a**) Aldol cleavage/condensation of a substrate with the OH group in 3 position relative to the carbonyl group. The carbonyl group induces a partially positively charged carbon in β (3-) position. Darker regions of covalent bond indicate higher electron density, here at the C2, while lighter grey indicates less electron density, here C3. The hydroxy oxygen of the 3-carbon assists in C–C cleavage by forming a double bond. The cleavage products are a new aldehyde and an enolate anion, which becomes protonated. In the reverse reaction, the carbonyl group forms an enolate anion that performs a nucleophilic attack on the partially positively charged aldehyde carbon, forming a new C–C bond. (**b**) Aldol cleavage/condensation of a thioester as in acetoacetyl-CoA. The carboxyl carbon of a thioester, an activated carboxyl group, has similar reactivity characteristics as a carbonyl carbon in (**a**) although the carbon is at a different oxidation state. This allows to perform carbonyl chemistry with a (n activated) carboxyl group. The thioester group of acetoacetyl-CoA similarly induces a partially positively charged carbon in the 3 position. In *β-ketothiolase*, the thiolate of a free coenzyme A assists in cleaving the carbon–carbon bond. In the reverse reaction, the cationic character of the activated carboxy carbon is a good nucleophile. *β-Ketothiolase* catalyzes the reversible condensation of 2 acetyl-CoA to acetoacetyl-CoA plus $^{-}$S-CoA in catabolic butyrate synthesis as well as in fatty acid biosynthesis. See text for details

anabolic pathways (e.g., lipid biosynthesis), because of the hydrolytic loss of the thioester bond of acetyl-CoA. On the other hand, ***citrate lyase*** involved in citrate degradation/fermentation in *E. coli* and *Klebsiella species* catalyzes the aldolytic cleavage of citrate into oxaloacetate and acetate. The α subunit of *citrate lyase* activates citrate with an acetyl-ACP thioester from the previous reaction cycle to citryl-ACP. Citryl-ACP is then cleaved aldolytically at the β subunit to oxaloacetate and acetyl-ACP (Fig. 7.9) which activates the citrate of the next reaction cycle to citryl-ACP. The overall reaction of *citrate lyase* is, therefore, the aldolytic cleavage of citrate to acetate and oxaloacetate.

Related prototypic aldolytic reactions in catabolic pathways are catalyzed by *isocitrate lyase* in the glyoxylate shunt in many microbes, *methylcitrate lyase* in propionate degradation by *E. coli* sp. and *Salmonella* sp., and *citramalate lyase* in glutamate fermentation in several *Clostridium* species. Notably, in the thermophilic *Desulfurella acetivorans* and *Thermosulfidibacter takaii* a *citrate synthase* is

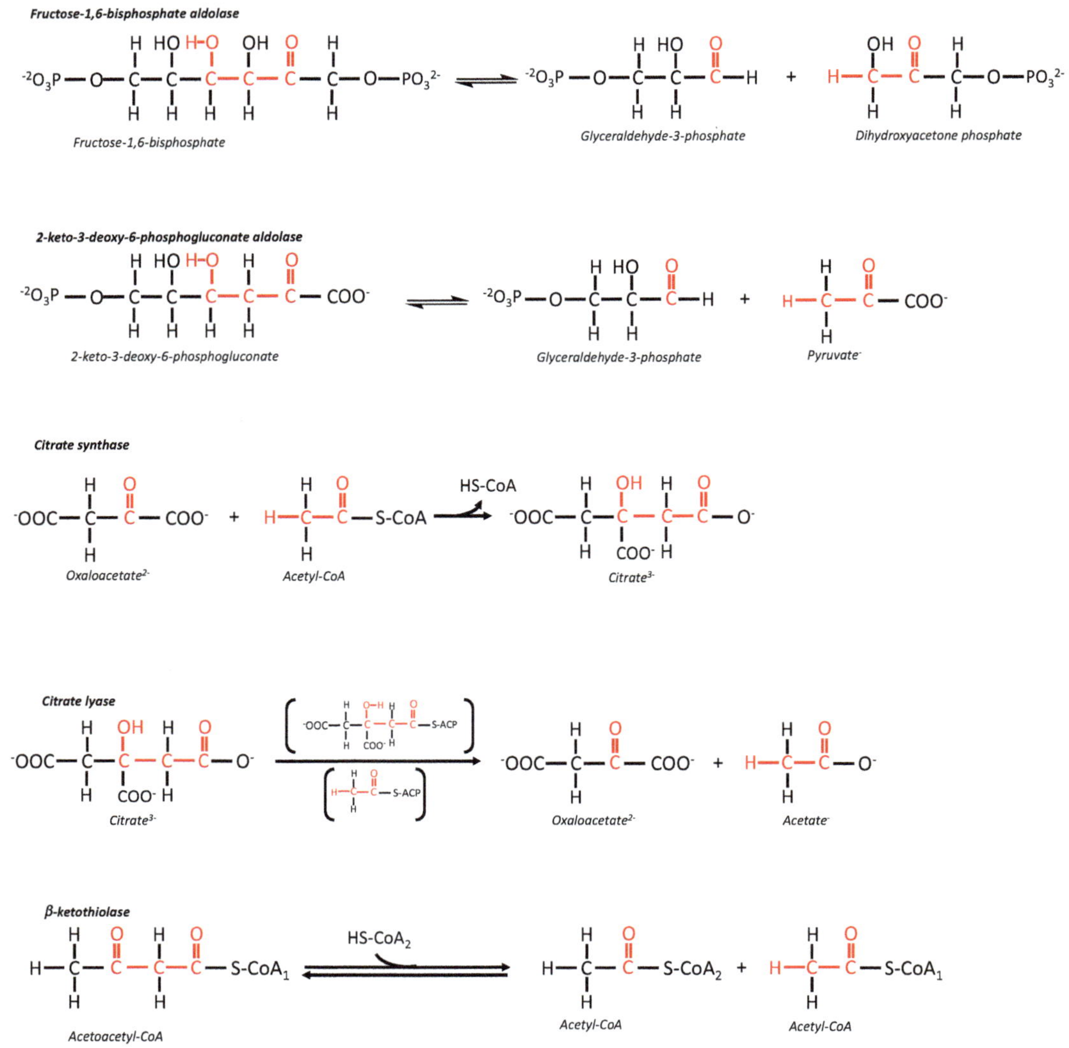

Fig. 7.9 Examples of important aldolytic cleavage reactions in diverse catabolic pathways. Highlighted in red are the structural requirements for aldol cleavage. See text for details

involved both in acetyl-CoA oxidation via the oxidative and in CO_2 fixation via reductive TCA cycle. In anaerobic acetate oxidation via the TCA cycle by *Desulfobacter postgatei,* a catabolic *ATP citrate lyase* is employed, which couples the condensation of oxaloacetate with acetyl-CoA to citrate and ATP. This enzyme enables energy conservation via substrate-level phosphorylation in acetate oxidation.

Finally, *β-ketothiolase* catalyzes the reversible cleavage of an 3-oxo acyl-CoA (acetoacetyl-CoA) in β-oxidation of fatty acids into two CoA thioesters (2 acetyl-CoA). Mechanistically, this *β-ketothiolase* reaction can be understood as a thiolate-assisted aldolytic cleavage (Figs. 7.8 and 7.9).

7.3.2 Ketol Cleavage

Ketolases catalyze the prototypic, ketol-cleavage of C-C bonds in diverse catabolic pathways. The structural requirement for an organic molecule to undergo a ketol cleavage is that of a hydroxy group in 2-position to a functional carbonyl (Fig. 7.10a). An important variation are enzymes

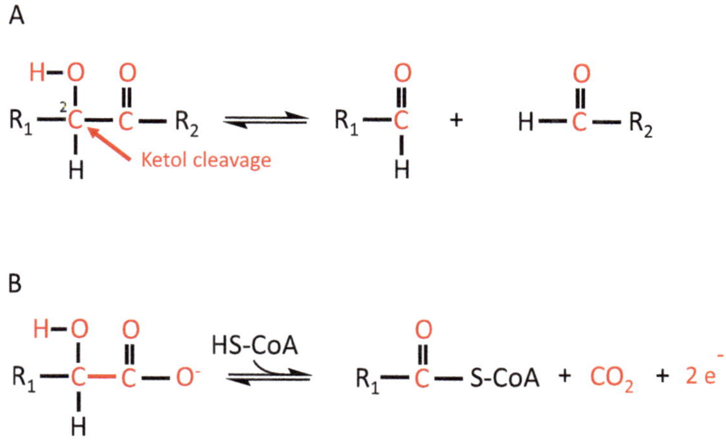

Fig. 7.10 Common catabolic ketol cleavage reactions. (**a**) Ketolytic cleavage of a 2-hydroxy-carbonyl into two aldehydes; (**b**) Ketolytic cleavage of a 2-keto carbonic acid coupled to subsequent oxidation of the carbonyl carbon (oxidative decarboxylation). Relevant bonding electron pairs are shown in red. In (**b**) the bonding electron pair is oxidatively eliminated from the cleaved product

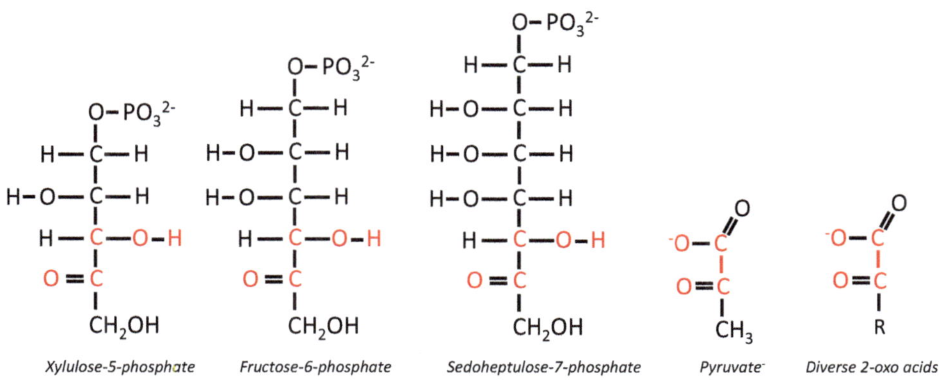

Fig. 7.11 Common substrates of catabolic *ketolases*. The structural features necessary for ketol cleavage are highlighted in red. Pyruvate and other 2-oxo acids can be oxidatively decarboxylation (electrons removed in red). See text for details

that catalyze a ketolytic C–C cleavage in conjunction with a net oxidation of the ketone carbon (= oxidative decarboxylation) such as by *2-oxo acid dehydrogenases* (e.g., *pyruvate dehydrogenase* or *2-oxo-glutarate dehydrogenase*) in aerobic degradation of organic matter (e.g., carbohydrates, lipids, and amino acids), or *2-oxo acid:ferredoxin oxidoreductases* (e.g., as *pyruvate:ferredoxin oxidoreductase*) in many anaerobic pathways of carbohydrate and amino acid fermentations (Figs. 7.10b and 7.11).

All *ketolases* contain thiamine pyrophosphate (TPP) as prosthetic group (Fig. 7.12). Of catalytic importance in TPP is the thiazole ring which is deprotonated in the active enzyme as a carbanion (TPP carbanion, Fig. 7.12b). In the catalytic cycle, the nucleophilic thiazole carbanion attacks the electrophilic carbonyl carbon of the substrate molecule, forming transiently a new C–C bond between the substrate and the prosthetic group (Fig. 7.12b). This thiazole-substrate intermediate is rearranged to an enamine with concomitant release of the first product molecule. Further transformation of the enamine is tightly controlled by the specific active site in diverse *ketolases*. The actual ketolytic cleavage of the

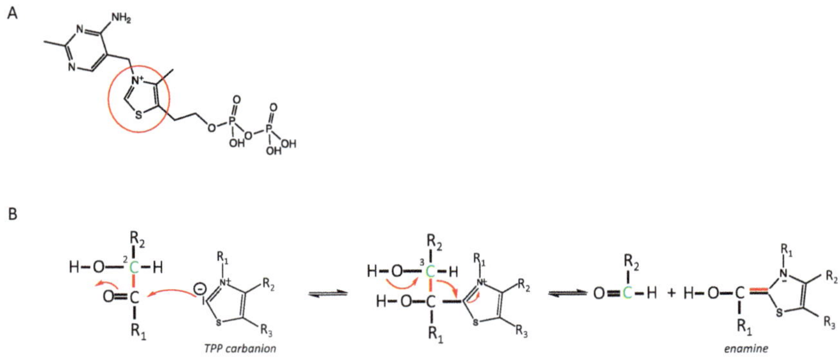

Fig. 7.12 Thiamine pyrophosphate (TPP) prosthetic group (**a**) and simplified scheme of the initial steps of a *ketolase* reaction mechanism. (**b**) The TPP carbanion (red circled in **a**) performs a nucleophilic attack on the carbonyl carbon and forms a substrate-TPP intermediate. The 2 carbon of the substrate molecule is now in 3 position relative to the iminium nitrogen. The substrate molecule is cleaved, similarly to an aldol cleavage; an aldehyde is released and a TTP-bound enamine is formed. The resolution of this enamine can proceed via different paths (Fig. 7.13). See text for details

C–C bond is reversible, while the reversibility of the overall reaction depends on the fate of the enamine.

Figure 7.13 shows the resolutions of the enamine-TPP in 5 different *ketolases*. It outlines how a single enzyme-substrate intermediate (here the enamine-TPP) can have multiple fates, depending on the active site environment of a specific enzyme. In *transketolases*, xylulose-5-phosphate is cleaved into glyceraldehyde-3-phosphate (Fig. 7.13) which leaves the enzyme. Ribose-5-phosphate enters the active site and the enamine attacks the ribose aldehyde to form a C_7 carbohydrate-TPP adduct, which resolves into sedoheptulase-7-phosphate plus the thiazole carbanion. The overall consequence is that a C_2 (hydroxyethyl) group is transferred from a C_5 ketose (xylulose-5-phosphate) to a C_5 aldose (ribulose-5-phosphate) releasing a C_7 ketose and a C_3 aldose:

$$2 \times C_5 \rightarrow 1\ C_7 + 1\ C_3.$$

This *transketolase* is a key enzyme in the "Bifidobacterium shunt" in *Bifidobacterium sp.* fermenting glucose to lactate and acetate, in the oxidative pentose phosphate cycle, and in the Archaea *Pyrococccus horikoshii* and *Thermococcus kodakarensis* in aerobic degradation of glucose, as well in anabolic reactions of the pentose phosphate cycle.

Phosphoketolase is a physiologically and biotechnologically important *ketolase* which catalyzes both a ketol cleavage and an intramolecular redox reaction. Substrates for different *phosphoketolases* are xylulose-5-phosphate or fructose-6-phosphate that are phosphorylytically cleaved into glyceraldehyde-3-phosphate and acetyl-P, or into erythrose-4-phosphate and acetyl-P, respectively (Fig. 7.13). In xylulose-5-phosphate-specific *phosphoketolase*, xylulose-5-phosphate is cleaved into glyceraldehyde-3-phosphate and a dihydroxyethyl-TPP intermediate. The electron pair from the C–C double bond of the dihydroxyethyl-TPP adduct is not directed toward a second substrate, as in the above-introduced *transketolases*, but attacks the vicinal carbon, supported by the hydroxy group as a good leaving group. The enol of the adduct resolves into the keto form, which is then released from TPP by the attack of a hydroxylate from inorganic phosphate (Fig. 7.13). The product released is acetyl-P, carrying an energy-rich bond. The physiologically important consequence of *phosphoketolase* is that **an energy-rich bond is formed without the net release of reducing equivalents**. In glyceraldehyde-3-phosphate oxidation, in contrast, the formation of the energy-

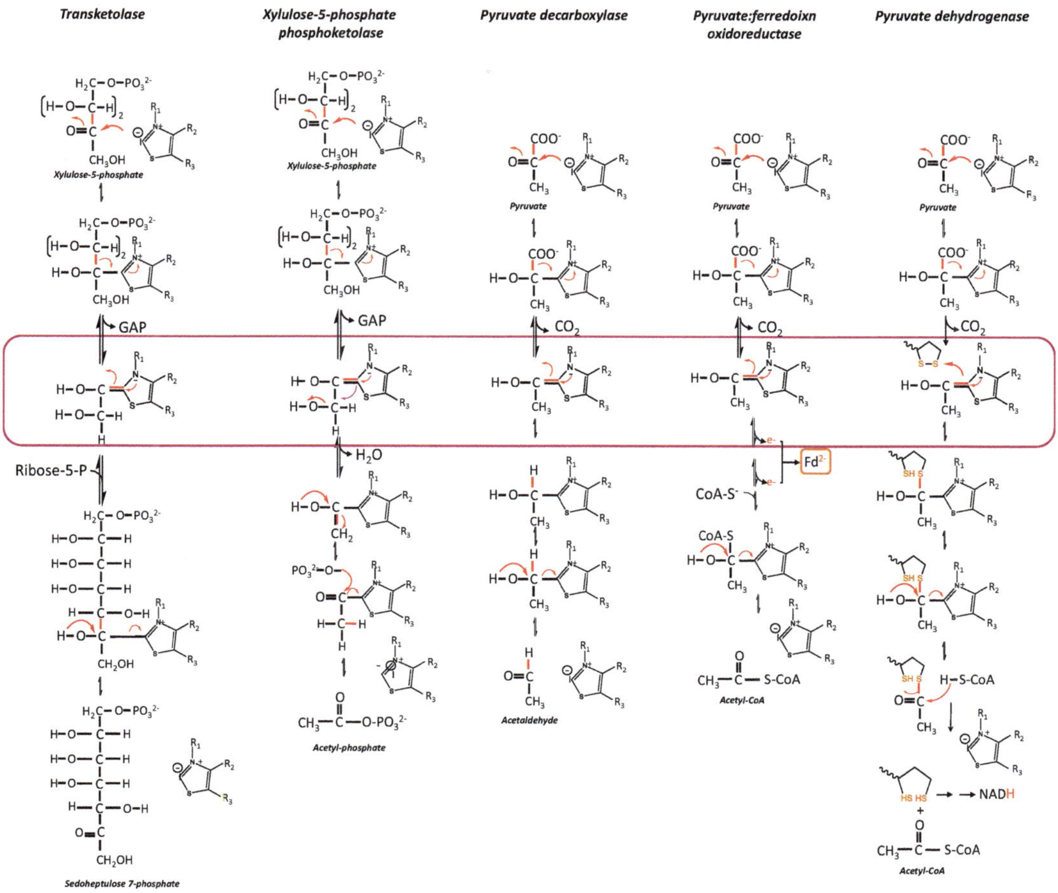

Fig. 7.13 Major catabolic ketol-cleaving enzymes. All enzymes have a common enamine-TPP intermediate (bracketed in magenta) that is differently resolved in the different enzymes: *transketolase*, *xylulose phosphoketolase*, *pyruvate decarboxylase*, *pyruvate:ferredoxin oxidoreductase*, *pyruvate dehydrogenase*. See text for details

rich bond of 1,3-bisphosphoglycerate is stoichiometrically coupled to NADH formation. Thus, for carbohydrate-fermenting microbes, which derive their cellular energy via substrate-level phosphorylation, the involvement of *phosphoketolase* is advantageous, because no substrate resources are diverted to form an electron acceptor (see Sects. 9.3.2 and 9.3.4). Furthermore, no enzyme for a reductive pathway needs to be expressed; thereby, enabling a relatively small catabolic proteome.

Pyruvate decarboxylase is an example for addition of a H^+ to the enamine-TPP intermediate (Fig. 7.13). *Pyruvate decarboxylase* reacts with pyruvate to form a lactyl-TPP intermediate that rearranges and irreversibly decarboxylates to form a hydroxyethyl-TPP adduct plus CO_2 (Fig. 7.13). Hydroxyethyl-TPP rearranges to release acetaldehyde and the thiazole carbanion. *Pyruvate decarboxylase* is the key enzyme in ethanol-forming fermentation from glucose in many microorganisms, where it is the cause for the loss of 1/3 of the glucose carbon as CO_2 in this fermentation. Furthermore, the presence and activity of *pyruvate decarboxylase* commits pyruvate irreversibly to acetaldehyde and, there by, into ethanol formation.

Pyruvate:ferredoxin oxidoreductase and ***NAD⁺-dependent pyruvate dehydrogenase***. A variation of *ketolases* is oxidative decarboxylation involving both a ketol cleavage and a net oxidation reaction (see Figs. 7.10 and 7.13). In ferredoxin-dependent *2-oxo acid:ferredoxin oxidoreductase* and in NAD^+-dependent *2-oxo acid dehydrogenase*, the substrate carbonyl carbon is oxidized after ketolytic cleavage, and the electrons are transferred to either Fd or NAD^+ (Figs. 7.10 and 7.13), respectively. Both classes of enzymes are coenzyme A-dependent and release a CoA thioester as end product; the mechanisms of oxidation, however, are different. While *2-oxo acid:ferredoxin oxidoreductase* catalyzes two single-electron oxidations of the enamine, the enamine in NAD^+-dependent *2-oxo acid dehydrogenases* attacks a lipoamide, which is a second prosthetic group, and reduces its disulfide during which the hydroxyethyl group is transferred from TPP to lipoamide. The thiolate of a HS-CoA releases the hydroxyethyl group from lipoamide as acetyl-CoA (Fig. 7.13). In *2-oxo acid:ferredoxin oxidoreductase*, the oxidized hydroxyethyl group is thiolytically ($^-$S-CoA) released from TPP, forming acetyl-CoA and Fd^{2-}. Since the electrons from enamine oxidation ($E^{0\prime} = -500$ mV) in NAD^+-dependent *2-oxo acid dehydrogenases* are transferred to NAD^+ ($E^{0\prime} = -320$ mV), a *2-oxo-acid dehydrogenase* catalyzed reaction is physiologically irreversible.

Relevant *2-oxo acid:ferredoxin oxidoreductases* include *pyruvate:ferredoxin oxidoreductase*, *2-oxo-glutarate:ferredoxin oxidoreductase*, *2-oxo-valerate:ferredoxin oxidoreductase* and *indolpyruvate:ferredoxin oxidoreductase* in pyruvate and amino acid fermentations. Numerous intestinal pathogens, such as *Clostridium difficile* use *2-oxo-glutarate:ferredoxin oxidoreductases* in Stickland fermentation in pathogenicity-induced amino acid fermentation (Fig. 9.20).

An interesting TPP-dependent enzyme is *oxalate:ferredoxin oxidoreductase* in oxalate catabolism of *Moorella thermoacetica*, where this CoA-independent enzyme catalyzes the disproportionation of oxalate into 2 CO_2 and Fd^{2-} (Fig. 7.14).

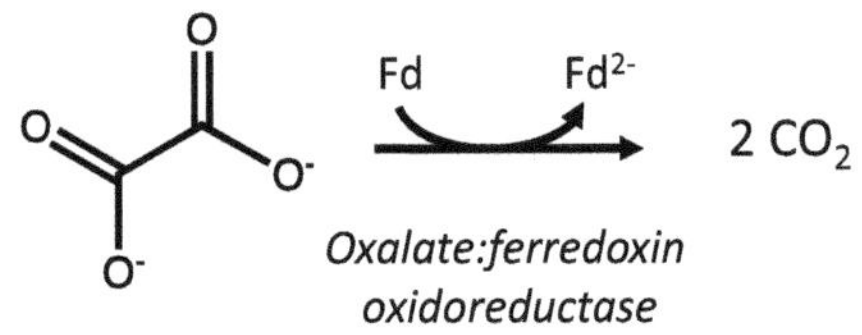

Fig. 7.14 *Oxalate:ferredoxin oxidoreductase* in *Moorella thermoacetica* catalyzes the ketolytic cleavage and oxidation of oxalate into 2 CO_2 and Fd^{2-}

7.3.3 Major Catabolic Pyruvate-Transforming Enzymes

Pyruvate is a key intermediate and metabolic node in numerous oxidative pathways of chemoorganotrophic microbes, in particular in carbohydrate catabolism. The most frequently found pyruvate-transforming, non-reducing enzymes are summarized in Fig. 7.15. With the exception of *pyruvate:formate lyase*, all pyruvate-transforming enzymes are TPP-dependent. Each enzyme class has a distinctive metabolic and ecophysiological consequence to the microorganism. Three of the six convert pyruvate into acetyl-CoA and one into acetyl-phosphate, indicating their involvement in energy conservation.

Pyruvate dehydrogenase is NAD^+-dependent (Fig. 7.15) and catalyzes pyruvate oxidation to acetyl-CoA plus CO_2. This oxidative decarboxylation is thermodynamically irreversible as the hydride from pyruvate oxidation ($E^{0\prime} = -500$ mV) is transferred to the more positive electron acceptor NAD^+ ($E^{0\prime} = -320$ mV). *Pyruvate dehydrogenase* contains TPP and is present in aerobic and some anaerobic microbes, where NADH is re-oxidized via a respiratory chain. Thus, the release of redox energy as heat in the *pyruvate dehydrogenase* reaction funnels pyruvate irreversibly into acetyl-CoA-dependent pathways.

Pyruvate:ferredoxin oxidoreductase (Fig. 7.15) catalyzes the same reaction as above, except that electrons are transferred to ferredoxin ($E' = -500$ mV), which makes this reaction reversible. In *pyruvate:ferredoxin oxidoreductase*-containing microbes, Fd^{2-} is often

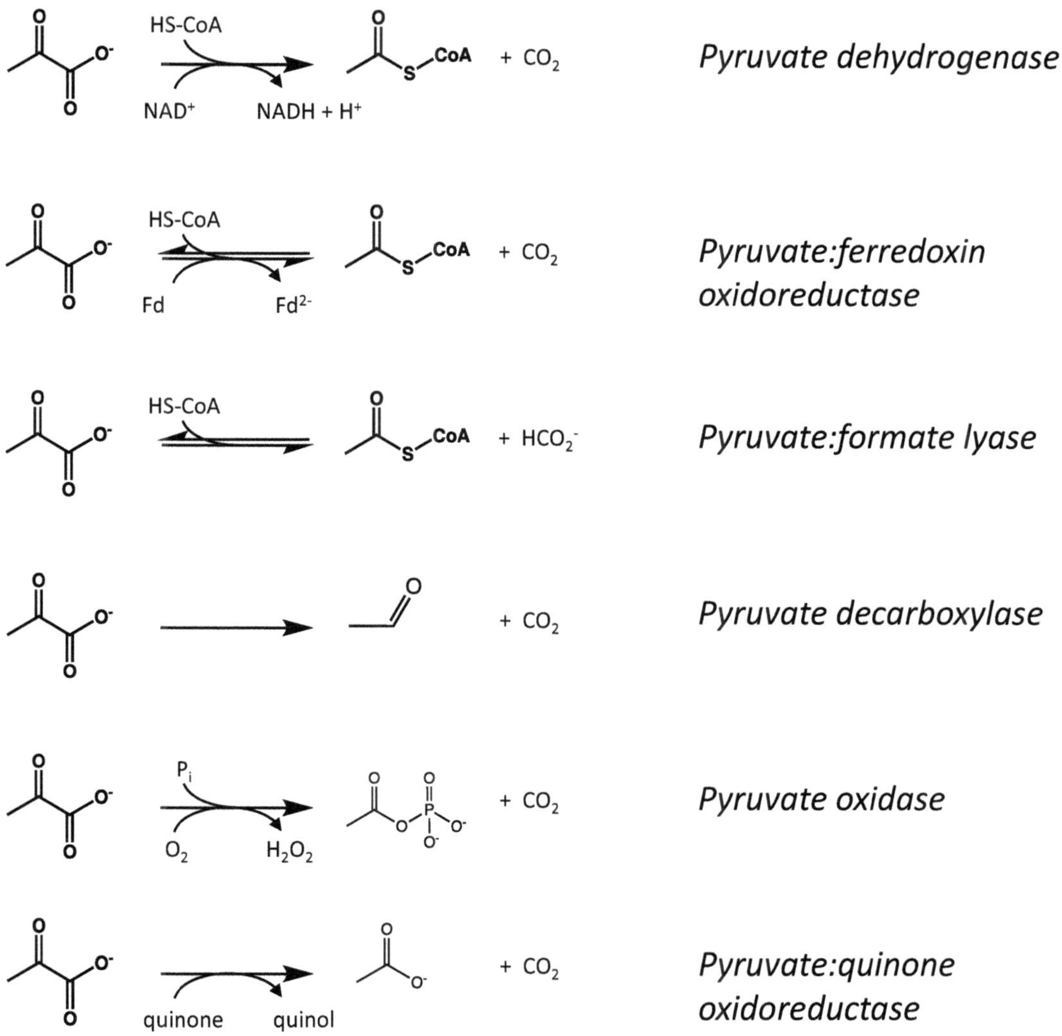

Fig. 7.15 Important pyruvate-transforming enzymes involved in microbial catabolism. See text for details

re-oxidized by a *Fd-dependent hydrogenase*, including Ech *hydrogenase*, to form H_2. The combination of these two enzymes enables cells to conserve the energy-rich bond of acetyl-CoA as ATP while the electrons "escape" cellular redox balancing as gaseous H_2. *Pyruvate:ferredoxin oxidoreductase* contains TPP and is found, among others, in strictly anaerobic fermenting microbes, such as *Clostridia* that ferment carbohydrates, amino acids, as well as nucleic acids. *Pyruvate:ferredoxin oxidoreductase* is oxygen-sensitive and, therefore, incompatible with an aerobic catabolism. *Pyruvate synthase* catalyzes the same reaction, however, functions in anabolism of anaerobes growing on C_2 compounds. In genome annotations, both names are often used synonymously without distinction between involvement in catabolism or anabolism.

Pyruvate:formate lyase (Fig. 7.15) does not contain TPP, and no cellular electrons are released in this pyruvate metabolism to acetyl-CoA and formate. Rather, the electrons from pyruvate remain associated with the C1 carbon as formate, in contrast to CO_2. *Pyruvate:formate lyase* is a reversible radical enzyme that is irreversibly inactivated by molecular oxygen.

Although not acetyl-CoA forming, *pyruvate decarboxylase* (Fig. 7.15) catalyzes the simple decarboxylation of pyruvate to acetaldehyde. The TPP-containing enzyme is found in mainly anaerobic ethanol-forming microbes, where cytotoxic, electrophilic acetaldehyde is eliminated by a rapid NADH-dependent *ethanol dehydrogenase* as ethanol.

Pyruvate oxidase (Fig. 7.15) catalyzes the O_2-dependent oxidation of pyruvate to acetyl-phosphate, CO_2, and H_2O_2. This TPP-containing flavoenzyme, as found in *E. coli, Lactobacillus plantarum,* and *L. delbrückii,* functions in the context of fermentation of substrates such as carbohydrates in the presence of (mostly limited) O_2. Such direct O_2 reduction by *pyruvate oxidase* without respiratory energy conservation is of advantage in these microbes (1) because electrons recovered in the oxidative (mostly glycolytic) pathway do not need to be balanced by the use of substrate-derived intermediates and (2) because this "metabolic shortcut" releases as heat – 254 kJ/mol ($\Delta E = ((+820\ \text{mV}) - (-500\ \text{mV}))$, and, therefore, enables these microbes to express a rate phenotype. Moreover, because "disposal" of electrons by *pyruvate oxidase* requires less enzymatic steps, a smaller catabolic proteome can enable a high flux of fermentative ATP synthesis. *Pyruvate oxidase* in eukaryotes forms acetyl-CoA rather than acetyl-phosphate.

Pyruvate:quinone oxidoreductase (Fig. 7.15) in *E. coli* and *Corynebacterium glutamicum* is also a TPP-containing flavoprotein and catalyzes the non-acetylating pyruvate oxidation to acetate, CO_2, and reduced quinone, which is ubiquinone in *E. coli* and most likely menaquinone in *C. glutamicum.*

7.3.4 Comparison Between Aldol and Ketol Cleavage

Although the structural configuration of functional groups in a substrate required for ketol cleavage is different from that of an aldol cleavage (aldol cleavage: hydroxy group in 3-position to carbonyl; ketol cleavage: hydroxy group in 2-position of carbonyl), the (bio)chemistries of C–C cleavage are, in a sense, quite similar. In *aldolase*, the C–C cleavage is initiated by the strongly polarized bond of the carbonyl group and supported by the hydroxy group at the β carbon forming a C=O double bond. In *ketolase*, the iminium ion of TPP functionally substitutes for that carbonyl group in aldolytically cleaved substrates: the original C2 carbon is now in 3-position to the iminium ion in the substrate-TPP adduct (Fig. 7.16).

Aldolases and *ketolases* with substrate specificities to diverse organic compounds are present in many degradation pathways, including carbohydrates, amino acids, nucleic acids, and of aromatic and aliphatic compounds. Common end products of such cleavage reactions are acetate, acetyl-CoA, and acetyl-phosphate, as well as pyruvate.

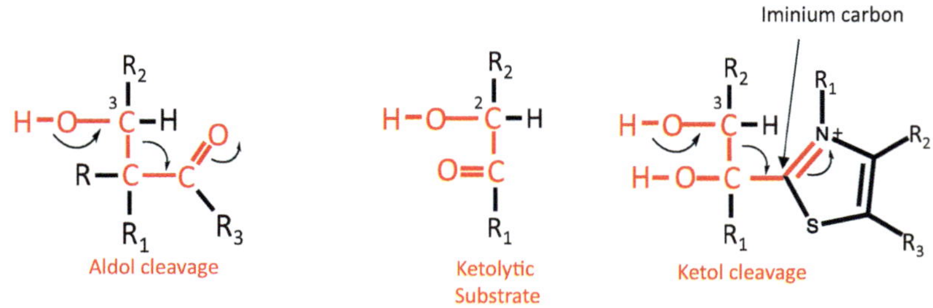

Fig. 7.16 Comparision between aldolytic and ketolytic enzymatic C–C cleavage. The key feature of aldolytic cleavage (left) is the electron-withdrawing property of the carbonyl function. This is functionally replaced in ketol cleavage (right) by the iminium in TPP, which places the carbon of the 2-position in the substrate molecule in 3-position to this iminium group. See text for details

7.3.5 Radical Enzyme-Dependent Decarboxylations

Some aromatic amino acid-fermenting microbes, such as *Clostridium difficile*, *Clostridium aerofoetidum*, or *Clostridium scatologenes* contain *decarboxylases* that catalyze the chemically challenging decarboxylation of phenylacetate, *p*-hydroxyphenylacetate, or indolacetate (Fig. 7.17). These compounds are the end products of the oxidative branch of tyrosine, phenylalanine, or tryptophane fermentation (see Sect. 9.5.3). Phenylacetate, *p*-hydroxyphenylacetate, or indolacetate are further transformed by non-oxidative *decarboxylases* forming toluene, *p*-cresol, or scatole. Although these decarboxylations do not function directly in catabolic pathways or are essential for energy conservation, the consequence of the activity of such glycyl radical-containing *decarboxylases* can be ecologically advantageous. For example, *p*-cresol, the decarboxylation product of *p*-hydroxyphenylacetate in tyrosine fermentation of *C. difficile*, is bacteriostatic on γ-proteobacteria as well as *Bacteroides thetaiotaomicron* and, thereby, disrupts a stable commensal microbiota of the human gut, which enables *C. difficile* to invade the thus-vacated niche and establish a pathogenic colonization. In addition to amino acid-fermenting *Clostridia*, such *decarboxylases* are present also in phylogenetically diverse microbes including *Pseudomonads* and *Lactobacilli*.

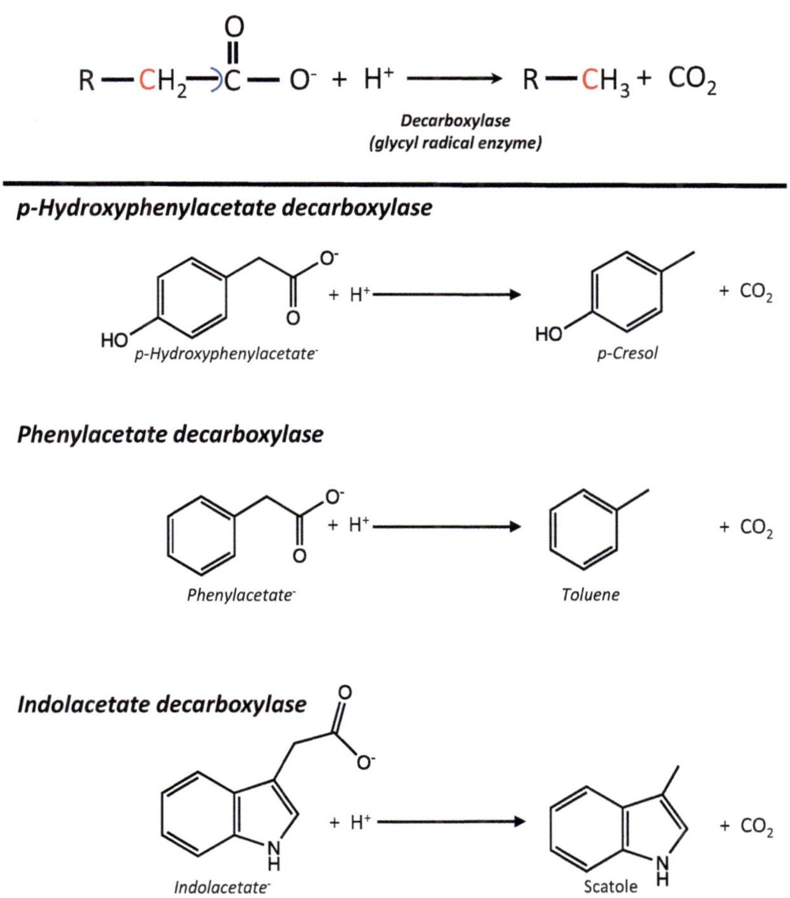

Fig. 7.17 Glycyl radical *decarboxylases*. Top line shows the general reaction, and below are three examples in fermentative amino acid metabolism: *p-hydroxyphenylacetate decarboxylase*, *phenylacetate decarboxylase*, and *indolacetate decarboxylase*. See text for details

7.4 Auxiliary, Extension, and Funneling Reactions

As the above section showed, prototypic reactions require in a substrate molecule the presence of alcohol, aldehyde/ketone, or carboxyl groups and their specific spatial arrangement. Substrate molecules that lack such structural organization are funneled into metabolism by **auxiliary enzymatic reactions** that (1) introduce a functional group, (2) rearrange the carbon–carbon skeleton, or (3) mediate group substitutions. Therefore, auxiliary enzymes extend the substrate range of a core pathway to otherwise structurally inaccessible compounds and enable microbes to metabolically occupy novel niches. Auxiliary enzymatic reactions are more restricted than prototypic reactions and functions only on specific compound classes and under specific environmental conditions, where they can be ubiquitously present nevertheless. The following sections introduce auxiliary reactions that often occur in combinations and function as upstream extensions of core metabolic pathways (Chap. 8).

7.4.1 *Mono-* and *Dioxygenases*

Hydrocarbons (aromatic and aliphatic) do not contain any functional group. There are two general metabolic strategies for activating hydrocarbons; one requiring the presence and the other one the absence of O_2.

Oxygenases catalyze in aerobic microbes the O_2-dependent oxidation of a C–H bond in alkanes and aromatic hydrocarbons (as well as of ammonia) (Fig. 7.18). In *oxygenase* reactions, the hydroxy group introduced into the substrate is derived from molecular oxygen (O_2) and not from H_2O. *Oxygenases* are physiologically irreversible. The stoichiometric dependence on molecular oxygen restricts the presence of *oxygenases* to aerobic microbes with exceptions of those that can form O_2 anaerobically as an intermediate from nitrite (methanotrophic *Planctomyces*, see below) or chlorate (chlorate-respiring bacteria). There are two types of *oxygenases*: ***monooxygenases*** and ***dioxygenases*** (Fig. 7.18). They differ by the number of oxygen atoms they incorporate into a substrate molecule: *monooxygenases* incorporate one atom of oxygen from O_2 into the substrate, while in dioxygenases both oxygen atoms are incorporated at vicinal carbons.

Since molecular oxygen contains two atoms of oxygen and each atom is reduced by one pair of electrons, a total of four, substrate-derived electrons are consumed both in *mono-* and *di-oxygenases*. Consequently, these electrons are unavailable for respiratory energy conservation. In *dioxygenases*, the bonding electron pair from two vicinal C–H bonds reduce one oxygen atom each; in *monooxygenases*, an additional pair of electrons (reducing equivalents) is required per reaction. These electrons are provided via oxidation reactions in the downstream metabolic pathway, and are typically in the form of NADH, although some *monooxygenases* use also NADPH, $FADH_2$, reduced ferredoxin, or rubredoxin as co-substrate, respectively.

Methane is oxidized to methanol by *methane monooxygenase* (MMO) (Fig. 7.18b). MMOs are metalloenzymes and based on the metal cofactor, two genetically unrelated types of MMOs are known: a membrane-(particulate)-associated *methane monooxygenase* (pMMO) which contains Cu at the active site and a Fe-dependent soluble *methane monooxygenase* (sMMO) which is expressed under Cu^{2+}-limitation. pMMO generally has a higher apparent affinity for methane than sMMO and a wider substrate range, including the capacity for co-metabolic transformations (e.g., oxidation of ammonia or halogenated alkenes). However, the catalytic mechanism of sMMO is currently better understood. In sMMO, NAD(P)H are co-substrate, while the physiological source of reductant in pMMO is unclear and can involve a reduced quinone or cytochrome. For both MMOs to function, the reducing equivalents of the MMO reaction are recovered from downstream oxidations of methanol.

The reaction catalyzed by di-iron center-containing sMMO follows a radical mechanism of H atom abstraction from CH_4, which is

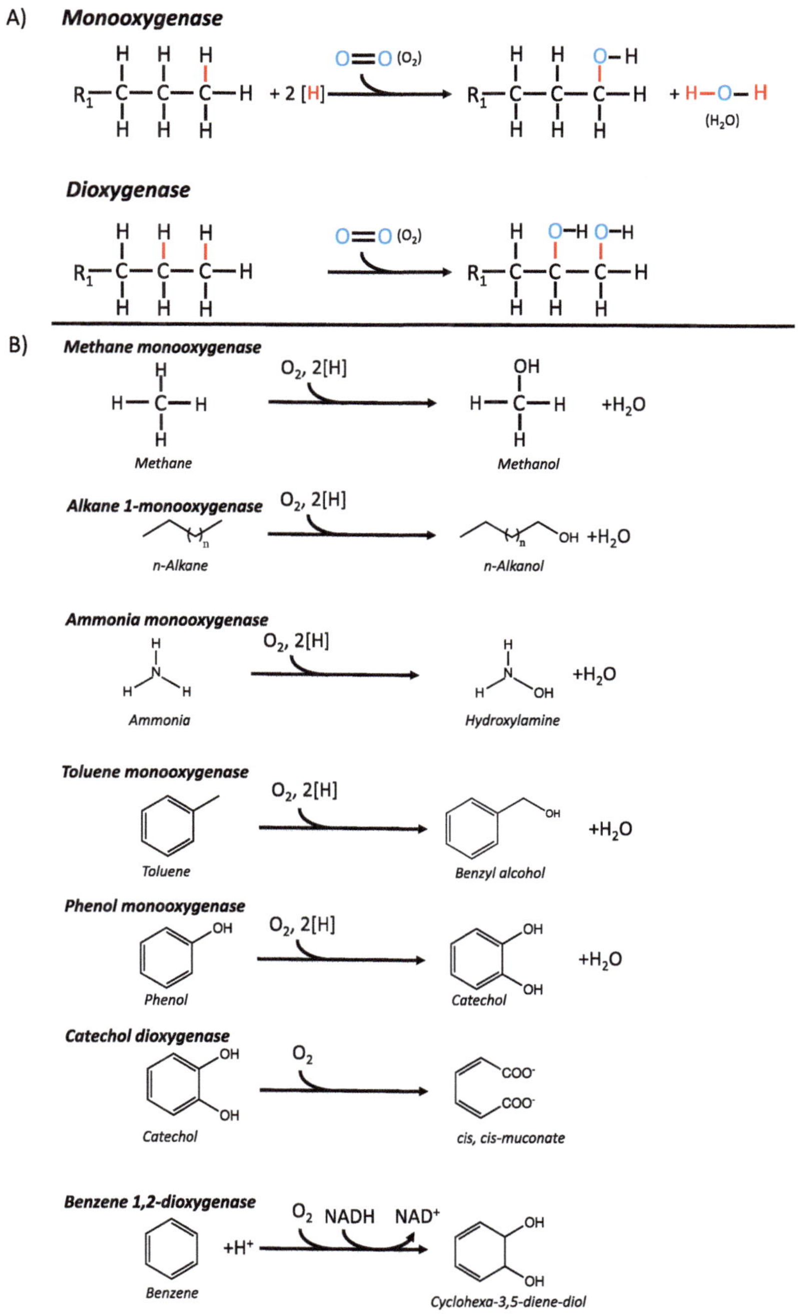

Fig. 7.18 *Mono-* and *di-oxygenases*. The formal fate of electrons being removed from the carbon (oxidized) rather than mechanistic details are highlighted. (**a**) In both *oxygenases*, one atom of molecular oxygen (blue) is incorporated into the substrate leading to an oxidation of the respective carbon (red: bond electron pair). In *monooxygenases*, the second O atom is reduced to H_2O by a pair of electrons from a co-substrate, while in *dioxygenases*, the second O atom oxidizes a C–H bond of a vicinal carbon. (**b**) Examples of common *mono-* and *di-oxygenases*. See text for details

ultimately based on O_2 being a biradical. Upon binding of O_2 to the active site, the two Fe(II) of the di-iron center reduce the O_2 biradical to a peroxide species. Further oxidation of 2 Fe(III) by the Fe-coordinated superoxide and rearrangement into two μ oxo bridges yields 2 Fe(IV). A substrate H atom is abstracted by the μ-oxo-Fe (IV) complex to form the bridging OH ligands, which reacts with the methyl radical to methanol and two Fe(III). The latter Fe(III) is reduced back to Fe(II) by the two electrons of the 2[H] co-substrate.

7.4.2 Fumarate Additions

The equivalent to an *oxygenase* reaction in anaerobic microbes metabolizing aromatic and aliphatic hydrocarbons are **fumarate addition reactions**, catalyzed by ***alkyl–/benylsuccinate synthases*** (Fig. 7.19). *Alkylsuccinate synthase* and *benzylsuccinate synthase* catalyze radical reactions and belong to the class of glycyl radical enzymes (GRE). In these reactions, a hydrogen atom is abstracted from the subterminal carbon of an aliphatic hydrocarbon or from the methyl carbon of methyl-substituted benzenes generating an *n*-alkyl or a benzyl radical, respectively (Fig. 7.19a). The organic substrate radical adds to the double bond of fumarate forming an alkylsuccinyl or benzylsuccinyl radical. The abstracted hydrogen atom is transferred back to these radical intermediates to form an alkylsuccinate and benzylsuccinate. In both products, the previously unreactive carbon (i.e., the subterminal carbon of an aliphatic hydrocarbon or the methyl group of a methyl-substituted benzene) is placed in β (3-) position to a fumarate-derived carboxyl group (Fig. 7.19b). Upon activation to an acyl-CoA derivative, alkylsuccinyl-CoA and benzylsuccinyl-CoA are oxidized via the previously discussed prototype β-oxidation reactions (see Sects. 7.2 and 8.3.2). Catabolic pathways involving fumarate addition reactions are inherently cyclic pathways as the initial

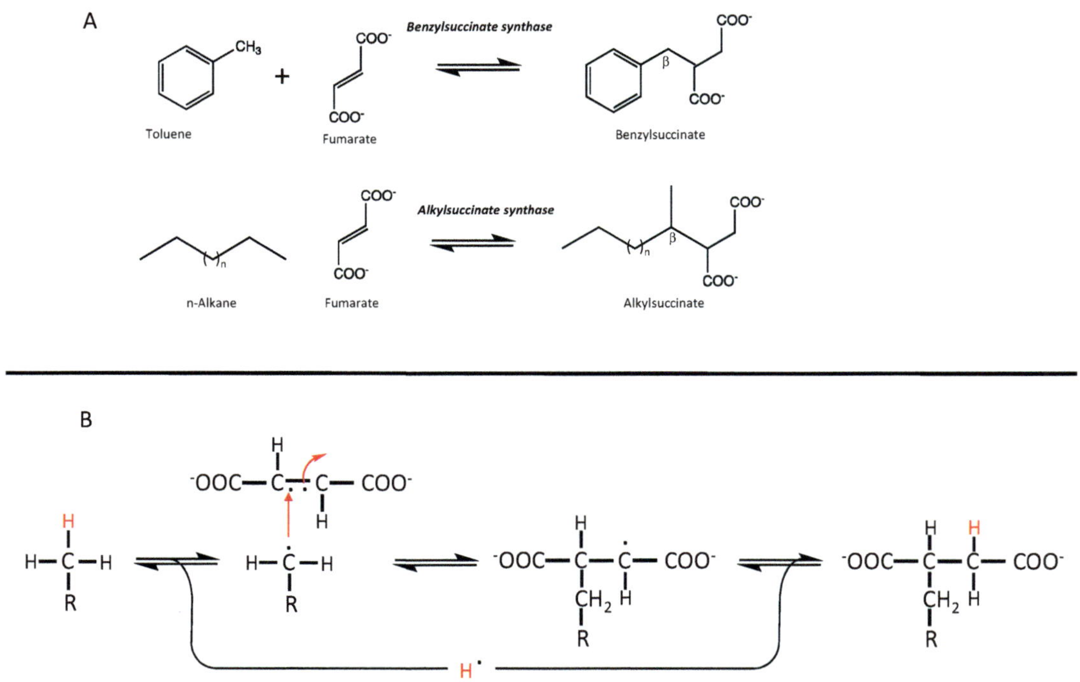

Fig. 7.19 Radical fumarate addition reactions in anaerobic degradation of hydrocarbons. (**a**) *Benzylsuccinate synthase* catalyzing the activation of methyl-substituted benzenes forming benzylsuccinate, and *alkylsuccinate synthase* catalyzing the activation of the subterminal carbon of an *n*-alkane. The metabolic strategy is that an unactivated carbon atom is now in β position to a carboxyl group and, therefore, activated once the CoA thioester is formed. (**b**) Simplified mechanism of methyl-substituted benzene activation by *benzylsuccinate synthase*: a hydrogen atom is abstracted from the hydrocarbon substrate, generating a substrate radical, which adds to the double bond of fumarate. The H· reacts with a glycyl radical at the enzyme. The hydrocarbon-succinyl radical accepts the H atom previously abstracted from the hydrocarbon substrate to form a C4 dicarboxylate adduct

acceptor molecule fumarate is regenerated in further downstream degradation reactions.

7.4.3 Benzoyl-CoA Reduction

Benzoyl-CoA reductase is a funneling enzyme in anaerobic degradation of diverse aromatic compounds (Fig. 7.20). *Benzoyl-CoA reductase* catalyzes the reduction of benzoyl-CoA, which is a central intermediate in anaerobic degradation of aromatic compounds, to cyclo-1,5-diene-1-carbonyl-CoA. For this reductive dearomatization, very low redox potential electrons ($E^{0\prime} < -$ 620 mV) are required to overcome the high resonance energy of the delocalized electrons of the aromatic ring of benzoyl-CoA. In *benzoyl-CoA reductase*, Fd^{2-}-derived electrons are activated by further lowering their redox potential. The two known classes of *benzoyl-CoA reductase* both use Fd^{2-} as cellular electron donor; the two classes differ in how the low-potential electrons are mechanistically activated from Fd^{2-} (Fig. 7.20). Class I *benzoyl-CoA reductases* (Bcr) use Fd^{2-} and hydrolyze 2 ATP, one ATP per e^- in Fd^{2-}, for activation, whereas class II *benzoyl-CoA reductases* (Bam) use a HdrA-type flavin-based electron bifurcation (Sect. 6.5.1, Fig. 7.20). Current evidence suggests that class II enzymes mediate a double electron bifurcation: in the first bifurcation, electrons from Fd^{2-} are bifurcated to low-potential electrons at ~ − 620 mV and a [4Fe-4S] cluster in BamDEF (~ − 200 mV). Electrons from this latter [Fe-S] cluster are further bifurcated to NADH/NADPH and a metabolic high potential electron acceptor such as menaquinol ($E^{0\prime}$ = − 74 mV) in *Geobacter metallireducens* or adenosinephosphosulfate (− 60 mV) or sulfite (− 110 mV) in *Desulfosarcina cetonica.*

These molecular differences in benzoyl-CoA activation by class I and class II BCR can explain why the more energy-intensive class I BCR is abundant in facultative anaerobic bacteria, while class II BCRs occur in strictly anaerobic sulfate-, metal-oxide-reducing or syntrophic bacteria, which are more energy-limited.

7.4.4 Carbon–Carbon Rearrangements

The most prevalent carbon-skeleton rearrangement reaction in anaerobic prokaryotes is the exchange of a H substituent with a C substituent between two vicinal carbon atoms via a substrate-derived radical as intermediate (Fig. 7.21). All carbon-skeleton rearranging ***mutases*** (Fig. 7.22),

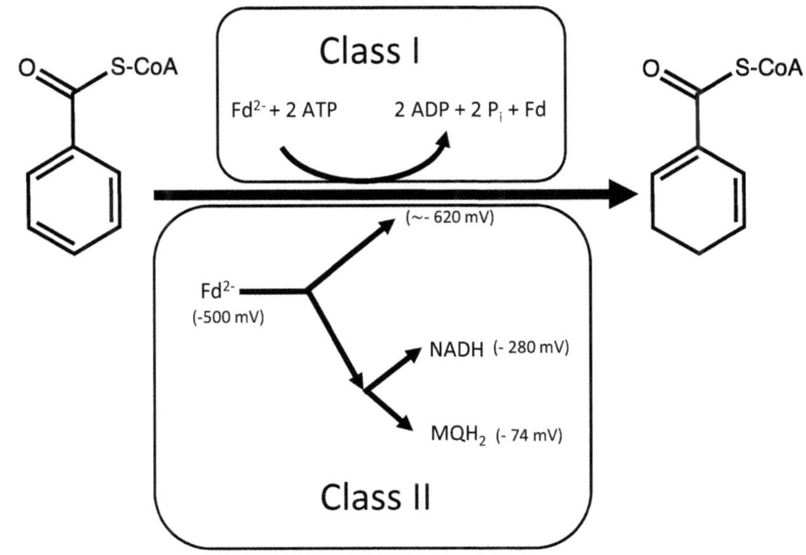

Fig. 7.20 Dearomatizing *benzoyl-CoA reductase* in anaerobic prokaryotes. Benzoyl-CoA is reduced via a radical ketyl to cyclo-1,5-diene-1-carbonyl-CoA. Class I (Bcr) and class II (Bam) are phylogenetically and mechanistically distinct. Class I enzymes use ATP-dependent electron activation; class II are proposed to use a double electron bifurcation as shown here for *Geobacter metallireducens* Bam. See text for details

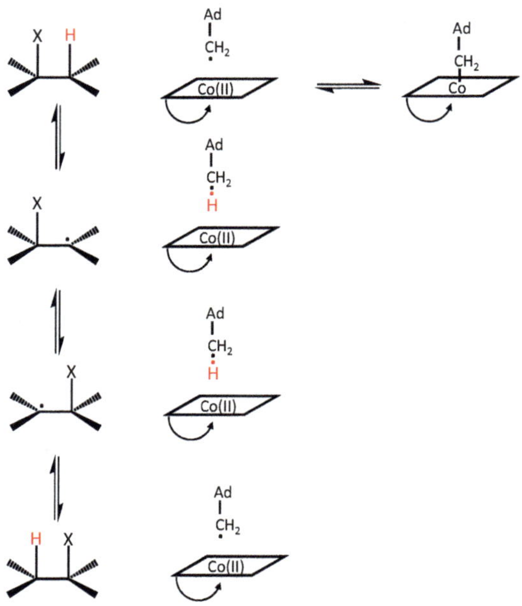

Fig. 7.21 General scheme of coenzyme B_{12}-dependent carbon rearrangements by *mutases* via radical-induced migration of substituents between vicinal carbons. See text for details

with the exception of *glutamate mutase*, have a C=C or C=O group in one of the substituents, leading to a cyclopropane-type intermediate in the migration cycle. Moreover, these carbon-skeleton rearranging *mutases* are coenzyme B_{12}-dependent radical enzymes. In thermodynamic equilibrium, the concentration of the less-branched component is slightly higher.

Two well-studied coenzyme B_{12}-dependent *mutases* are *methylmalonyl-CoA mutase (Acyl-CoA mutase)*, which is also found in aerobic microbes, and *glutamate mutase* (e.g., in *Clostridium cochlearium*) (Fig. 7.22). *Methylmalonyl-CoA mutase* is involved in one type of propionate-forming fermentation to lactate (see Figs. 9.9 and 9.10) as well as in the oxidation of propionate in the absence of a methylcitrate cycle. *Glutamate mutase* in *Clostridium tetanomorphum* is involved in glutamate fermentation to butyrate and acetate and converts glutamate to methylaspartate (see Fig. 9.23). Methylaspartate is converted via two non-redox reactions to pyruvate and acetate for energy conservation. Other important carbon-skeleton rearranging *mutases* are *methyleneglutarate mutase* (e.g., *Eubacterium barkeri*) isomerizing 2-methyleneglutarate to 3-methylitaconate in nicotinate fermentation to propionate, acetate, and NH_3, and *isobutyryl-CoA mutase* (e.g., in *Steptomyces cinnamonensis*), converting isobutyryl-CoA to butyryl-CoA in valine degradation. *Acyl-CoA mutases* are also involved in anaerobic degradation of alkanes and methyl-substituted benzenes.

Fig. 7.22 Mechanism of coenzyme B_{12}-dependent enzymatic carbon–carbon rearrangement in *methylmalonyl-CoA mutase* and *glutamate mutase*. See text for details

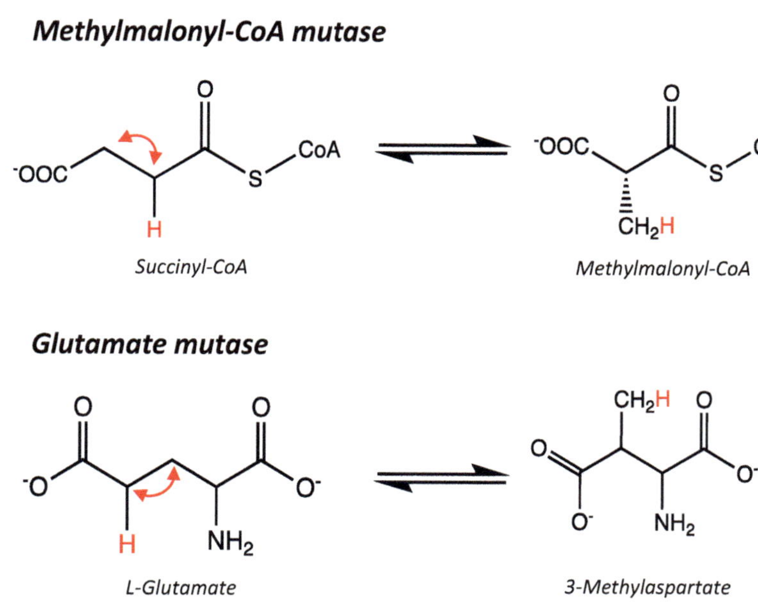

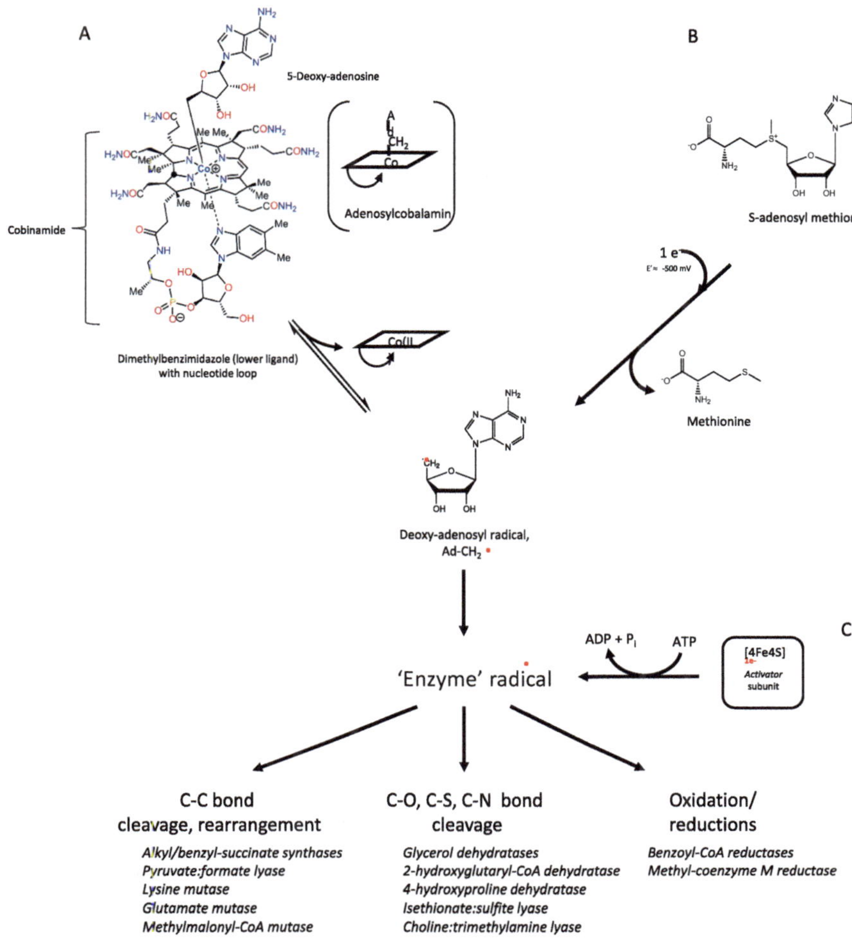

Fig. 7.23 Three modes of generating enzyme/substrate radicals and the catabolic radical reactions they catalyze. "Enzyme" denotes that either an enzyme radical, containing a radical at an amino acid residue such as glycine, cysteine, or tyrosine, or a substrate radical as enzyme-bound intermediate, can be generated. An initiating radical can be formed by different modes for enzymes catalyzing the same reaction. (**a**) Coenzyme B_{12}-dependent. The deoxy-adenosyl radical is generated by homolytic cleavage of the carbon-cobalt bond of coenzyme B_{12}. Coenzyme B_{12} consists of a central cobinamide, containing four pyrrole rings plus a central cobalt atom. In contrast to hemes, which also are tetrapyrroles, one methine bridge between two pyrrole rings in B_{12} forms a direct C–C bond between two pyrrole ring carbons. Next to the four planar imidazole-nitrogen ligands, a lower ligand (in this case a dimethylbenzimidazole nucleotide) and an upper deoxy-adenosyl ligand are important for catalysis. The C5 carbon of the upper ligand deoxy-ribose of adenosine is in a covalent bond to cobalt forming a unique C–Co bond. A lower ligand of the cobinamide controls the reactivity of Co by polarization of the C–Co bond. (**b**) SAM-dependent. The S–CH_2 bond of SAM is reductively cleaved by a single electron to form methionine and the deoxy-adenosyl radical. (**c**) ATP-dependent. In *2-hydroxyglutaryl-CoA dehydratase* from *Acidaminococcus fermentans*, the enzyme radical is formed by ATP-dependent activation and transfer of a single electron from the [4Fe4S] cluster of the *activase* subunit to the catalytically active *dehydratase* subunit. Once activated the *dehydratase* can go through thousands of reaction cycles before another ATP-dependent activation is required. See text for details

Carbon–carbon rearrangements by *mutases* or *lyases*. Homolytic cleavage of the C–Co bond of adenosylcobalamine forms a deoxyadenosyl radical and a Co(II) in *mutases* or *lyases* (Fig. 7.21). This deoxy-adenosyl radical then abstracts a hydrogen atom from the substrate molecule to form a substrate radical. This substrate radical rearranges and a substituent "X" migrates between two vicinal carbon atoms. At the end of the reaction cycle, the abstracted H atom is transferred back to the substrate radical, regenerating the deoxy-adenosyl radical (Fig. 7.21). Coenzyme-B_{12}-dependent enzymes catalyze diverse radical reactions, some of which are also catalyzed by SAM-dependent GREs.

Radical Reactions, Radical Enzymes, and O_2-Sensitivity

Radical enzymes catalyze reactions that are difficult to catalyze by acid-base-type enzyme chemistries, and involve an enzyme or a substrate radical. Since organic radicals are highly reactive, most radical enzymes function only in strictly or facultative anaerobic microbes, in particular in those fermenting amino acids, taurine, choline, or other compounds including those present in the human gut microbiota. The steps in catalysis of radical enzymes involve (1) generating an initiating enzyme-based radical, (2) reaction of this radical with the substrate, typically by abstracting a H atom to form a substrate radical, (3) transformation of the substrate-derived radical into a product-related radical, and (4) conversion of the product-related radical into the product by regenerating the initiating radical (see Fig. 7.18). In other radical enzymes, such as *lactoyl-CoA dehydratase,* a single electron is added from the enzyme to generate a substrate radical.

Initiating radicals can be generated involving (1) **coenzyme B_{12}**, (2) **S-adenosyl methionine (SAM)**, or (3) **ATP-dependent single electron activation** (see Fig. 7.23).

Coenzyme B_{12}. Homolytic cleavage of the C–Co bond of adenosylcobalamine forms a deoxyadenosyl radical and a Co (II) as found in many *lyases* and *mutases* (Fig. 7.23a). The deoxyadenosyl radical generates the enzyme or substrate radical in coenzyme B_{12}-dependent enzymes. Coenzyme B_{12} biosynthesis is costly as ~30 genes are required, and the biosynthetic genes are unevenly distributed in different microbial phyla. About 60% of microorganisms lack biosynthetic genes for the de novo synthesis although a microbe's catabolism and/or anabolism requires coenzyme B_{12} or related corrinoids. It is assumed that in microbial communities, B_{12} or derivatives are actively released or passively escape from decaying cells and taken up by B_{12} auxotrophs via several mechanisms, including a TonB-type uptake. The capacity for B_{12} uptake is widely distributed, including the incorporation of diverse lower ligands.

S-adenosyl methionine (SAM). A deoxy-adenosyl radical is also generated from SAM, which is structurally simpler than coenzyme B_{12} (Fig. 7.23b). Therefore, SAM is often described as the 'poor man's version' of adenosylcobalamine. In SAM, homolytic cleavage of the S–CH_2 bond by a single electron generates a deoxy-adenosyl radical plus methionine. Formation of this radical is irreversible. Interestingly, SAM is also the biosynthetic precursor of the signaling molecule AI-2, a furanosyl borate diester.

Some SAM- and coenzyme B_{12}-dependent radical enzymes catalyze the same overall reaction while the deoxy-adenosyl radical is generated via different modes. There are radical enzymes that require SAM stoichiometrically to the substrate, and there are GRE that require SAM to

(continued)

initiate a stable enzyme radical for multiple turnovers.

The third mode of radical generation is via *activases*, or "*archerases*," that were discovered in the context of anaerobic *dehydratases* (see below). As shown for *2-hydroxygutaryl-CoA dehydratase*, *activases* generate an enzyme radical by ATP-dependent addition of a single electron from the [4Fe-4S] cluster of the *activase* that is transferred and stored in the flavin of the catalytically active *dehydratase* subunit.

Radical enzymes can be classified either based on the mode of radical generation or on the type of reaction they catalyze; however, there is no strict correlation between both. For example, *dehydratases* can receive an activating radical from SAM or via ATP-dependent activation. However, each radical enzyme has its specific, non-exchangeable mode of formation of the initiating radical. Radical enzymes catalyze diverse reactions in catabolic pathways including (Fig. 7.23)

- cleavage, formation, and rearrangement of C–C bonds, described in this chapter in detail,
- cleavage of C–O, C–S, and C–N bonds, and
- oxidation/reduction reactions.

One emerging class of SAM radical enzymes is GRE that contains an active site glycyl radical such as in *benzylsuccinate synthase* (Fig. 7.19). At the beginning of a reaction cycle, the glycyl radical of GREs abstracts a H atom from a vicinal cysteine residue in the enzyme to form a thiyl radical. This thiyl radical abstracts a H atom from the substrate molecule, thereby forming a substrate radical to initiate the radical substrate transformation. The glycyl radical of GREs is generated post-translationally by a *GRE activase*, catalyzing the reductive cleavage of S-adenosylmethionine with a single electron from a reduced [4Fe-4S] cluster to form a deoxyadenosyl radical (and methionine), or via the ATP-dependent mode (Fig. 7.23). The glycyl radical in GREs is stable and can initiate thousands of reaction cycles.

Based on amino acid sequence similarity and substrate specificity, there are currently several classes of catabolic GREs:

C–C bond cleaving and forming GREs e.g., *pyruvate:formate lyase* (see Sect. 7.3.3), *alkyl-* and *aryl-succinate synthases* (Fig. 7.19), and *4-hydroxyphenylacetate decarboxylases*

C–O bond cleaving e.g., *propanediol* and *glycerol dehydratases* and *4-hydroxyproline dehydratases*;

C–S bond cleaving e.g., *isethionate:sulfite lyases*, and

C–N bond cleaving e.g., *choline: trimethylamine lyases*.

While the C–C and C–O bond-cleaving/forming GREs are described below in more detail, C–S and C–N bond-cleaving GREs function in pathways of anaerobic microbes to generate an inorganic electron acceptor from a fermentable substrate. *Isethionate: sulfite lyase* in *Bilophila wadsworthia*, for example, catalyzes cleavage of isethionate, which is derived from dietary or host taurine (2-aminoethanesulfonate), into SO_3^{2-} and acetaldehyde. In *B. wadsworthia* and selected other microbes, sulfite is used as a unique electron acceptor in an otherwise electron acceptor-limited host environment and reduced to H_2S. Such metabolic consequences are significant in the intestinal microbiota and microbial pathogenesis. Hydrogen sulfide at higher concentration has been found to be associated with irritable bowel disease and colorectal cancer in men; diseases that further destabilize an

(continued)

intact commensal gut microbiota. Acetaldehyde can be further catabolized in this microbe and disproportionated to acetate and ethanol in analogy to ethanolamine fermentation by *Salmonella typhimurium*. *Choline:trimethylamine lyase* of *Desulfovibrio desulfuricans* catalyzes cleavage of dietary- and host-derived choline (2-hydroxyethyl)-trimethylammonium) into trimethylamine and acetaldehyde. Also here, acetaldehyde is catabolically disproportionated into acetate and ethanol. Trimethylamine is either used by methanogens for catabolic methane formation (the methyl groups of TMA are at the oxidation level of methanol (Chap. 10) or metabolized by the human host to trimethylamine-N-oxide. The latter is a unique electron acceptor to several intestinal microbes. Many GREs, including *choline:trimethylamine lyase* from *Clostridium botulinum, ethanolamine ammonia lyase,* as well as *propanediol dehydratase* in *Vibrio* sp. and *Shewanella putrefaciens*, are enclosed in bacterial microcompartments (see Chap. 2, Fig. 2.2). Such segregation and "privatization" of selected catabolism from the bulk cytoplasm is advantageous for preventing toxicity (e.g., of aldehydes in ethanolamine and glycerol utilization), inactivation (e.g., of the radical enzyme), or by decreasing diffusional distance between enzymes and substrates to enhance the rate of catabolism.

Important examples of radical, single electron-transferring oxidation/reduction enzymes are

- *methyl-coenzyme M reductase* (see Sect. 10.2.3.3) in anaerobic reduction of a methyl carbon to methane and in anaerobic methane oxidation,
- *mono/di-oxygenases* (see Sect. 7.4.1) catalyzing O_2-dependent oxidation of hydrocarbons,
- *benzoyl-CoA reductases* catalyze radical substrate reduction by two low redox potential single electrons (see Fig. 7.20), and
- molybdopterine-dependent *dehydrogenases* or *oxidases*, such as *formate dehydrogenase* or *ethylbenzene dehydrogenase*. While not radical enzymes as defined above, enzymes of this later group oxidize the substrate via two single electron oxidations rather than a hydride abstraction involving the reversible transitioning of Mo from the oxidation state of +6 to +4.

The sensitivity of radical enzymes as well as of other low-potential redox enzymes toward molecular oxygen is due to O_2 being a biradical in a triplet state. The O_2 biradical can be reduced rapidly by a single electron, such as an electron from a radical enzyme, to a superoxide anion radical ($O_2^{\cdot}$) at a redox potential of $E^{0\prime} = -160$ mV, and upon further single electron reduction to H_2O_2 ($O_2^{\cdot}/O_2^{2-}$; $E^{0\prime} = +890$ mV). Therefore, single electrons at redox potentials of < -400 mV, as present in reduced Fe–S cluster enzymes, radical enzymes, reduced ferredoxin, or Ni(I) of F_{430}, can be easily scavenged by O_2 and $O_2^{\cdot}$. Peroxide (O_2^{2-}) is a strong oxidant and can irreversibly inactivate enzymes. Its destructive reactivity is mitigated by *superoxide dismutase*, disproportionating 2 O_2^{2-} into O_2 and H_2O_2, and further by *catalase* converting H_2O_2 into H_2O and $1/2O_2$.

7.4.5 Addition/Elimination Reactions

Addition reactions to activated double bonds in β position to a functional carbonyl or carboxyl are important "auxiliary" reactions. These reactions are different from the addition reactions of Fig. 7.19 as they are not radical additions and do not involve the formation of a new C–C bond.

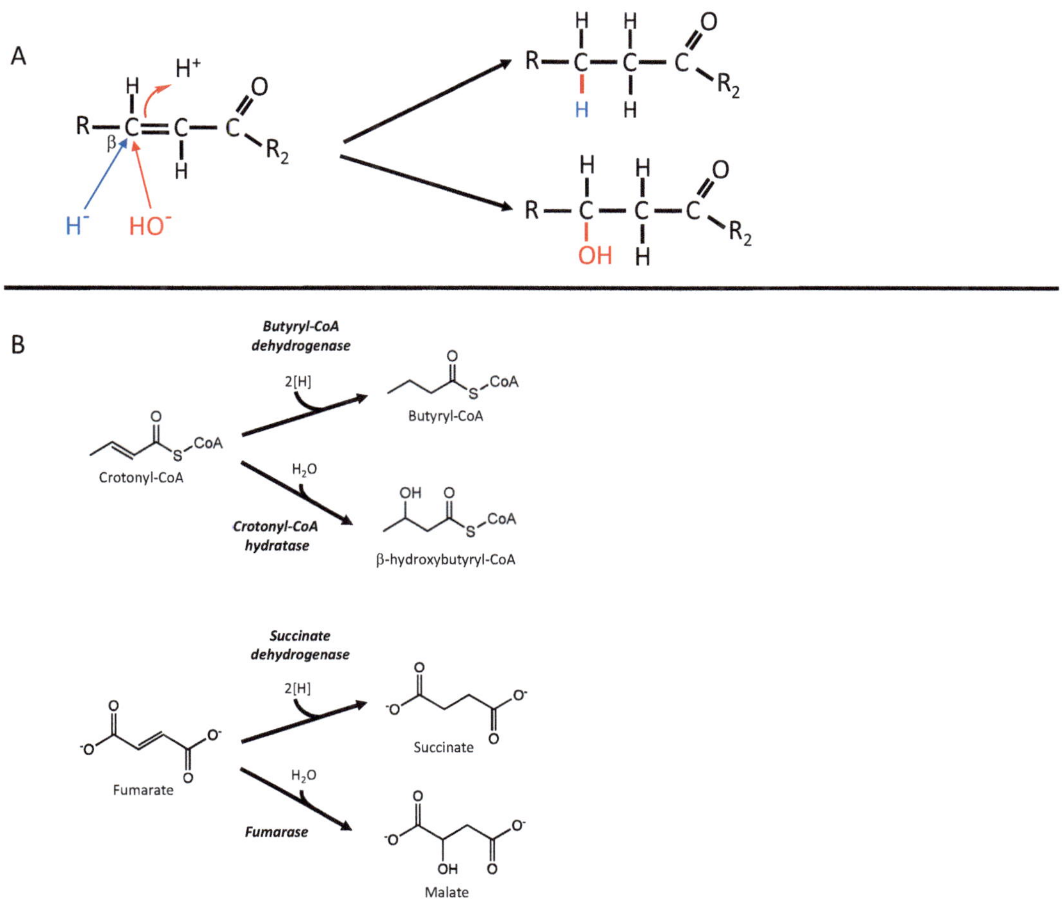

Fig. 7.24 Common enzymatic nucleophilic addition reactions to carbon–carbon double bonds. (**a**) The two most common types are hydride or $^{-}$OH addition. (**b**) Examples of hydride and $^{-}$OH additions to (1) crotonyl-CoA resulting in butyryl-CoA or β-hydroxybutyryl-CoA formation, and (2) fumarate, forming succinate or malate, respectively

Figure 7.23 shows the principle of these nucleophilic addition reactions. As introduced above, addition by a nucleophile occurs at the β (3-position) carbon (Fig. 7.24). The most common nucleophiles are a hydride (H^-) or a $^{-}$OH group. A functional carbonyl is in the thioester bond of a fatty acid with HS-CoA, or present in the dicarboxylate fumarate. Figure 7.24 shows two examples of hydride and $^{-}$OH addition as found in many metabolic pathways. Crotonyl-CoA, for example, can be either reduced to butyryl-CoA or hydrated to β-hydroxybutyryl-CoA; fumarate can be reduced to succinate or hydrated to malate, respectively. These enzymatic addition reactions are present in diverse ($C > 4$) fatty acid degradation pathways as well as in the TCA cycle.

7.4.6 Hydroxy/Amino Eliminations

Hydroxy or amino groups in non-acyl-CoA compounds can be eliminated by coenzyme B_{12}-dependent or SAM-dependent radical enzymes. Coenzyme B_{12}-dependent hydroxy eliminations are catalyzed by *glycerol dehydratase* in enteric bacteria and *Clostridia,* and mediate the elimination of a hydroxy group resulting as water (Fig. 7.24). Notably, *glycerol dehydratase* in *Clostridium butyricum* is a SAM-dependent

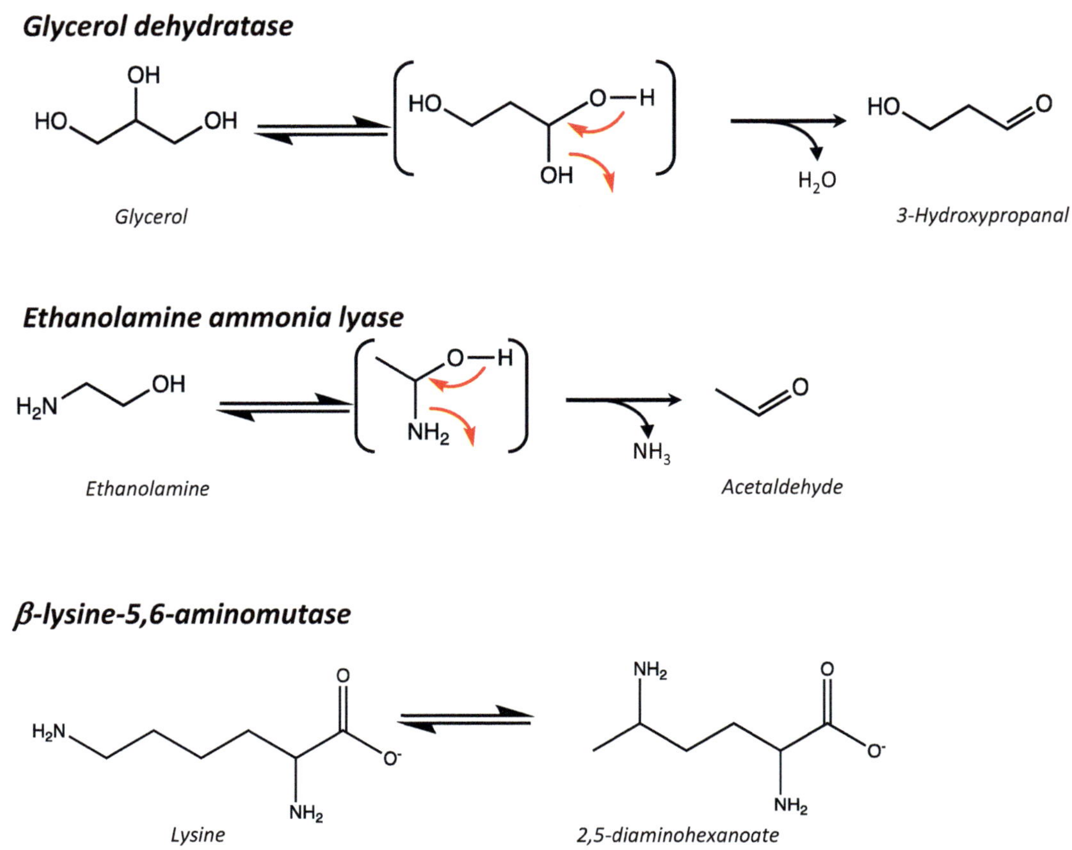

Fig. 7.25 Examples radical enzyme reactions of hydroxy/amino group rearrangements and eliminations in anaerobic catabolism. See text for details

GRE. In glycerol fermentation, where glycerol is both electron donor and acceptor, glycerol itself is not a good electron acceptor as neither of the three carbon atoms can be reduced by any of the above prototype oxidation/reduction reactions. *Glycerol dehydratase* converts glycerol in a single step into hydroxypropionaldehyde, where the aldehyde group now is an excellent electron acceptor (Fig. 7.25) and can serve as electron sink and be reduced to 1,3-propanediol.

Aminomutases catalyze the vicinal migration of an amino group, which is subsequently eliminated as NH_3. ***Ethanolamine ammonia lyase*** in *Clostridia* sp. and *Salmonella typhimurium* catalyzes the coenzyme B_{12}-dependent conversion of ethanolamine to acetaldehyde and NH_3 (Fig. 7.25). *Ethanolamine ammonia lyase* together with acetylating *acetaldehyde dehydrogenase*, *ethanol dehydrogenase*, and *phosphotransacetylase* are typically enclosed in a bacterial microcompartment, presumably to mitigate the toxicity of acetaldehyde (see Chap. 2, Fig. 2.2). This overall short fermentation pathway yields 1 ATP per acetate formed.

β-lysine-5,6-aminomutase catalyzes amino group migration in D-lysine fermentation to acetate, butyrate, and NH_3 in *Clostridium sticklandii* (Fig. 7.25). In addition to the coenzyme B_{12} cofactor, this enzyme also contains pyridoxal phosphate as prosthetic group. SAM-dependent ***lysine 2,3-aminomutase*** of *Clostridium subterminale* catalyzes the first step in lysine fermentation to acetate and butyrate.

Hydroxy or amino groups in a position other than the β-position to a functional carbonyl can be eliminated by other radical enzymes. Ketyl radical ***dehydratases*** in anaerobic amino acid fermentations act on activated 2-hydroxy acids (Fig. 7.26a). The metabolic context of these *dehydratases* is mostly in reductive fermentative

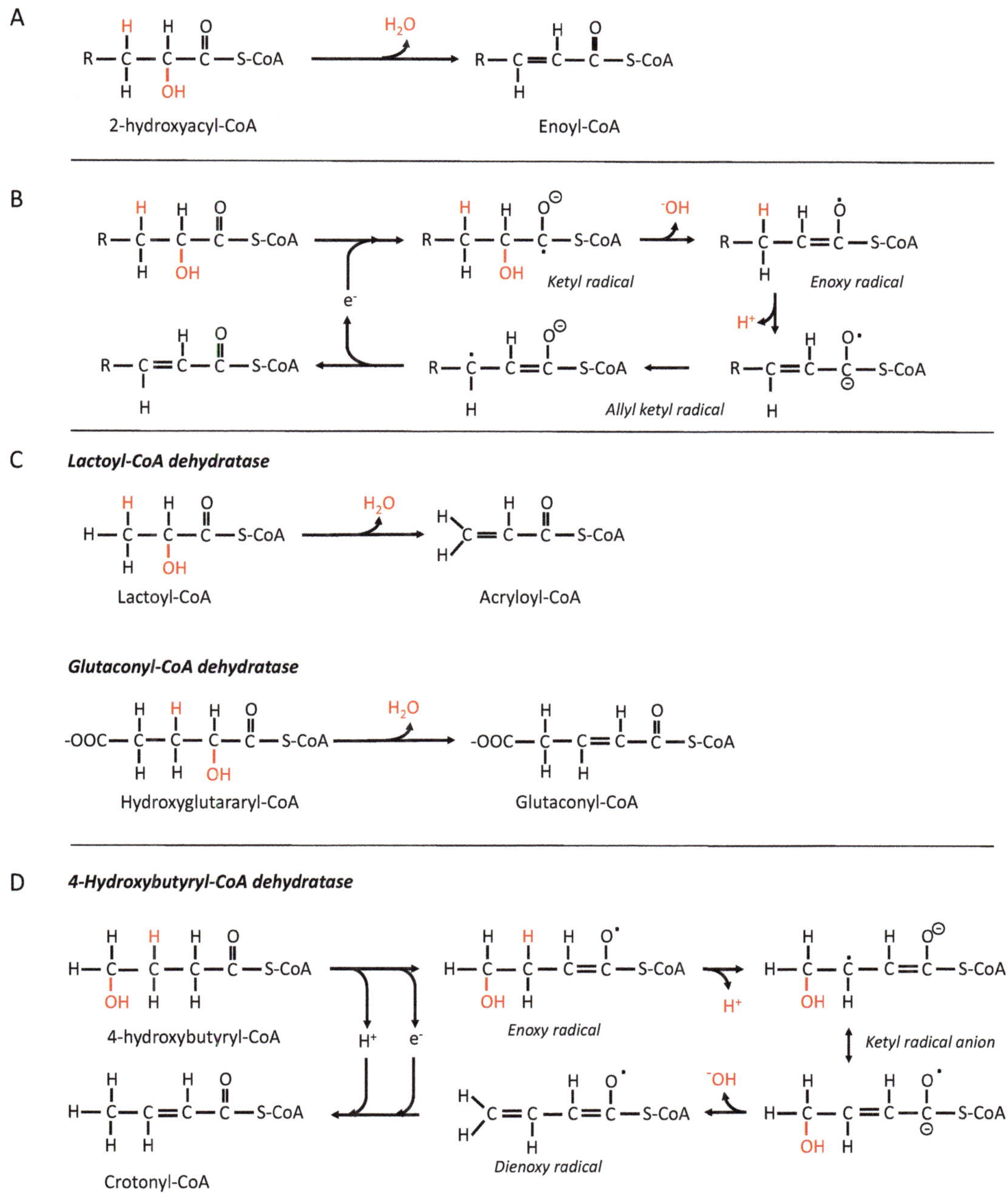

Fig. 7.26 2-and 4-Hydroxy (α- and γ-) eliminations. (**a**) Overall reaction of a *2-hydroxyacyl-CoA dehydratase* to form an enoyl-CoA. (**b**) A ketyl radical-based reaction mechanism. Addition of a single electron, stored in the *dehydratase*, to the substrate carbonyl generates a ketyl radical that leads to an "Umpolung" at the α carbon and a rapid elimination of the hydroxy group, forming an enoxy radical, which rearranges after deprotonation to an allyl ketyl radical. The radical electron is transferred back to *dehydratase* and the enoyl-CoA end product is released. (**c**) Examples of *dehydratases* in fermentations for producing an enoyl-CoA electron acceptor: *lactoyl-CoA dehydratase* in propionate-forming fermentation via the acrylate pathway by *Clostridium propionicum* and *Megasphera elsdenii* (see Fig. 9.21) and *glutaconyl-CoA dehydratase* in glutamate fermentation by *Acidaminococcus fermentans* and *Fusobacterium nucleatum* (see Fig. 9.22). (**d**) *4-Hydroxybutyryl-CoA dehydratase* from *Clostridium aminobutricum*. In contrast to the elimination of a hydroxy group in 2-position (**a**, **b**, **c**), in *4-hydroxybutyryl-CoA dehydratase* a single electron is abstracted from the substrate molecule (after deprotonation) to generate the substrate ketyl radical anion radical for HO^- elimination in the 4- (γ) position

pathways to convert a 2-hydroxy acid into an electron acceptor by eliminating the hydroxy group under the formation of an enoyl compound, such as crotonyl-CoA (see Sect. 9.5.3, Fig. 9.22). In *2-hydroxyacyl-CoA dehydratase,* a single electron at a redox potential of ~ −900 mV is added by the enzyme to the functional carbonyl of the activated 2-hydroxy acid to form a ketyl radical, leading to an "Umpolung" of the activated 2-hydroxy acid such that the elimination of water is possible (Fig. 7.26b). Due to the reactivity and low redox potential of the active site radical, *2-hydroxyacyl-CoA dehydratases* are rapidly inactivated by O_2 and, therefore, found only in some strictly anaerobic, in particular in amino acid-fermenting prokaryotes such as *Clostridium propionicum, C. diphteriae, C. difficile, Acidaminococcus fermentans,* or *Fusobacterium nucleatum.* Other *dehydratases* include *lactoyl-CoA dehydratase* in lactate fermentation to propionate by the acryloyl-CoA pathway and *glutaconyl-CoA dehydratase* in glutamate fermentation (Figs. 7.26c and 9.22). Other important *dehydratases* in amino acid fermentations are *2-hydroxyglutaryl-CoA dehydratase* in glutamate, glutamine, and histidine fermentation of *Acidaminococcus fermentans* (see also Fig. 9.22) and *Clostridium symbiosum, 2-hydroxyisocaproyl-CoA dehydratase* in *Clostridium difficile* in fermentation of leucine (see also Fig. 9.24), and *phenyllactyl-CoA dehydratase* in phenylalanine, tyrosine, and tryptophane fermentation of *Clostridium sporogenes.*

An interesting variation of radical-based *dehydratases* is *4-hydroxybutyryl-CoA dehydratase* in fermentation of γ-aminobutyrate to acetate, butyrate and ammonia by *C. aminobutyricum* (Fig. 7.26d). In contrast to *2-hydroxyacyl-CoA dehydratases*, which add a single electron to the substrate carbonyl, *4-hydroxybutyryl-CoA dehydratase* abstracts (after deprotonation) a single electron from the substrate to generate a substrate ketyl radical. The abstracted electron is transiently stored in *hydroxybutyryl-CoA dehydratase* in an enzyme-bound flavin as a semiquinone.

Under oxic conditions, 2-hydroxy acids are generally oxidized by an *alcohol dehydrogenase* to 2-oxo acids, which are then oxidatively decarboxylated as described above.

Another interesting $^{-}$OH elimination reaction that is, however, not found so far in catabolic pathways is the ATP-dependent elimination of a hydroxy group (Fig. 7.27). *Methylglyoxal synthase* in glycerol fermentation catalyzes the dehydration of dihydroxyacetone phosphate to methylglyoxal as electron acceptor in reductive

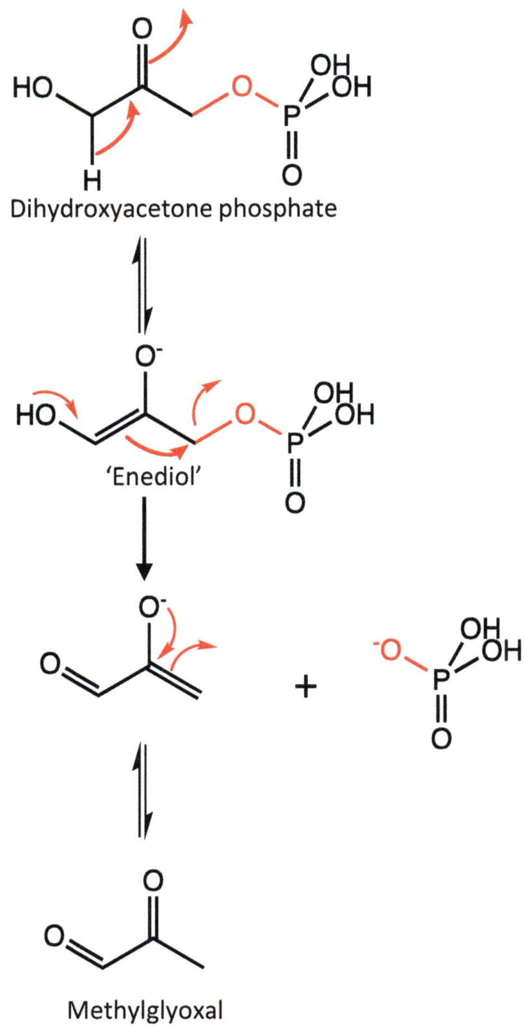

Fig. 7.27 Water elimination by *methylglyoxal synthase.* Dihydroxyacetone phosphate is dephosphorylated to methylglyoxal with formation of an enediol as intermediate. The net result of this intramolecular redox reaction is that the terminal alcohol carbon (OH indicated in red) is reduced to a methyl carbon as other terminal alcohol carbon is oxidized of to an aldehyde

glycerol fermentation. The phosphate-activated hydroxy group of dihydroxyacetone phosphate is eliminated by formation of an enediol as intermediate, resulting in the leaving of the phosphate group from the organophosphate and concurrent elimination of the OH group of the glycerol (Fig. 7.27)! A phosphoryl group-dependent $^{-}$OH elimination is also present in anabolic isopentenyl pyrophosphate biosynthesis of the mevalonate pathway. Such ATP-dependent dehydrations, however, are rare in catabolic pathways because this step consumes stoichiometrically one ATP for the phosphorylation. ATP-consuming steps are minimized or absent in particular in pathways of anaerobes with an overall small ΔG and low n_{ATP}. Nevertheless, this ATP-dependent dehydration is an interesting reaction as an intramolecular redox reaction converting a polyalcohol, after phosphorylation, to an aldehyde/ketone (see also *glycerol dehydratase*).

The seven prototypic and six auxiliary reactions discussed in this chapter are summarized in Table A.3 in the Appendix.

Further Reading

Metzler D (2003) Biochemistry: the chemical reactions of living cells, 2nd edn. Academic

Buckel W, Zhang J, Friedrich P, Parthasarathy A, Li H, Djurdjevic I, Dobbek H, Martins BM (2012) Enzyme catalyzed radical dehydrations of hydroxy acids. Biochim Biophys Acta 1824(11):1278–1290

Huwiler SG, Loffler C, Anselmann SEL, Stark HJ, von Bergen M, Flechsler J et al (2019) One-megadalton metalloenzyme complex in Geobacter metallireducens involved in benzene ring reduction beyond the biological redox window. Proc Natl Acad Sci U S A 116:2259–2264

Rabus R, Boll M, Heider J, Meckenstock RU, Buckel W, Einsle O et al (2016) Anaerobic microbial degradation of hydrocarbons: from enzymatic reactions to the environment. J Mol Microbiol Biotechnol 26:5–28

Metabolic Modules, Pathways, and Nodes of Intermediates

8

Contents

8.1 Introduction

"Metabolic modules" are *conceptual* and exist primarily in the eye of the beholder. A module is defined as a set of prototypic enzymatic reactions that catalyzes a short transformation of a substrate or metabolite. Reactions of a module occur in a specific order and are connected by common metabolic intermediates. Modules combine and recur often in different metabolic pathways (Fig. 8.1), and reflect the logic of pathways. With an invariant overall reaction for transformation of an organic compound, there are frequent variations of a module, including the

The original version of the chapter has been revised. A correction to this chapter can be found at https://doi.org/10.1007/978-3-031-28218-8_11

A. M. Spormann, *Principles of Microbial Metabolism and Metabolic Ecology*,
https://doi.org/10.1007/978-3-031-28218-8_8

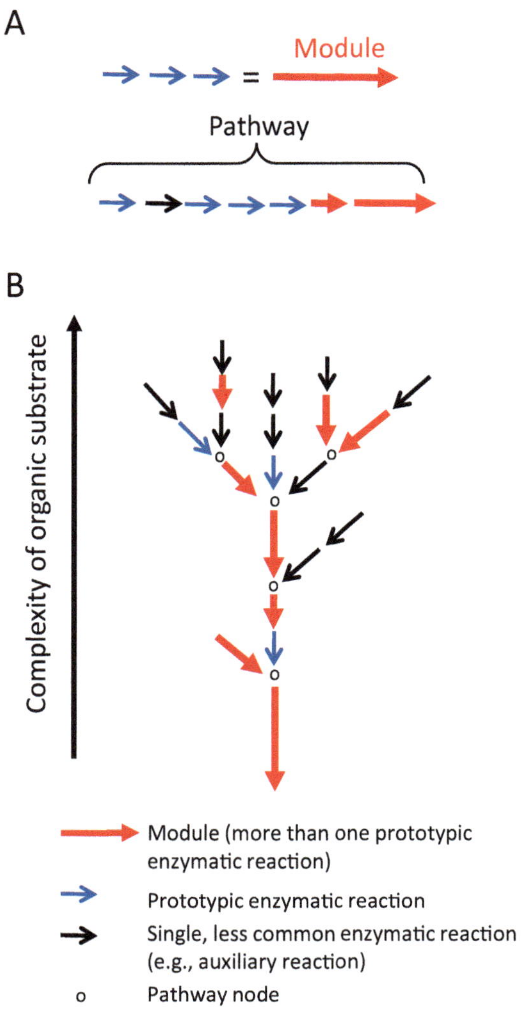

Fig. 8.1 Microbial catabolism as a mosaic of modules and prototypic enzymatic reactions. (**a**) Modules include enzymes catalyzing prototypic reactions and acting consecutively on structurally-related intermediates. Length of a module arrow (red) is proportional to the number of prototypic reactions included. Pathways may consist of modules plus non-module, prototypic reactions (blue, e.g., auxiliary reaction) plus a few less common, enzymatic funneling reactions (black). The pathway shown in **a**) is the modular deconvolution of the citric acid cycle (see text). (**b**) Organization of general catabolic pathways. Organic substrates with increasing structural complexity converge in oxidative degradations to simpler intermediates in nodes that are common to diverse pathways with unrelated parental substrates, reflecting the evolution of pathways by upstream extension. See text for details

type of electron carriers involved, which lead to different pathway n_{ATP}.

The primary purpose of the "module" concept is a metabolic rather than a chemical one: to deconvolute the apparent complexity of extant pathways into organizing principles. Module enzymes accommodate structural variation of an organic substrate molecule in regions that are not relevant for the transformation chemistry. For example, the β-oxidation module acts on the C_6-activated fatty acid caproyl-CoA as well as on the C_4-activated butyryl-CoA, where the additional C_2-unit in caproyl-CoA is inconsequential for the functioning of the β-oxidation module; however, the respective enzymes of the C_4 and C_6 specific module have different substrate specificities.

8.2 Metabolic Modules, Nodes, Pathways

8.2.1 Frequently-Occurring Modules

Catabolic pathways consist of one or more modules plus a few auxiliary steps (Fig. 8.1a, Table A.4 in the Appendix). For example, the citric acid cycle contains a β-oxidation/succinate ↔ oxaloacetate module as well as prototypic reactions of aldol condensation, alcohol oxidation, and oxidative decarboxylation (see below, Fig. 8.1a). Prototypic *aldolase* reactions and β-oxidation modules also occur independent of each other in many other pathways (see below).

Substrates, Metabolites, and End Products

The term (metabolic) "end product" is used here to describe those compounds that are at the end of a pathway and are released into the extracellular environment. The term "intermediate" is used to describe intracellular organic compounds derived from catabolic substrates by module/pathway transformations and that are transformed further. For example, lactic acid is the end product of glucose

(continued)

fermentation, but glyceraldehyde-3-phosphate or pyruvate are intermediates as they are further transformed enzymatically. Finally, a “catabolic substrate” is an environmentally available organic compound that is used as electron donor or acceptor, while the term “substrate” alone is often used in the context of a reactant of an enzyme.

Most enzymatic reactions are mechanistically reversible and so are modules, as they can operate in the forward or reverse direction depending on the pathway context. If a module is involved in energy conservation or, in reverse, contains under standard conditions a large endergonic step, there are key variations of a module to accomplish the physiologically important flux. Such differences are often, but not exclusively, in the energetic coupling and/or type of electron carriers employed.

In the first sections of this chapter, catabolic pathways are generally drawn and labeled in the oxidative, degradative, and/or thermodynamically exergonic direction, unless otherwise indicated. Such representation is with the understanding that individual reactions and modules are reversible and can function in the reverse direction depending on the overall metabolic context. Therefore, reaction arrows in the Figures show a physiological direction and will neglect the chemical convention of reversibility. How the same module functions physiologically in the reverse direction is discussed, for example, in the reductive fermentative pathway (Sect. 8.5).

In contrast to Chap. 7, this chapter will not go into mechanistic details. For example, chemical leaving groups or migration of electrons are not specifically indicated as such for the sake of clarity and focus. The specificity of electron carriers (e.g., NAD^+, $NADP^+$, F_{420}, Fd, quinone) is relevant and shown in the context of different versions of modules. Also, the overall reaction of a module (or a pathway) shown at the top of each paragraph is not balanced in order to highlight the overall metabolic consequence of a module or pathway for carbon transformation, electron balance, and energy-rich bonds (e.g., ATP, acetyl-CoA). Correctly-balanced pathway equations are provided in the summary Tables A.4, A.5 in the Appendix, and A.6 at the end of this book.

Finally, the list of modules and pathways introduced here covers the most common ones and is sufficient to reveal the organizing principles of metabolic pathways. With an understanding of metabolic modularity (Chaps. 7–10) in conjunction with the (eco)physiology of a microbe (Chaps. 2–6), catabolic pathways not listed here can be easily comprehended.

8.2.2 Nodes

Pathway branch points are considered as “nodes” where an intermediate is common to more than one module or pathway as emphasized in Fig. 8.1. In the context of catabolic degradation pathways, such nodes represent funneling intermediates which are both endpoints of specific modules and starting intermediates of more common pathways acting on simpler compounds. For example, in aerobic chemotrophic microbes, acetyl-CoA and the citric acid cycle intermediates oxaloacetate, succinate, fumarate, or 2-oxo-glutarate, are nodes and funneling metabolites from degradations of diverse compounds, such as carbohydrate, fatty acid, and aromatic compound. Other common metabolite nodes are pyruvate, benzoate/benzoyl-CoA, or catechol.

8.2.3 Pathway Architecture

A metabolic pathway is composed of one or more metabolic modules, and is either **linear** or **cyclic**. Cyclic pathways and the associated idiosyncratic substrate activations are a consequence of the limited reactivity of a substrate molecule and the limited number of prototypic enzymatic reactions. In cyclic pathways, the degradation of the substrate molecule is initiated by covalent bonding to an (adaptor) metabolite, that is derived (regenerated) from downstream degradation of that primary substrate. The metabolic logic of activation by an "adaptor" metabolite is that an (unreactive) substrate molecule is placed into a configuration so that it can be transformed by existing modules and prototypic reactions. Examples for cyclic pathways are the citric acid cycle for acetyl-CoA oxidation (acceptor molecule: oxaloacetate) or the *acyl/benzyl-succinate synthase* pathways for anaerobic degradation of long-chain alkanes and methyl-substituted aromatic benzenes (acceptor molecule: fumarate). A consumption of node metabolites from a cyclic pathway, e.g., toward biosynthesis, renders the pathway depleted of the primary accepting adaptor metabolite, which needs to be generated in downstream **anaplerotic** bypass reactions (see below).

8.2.4 Modules and Pathway Evolution

Evolution of new modules is rare and even more rare is the evolution of novel prototypic enzyme reactions based on new chemistries. Gene/operon duplications, point mutations, and recombination that alter substrate specificity of modules are relatively frequent. Thus, new catabolic pathways for degradation of novel and/or complex compounds are more likely to arise by combination of extant common modules and enzymes with altered substrate specificity (see Fig. 8.1) than by *de novo* evolution of novel enzymes with novel chemistries. Evidence thereof can be found in virtually all microbial genomes.

8.3 Common Modules and Pathways in Organic Acid Oxidation/Degradation

This section describes first a common module for organic acid activation, followed by common key reactions cleaving (or forming) carbon–carbon bonds, followed by auxiliary reactions connecting modules and funneling organic compounds into diverse modules.

8.3.1 Activation of Organic Acids and the Acylate ↔ Acyl-CoA Module

One of the most frequent modules is the **Acylate ↔ acyl-CoA Module** for activation of organic acids for degradation and, in reverse, for recovery of energy from a CoA thioester (Fig. 8.2a). CoA thioesters of fatty acids are central as they enable oxidation and carbon–carbon cleavage reactions (see Sects. 7.2.4 and 7.3.1). There are at least four different versions of the Acylate ↔ acyl-CoA Module (Fig. 8.2). All versions have in common that in the forward direction, metabolic energy is required for conversion of an acylate into a thioester. The Gibbs' free energy of formation of a thioester is $\Delta G^{0\prime} = +\ 36$ kJ/mol, indicating that hydrolysis of one ATP ($\Delta G^{0\prime} = -\ 32$ kJ/mol) or equivalent is sufficient to provide the required energy.

In the ***acetate kinase/phosphotransacetylase*** (AK/PTA) version of the Acylate ↔ acyl-CoA Module, acetate is first activated to acetyl-phosphate ($\Delta G^{0\prime} = -\ 33$ kJ/mol) via an ATP-consuming, ADP-forming *acetate kinase*. Acetyl-phosphate is subsequently converted to acetyl-CoA via a *phosphotransacetylase* (Fig. 8.2b). Such activation of a carboxylic acid is frequently found in facultative and anaerobic microbes metabolizing acetate, propionate, or butyrate. This version operates also in these microbes in the direction of recovering ATP from acetyl-CoA, propionyl-CoA, and butyryl-CoA.

In the ***CoA transacetylase*** version, activation of a carboxylic acid is coupled to transfer of HS-CoA from acetyl-CoA (or other short-chain fatty acid thioester) in an equilibrium reaction. Such activation often couples stoichiometrically diverse parts of an overall degradation pathway (for example in propionate formation via the methylmalonyl-CoA pathway), which is important for the overall fluxes through a catabolic pathway. The *CoA transacetylase* version of the

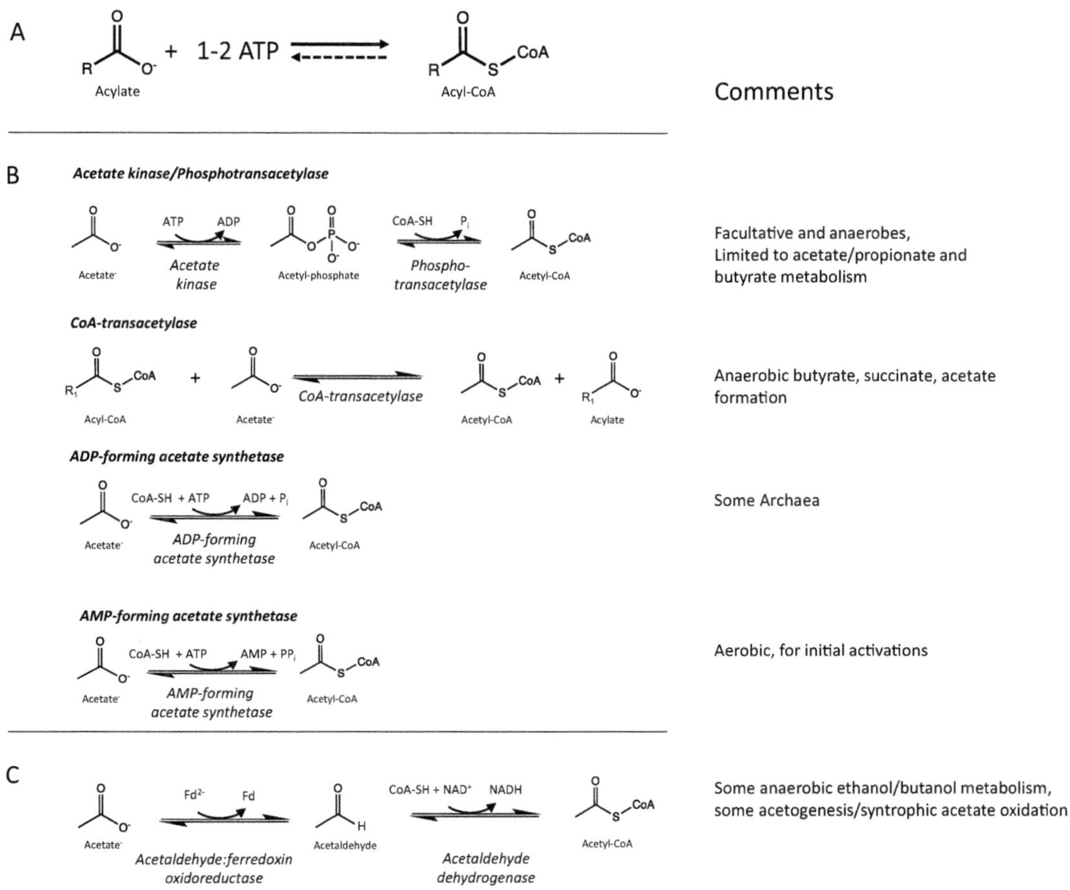

Fig. 8.2 The Acylate ↔ acyl-CoA Module and variations thereof for activation of organic acids to a CoA thioester. (**a**) shows the overall reaction of the Acylate ↔ acyl-CoA Module. R is an organic rest. (**b**) Variations of the Acylate ↔ acyl-CoA Module. The AK/PTA version is frequently found in facultative and anaerobic Prokaryotes, e.g., in acetate, propionate, or butyrate formation or activation. The *CoA-transacetylase* version is associated mostly with anaerobic acetate, butyrate, or succinate catabolism. *ADP-forming acetate synthetase* is found in some Archaea, while *AMP-forming acetate synthetase* is found in aerobic microbes, primarily for the activation of long-chain fatty acid for oxidative degradation. (**c**) Postulated acetate activation to acetyl-CoA via a Fd^{2-}-dependent reduction by *acetaldehyde:ferredoxin oxidoreductase* first, followed by NAD^+-dependent oxidation of acetaldehyde to acetyl-CoA. Here, the energy for thioester formation is not derived from ATP but from the difference of redoxpotentials of the Fd/Fd^{2-} and NAD^+/NADH couples. Abbreviations: AK/PTA: acetate kinase/ phosphotransacetylase. See text for details

Acylate ↔ acyl-CoA Module is often found in anaerobic acetate-, propionate-, or butyrate-forming microbes. For example, in *Clostridium kluyveri* fermenting ethanol plus acetate to butyrate, the free energy of the butyryl-CoA thioester is transferred to acetate via a *acetyl-CoA:butyrate CoA transferase* forming butyrate and acetyl-CoA (Fig. 9.18).

Acetate synthetases activate a carboxylate under the consumption of ATP to the respective thioester without a free phospho-anhydride as intermediate as in the AK/PTA version. There are two *acetate synthetase* versions: an **ADP-forming** and an **AMP-forming *acetate synthetase***. ADP-forming *acetate synthetase* hydrolyzes one ATP equivalent and is found in some Archaea. The AMP-forming *acetate synthetase* hydrolyzes ATP to AMP + pyrophosphate (PP_i), where PP_i is subsequently hydrolyzed by a *pyrophosphatase* to two P_i. The energy released in the *pyrophosphatase* reaction is not recovered. Therefore, AMP-forming *acetate synthetase* effectively consumes 2 ATP and is, in conjunction with *pyrophosphatase*, thermodynamically irreversible. This irreversibility is physiologically used to either commit acetyl-CoA into a specific pathway, such as in aerobic alkane degradation, or for sequestration of acetate as catabolic substrate at very low environmental concentration. Acetoclastic methanogen *Methanotrix soehngenii (Methanosaeta sp.)*, or *E.coli* in an acetate-limited chemostat (see Julian Adams experiment, Chaps. 1 and 5) are examples of metabolic use of AMP-forming *acetate synthetase*.

Acetate activation via ***aldehyde:ferredoxin oxidoreductase*** and ***acetaldehyde dehydrogenase*** (Fig. 8.2c) involves a reduction-oxidation loop. In anaerobic microbes, reduced ferredoxin (Fd^{2-}, $E^{0\prime} = -500$ mV) can function as electron donor for *acetaldehyde:ferredoxin oxidoreductase* to reduce the protonated form of acetate, acetic acid, to acetaldehyde ($E^{0\prime} = -580$ mV) in a reaction that is endergonic under standard conditions and pH 7. Under slightly acidic condition (e.g., intracellular pH = 6) and at high internal acetic acid concentration (pKa = 4.7), this reaction becomes thermodynamically favorable because 3 H^+ are consumed in this reaction (acetate$^-$ + 3 H^+ $\rightarrow$ acetaldehyde +2 H_2O), which pushes this reaction toward acetaldehyde formation. An *acetaldehyde dehydrogenase* rapidly re-oxidizes acetaldehyde, however, not with Fd but with NAD^+ as electron carrier and CoA-SH to form acetyl-CoA. Thus, a high acetic acid and a low acetaldehyde concentration make this reaction exergonic under slightly acidic conditions in vivo. For the overall reaction, the energy of acetate activation via this reaction sequence is not derived from ATP but from the difference in redox potential between Fd and NAD^+ ($\Delta E^{0\prime} = 180$ mV), which is equivalent to about ½ ATP. Notably, the same overall reaction of Fd^{2-} oxidation to NADH coupled to energy conservation is also catalyzed by Rnf; however, in completely different steps. Thus, acetate activation via the *acetaldehyde:ferredoxin oxidoreductase/acetaldehyde dehydrogenase* reaction sequence consumes less energy than via the ATP-dependent activation above. Such proposed activation would be important in some acetogenic microbes, in anaerobic, syntrophic acetate oxidation, or other microbes where catabolism involving fatty acids is minimized for ATP consumption.

Molecular Rate-Yield Trade-Off in Module Versions

The difference in acetate activation to acetyl-CoA via AK/PTA versus AMP-forming *acetate synthetase* is one example of a molecular rate-yield trade-off in catabolism. While acetate activation via the AK/PTA module costs one ATP, activation by AMP-forming *acetate synthetase* costs two ATP equivalents. This loss in thermodynamic efficiency enables *acetate synthetase*-containing *Methanotrix sp.* to grow at a significant rate at low acetate concentration, at which AK/PTA module-containing *Methanosarcina sp.*, e.g., is kinetically inferior. This is reflected in different apparent K_M values for acetate (>5 mM for *Methanosarcina* sp. vs 0.5 mM for *Methanotrix* sp.). Thus, the capacity of *Methanotrix* sp. to scavenge acetate effectively (in a kinetic sense) at low concentration comes at a trade-off in the form of a lower molar growth yield relative to *Methanosarcina* sp. The choice between these two versions of the Acylate $\leftrightarrow$ acyl-CoA Module reflects the different ecological strategies of these two methanogenic Archaea: *Methanosarcina* sp. are "generalists" with relatively fast growth on multiple substrates including H_2/CO_2, acetate, methanol, or methylamines, while *Methanotrix* sp. appears to be restricted to catabolic acetate utilization at even very low concentration. This dissipation of metabolic energy as heat at low substrate concentration is

(continued)

reminiscent of the variation of energetic coupling (n_{ATP}) of *NADH dehydrogenases* and *quinol oxidases* in the respiratory chains of *E. coli* and other enteric microorganisms at low oxygen. As discussed in Chap. 3, expression of a less-coupling (low n_{ATP}) *cytochrome bd oxidase* (no net H^+ pumping) under low pO_2 is more beneficial to the microbe than expressing an efficient cytochrome *bo* oxidase (Fig. 3.4). In halophilic archaea, such as *Halococcus saccharovorum* or *Haloferax volcanii* when metabolizing glucose, acetate is formed via an ADP-forming *acetate synthetase*. When these microbes consume the secreted acetate in stationary phase, AMP-forming *acetate synthetase mediates* acetate activation.

8.3.2 β-oxidation Module

Perhaps the most common **oxidation** module is the **β-oxidation module** or the β-oxidation pathway. The overall transformation of a β-oxidation module is:

$$\text{Acyl}_{(n)} - \text{CoA} \rightarrow \text{Acyl}_{(n-2)} - \text{CoA} + \text{Acetyl} - \text{CoA} + 4\,[\text{H}]$$

Chapter 7 introduced the prototype oxidation reactions, *acyl-CoA dehydrogenase*, *3-hydroxyacyl-CoA dehydrogenase*, as well as *enoyl-CoA hydratase* and *β-ketothiolase*. The β-oxidation module combines these four prototypic reactions into a module resulting in the oxidation AND shortening of a fatty acid by a C_2 ("acetate") unit (Fig. 8.3a). Prerequisite for oxidation of a fatty acid by β-oxidation via *acyl-CoA dehydrogenase* is the activation of the fatty acid via the Acylate ↔ acyl-CoA Module (Fig. 8.2). Thus, the β-oxidation module often operates in conjunction with the Acylate ↔ acyl-CoA Module, unless the CoA thioester substrate is formed via some different metabolic reaction sequence, such as in anaerobic degradation of coumarate, phenylacetate, phenol, anthranilate or toluene, respectively. The example of the

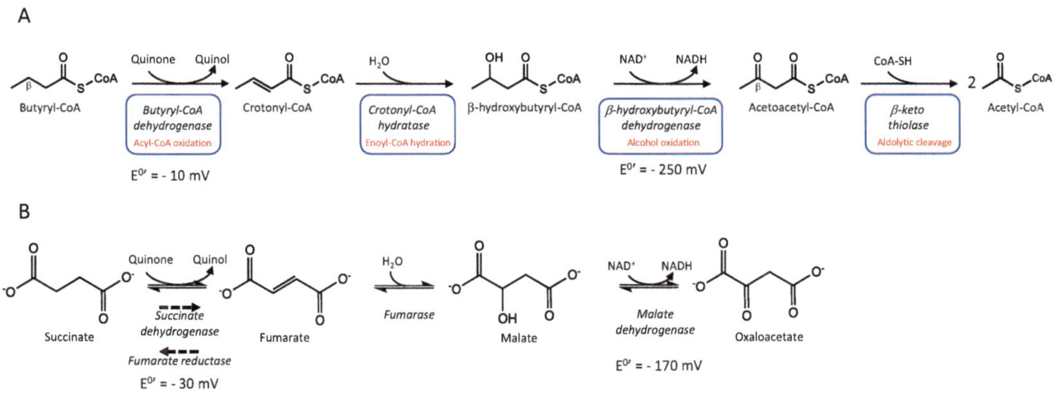

Fig. 8.3 β-oxidation module and the related succinate ↔ oxaloacetate variation. β-oxidation module enzymes catalyze analogous reaction sequences of (left to right) acyl-CoA oxidation, enoyl-CoA hydration, and alcohol oxidation. (**a**) For illustrative purposes, the β-oxidation module is shown here for oxidation of butyryl-CoA to 2 acetyl-CoA. This module can, with enzymes of appropriate substrate specificity, act iteratively on longer-chain fatty acids. When the β-oxidation module acts on odd-chain fatty acids, propionyl-CoA is the final product. This module functions also on organic acids other than fatty acids if the structural requirement for each prototypic reaction is met, for example as the succinate ↔oxaloacetate variation (**b**). (**b**) The succinate ↔ oxaloacetate variation of the β-oxidation module. This module represents 2/3 of the oxidative reactions of the Citric Acid Cycle. Moreover, functioning in the reverse direction, the succinate ↔ oxaloacetate variation is found in reductive fermentative pathways. See text for details

β-oxidation module for butyryl-CoA oxidation in Fig. 8.3 is for illustrative purposes. The two oxidation reactions are the *butyryl-CoA dehydrogenase* and *β-hydroxybutyryl-CoA dehydrogenase* reactions with redox potentials ($E^{0\prime}$) at –10 mV and −250 mV, respectively. The cellular electron acceptor for *β-hydroxybutyryl-CoA* oxidation is NAD^+, while the more positive electrons from butyryl-CoA oxidation are transferred to a quinone.

This β-oxidation module operates in catabolic pathways in both the oxidative direction, e.g., for fatty acid ($C_{>3}$) oxidation in respiring or in anaerobic fatty acid-oxidizing syntrophic microbes, or, with some modifications, in the reductive direction in, e.g., butyrate-forming, carbohydrate- or amino acid-fermentations. The kinetic parameters of the involved enzymes reflect the idiosyncratic solution of the Haldane equation in terms of K_M and V_{max} values (see *β-ketothiolase*, *fumarase*) depending on the physiological direction. As discussed in Sect. 5.4.2, *fumarase,* for example, catalyzes hydration of fumarate to malate in the oxidative TCA cycle, and dehydration of malate to fumarate in reductive TCA for CO_2 fixation. *Fumarase* functioning in the former pathway has a higher k_{cat} and K_M in the direction of water addition to fumarate than in the direction of malate dehydration to fumarate, while in the latter pathway, the kinetic parameters of the corresponding *fumarase* are reverse. Also, when operating in the direction of butyryl-CoA formation, *β-ketothiolase* is expressed at a very high cellular level (large fraction of proteome) to kinetically compensate for the unfavorable thermodynamics ($\Delta G^0 = +25$ kJ/mol) in butyrate synthesis (see Sect. 5.4.2).

8.3.3 Succinate ↔ Oxaloacetate Module

A frequently found variation of the catabolic β-oxidation module is the succinate ↔ oxaloacetate module (Fig. 8.3b). In this module, the reaction sequence of succinate oxidation to fumarate, hydration of fumarate to malate, and malate oxidation to oxaloacetate by *succinate dehydrogenase*, *fumarase*, and *malate dehydrogenase* is analogous to the sequence catalyzed by *butyryl-CoA dehydrogenase*, *crotonyl-CoA hydratase*, and *β-hydroxybutyryl-CoA dehydrogenase*. This three-enzyme succinate ↔ oxaloacetate module, for example, represents the β-oxidation part of the TCA cycle. The succinate ↔ oxaloacetate module functions frequently also in reverse in numerous reductive fermentation pathways (see Sect. 8.5.1). In a reductive fermentative pathway, either a cytoplasmic or a membrane-associated *fumarate reductase* uses NADH as electron donor for fumarate reduction to succinate ($E^{0\prime} = +30$ mV). The redox energy in the latter reaction can be conserved in a proton motive force or can drive electron bifurcation.

Outside the succinate ↔ oxaloacetate module, *fumarate reductase* can also be an important stand-alone reductase in anaerobic respiration of H_2 or formate, or in fermentation, in particular in conjunction with *aspartase* that deaminates aspartate to fumarate + NH_3. By the single *aspartase* step, the electron acceptor fumarate can be generated simply from the abundant amino acid aspartate in anoxic organic-rich environments.

8.3.4 *Acyl-CoA Mutase* Module

A common, carbon skeleton rearrangement reaction/module in anaerobic Prokaryotes is the ***acyl-CoA mutase* module** (Fig. 8.4):

$$\text{Acyl}_n\text{-CoA} \rightleftharpoons \text{2-Methylacyl}_{n-1}\text{-CoA}$$

The *methylmalonyl-CoA mutase* module is a common version in numerous catabolic and anabolic pathways and is involved in a reaction sequence interconverting propionyl-CoA to succinyl-CoA:

$$\text{Propionyl-CoA} + \text{ATP} + CO_2 \rightarrow \text{Succinyl-CoA}$$

Methylmalonyl-CoA mutase catalyzes the reversible carbon–carbon rearrangement between (*2R*)- methylmalonyl-CoA and succinyl-CoA (Fig. 7.22). In the overall reaction sequence,

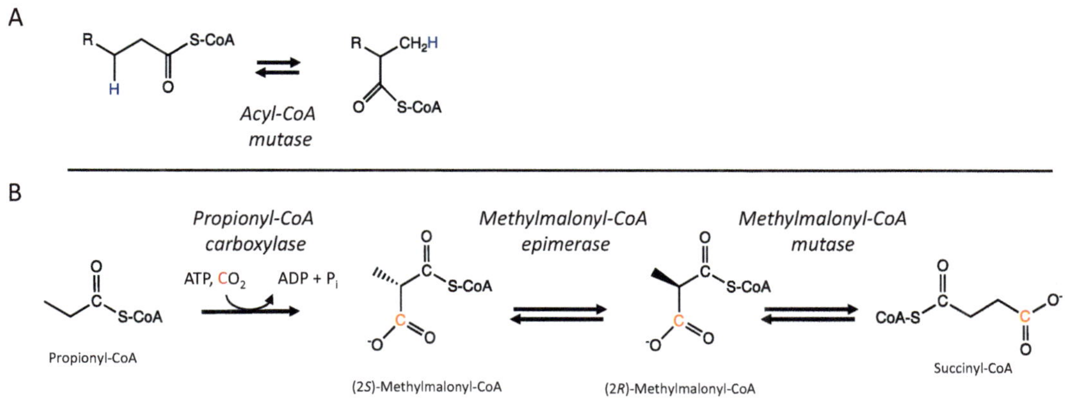

Fig. 8.4 The *methyl malonyl-CoA mutase* module for interconversion of propionyl-CoA (plus CO_2) to succinyl-CoA. See text for details

propionyl-CoA is first carboxylated to 2*S*-methylmalonyl-CoA in an ATP- and CO_2-dependent reaction by *propionyl-CoA carboxylase*. This reaction is the reverse of an aldolytic decarboxylation of a 2-oxo acid (see Sect. 7.3.1). Since such decarboxylation is exergonic, energy in the form of 1 ATP is required in the forward direction of *propionyl-CoA carboxylase*. 2*S*-methylmalonyl-CoA is subsequently epimerized to 2*R*-methylmalonyl-CoA, which is the substrate of *methylmalonyl-CoA mutase*. The product of this carbon–carbon rearrangement is the common intermediate (node) succinyl-CoA that can be metabolized further, e.g., via the Citric Acid Cycle.

When this module operates in the reverse direction as in some reductive fermentative pathways, there is a variation of the *propionyl-CoA carboxylase* reaction that involves either a *methylmalonyl-CoA transcarboxylase* or a *methylmalonyl-CoA decarboxylase*; the latter conserves 1/3 ATP via a membrane-bound, Na^+-translocating *decarboxylase* (see below) in a non-redox reaction.

In some pathways, the *acyl-CoA mutase* reaction is present in conjunction with other modules, such as β-oxidation module. For simplification, even a single *acyl-CoA mutase* reaction in a pathway is considered here as a "module" as shown in the following examples.

8.3.5 Examples of β-Oxidation Modules in Different Pathways

8.3.5.1 Aerobic *n*-Alkane Degradation

n-Alkane oxidation can be thought of as the acyl-CoA-oxidizing β-oxidation module upstream extended by a *monooxygenase* and an *alcohol/aldehyde dehydrogenase* (Fig. 8.5):

$$\text{Octane} + O_2 \rightarrow 4\ \text{Acetyl-CoA} + 14\ (e^- + H^+)$$

After the initial hydroxylation and oxidation to the corresponding fatty acid, an *acyl-CoA synthetase* activates the fatty acid to the CoA thioester. Each round of subsequent β-oxidation releases one acetyl-CoA, an acyl-CoA that is shorter by a C_2 unit, and 4 e^- (Fig. 8.5). A *monooxygenase* in *n*-alkane degradation pathway catalyzes the initial, O_2-dependent hydroxylation of a terminal methyl group (see Sect. 7.4.1).

One physiological consequence of substrate activation by a *monooxygenase* is that neither the electrons from carbon hydroxylation nor the energy released in the highly exergonic *monooxygenase* reaction ($\Delta G^0 = -361$ kJ/mol CH_4 with H_2 as co-substrate) are conserved. Furthermore, *monooxygenases* require 2 e^- as a co-substrate (Fig. 7.18). These electrons are recovered in subsequent oxidation reactions and are also unavailable for energy conservation!

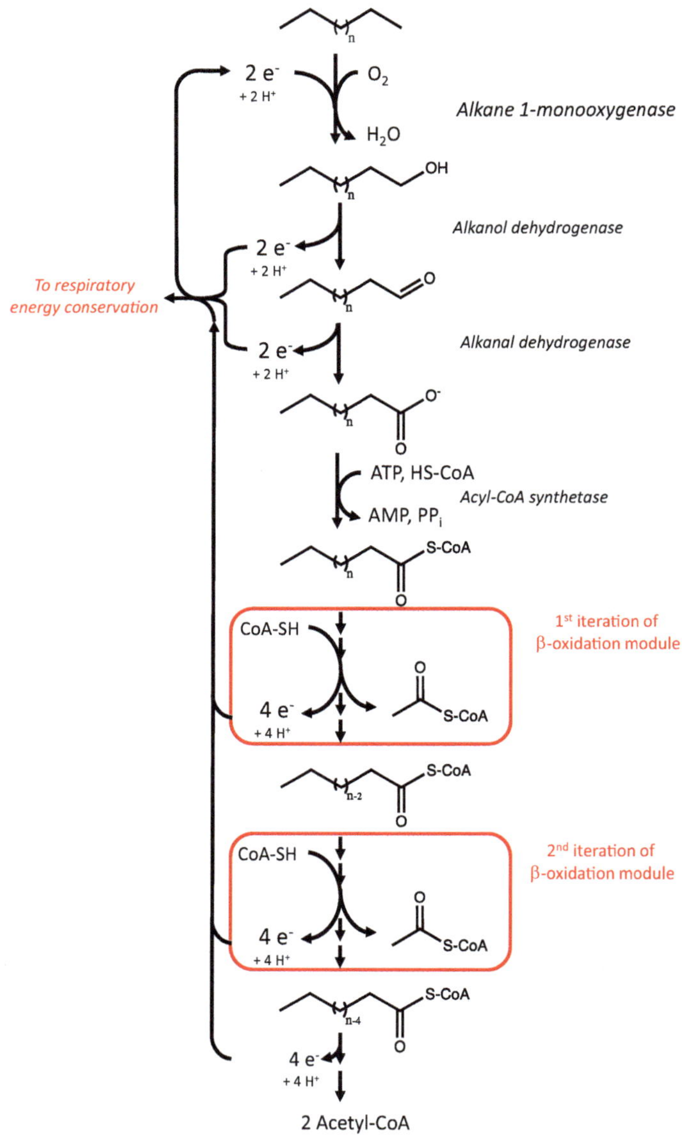

Fig. 8.5 Metabolic modules involved in aerobic, oxidative degradation of even chain *n*-alkanes. A *monooxygenase* activates the *n*-alkane by oxidative hydroxylation, followed by oxidation via *alcohol/aldehyde dehydrogenase*. The fatty acid is activated by a *acylate synthetase* to the corresponding CoA thioester (see Fig. 8.2), and further degradation proceeds by iterative action of the β-oxidation module. While the type of transformation is modular, enzymes with substrate specificity appropriate to the chain length of an acyl-CoA intermediate catalyze the respective reaction in each module. 2 e^- as co-substrate for the initial *monooxygenase* are retrieved from a downstream oxidation reaction. *n* is an even number. Note, that no electrons are released in the *monooxygenase* reaction; therefore, these electrons are unavailable for respiratory energy conservation

Odd-chain alkanes are degraded via β-oxidation module to propionyl-CoA and then via the *acyl-CoA mutase* module or the methylcitrate cycle.

Aerobic hydrocarbon degradation is mostly affiliated with α-proteobacteria, such as *Rhodospirillum* and *Roseobacter* species, and with γ-proteobacteria, such as *Oceanospirillum*, *Colwellia*, *Alcanivorax*, and *Oleispira* species.

8.3.5.2 Anaerobic *n*-Alkane Degradation

Because of the absence in anoxic environments of the strong oxidant O_2 for C–H bond activation, a different metabolic strategy is required for anaerobic activation of alkanes. The example is again the oxidation of octane:

$$\text{Octane} + O_2 \rightarrow 4\text{Acetyl-CoA} + 14(e^- + H^+)$$

The current evidence from denitrifying prokaryotes indicates that anaerobic degradation of *n*-alkanes with $C_{>4}$ is initiated by a fumarate addition reaction catalyzed by an *alkylsuccinate synthase* (Figs. 7.18 and 8.6). Using fumarate as co-substrate *alkylsuccinate synthase* catalyzes the radical addition of the subterminal C2 carbon to fumarate to form an alkylsuccinate. *n*-Alkylsuccinate is activated presumably in a *CoA transferase* reaction (see Acylate ↔ acyl-CoA Module) to *n*-alkylsuccinyl-CoA. An *n-alkylsuccinyl-CoA mutase* (i.e., *acyl-CoA mutase* module) rearranges the thioester-activated carboxyl group at the vicinal carbon. The unreactive subterminal C2 carbon of octane is now activated as it is in β position to a carbonyl function (carboxyl-CoA). Therefore, subsequent reactions are straight β-oxidation module reactions generating acetyl-CoA and propionyl-CoA (Fig. 8.6). Propionyl-CoA is converted via the *methylmalonyl-CoA mutase* version of the *acyl-mutase* module to succinyl-CoA. The activated CoA thioester is transferred via the acylate ↔ acyl-CoA (*CoA transferase*) module, and succinate is oxidized to regenerate fumarate consumed in the initial addition reaction. Thus, the cyclic pathway for anaerobic *n*-alkane oxidation consists of several β-oxidation modules, an *acyl-CoA mutase* module, *succinate dehydrogenase* of the succinate ↔ oxaloacetate module, Acylate ↔ acyl-CoA (*CoA transferase*) Module, and a single, specific key funneling reaction: the *alkylsuccinate synthase* reaction. The red-framed modules in Fig. 8.6 visualize the large extent to which this pathway is simply a mosaic of several modules.

While the overall oxidation of n-alkanes (acyl-CoA) in aerobic and anaerobic microbes via β-oxidation modules is similar, formation of the first acyl-CoA intermediate from *n*-alkane proceeds by substantially different metabolic strategies: anaerobic activation, using *alkylsuccinate synthase* and C–C rearrangement reaction, versus oxidative *monooxygenase* and alcohol/aldehyde oxidations in aerobic pathways.

8.3.5.3 Anaerobic Degradation of Alkyl-Substituted Benzenes

The anaerobic pathway for degradation of methyl-substituted benzenes, such as of toluene or xylenes, in denitrifying and sulfate-reducing bacteria is another example of mosaic and iterative combinations of the β-oxidation module. The benzylsuccinate pathway (Fig. 8.7) is an upstream extension of the reductive benzoyl-CoA pathway (see Fig. 8.8) as toluene is metabolized to benzoyl-CoA as intermediate (node). The overall metabolic consequence of the benzylsuccinate pathway is the oxidation of the toluene methyl group to benzoyl-CoA:

$$\text{Toluene} \rightarrow \text{Benzoyl-CoA} + 6\,(e^- + H^+)$$

In toluene degradation, ***benzylsuccinate synthase*** catalyzes the radical addition of the methyl group of toluene to fumarate to form benzylsuccinate. Benzylsuccinate is converted to the CoA-thioester by a *CoA transferase* (Acylate ↔ acyl-CoA Module) to form benzylsuccinyl-CoA, which is oxidized to benzoyl-CoA by a β-oxidation module: *benzylsuccinyl-CoA dehydrogenase* catalyzes oxidation to phenylitaconyl-CoA, which is hydrated by an *E-phenylitaconyl-CoA hydratase* to 2-(hydroxymethylphenyl)-succinyl-CoA. The latter is oxidized by *(hydroxymethylphenyl)-succinyl-CoA dehydrogenase* to benzoylsuccinyl-CoA, which is thiolytically cleaved to benzoyl-CoA and succinyl-CoA. Succinyl-CoA serves as the CoA donor for the *CoA transferase* reaction with benzylsuccinate, and succinate is oxidized by a *succinate dehydrogenase* to regenerate fumarate.

Similar as in anaerobic *n*-alkane oxidation, the use of the β-oxidation module for substrate oxidation was enabled by placing the unreactive methyl group of toluene in β-position to a thioester of benzylsuccinyl-CoA leading to a perfect structural configuration for subsequent β-oxidation. Thus, six of the seven reactions in

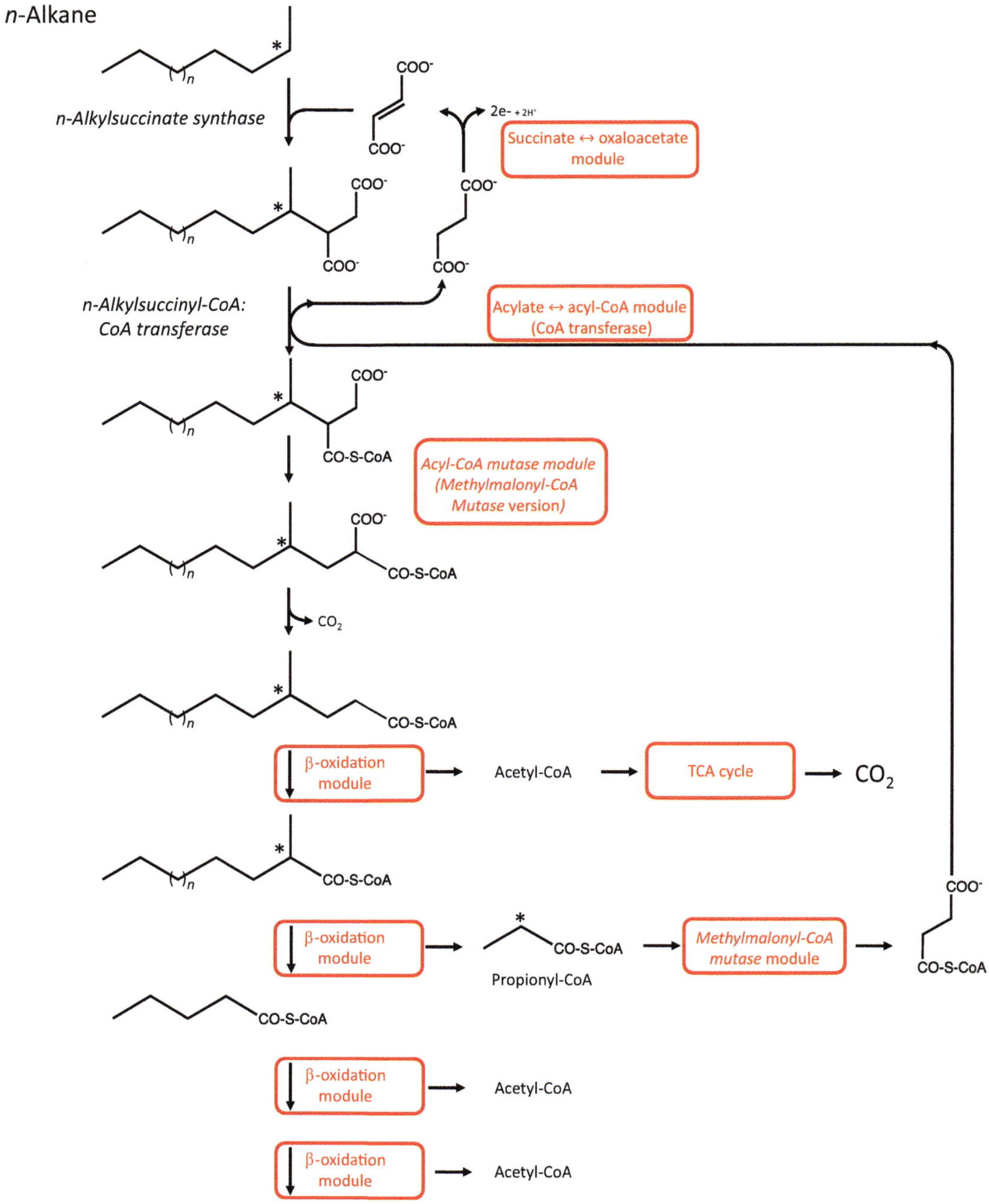

Fig. 8.6 Proposed anaerobic *n*-alkane degradation pathway in denitrifying bacteria. The overall degradation of even- or odd-chain *n*-alkanes to acetyl-CoA or propionyl-CoA, respectively, involves several modules (indicated in red): β-oxidation module, *CoA transferase* module, *acyl-CoA mutase* module. Electrons recovered from oxidation reactions are omitted to emphasize the carbon reactions as they are discussed in the modules. Only *alkylsuccinate synthase* is a unique, non-modular key enzyme. See text for details

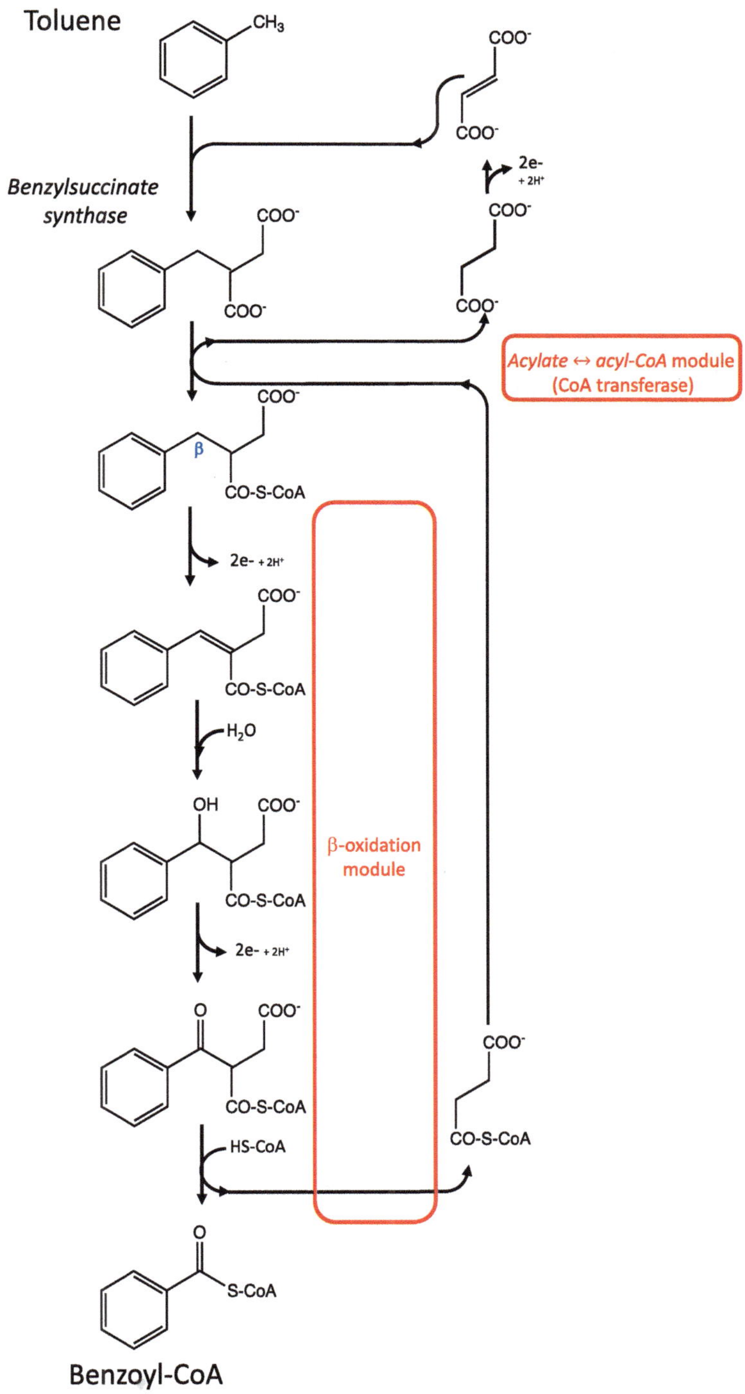

Fig. 8.7 *Benzylsuccinate synthase* pathway of anaerobic oxidation of toluene to benzoyl-CoA. The pathway consists of the modules (boxed in red): β-oxidation, Acylate ↔ acyl-CoA Module (CoA transfer), and *succinate dehydrogenase*, as well as an auxiliary fumarate addition reaction. The metabolic strategy of this pathway is to place the unreactive methyl group of toluene in β (3-) position to a CoA thioester (benzylsuccinyl-CoA) for β-oxidation. See text for details

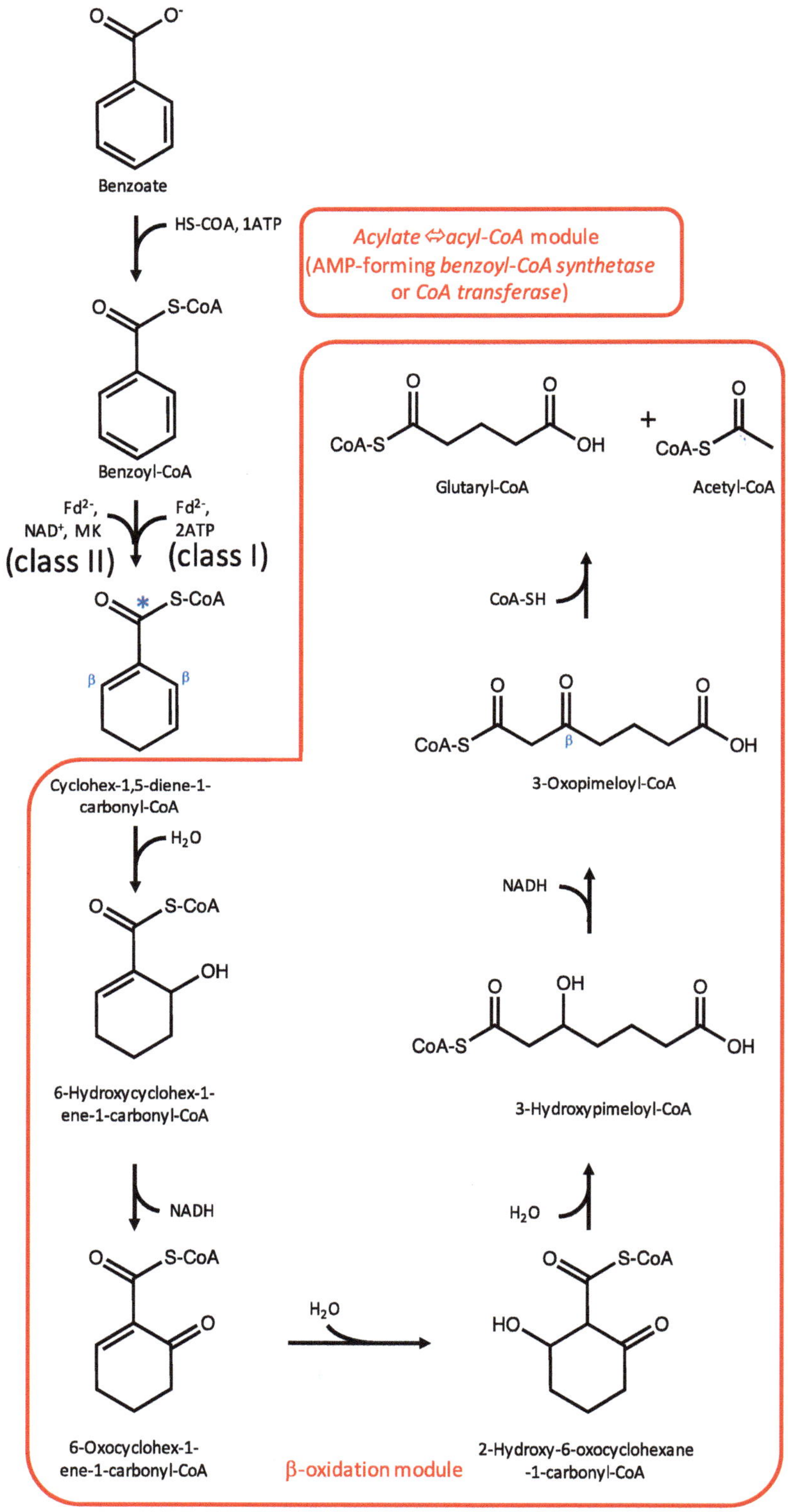

Fig. 8.8 Reductive benzoyl-CoA pathway for anaerobic degradation of benzoate/benzoyl-CoA. Upon formation of benzoyl-CoA by a *benzoyl-CoA ligase*, benzoyl-CoA is reductively dearomatized by either a class I *benzoyl-CoA reductase*, consuming 1 ATP per Fd^{2-}-derived electron, or a class II *benzoyl-CoA reductase*, bifurcating Fd^{2-}-derived electrons with NAD^+ and menaquinone as electron acceptor (see Sect. 7.4.3). The subsequent β-oxidation module reactions are framed in red. The enzymes involved are 1) *cyclohexa-1,5-diene-1-carbonyl-CoA hydratase*, 2) *6-hydroxycyclohex-1-ene-1-carbonyl-CoA dehydrogenase*, 3) *6-oxocyclohex-1-ene-1-carbonyl-CoA hydrolase*, 4) *6-oxocyclohex-1-ene-1-carbonyl-CoA hydrolase*, 5) *3-hydroxypimelyl-CoA dehydrogenase*, and 6) *3-oxopimelyl-CoA thiolase*, respectively

this cyclic pathway are modules; the seventh (*benzylsuccinate synthase*) is a recurrent auxiliary enzymatic reaction.

8.3.5.4 Anaerobic Degradation of Benzoate

Another illustrative example of the repetitive use of the β-oxidation module and its combination with diverse other modules is the pathway of anaerobic benzoate degradation (Fig. 8.8). Benzoate/benzoyl-CoA is a funneling intermediate (node) in anaerobic degradation of various aromatic hydrocarbons, including methyl-substituted benzenes (see above). Benzoyl-CoA is oxidized via the reductive benzoyl-CoA pathway:

$$\text{Benzoate/Benzoyl-CoA} \rightarrow \text{Acetyl-CoA} + \text{Glutaryl-CoA} + 2[\text{H}]$$

Benzoate is activated to benzoyl-CoA via the Acylate ↔ acyl-CoA Module by an AMP-forming *benzoyl-CoA synthetase* or via a *CoA transferase* (Fig. 8.8). Benzoyl-CoA is subsequently reduced by either a class I or a class II *benzoyl-CoA reductase* to cyclohex-1,5-diene-1-carbonyl-CoA (see Sect. 7.4.3). In anaerobic benzoate-utilizing *Thauera* and *Azoarcus* species, a class I *benzoyl-CoA reductase* utilizes 2 ATP per 2 ferredoxin-derived electron for reductive dearomatization, while in *Geobacter metallireducens* a membrane-associated electron-bifurcating class II enzyme is employed. After ring reduction, cyclohex-1,5-diene-1-carbonyl-CoA is oxidized and cleaved to linear glutaryl-CoA and acetyl-CoA by β-oxidation module reactions, *enoyl-CoA hydratases*, *alcohol dehydrogenases*, and an *aldolase,* using OH^- of water or the thiolate of CoA-SH (*thiolase*) as nucleophile.

8.3.5.5 Citric Acid Cycle

Although introduced as a separate module/pathway here, the **citric acid cycle** (or Tricarboxylic Acid Cycle, TCA cycle) (Fig. 8.9) represents another variation of the β-oxidation/succinate ↔ oxaloacetate module rather than a set of different types of enzymatic reactions. The TCA cycle catalyzes the complete oxidation of acetate (via acetyl-CoA) to 2 CO_2:

$$\text{Acetate}^-/\text{Acetyl-CoA} \rightarrow 2\text{CO}_2 + 8\,(\text{e}^- + \text{H}^+)$$

As the previous paragraphs showed, acetyl-CoA is an intermediate (node) in numerous degradation pathways. Further oxidation of acetyl-CoA by simple dehydrogenation of the methyl group is chemically difficult because of the carbanion character of the methyl carbon (see Sect. 7.2.3). Thus, the methyl group needs to be "activated" or placed into a structural configuration so that the above-outlined modules can act on the (former) acetate (acetyl-CoA) molecule. So far, there are only two major catabolic pathways known for complete oxidation of acetyl-CoA to 2 CO_2: the tricarboxylic acid cycle (see below) and the Wood–Ljungdahl pathway (see Sects. 8.5.5 and 10.3). While the TCA cycle functions in many aerobic and anaerobic prokaryotes, the oxidative Wood–Ljungdahl pathway follows a fundamentally different metabolic strategy and is so far found only in some strictly anaerobic sulfate-reducing and some syntrophic prokaryotes. However, when operating in reverse, the Wood–Ljungdahl pathway is a major catabolic pathway for using CO_2 as electron acceptor. 2 CO_2 are reduced to acetate (see below and Chap. 10).

In chemoorganotrophic microbes, the oxidative TCA cycle also couples catabolic substrate oxidation to anabolic pathways via a flux of TCA cycle intermediates as well as of electrons to biosynthetic pathways (Chap. 4). For example, about half of all amino acids are biosynthetically derived from intermediates of the TCA cycle. The TCA cycle, when operating in the reductive direction, is an anabolic pathway for autotrophic CO_2 fixation in some Prokaryotes (reverse TCA cycle).

The overall "strategy" of the oxidative TCA cycle is to place the methyl carbon of acetate functionally into β position of a carboxyl group to enable oxidation via the β-oxidation/succinate ↔ oxaloacetate module. Such strategy is similar

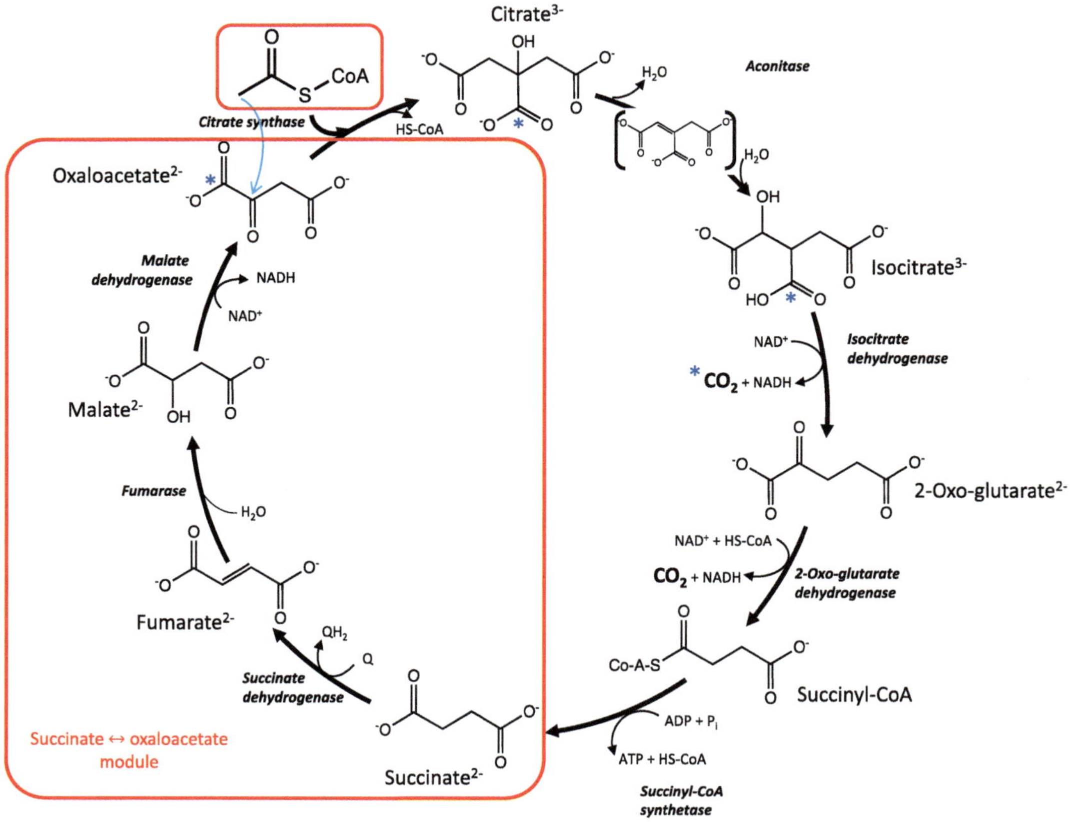

Fig. 8.9 The citric acid cycle (TCA) for complete oxidation of acetyl-CoA to 2 CO_2 and 8 e^-. The TCA cycle is written in the oxidative direction and is driven thermodynamically by the *citrate synthase* and *2-oxo-glutarate dehydrogenase* reactions, which are thermodynamically irreversible. The TCA cycle consists of succinate ↔ oxaloacetate module and prototypic aldol condensation (*citrate synthase*), alcohol oxidation (*isocitrate dehydrogenase*) and oxidative decarboxylation (*2-oxo-glutarate dehydrogenase*), and Acylate ↔ acyl-CoA (ADP-forming *succinyl-CoA synthetase*) Module. The instable oxalosuccinate is not shown. Abbreviations: Q, quinone; QH_2, quinol. See text for details

to anaerobic hydrocarbon degradation and requires an acceptor metabolite, in this case. Consequently, the TCA cycle is a cyclic pathway. The first reaction of the oxidative TCA cycle (Fig. 8.9), catalyzed by ***citrate synthase***, is the aldol condensation of acetyl-CoA with the 2-oxo carbon of oxaloacetate to form citrate (*si*/*re*) with concomitant release of HS-CoA. The alcohol carbon of citrate is a tertiary alcohol and cannot be oxidized by a prototypic, hydride-abstracting *dehydrogenase*. Thus, ***aconitase*** catalyzes the interconversion of citrate to isocitrate, which carries the alcohol function at a secondary carbon. ***Isocitrate dehydrogenase*** oxidizes that carbon to an instable oxalosuccinate intermediate, which spontaneously decarboxylates to 2-oxo-glutarate plus CO_2 (this is formally a decarboxylation of a 2-oxo acid). 2-oxo-glutarate is subsequently oxidatively decarboxylated by ***2-oxo-glutarate dehydrogenase*** to succinyl-CoA in a reaction analogous to the *pyruvate dehydrogenase* reaction (see Sects. 7.3.2 and 7.3.3). Succinyl-CoA is converted to succinate by ***succinyl-CoA synthetase*** plus ATP (or equivalent). ***Succinate dehydrogenase*** oxidizes succinate to fumarate, which is hydrated to malate by ***fumarase***. Finally, ***malate dehydrogenase*** oxidizes malate to oxaloacetate, regenerating the initial acceptor oxaloacetate for condensation with acetyl-CoA.

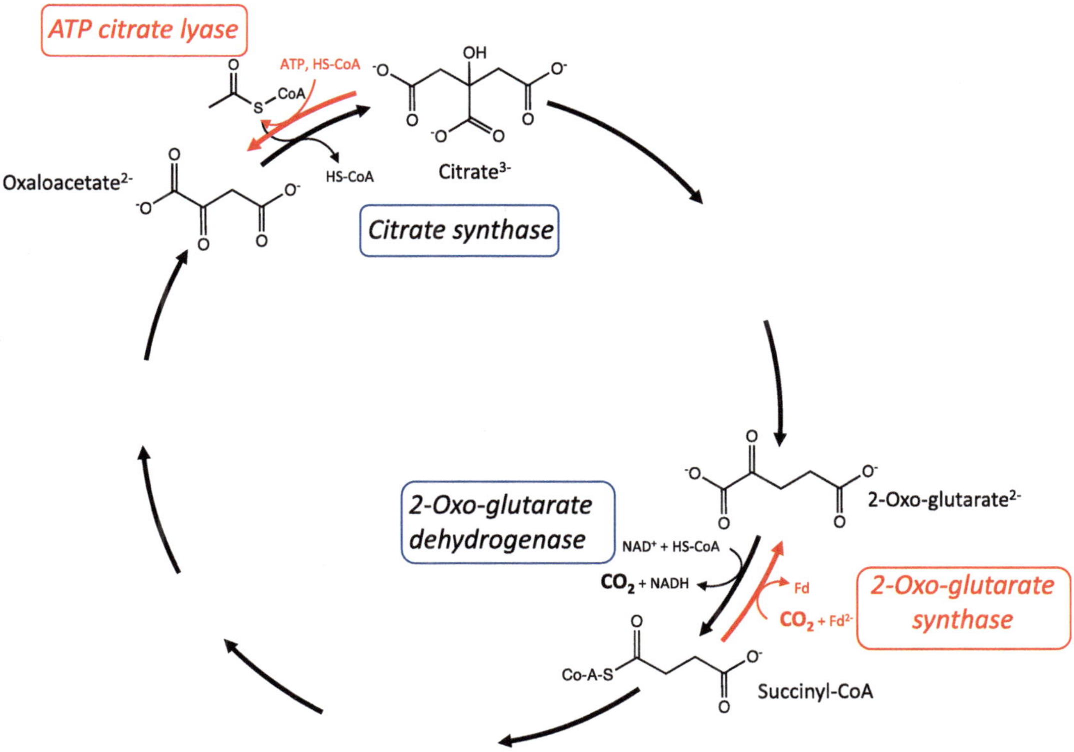

Fig. 8.10 Direction-defining enzymes of the oxidative (black) and reductive (red) TCA cycle. While all enzyme reactions of the TCA cycle other than those catalyzed by *citrate synthase* and *2-oxo-glutarate dehydrogenase* are reversible, the oxidative direction is determined by the activity of *citrate synthase* and *2-oxo-glutarate dehydrogenase* (both in black and black arrowed reactions), while operation in the reductive direction requires instead *ATP citrate lyase* and *2-oxo-glutarate synthase* (both in red and red arrowed reactions). Acetyl-CoA is substrate for *citrate synthase* and product of *ATP citrate lyase*, respectively. Note, that for the reductive carboxylation of succinyl-CoA, reduced ferredoxin is the electron donor, while typically NAD^+ is electron acceptor in the oxidative directions. See text for details

Figure 8.9 shows that all reactions of the TCA cycle, with the exception of the *aconitase* reaction, can be understood as previously introduced modules or prototype reactions; *citrate synthase* reaction, as an aldol condensation reaction, is similar to the reverse direction of a *β-ketothiolase* reaction in the β-oxidation module while *isocitrate dehydrogenase* is a bonified hydride-abstracting *alcohol dehydrogenase.*

In the anabolic (reverse) direction, the TCA cycle is used for autotrophic CO_2 fixation, however with modifications. The reductive TCA cycle is found, for example in phototrophic *Chlorobium limicola*, chemotrophic *Desulfobacter hydrogenophilus*, hyperthermophilic *Aquifex pyrophilus, Hydrogenobacter thermophilus*, ε-Proteobacterium *Thiomicrospira denitrificans*, and sulfur-reducing Crenarchaeota *Thermoproteus neutrophilus* and *Pyrobaculum islandicum.* Two pathway steps determine whether the TCA cycle operates in the forward oxidative or reverse reductive direction: *citrate synthase vs ATP citrate lyase*, and *2-oxo-glutarate dehydrogenase vs 2-oxo-glutarate synthase,* respectively (Fig. 8.10). The *citrate synthase* reaction in the direction of citrate formation is exergonic under standard conditions by −32.2 kJ/mol, thereby providing a thermodynamic pull for kinetically "trapping" acetate (acetyl-CoA) into the TCA cycle. Many *citrate synthases* appear irreversible in forming a thioester bond from HS-CoA and citrate in the absence of ATP. For a TCA cycle operating in the reverse, reductive direction, such kinetic bias in the oxidative direction can be overcome in two ways. Mesophilic microbes using the anabolic reductive TCA cycle employ an *ATP citrate lyase* for citrate cleavage into oxaloacetate and acetyl-CoA. This reaction is the end point of fixing 2 CO_2 into acetate and releases oxaloacetate as the acceptor for a new round of carboxylation reactions. Acetyl-CoA is assimilated into biomass. Interestingly, *Desulfobacter postgatei* contains a catabolic *ATP citrate lyase* in its oxidative TCA cycle.

On the other hand, thermophilic *Desulfurella acetivorans* and *Thermosulfidibacter takaii* utilize a catabolic oxidative TCA cycle for acetyl-CoA oxidation and an anabolic reductive TCA cycle when growing autotrophically. In both modes, the TCA cycles use a reversible *citrate synthase* as opposed to a *ATP-citrate lyase* during CO_2 reduction. It is assumed that dissipation of sufficient metabolic energy in the reductive TCA cycle, e.g., by employing Fd^{2-} dependent *reductases*, enhances the overall carbon flux through this anabolic cycle in the direction of CO_2 fixation. In addition, the loss of energy as heat, or irreversibility, is ecologically beneficial as this "locks" carbon into newly formed biomass, which is ecological success. Nevertheless, based on the Haldane equation and assuming similar K_M values in the forward and reverse reaction, a reversible *citrate synthase* would need to be present in several thousand-fold excess relative to the other anabolic enzymes of the reverse citric acid cycle so that carbon flux to acetyl-CoA is not kinetically limited by the unfavorable *citrate synthase* reaction. A requirement for a significant proteome space of *citrate synthase* in an anabolic proteome can be easier accommodated than in a catabolic proteome because its lower specific activity, adjusted by a lower *citrate synthase* concentration, is sufficient for an anabolic function.

The second enzyme defining the direction of the TCA cycle is *2-oxo-glutarate dehydrogenase* catalyzing 2-oxo-glutarate oxidation to succinyl-CoA plus CO_2 (Fig. 8.10). Oxidation of 2-oxo-glutarate releases electrons at a redox potential of $E^{0\prime} \approx -500$ mV. Transfer of these electrons to NAD^+ at $E^{0\prime} = -320$ mV via *NAD^+-dependent 2-oxo-glutarate dehydrogenase* renders such transfer thermodynamically irreversible, in particular in a respiratory microorganism with O_2 or NO_3^- as terminal electron acceptor. Microbes that fix CO_2 via the reductive TCA cycle contain and express a ***2-oxo-glutarate synthase*** that uses Fd^{2-} as electron donor for the reduction of succinyl-CoA plus CO_2 to 2-oxo-glutarate. The low redox potential of Fd^{2-} of $E^{0\prime} = -500$ mV is sufficient for Fd^{2-} to provide electrons for *2-oxo-glutarate synthase*. *2-Oxo-glutarate synthase* is mechanistically similar to *pyruvate synthase*, also a *2-oxo acid synthase*, catalyzing the assimilation of the C_2 compound acetyl-CoA into the C_3 pyruvate for biosynthesis (see below). A *ferredoxin-dependent 2-oxo-glutarate dehydrogenase* (which effectively is a *2-oxo-glutarate synthase*) operates in the oxidative direction of the TCA cycle of some carbohydrate-oxidizing archaea and in *Thauera aromatica* when oxidizing benzoate anaerobically.

In *Hydrogenobacter thermophilus* TK-6, there is a third enzyme defining the direction of the TCA cycle, namely an *ATP-dependent isocitrate dehydrogenase*, which contains biotin.

As previously discussed, *fumarase* has different kinetic properties, depending on whether the enzyme functions in the oxidative or reductive direction of the cycle. *Fumarase* in the oxidative direction catalyzes fumarate hydration to malate with a higher k_{cat} and K_M than malate dehydration to fumarate, while the kinetic parameters of *fumarase* in a reductive TCA cycle are reverse (see Sect. 5.4).

8.3.5.6 Methylcitrate Cycle

A variation of the TCA cycle and its modules is the **methylcitrate cycle** for oxidation of propionyl-CoA as found in *E. coli*, *Salmonella* sp., *Cupriavidus necator*, *Mycobacteria* sp., *Neisseria* sp., and *Bacillus subtilis* among others (Fig. 8.11). The overall reaction mediated by the methylcitrate cycle is the oxidation of propionate (via propionyl-CoA) to pyruvate:

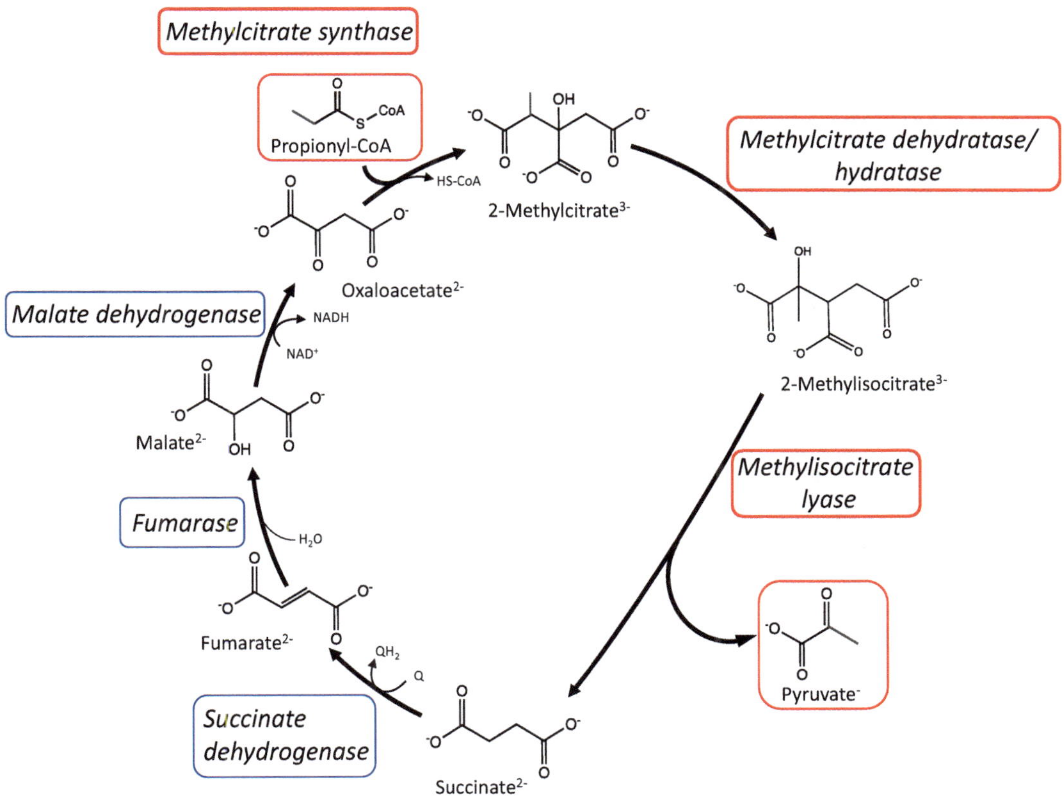

Fig. 8.11 The Methylcitrate Cycle for oxidation of propionyl-CoA to pyruvate. The pathway-specific enzymes are (framed in red) *methylcitrate synthase*, *methylcitrate dehydratase/hydrolase*, and *methylisocitrate lyase*. All other enzymes (framed in blue) are also present in the TCA cycle. See text for details

$$\text{Propionate}^- / \text{Propionyl-CoA} \rightarrow \text{Pyruvate}^- + 4\,(e^- + H^+)$$

An aldol condensation-type *methylcitrate synthase* catalyzes the condensation of the C2 carbon of propionyl-CoA to oxaloacetate to form 2-methylcitrate. A *methylcitrate dehydratase/hydratase* dehydrates 2-methylcitrate to 2-methylaconitate and subsequently hydrates the latter to 2-methylisocitrate in reactions analogous to *aconitase*. 2-Methylisocitrate is aldolytically cleaved by a *2-methylisocitrate lyase* to pyruvate and succinate. Succinate is oxidized by the succinate ↔ oxaloacetate module to oxaloacetate, which is the propionyl-CoA acceptor molecule. Pyruvate is metabolized further by several pathways as discussed below.

The modular architecture of the methylcitrate cycle (Fig. 8.11) again reemphasizes that for prototypic reactions and modules to function only certain structural features in organic substrates need to be met, while the chemical nature of substituents that are less important for a given transformation can be less restrictive. Specifically (Figs. 8.9 and 8.11), the CoA thioester in propionyl-CoA and acetyl-CoA induces a carbanionic C2 carbon of acetyl-CoA and propionyl-CoA, respectively. Therefore, these carbons are good nucleophile regardless of having a methyl group as one of the three substituents as in propionyl-CoA.

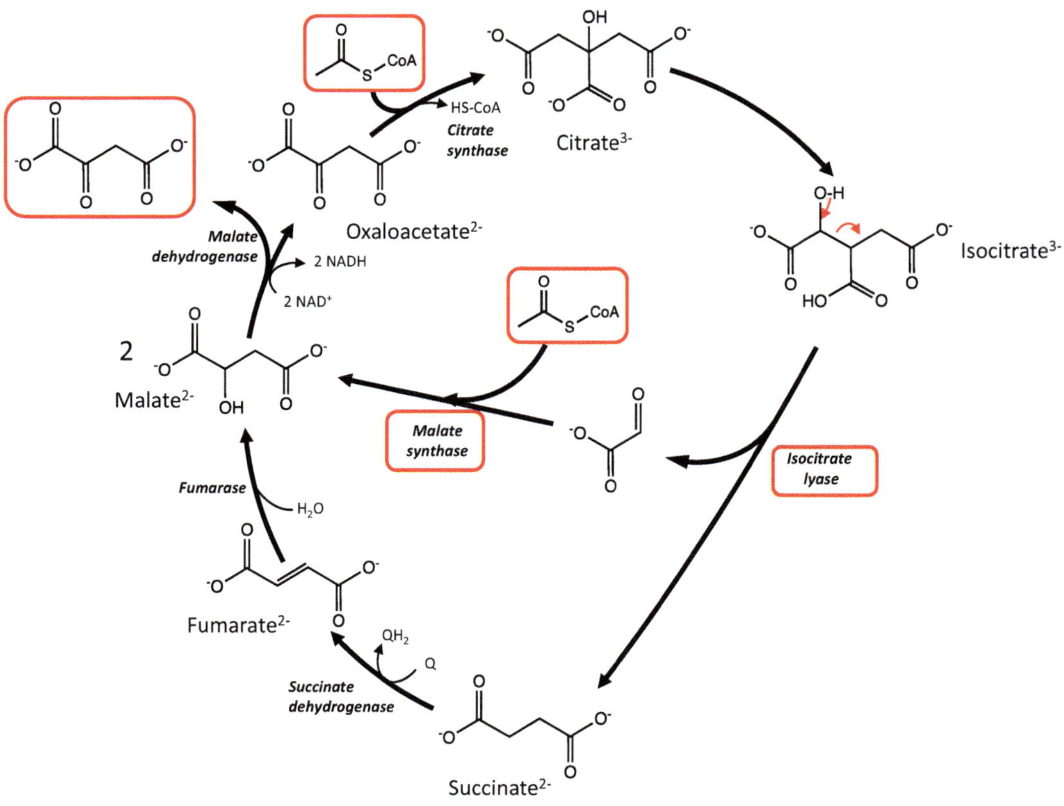

Fig. 8.12 The glyoxylate shunt for anaplerotic formation of oxaloacetate from acetyl-CoA. The enzymes that distinguish the glyoxylate shunt from the oxidative TCA cycle are *isocitrate lyase*, in lieu of *isocitrate dehydrogenase*, and *malate synthase*, which is absent in the TCA cycle. Boxed in red are the pathway-defining enzymes (*isocitrate lyase* and *malate synthase*). The glyoxylate shunt lacks *isocitrate dehydrogenase*, *2-oxo-glutarate dehydrogenase* and *succinyl-CoA synthetase*, but contains the succinate ↔ oxaloacetate module (*succinate dehydrogenase*, *fumarase*, and *malate dehydrogenase*). See text for details

8.3.5.7 Anaplerotic Reactions

Anaplerotic reactions are reaction sequences necessary in cyclic catabolic pathways to compensate for consumption of intermediates in anabolic pathways. For example, nine out of the 20 essential amino acids are biosynthetically derived from TCA cycle intermediates (e.g., oxaloacetate, 2-oxo-glutarate), and tetrapyrrole biosynthesis via the δ-aminolevulinate pathway starts with succinyl-CoA in chemoorganotrophic, acetate-utilizing microbes. Thus, oxaloacetate and 2-oxo-glutarate, for example, are nodes that interconnect catabolic and anabolic pathways. Anaplerotic reaction sequences replenish the acceptor metabolite of a cyclic pathway, here oxaloacetate, via the glyoxylate shunt (Fig. 8.12) or the synthesis of oxaloacetate from pyruvate or PEP.

In the overall reaction, the glyoxylate shunt converts 2 acetyl-CoA into 1 oxaloacetate. The modules of the glyoxylate shunt include *citrate synthase* and *aconitase* as the first steps and the succinate ↔ oxaloacetate module, which is also the second half of the TCA cycle. Distinguishing the TCA cycle and the methylcitrate cycle from the glyoxylate shunt is aldolytically cleaving *isocitrate lyase* (vs. *isocitrate dehydrogenase* in the TCA cycle and *2-methylisocitrate lyase* in the methylcitrate cycle) forming succinate and glyoxylate, and *malate synthase*, which catalyzes the aldol condensation of a second acetyl-CoA with the carbonyl carbon of glyoxylate. The latter reaction is analogous to *citrate synthase* reaction for condensation of acetyl-CoA to the carbonyl carbon of oxaloacetate.

This comparison of these three TCA cycle-related pathways reemphasizes the modularity of and recurrence of modules in pathways and the different metabolic outputs of each variation. Despite the similarity in prototypic reactions, the high substrate specificity of enzymes involved in each pathway guarantees that there is only controlled metabolic cross-talk between them.

There are other important enzymatic reactions, based on the introduced prototype reactions and modules, for generating the metabolic nodes oxaloacetate and phosphoenolpyruvate in chemoorganotrophic microorganisms. Although not catabolic, such reactions connect metabolic nodes for balancing cellular carbon flux by interconverting C_2 compounds to C_3 and C_6 biosynthetic precursors such as 2-phosphoglycerate, phosphoenolpyruvate, or glucose-6-phosphate. Figure 8.13 shows some important enzymes and metabolic connections.

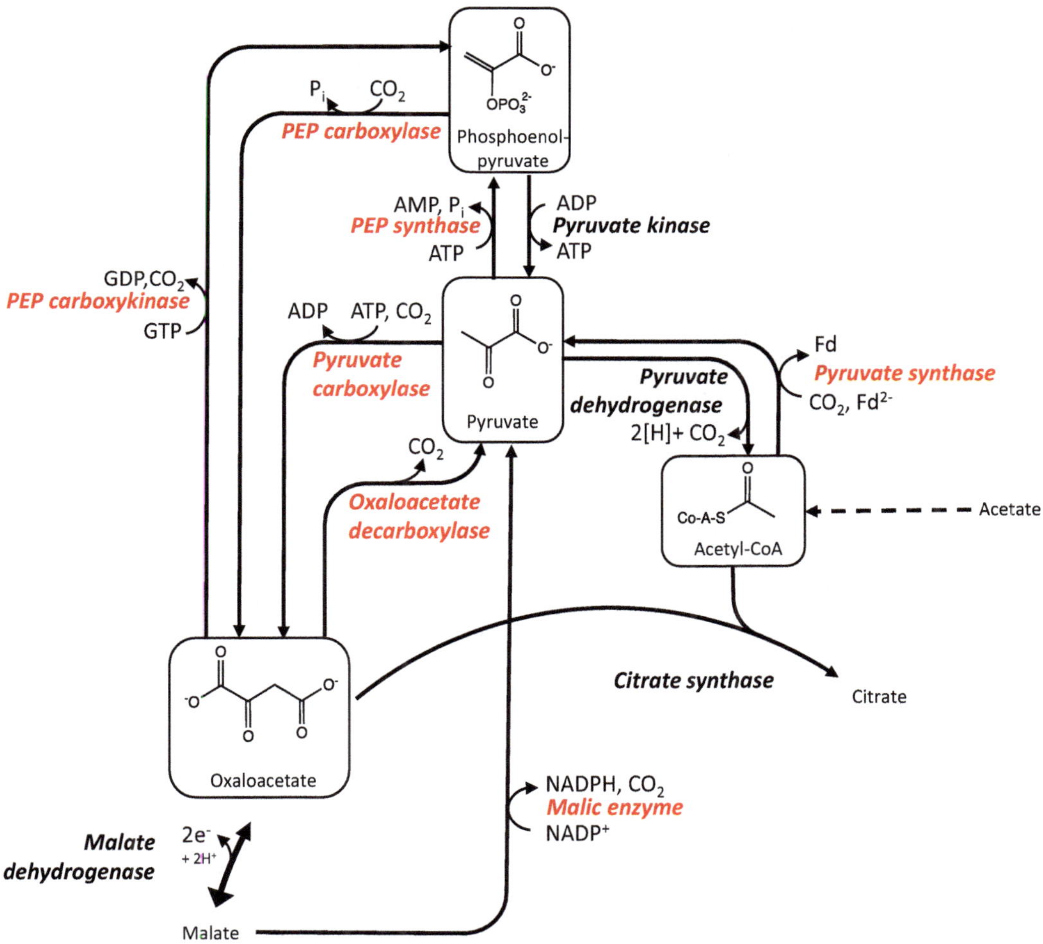

Fig. 8.13 Enzyme reactions connecting metabolism of C_2 to C_6 carbon intermediates in organotrophic microorganisms. These enzyme reactions at the intersection of pyruvate-phosphoenolpyruvate-acetyl-CoA-oxaloacetate are primarily anabolic to balance carbon flux between catabolism of organic compounds and assimilation by anabolism. Indicated in red are anaplerotic enzymes; indicated in black are catabolic enzymes. See text for details

Phosphoenolpyruvate (PEP) carboxykinase catalyzes the decarboxylation and phosphorylation of oxaloacetate to PEP. Phosphoryl group donor can be ATP (*Planctomyes maris, Formosa agariphila*) or GTP (*Thioalcalivibrio nitratireducens, Actinomyces graevenizii*), and the reaction is kinetically biased toward PEP formation. In *Propionibacterium freudenreichii* while using PP_i as phosphoryl donor, the reaction is biased in the direction of oxaloacetate formation.

Phosphoenolpyruvate (PEP) carboxylase, such as in *E. coli*, catalyzes the carboxylation and dephosphorylation of PEP to oxaloacetate. This reaction is physiologically irreversible because the energy released in PEP dephosphorylation ($\Delta G^{0\prime} = -52$ kJ/mol) is larger than the energy required for carboxylation ($\Delta G^{0\prime} = +20$ kJ/mol).

Phosphoenolpyruvate (PEP) synthetase catalyzes formation of PEP from pyruvate. This reaction is driven by the hydrolysis of ATP to AMP + P_i, which is sufficient for phosphorylating pyruvate ($\Delta G^{0\prime} = +52$ kJ/mol). Interestingly, *Pyrococcus furiosus* contains a PEP *synthetase* that may substitute for the absence of *pyruvate kinase* in catabolic carbohydrate degradation by the Embden–Meyerhof pathway. When catalyzing the conversion of PEP to pyruvate using PEP *synthetase,* 1 AMP plus P_i plus the phosphoryl group of PEP are converted to 1 ATP; thereby effectively catalyzing formation of 2 ATP equivalents from 1 PEP.

Pyruvate carboxylase is a biotin-containing enzyme and catalyzes the irreversible carboxylation of pyruvate with bicarbonate to oxaloacetate with concomitant hydrolysis of 1 ATP to 1 ADP plus P_i.

Pyruvate synthase catalyzes carboxylation and Fd^{2-}-dependent reduction of acetyl-CoA to pyruvate. This reaction is reversible and functionally equivalent to *pyruvate:ferredoxin oxidoreductase* as found in catabolism of anaerobic fermenting microbes (see Chap. 9). *Pyruvate synthase* is important for anaerobic microbes metabolizing C_2 compounds, such as acetate, for biosynthesis of C_3, C_5, and C_6 metabolites, and functions often in conjunction with *PEP synthetase*, e.g., for gluconeogenesis.

Oxaloacetate decarboxylase catalyzes the irreversible decarboxylation of oxaloacetate to pyruvate ($\Delta G^{0\prime} = -20$ kJ/mol). When functioning as anaplerotic/anabolic enzyme, *oxaloacetate decarboxylase* is cytoplasmic and not coupled to energy conservation as in catabolic *oxaloacetate decarboxylase*.

Malic enzyme catalyzes $NADP^+$-dependent oxidative decarboxylation of malate to pyruvate. The overall reaction can be considered as a combined *NAD^+-dependent malate dehydrogenase* plus an *oxaloacetate decarboxylase* reaction.

8.4 Common Modules and Pathways of Carbohydrate Degradation

Most hexose- and pentose-degrading pathways have in common the module of *glyceraldehyde-3-phosphate dehydrogenase (GAP-DH)* converting glyceraldehyde-3-phosphate to pyruvate (GAP → Pyruvate module, Fig. 8.14a). Prototypic *aldolases or ketolases* convert phosphorylated C_5 or C_6 carbohydrate substrate into glyceraldehyde-3-phosphate for entry into the GAP → Pyruvate module. This module is discussed first, followed by prototypic *aldolase* and *ketolase* reactions, followed by their functioning in diverse carbohydrate-transforming pathways. All carbohydrate-degrading pathways have in common that the carbohydrate substrate is degraded via pyruvate or acetyl-CoA/acetyl-phosphate as intermediates.

8.4.1 The GAP → Pyruvate Module

The GAP → Pyruvate module mediates the overall oxidation of glyceraldehyde-3-phosphate (GAP) to pyruvate (Fig. 8.14a) with energy conservation according to:

$$\text{Glyceraldehyde-3-phosphate} \rightarrow \text{Pyruvate}^- + 2\,(e^- + H^+) + 1\text{-}2\ \text{ATP}$$

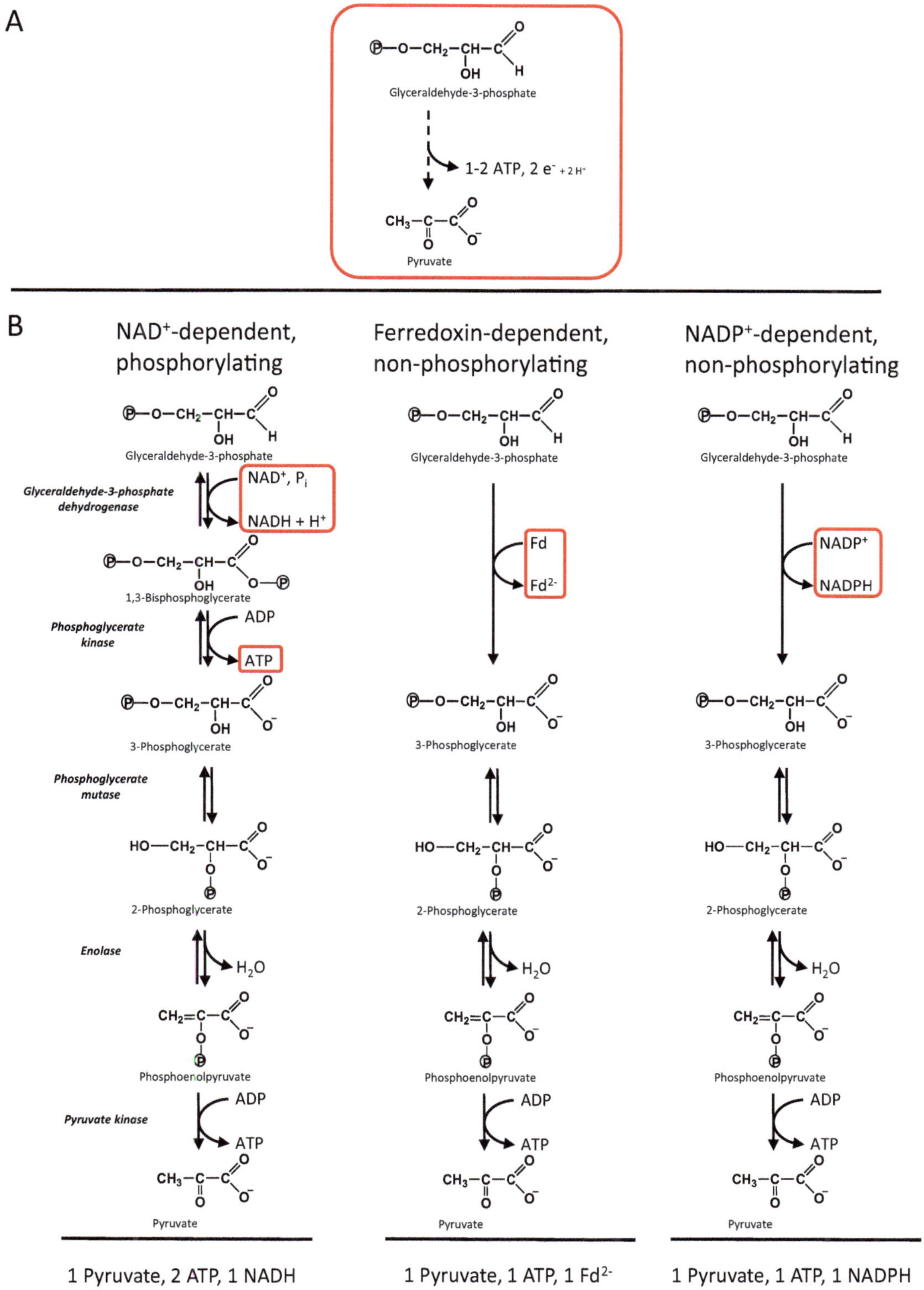

Fig. 8.14 The GAP → Pyruvate module (**a**) and different versions thereof (**b**): *NAD⁺-dependent, phosphorylating glyceraldehyde-3-phosphate dehydrogenase*, *non-phosphorylating glyceraldehyde-3-phosphate:ferredoxin oxidoreductase*, and *NADP⁺-dependent, non-phosphorylating glyceraldehyde-3-phosphate dehydrogenase*. Electrons from glyceraldehyde-3-phosphate oxidation are transferred to either NAD^+, Fd, or $NADP^+$. ATP recovery (n_{ATP}) varies between 1–2 ATP per pyruvate formed. Fd^{2-} oxidation with NAD^+ via Rnf recovers 0.5 ATP. See text for details

There are at least three variations of this module that differ in n_{ATP} (ATP yield) and electron carriers involved (i.e., (NAD^+, Fd, $NADP^+$). These differences are based on different versions of the *glyceraldehyde-3-phosphate dehydrogenase* reaction as discussed in Sect. 7.2.3. Figure 8.14b shows the three GAP → Pyruvate modules and their net metabolic effects. These modules vary only in the *GAP dehydrogenase* reaction and are identical in the overall conversion of 3-phosphoglycerate to pyruvate, although there can also be variations in PEP conversion to pyruvate (not shown here). The first version of the GAP → Pyruvate module (Fig. 8.14b) is catalyzed by ***NAD⁺-dependent, phosphorylating glyceraldehyde-3-phosphate dehydrogenase*** oxidizing the aldehyde of glyceraldehyde-3-phosphate to 1,3-bisphosphoglycerate. In this version, the energy of hydrolysis of 1,3-bisphosphoglycerate is conserved as 1 ATP by ***phosphoglycerate kinase*** releasing 3-phosphoglycerate. The two other versions of this module are based on a ***non-phosphorylating glyceraldehyde-3-phosphate:ferredoxin oxidoreductase*** and a ***NADP⁺-dependent, non-phosphorylating GAP dehydrogenase*** (Fig. 8.14b). In these two versions, 3-phosphoglycerate and not 1,3-bisphosphoglycerate is the product of the respective *GAP dehydrogenase* reactions. *NADP⁺-dependent, non-phosphorylating GAP dehydrogenase* (commonly referred to as GAPN) as well as *non-phosphorylating glyceraldehyde-3-phosphate:ferredoxin oxidoreductase* (GAPOR) are irreversible. All three versions have in common the subsequent steps from 3-phosphoglycerate to pyruvate: ***phosphoglycerate mutase*** converts 3-phosphoglycerate to 2-phosphoglycerate, from which H_2O is abstracted by an ***enolase*** to form PEP. ***Pyruvate kinase*** catalyzes the transfer of the phosphoryl group of PEP to ADP, forming pyruvate and ATP. Variations in PEP conversion to pyruvate involving either a *pyruvate kinase* or a *PEP synthetase* are not discussed here. The differences between the three variations in n_{ATP} and the involvement of different electron carriers have physiological and ecological implications.

NAD^+-dependent, phosphorylating GAP → Pyruvate module is found in diverse pathways such as the Embden–Meyerhof ("lower glycolysis"), Entner–Doudoroff Pathway, Phosphoketolase Pathway, Bifidobacterium pathway, and in diverse Bacteria and Archaea. The Fd-dependent GAP → Pyruvate module is found in hyperthermophilic Archaea, such as *Pyrococcus furiosus*, *Pyrobaculum* sp., and the $NADP^+$-dependent module is found in *Thermococcus kodakarensis, Thermoproteus tenax*, and *Sulfolobus solfataricus.*

8.4.2 *Aldolase* and *Phosphoketolase* Reactions

For converting hexoses and pentoses to GAP as the substrate for the nearly ubiquitous GAP → Pyruvate module, carbon–carbon bonds of C_5 and C_6 carbohydrates are cleaved in diverse pathways via **prototypic aldolytic** or **ketolytic cleavage reactions** (Fig. 8.15). As described in Sects. 7.3.1 and 7.3.2, the essential structural configuration of substrates of ***aldolases*** is that of a hydroxy group in 3-position to a carbonyl function. *Aldolases* are reversible and can also act on non-carbohydrate compounds. Figure 8.15a shows the reaction of *fructose-1,6-bisphosphate aldolase*, the key enzyme in the Embden–Meyerhof pathway. Many other aldolases, such as *2-keto-3-deoxy-6-phosphogluconate aldolase, transaldolase*, *citrate synthase,* or *isocitrate lyase*, function in the Entner–Doudoroff pathway, the pentose phosphate pathway, the TCA cycle, or the isocitrate shunt.

The second important prototypic reaction shared by many carbohydrate-transforming pathways in anaerobes is catalyzed by ***phosphoketolase*** (Fig. 8.15b). As described in Sect. 7.3.2 (Fig. 7.12), *phosphoketolase* catalyzes the ketolytic cleavage of xylulose-5-phosphate to glyceraldehyde-3-phosphate and a thiamine pyrophosphate-bound dihydroxyethyl adduct, which undergoes a subsequent intramolecular redox reaction: the electron pair (circled orange in Fig. 8.15b) of the carbonyl carbon of xylulose-5-phosphate reduces the terminal alcohol to a

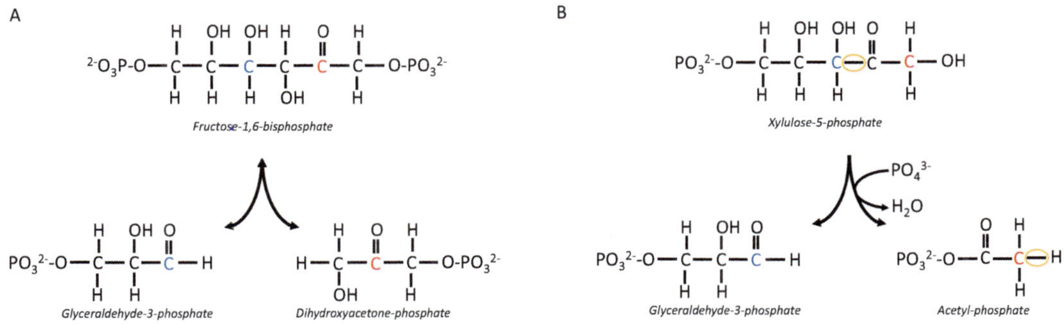

Fig. 8.15 Aldolytic and ketolytic enzymes in carbohydrate conversions. (**a**) The *fructose-1,6-bisphosphate aldolase* prototype reaction. Indicated in blue and red are the carbon atoms of relevant functional groups. (**b**) The *phosphoketolase* prototype reaction. Circled in yellow is the electron pair involved in the intramolecular redox reaction. While the *phosphoketolase* reaction (shown here) is physiologically irreversible, *transketolases* are reversible. See text for details

methyl group. Concomitant H_2O elimination makes this reaction physiologically irreversible. During the intramolecular redox reaction, the carboxylate carbon is phosphorylytically released from TPP by inorganic phosphate. The overall metabolic consequence of *phosphoketolase* is a **substrate-level-phosphorylation WITHOUT generation of a reduced cellular electron carrier (i.e., NADH)!**

Phosphoketolase is found in carbohydrate fermentations of pentoses as well as in Bifidobacterial fermentation of hexoses and in heterofermentative lactic acid bacteria metabolizing hexoses. *Ketolases* without acetyl-phosphate formation are found, for example, in the pentose phosphate cycle as *transketolases*.

8.4.3 Examples of GAP → Pyruvate Modules and *Aldolases/Ketolases* in Different Pathways

8.4.3.1 The Embden–Meyerhof (Glycolytic) Pathway

When the Embden–Meyerhof pathway uses a *NAD⁺-dependent, phosphorylating GAP dehydrogenase* in the GAP → Pyuvate module, glucose is oxidized to 2 pyruvate, 2 NADH, and 2 ATP (Fig. 8.16):

$$\text{Glucose} \rightarrow 2\ \text{Pyruvate}^- + 2\ \text{NADH} + 2\ \text{ATP}$$

This stoichiometry is different when the two other versions of *GAP dehydrogenase* are involved.

Entry point of the pathway is glucose-6-phosphate, which is formed from glucose by several routes depending on the mode of glucose uptake or intracellular release from oligosaccharides. In enteric bacteria such as *E. coli*, uptake of glucose proceeds via the PEP-dependent transferase system (PEP transferase system). In this system, transport of glucose is mechanistically coupled and thermodynamically driven by hydrolysis of PEP (from the GAP → Pyruvate module), resulting in glucose-6-phosphate formation as the first intracellular intermediate. Other carbohydrate-utilizing microbes take up glucose by a symport with H^+ or by an ATP-dependent ABC transporter, and activate glucose intracellularly by *hexokinase* to glucose-6-phosphate (Fig. 8.16). The *hexokinase* reaction is physiologically irreversible in vivo in fast-growing microbes as −18 kJ/mol glucose is released as heat under standard conditions; for glucose-6-phosphate synthesis from glucose and inorganic phosphate, only +14 kJ/mol is required while −32 kJ/mol are hydrolyzed from 1 ATP. One important metabolic implication of this type of phosphorylation is that the carbohydrate is

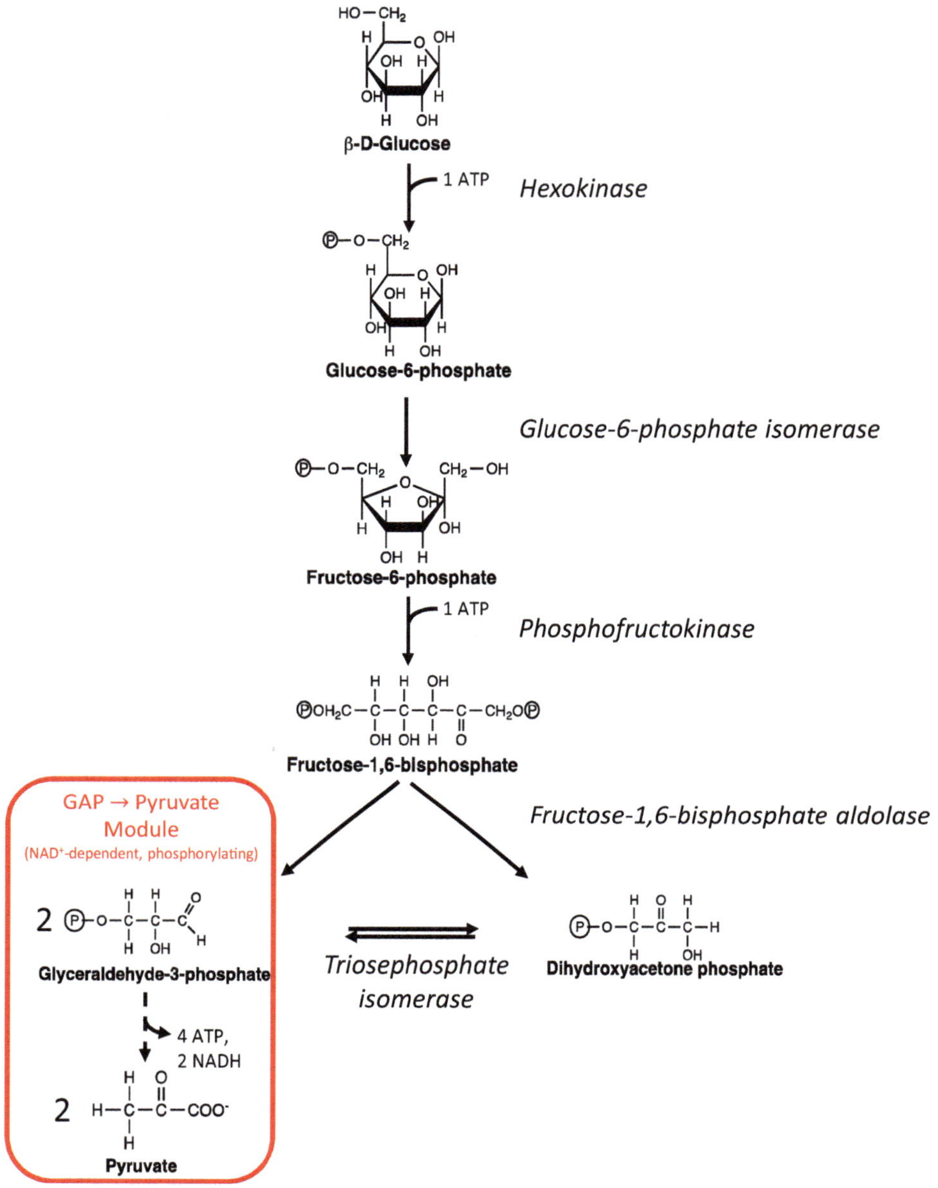

Fig. 8.16 The Embden–Meyerhof (glycolytic) pathway for glucose oxidation to 2 pyruvate, shown here with a *NAD^+-dependent, phosphorylating GAP dehydrogenase*. See text for details

irreversibly committed to the respective pathway for degradation in fast-growing microbes.

Glucose-6-phosphate is isomerized in an equilibrium reaction by ***glucose-6-phosphate isomerase*** to fructose-6-phosphate, which is subsequently phosphorylated at the C1 carbon to fructose-1,6-bisphosphate by ***phosphofructokinase*** (Fig. 8.16). This second phosphorylation reaction also is thermodynamically irreversible for the reason outlined. In the next step, ***fructose-1,6-bisphosphate aldolase*** catalyzes the reversible aldolytic cleavage of fructose-1,6-bisphosphate into glyceraldehyde-3-phosphate and dihydroxyacetone phosphate. A ***triosephosphate isomerase*** interconverts glyceraldehyde-3-phosphate and dihydroxyacetone phosphate in a reaction close to equilibrium. Glyceraldehyde-3-phosphate is subsequently oxidized to pyruvate by the GAP → Pyruvate module. Two glyceraldehyde-3-phosphate are oxidized and 2 pyruvate, 4 e^- and 2–4 ATP, are formed per glucose depending on the type of *GAP dehydrogenase* involved. Of these, 2 ATP equivalents are required in the Embden–Meyerhof pathway for hexose phosphorylation at the C1 and C6 positions.

Phosphorylation of Glucose

Why is the aldehyde of hexoses and pentoses not directly oxidized via Substrate Level Phosphorylation, and why are these carbohydrates cleaved first to GAP? One chemical rationale is that carbohydrates with carbon length of C > 4 in aqueous solution are present as the **cyclic hemiacetal** (e.g., 99% of glucose). Hemiacetals form between the aldehyde carbon in C1- or C2-position and the carbon in C5- or C4-position of hexoses or pentoses (Fig. 8.17), respectively. Thus, in aqueous solution, there is no functional carbonyl group at the C1 or C2 carbon of hexoses or pentoses (they are hemiacetals), and, therefore, no Substrate Level Phosphorylation can occur (see Sect. 6.3.1). In order to stabilize the carbonyl, in most degradation pathways hexoses and pentoses are first converted to a linear C_3 carbohydrate which cannot undergo hemiacetal formation. It is assumed that electrostatic repulsion of the negatively charged phosphoryl (e.g., in fructose-1,6-bisphosphate) or carboxyl and phosphoryl (e.g., 2-keto-3-deoxy-6-phosphogluconate) groups at each end of a hexose or pentose shift the equilibrium to the linear, non-hemiacetal form (Fig. 8.17).

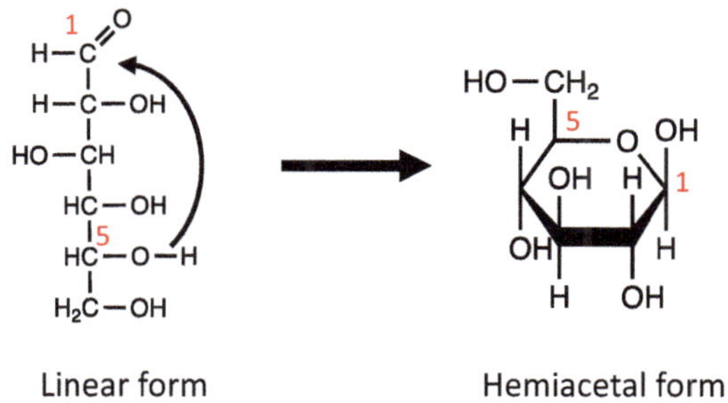

Fig. 8.17 Linear and hemiacetal form of glucose. In aqueous solution, glucose or fructose and other hexoses and pentoses are predominantly (99% for glucose) in the cyclic hemiacetal form

The Embden–Meyerhof pathway with its variations in the GAP → Pyruvate module operates in a large variety of microbes including enteric bacteria and primary fermenting bacteria, such as lactic acid bacteria and Clostridia.

Variations of the Embden–Meyerhof Pathway

There are interesting variations, predominantly in Archaea, of the glycolytic pathway as shown in Fig. 8.16. For example, in the Archaea *Pyrococcus furiosus*, *Thermococcus litoralis*, and *Archaeoglobus fulgidus*, glucose is not phosphorylated by an *ATP-dependent hexokinase*, but an *ADP-dependent glucokinase*. In *Thermoproteus tenax*, a PP_i*-dependent phosphofructokinase* rather than an *ATP-dependent phosphofructokinase is present*, using PP_i as the phosphoryl group donor. And *Pyrococcus furiosus* and *Archaeoglobus fulgidus* use an *ADP-dependent phosphofructokinase* instead.

As outlined above, an interesting variation is in the *GAP dehydrogenase* reaction. Anaerobic hyperthermophilic Archaea such as *Pyrococcus furiosus*, *Pyrobaculum aerophilum*, *Thermoproteus tenax*, *Archaeoglobus fulgidus*, and *Thermococcus kodakarensis* use an Embden–Meyerhof Pathway with a catabolic ***non-phosphorylating glyceraldehyde-3-phosphate: ferredoxin oxidoreductase*** which oxidizes GAP to 3-phosphoglycerate rather than to 1,3-bisphosphoglycerate with ferredoxin as electron acceptor (Fig. 8.14). The absence of ATP synthesis by SLP in this reaction is partially compensated in this microbe by re-oxidation of Fd^{2-} with NAD^+ via a membrane-associated Rnf forming an electrochemical ion gradient that, in turn, drives the phosphorylation of 0.5 ADP.

Some carbohydrate-fermenting Archaea and Bacteria carry a *NADP*$^+$*-specific, non-phosphorylating GAP dehydrogenase*.

8.4.3.2 Entner–Doudoroff Pathway

The Entner–Doudoroff pathway is a second pathway for glucose oxidation to 2 pyruvate:

$$\text{Glucose} \rightarrow 2\,\text{Pyruvate} + 4\,(e^- + H^+) + 0\text{-}1\,\text{ATP}$$

The stoichiometry of ATP/pyruvate (n_{ATP}) is different depending on the versions of *GAP dehydrogenase* involved.

The notable differences in hexose conversion to pyruvate by this pathway relative to the Embden–Meyerhof pathway are (1) that a ***2-keto-3-deoxy-6-phosphogluconate aldolase*** rather than a *fructose-1,6-bisphosphate aldolase* cleaves the phosphorylated hexose into C_3 units, and (2) that **only one pyruvate is formed** via GAP → Pyruvate conversion. Similar as in glycolysis, glucose is phosphorylated first to glucose-6-phosphate by intracellular ***hexokinase*** or coupled to uptake (Fig. 8.18). In the distinguishing next step, ***glucose-6-phosphate dehydrogenase*** oxidizes the hemiacetal C1 carbon without energy conservation to a carboxylate to form 6-phosphogluconate, which is dehydrated by ***6-phosphogluconate dehydratase*** at the C3 carbon to 2-keto-3-deoxy-6-phosphogluconate. Subsequent aldolytic cleavage of 2-keto-3-deoxy-6-phosphogluconate is catalyzed by ***2-keto-3-deoxy-6-phosphogluconate (KDPG) aldolase*** to form pyruvate and GAP. Thus, in the overall reaction, the Entner–Doudoroff pathway as shown in Fig. 8.18 converts glucose into 2 pyruvate, 2 NADH, and 1 ATP.

The Entner–Doudoroff pathway is predominantly found in Gram-negative Bacteria, such as *E. coli* or *Pseudomonas sp.* Many bacteria, including *E. coli* contain both a catabolic Emben–Meyerhof and an Entner–Doudoroff pathway.

Variations of the Entner–Doudoroff Pathway

Similar as in the Embden–Meyerhof pathway, the extent to which ATP (n_{ATP}) is formed in the Entner–Doudoroff pathway depends on which of the three types of *GAP dehydrogenases* are present in the GAP → Pyruvate module. If a *non-phosphorylating GAP:ferredoxin oxidoreductase* or a *NADP$^+$-dependent, non-phosphorylating GAP dehydrogenase* is involved, as in *Thermoplasma acidophilum*, only 1 ATP is conserved in the respective GAP → Pyruvate module. This ATP is, however, required in the Entner–Doudoroff pathway for hexose activation, and no net ATP is formed in the overall pathway from hexose to pyruvate. A **NADP$^+$-dependent**, ***non-phosphorylating glyceraldehyde-3-phosphate dehydrogenase*** operates in *Thermoproteus tenax*, *Aeropyrum pernix*, and *Sulfolobus solfataricus*.

8.4.3.3 Comparison Embden–Meyerhof and Entner–Doudoroff Pathway

Both pathways convert a hexose into 2 pyruvate; however, n_{ATP} is different even if both pathways contain a *NAD$^+$-dependent, phosphorylating GAP dehydrogenase* (Figs. 8.16 and 8.18). When employing this GAP → Pyruvate module, n_{ATP} is 1 in the Entner–Doudoroff pathway while in the Embden–Meyerhof pathway $n_{ATP} = 2$ at identical substrate and endproducts. This raises the question as to why microbes might operate and maintain an "inefficient" Entner–Doudoroff pathway for glucose oxidation to pyruvate. One explanation is that the difference in n_{ATP} enables a metabolic rate-yield trade-off: An inefficient glucose oxidation to 2 pyruvate by the Entner–Doudoroff pathway can proceed faster than via an efficient Embden–Meyerhof pathway. In fact, the above example of *NADP$^+$-dependent, non-phosphorylating GAP dehydrogenase* in *Thermoplasma acidophilum* shows that glucose conversion to pyruvate is at $n_{ATP} = 0$, enabling a faster flux of carbon to the TCA cycle and of electrons to the respiratory chain. Support for this explanation can be drawn also from the example of *Zymomonas mobilis* fermenting 1 glucose rapidly to 2 ethanol via the Entner–Doudoroff pathway, and outcompeting yeast, which uses the more efficient Embden–Meyerhof pathway. Moreover, for aerobic microbes under competitive conditions, glucose oxidation via the Entner–Doudorff pathway has the advantage that pyruvate is formed in fewer enzymatic steps (the first pyruvate is formed after 4 steps rather than 9 in glycolysis!).Therefore, pyruvate is formed

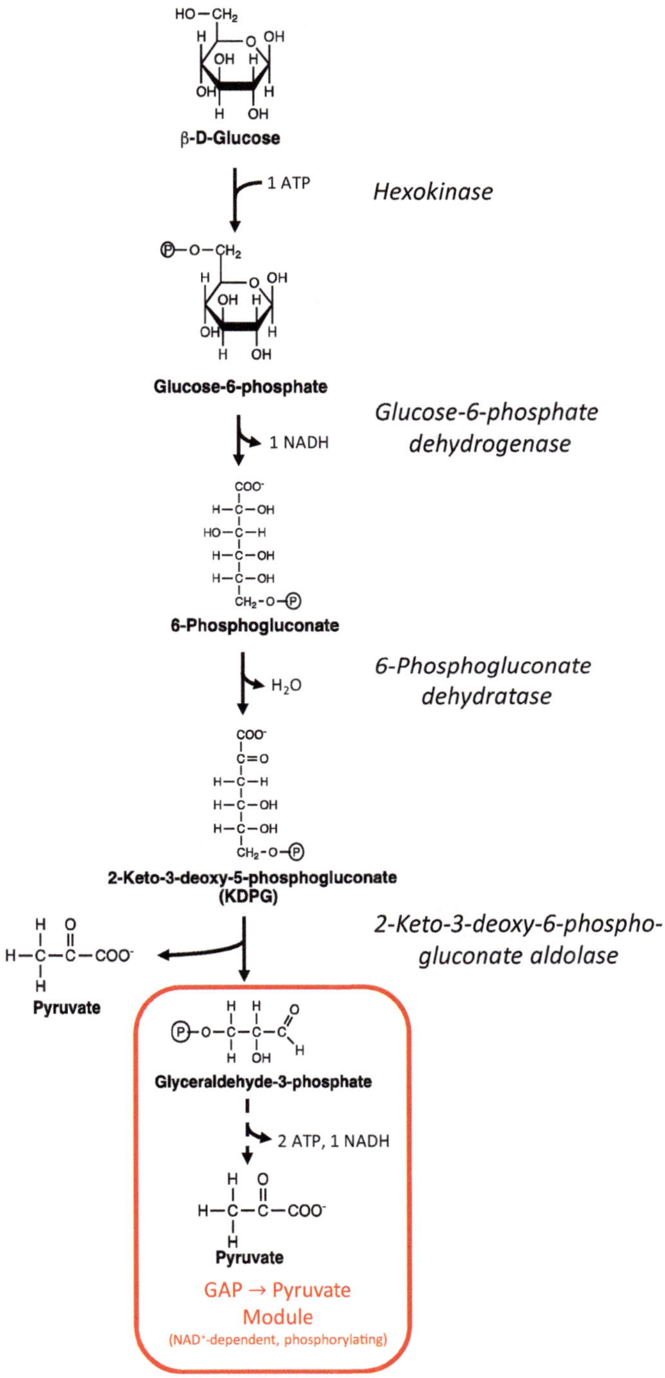

Fig. 8.18 The Entner–Doudoroff pathway for glucose oxidation to 2 pyruvate, shown here with a NAD^+-dependent, phosphorylating GAP → Pyruvate module. Key enzymes of this pathway are 6-*phosphogluconate dehydratase* and *2-keto-3-deoxy-6-phosphogluconate aldolase*. There are important variations of the Entner–Doudoroff pathway with respect to ATP yield (n_{ATP}) and electron carrier involved if *NAD^+-dependent, phosphorylating GAP dehydrogenase* is replaced by a *ferredoxin-dependent, non-phosphorylating GAP: ferredoxin oxidoreductase* or a *$NADP^+$-dependent, non-phosphorylating GAP dehydrogenase*. See text for details

faster for oxidation by *pyruvate dehydrogenase,* the TCA cycle, and respiration which are the pathways leading to most of the energy conservation in aerobic microbes. Indeed, many aerobic microorganisms, including *E. coli*, contain genes for both an Embden–Meyerhof and an Entner–Doudoroff pathway, enabling to switch between a rate and yield mode of growth.

8.4.3.4 *Phosphoketolase* Pathway

The Phosphokelolase Pathway (Fig. 8.19) is specific for anaerobic **pentose** degradation, and oxidizing a pentose to pyruvate and acetate:

$$\text{Pentose} \rightarrow 1\ \text{Pyruvate} + 1\ \text{Acetate} + 2\,(e^- + H^+) + 2\ \text{ATP}$$

The key enzyme is ***phosphoketolase***. Ribose is chosen here as an example for this pathway and, for simplification, the linear form of the pentose is shown (Fig. 8.19). Ribose is activated under consumption of 1 ATP by ***ribokinase*** to ribose-5-phosphate in an irreversible step for reasons outlined above. Ribose-5-phosphate is isomerized by ***ribose-5-phosphate isomerase*** to ribulose-5-phosphate, which is subsequently epimerized to xylulose-5-phosphate by ***ribulose-5-phosphate epimerase***. Xylulose-5-phosphate is the substrate for ***phosphoketolase***. After ketolytic cleavage into glyceraldehyde-3-phosphate and acetyl-phosphate, GAP is converted to pyruvate by a GAP → Pyruvate module, and acetyl-phosphate is converted to acetate by the Acylate ↔ acyl-CoA Module (Fig. 8.19). Depending on which variant of the GAP → Pyruvate module is operative, 1-2 ATP are conserved in the GAP → Pyruvate module and 1 ATP via the Acylate ↔ acyl-CoA Module. In the *phosphoketolase* pathway with *NAD⁺-dependent, phosphorylating GAP dehydrogena*se as shown in Fig. 8.19, 1 ATP is required for pentose activation, 2 ATP are net conserved by this pathway, and 1 NADH is recovered and released into the cellular NADH pool.

While illustrated above for ribose, the *phosphoketolase* pathway converts diverse pentoses into xylulose-5-phosphate through a combination of *isomerases* and *epimerases* with respective substrate specificities. The *phosphoketolase* pathway is found in anaerobic pentose-utilizing, heterofermentative lactic acid bacteria and in *Clostridium acetobutylicum*, and in anaerobic hexose-metabolizing *Bifidobacteria* in the bifido shunt, containing both a fructose-6-phosphate- and a xylulose-5-phosphate-specific *phosphoketolase*. Aerobic degradation of pentoses proceeds via pentose-to-hexose interconversion via the Pentose Phosphate Pathway (see below) and subsequent glycolytic oxidation of hexose.

8.4.3.5 The (Oxidative) Pentose Phosphate Pathway

The Pentose phosphate pathway is a pathway converting pentoses into hexoses and, *vice versa*:

$$2.5\ \text{Hexoses} \rightleftharpoons 3\ \text{Pentoses}$$

The 0.5 hexose is a triose ($0.5 \times C_6 = 1\ C_3$, e.g., GAP or dihydroxyacetone phosphate) which can condensate with a second GAP from the next cycle via *fructose-1,6-phosphate aldolase* to fructose-1,6-phosphate. *Transketolases* and *transaldolases* are the key enzymes of the Pentose Phosphate Pathway (Fig. 8.20).

There are **two catabolic versions** of the Pentose phosphate pathway: (1) The Pentose Phosphate pathway for converting pentoses into hexoses for aerobic degradation of pentoses (e.g., from hemicellulose). This version functions as a pathway extension in aerobic Prokaryotes to catabolize pentoses via the hexose-specialized glycolytic or Entner–Doudoroff pathways and (2) the oxidative Pentose Phosphate pathway as a means to oxidize a hexose to 6 CO_2 (see below) in the absence of a GAP → Pyruvate module. This version is only found under aerobic respiratory conditions. Both versions have also highly conserved **anabolic** functions (1) for generating (deoxy)pentoses or hexoses for nucleic acid or cell wall biosynthesis, and (2) for generating NADPH for biosyntheses (see below).

When C_5 carbohydrates are the substrate, pentose-5-phosphates are converted via *isomerases* and *epimerases* to one ribulose-5-phosphate and two xylulose-5-phosphate, respectively (Fig. 8.20). A ***transketolase*** cleaves xylulose-5-phosphate into GAP and a dihydroxyethyl group (light green Fig. 8.20) which is transferred at the same TPP-containing *transketolase* to ribose-5-phosphate to form sedoheptulose-7-phosphate. Sedoheptulose-7-phosphate is aldolytically cleaved by a ***transaldolase*** into erythrose-4-phosphate and a dihydroxyacetone group, which is

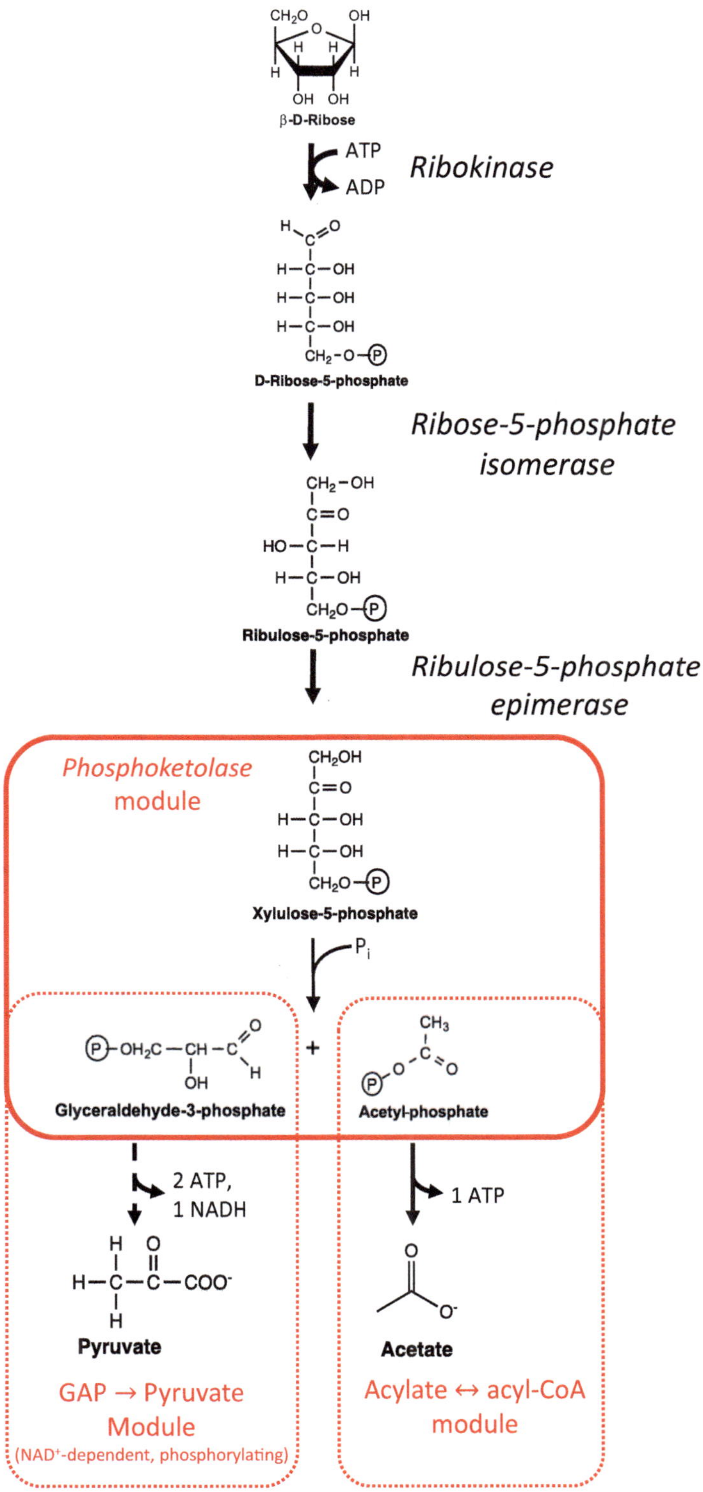

Fig. 8.19 The *phosphoketolase* pathway for anaerobic fermentation of pentoses, shown here with a NAD^+-dependent, phosphorylating GAP →-Pyruvate module. For simplification, the linear form of the pentose is shown. See text for details

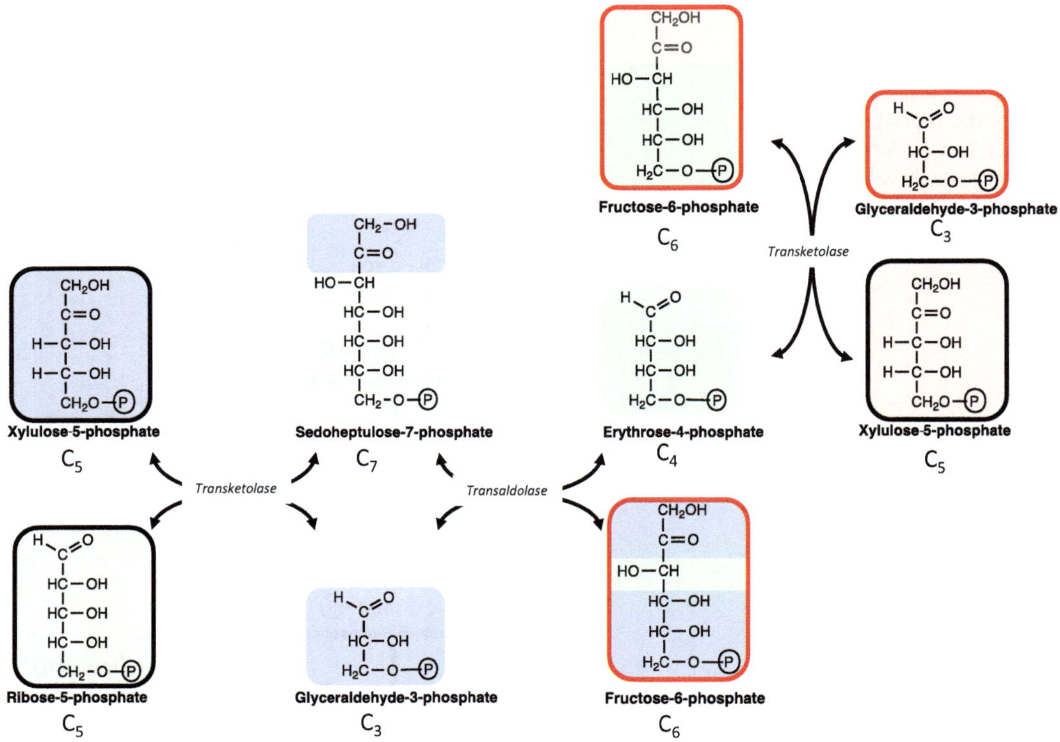

Fig. 8.20 The Pentose Phosphate Pathway for interconversion of pentoses and hexoses. Physiologically, this primarily anabolic pathway is fully reversible and catalyzed by *transketolases* and *transaldolases*. Framed in black are the three pentoses and framed in red are the 2.5 hexoses. The differently shaded sections highlight the fate of the carbon atoms in a conversion of pentose to hexose (left to right). See test for details

transferred at the same *transaldolase* to the previously released GAP to form fructose-6-phosphate. A second ***transketolase*** then transfers a dihydroxyethyl group from the second ketolytically cleaved xylulose-5-phosphate to erythrose-4-phosphate to form the second fructose-6-phosphate and a GAP (Fig. 8.20).

The oxidative Pentose phosphate pathway functions in conjunction with *glucose-6-phosphate dehydrogenase* and *6-phosphogluconate dehydrogenase* to oxidize hexoses in the absence of a GAP → Pyruvate module (Fig. 8.21), according to:

$$\text{Glucose} \rightarrow 6\ CO_2 + 24\ (e^- + H^+)$$

The oxidative Pentose phosphate pathway operates by iterative oxidation of the C1 carbon of glucose-6-phosphate to ribulose-5-phosphate plus CO_2, and subsequent conversion of 3 ribulose-5-phosphate to 2.5 hexose-phosphates. The latter is the substrate for the next iterative oxidative reactions (Fig. 8.21). In general, the oxidative pentose phosphate pathway uses **$NADP^+$** as cellular electron acceptor and is, therefore, an important anabolic pathway for generation of NADPH in carbohydrate-utilizing microbes. No ATP is formed via SLP during glucose oxidation via the oxidative pentose phosphate pathway. Thus, such a 5-step pathway can catabolically only operate in respiring microbes.

Comparison of the Entner–Doudorff and the oxidative pentose phosphate pathway for hexose (glucose) degradation reveals a common intermediate: 6-phosphogluconate. Two different enzymes, *6-phosphogluconate dehydratase* and *6-phosphogluconate dehydrogenase*, act on the

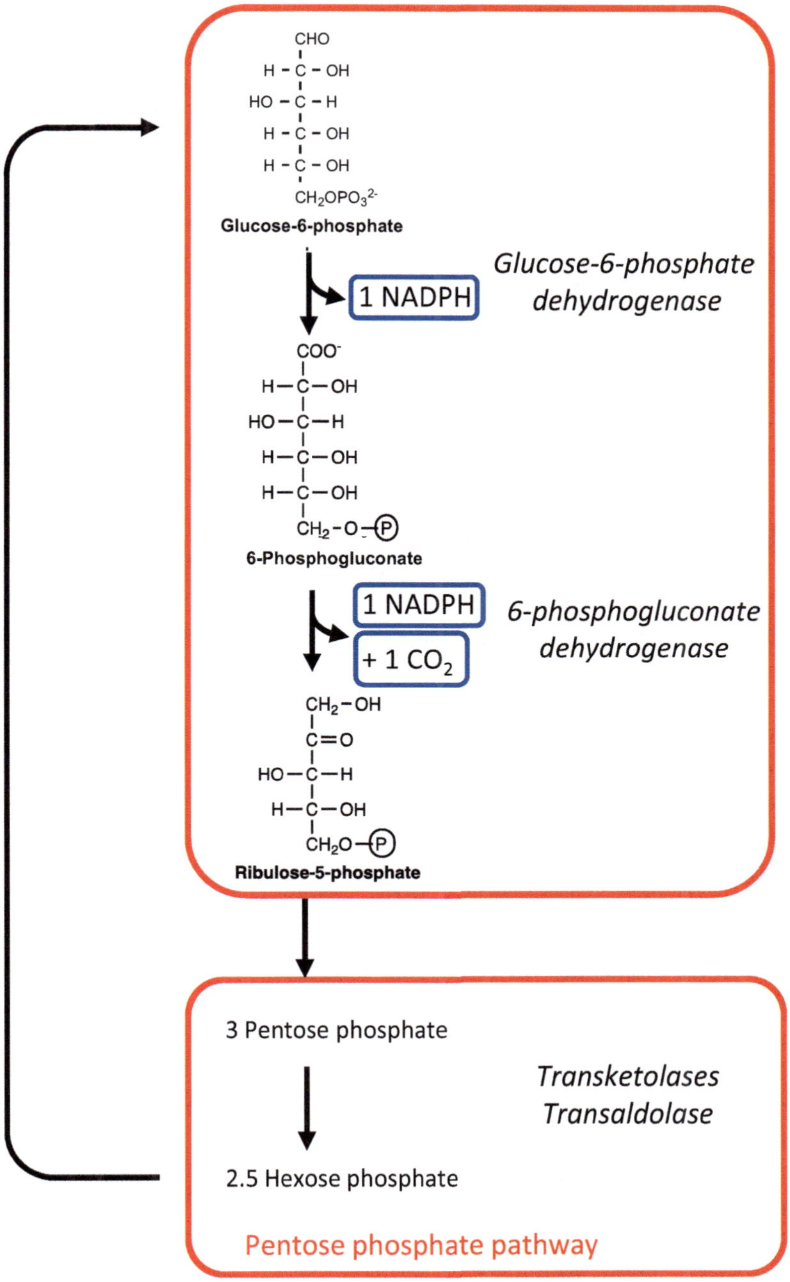

Fig. 8.21 Oxidative Pentose Phosphate pathway for complete oxidation of glucose to 6 CO_2. This pathway is a modular combination of the redox-neutral Pentose Phosphate Pathway plus the enzymes *glucose-6-phosphate dehydrogenase* and *6-phosphogluconate dehydrogenase* from the Entner–Doudoroff pathway and (upper red frame) from heterofermentative glucose oxidation, respectively. See text for details

same C3 carbon of 6-phosphogluconate, however, abstracing different substituents (Fig. 8.22). Activity of both enzymes is based on the (weak) inductive effect of the carboxyl group of the C3 carbon, making the hydroxy or hydride substituent a good leaving group for *6-phosphogluconate dehydratase* and *6-phosphogluconate dehydrogenase,* respectively. When water (i.e., the hydroxy group) is abstracted, the C_6 compound is committed to the Entner–Doudoroff pathway; when the C3 carbon is oxidized, the resulting 2-oxo acid decarboxylates; thereby, committing the C_6 compound into the *Phosphoketolase* pathway.

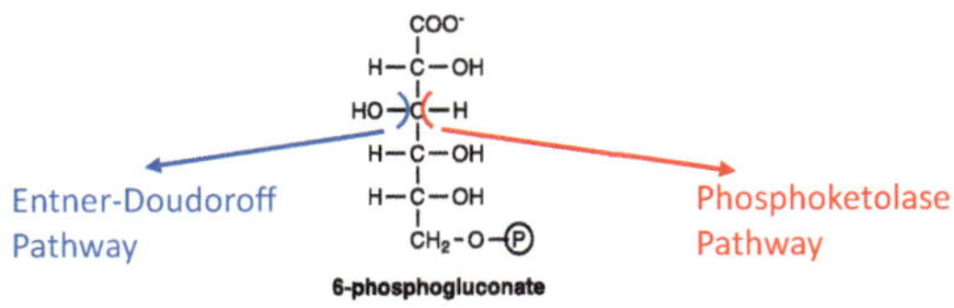

Fig. 8.22 Different transformations of 6-phosphogluconate at the C3 carbon which is in β position to the carboxyl group. Abstraction of water from the C3 carbon forms 2-keto-3-deoxy-6-phosphogluconate, funneling 6-phosphogluconate into the Entner–Doudoroff Pathway; abstraction of a hydride from the C3 carbon forms an instable 2-oxo acid that decarboxylates to ribulose-5-phosphate, funnel 6-phosphogluconate into the Pentose Phosphate Pathway. See text for details

8.5 Examples of Modules in Reductive Carbon Pathways

The above oxidative pathways have in common that the substrate is partially **oxidized** to **either pyruvate or acetyl-CoA** as intermediates. Thus, electrons recovered in the associated oxidations need to be transferred further to electron acceptors in order to balance the cellular pool of electrons in electron-carrying coenzymes. In respiring microbes, these electron acceptors are inorganic compounds, such as O_2, SO_4^{2-}, NO_3^-, NO_2^-, Fe^{3+}, etc., and only depend on the environmental availability. In fermenting microbes, the electron acceptors are organic (see Chaps. 4 and 9) (or H^+, which is ignored in this chapter) and derived from oxidized intermediates, such as pyruvate or acetyl-CoA, and are stoichiometric to the reduced electron carrier formed. In many fermenting microbes, pyruvate can be converted to acetyl-CoA by three important enzymes: *pyruvate dehydrogenase, pyruvate:ferredoxin oxidoreductase,* and *pyruvate:formate lyase* (see Sect. 7.3.3, Fig 7.15). The trade-offs between pyruvate as electron acceptor or being converted to acetyl-CoA, and between these three reactions are discussed in detail in Chap. 9.

In fermenting microbes, pyruvate and acetyl-CoA are the starting intermediates of reductive pathways for balancing cellular pools of NADH, NADPH, $F_{420}H_2$, or Fd^{2-}. Prerequisite for the overall function of such modules in the reductive direction is that the redox potential of the electron-accepting intermediates in a reductive pathway is more positive than that of the cellular electron carriers.

8.5.1 Butyrate-Forming Fermentation

One common reductive pathway is the reduction of 2 acetyl-CoA to butyryl-CoA by the **reverse β-oxidation module** (Fig. 8.23), which consumes 4 e^- according to:

$$2\text{Acetyl-CoA} + 4\,(e^- + H^+) \rightarrow 1\text{Butyryl-CoA}$$

Butyrate is formed as a reduced fermentative endproduct in microbial fermentations of carbohydrates and amino acids (Chap. 9) e.g., by *Clostridium pasteurianum*, *C. acetobutylicum*, *Eubacterium* sp., and *Fusobacterium* sp. The first reaction in this pathway is the condensation of 2 acetyl-CoA to acetoacetyl-CoA, which is a highly endergonic reaction (ΔG^0 = + 25 kJ/mol), requiring high cellular levels of *β-ketothiolase* in order to promote a sufficient forward flux in the catabolic pathway (see Haldane equation, Sect. 5.4.2). The two reduction reactions in this pathway are the *β-hydroxybutyryl-CoA dehydrogenase* and *butyryl-CoA dehydrogenase* reaction (Fig. 8.23). While the redox potential of acetoacetyl-CoA/β-hydroxybutyryl-CoA of $E^{0\prime} = -250$ mV is suitable for in vivo oxidation of NADH, the relatively positive redox potential of the butyryl-CoA/crotonyl-CoA redox couple of $E^{0\prime} = -10$ mV is exploited for energy conversion by ***electron bifurcating butyryl-CoA dehydrogenase*** in complex with electron-transferring flavoprotein (**Etf**), as in *C. pasteurianum*, *C. klyuveri*, or *C. difficile* (left branch, Fig. 8.23). The latter complex couples the bifurcation of 2 NADH with formation of Fd^{2-} and butyryl-CoA, where subsequently energy is conserved from Fd^{2-} oxidation via Rnf or Ech by transport-coupled phosphorylation. Alternatively, a non-bifurcating ***butyryl-CoA dehydrogenase***

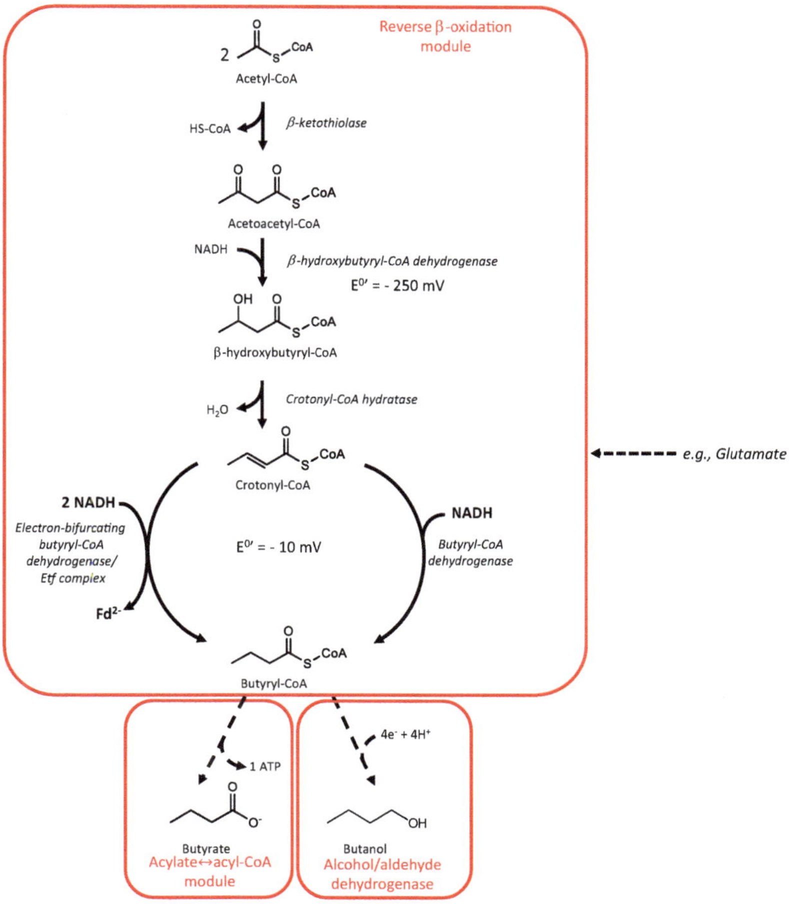

Fig. 8.23 Reductive butyryl-CoA-forming pathway in conjunction with the Acylate ↔ acyl-CoA or *alcohol/aldehyde dehydrogenase* modules. The butyrate pathway consists primarily of the β-oxidation module operating in reverse. In conjunction with an Acylate ↔ acyl-CoA module, butyryl-CoA is converted to butyrate via butyryl-CoA and AK/PTA, or a *butyryl-CoA:acetate CoA transferase* (not shown). In some anaerobic Bacteria, *butyryl-CoA dehydrogenase* is coupled to Fd reduction by electron bifurcation via an electron-transferring flavoprotein (Etf) (left branch of crotonyl-CoA reduction) and subsequent energy conservation via Ech or Rnf. In aerobes, crotonyl-CoA reduction proceeds via a non-bifurcating *butyryl-CoA dehydrogenase* (right branch of crotonyl-CoA reduction) See text for details

catalyzes crotonyl-CoA reduction with NADH in aerobic microbes (Fig. 8.23). Both pathway variants differ in their respective n_{ATP}. Crotonyl-CoA can also be formed during **glutamate fermentation**, thereby, funneling glutamate fermentation into the reductive β-oxidation module (see Fig. 9.22).

Butyryl-CoA is either converted to butyrate by a combination with the Acylate ↔ Acyl-CoA Module and 1 ATP is formed, or reduced further by combined NADH-dependent *alcohol/aldehyde dehydrogenase* such as in solvent fermentation (Figs. 8.23 and 9.17). In the latter pathway version, the butyryl-CoA thioester cannot be used for ATP synthesis. This different fate of a product of a reductive pathway, as exemplified here for butyryl-CoA, between energy conservation and use as electron acceptor is a common theme in fermentative pathways.

8.5.2 Propionate-Forming Fermentation

Two major reductive pathways result in **propionate** formation as a fermentation endproduct (Fig. 8.24): (A) the CO_2-dependent **methylmalonyl-CoA pathway**, and (B) the **acryloyl-CoA pathway**.

The methylmalonyl-CoA pathway, containing the *methylmalonyl-CoA mutase* module (Fig. 8.14a), converts pyruvate into an electron acceptor to accept 4 e^- according to:

$$1 \text{ Pyruvate} + 4(e^- + H^+) \rightarrow 1 \text{ Propionate}$$

Pyruvate is the endproduct of, e.g., glycolytic oxidation of glucose via the GAP → Pyruvate module (entry substrate: pyruvate, Fig. 8.24a) or is formed by oxidation of lactate, that is produced by lactic acid bacteria fermenting glucose (see Figs. 9.9 and 9.10).

The lower part of the methylmalonyl-CoA pathway can also function as a module for microbes fermenting succinate (from fumarate respiration) to propionate (Fig. 8.24):

$$1 \text{ Succinate} + 1H^+ \rightarrow 1 \text{ Propionate} + 1 \ CO_2$$
$$\Delta G^{0'} = -25 \ \text{J/mol}$$

Since this fermentation does not involve any net redox reaction, a membrane-bound, biotin-containing ***methylmalonyl-CoA decarboxylase*** conserves the energy of the exergonic decarboxylation of methylmalonyl-CoA to propionyl-CoA to

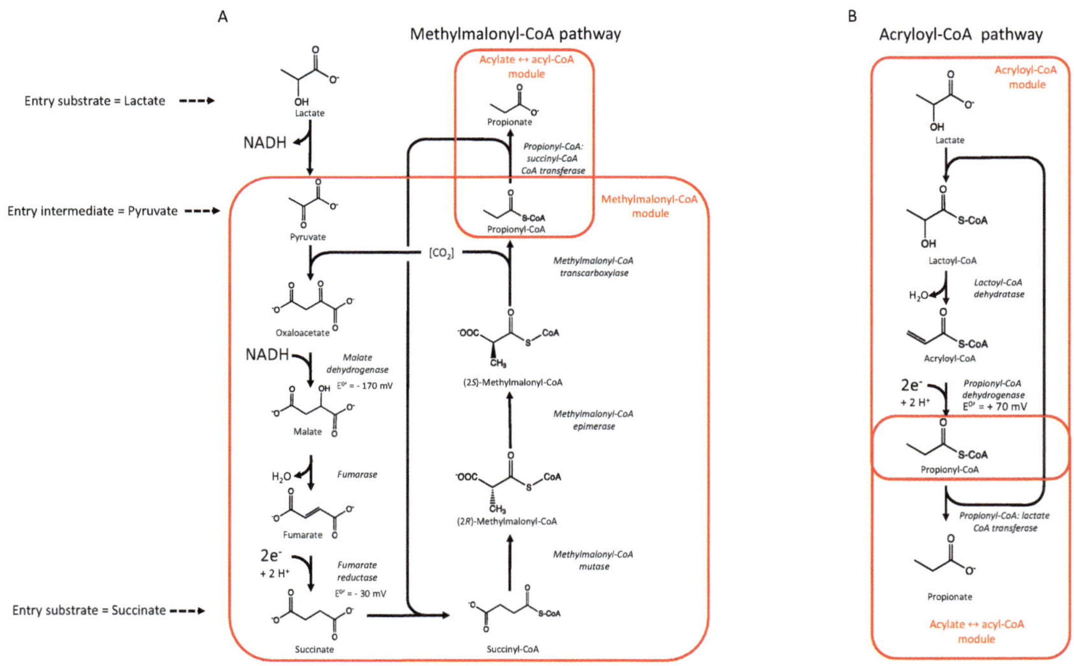

Fig. 8.24 Two reductive pathways of propionate-forming fermentation: (**a**) the methylmalonyl-CoA pathway, and (**b**) the acryloyl-CoA pathway. Modules are described in the text. (**a**) The methylmalonyl-CoA pathway uses i) pyruvate as entry intermediate, ii) lactate as entry substrate in fermentation of lactate, and iii) succinate as entry substrate for fermentation of succinate. In the latter case, *methylmalonyl-CoA transcarboxylase* is replaced by an energy-conserving, membrane-associated *methylmalonyl-CoA decarboxylase*. The reductive reactions of this pathway are reactions of *malate dehydrogenase* ($E^{0\prime}$ = − 170 mV) and *fumarate reductase* ($E^{0\prime}$ = − 30 mV). When lactate is used as substrate, only one reduction reaction net is available per lactate as electron acceptor. (**b**) The acryloyl-CoA Pathway has less metabolic flexibility and has been found so far only in some lactate-fermenting microorganisms. The electron-accepting reaction is the *propionyl-CoA dehydrogenase* reaction ($E^{0\prime}$ = + 70 mV). See text for details

generate a $\Delta\mu Na^+$ (see Fig. 6.24 and Sect. 6.4.4.4.1) as in *Propiogenium modestum* and *Veillonella parvula*. Energy-conserving and sodium-pumping decarboxylations are also found for oxaloacetate decarboxylation to pyruvate in *K. pneumoniae*, *Salmonella typhimurium*, and other enteric bacteria, and for glutaconyl-CoA decarboxylation to crotonyl-CoA in *Acidaminococcus fermentans* (Fig. 9.22), *Fusobacterium nucleatum*, and *Clostridium symbiosum*.

The **Acryloyl-CoA Pathway** (Fig. 8.24) is used as reductive pathway by microbes, such as *Clostridium propionicum* and *Megasphera elsdenii* (see Fig. 9.12) for fermentation of lactate to propionate:

$$1 \text{ Lactate} + 2[H] \rightarrow 1 \text{ Propionate}$$

This pathway is based on a prototypic *α-hydroxy acid dehydratase* (***lactoyl-CoA dehydratase***, Sects. 7.4.4 and 7.4.6), eliminating an α-hydroxy group with concomitant formation of a double bond, and a prototypic acyl-CoA dehydrogenase (***acryloyl-CoA reductase***) as the central electron-accepting reaction. Electron donor is typically NADH from lactate oxidation via *electron confurcating lactate dehydrogenase* (lactate + $Fd^{2-} \rightarrow$ 2 NADH, Fig. 9.12). Because formation of the electron acceptor acryloyl-CoA is directly derived from activated lactate, this pathway is typically not associated with glucose fermentation in a single microorganism. Acryloyl-CoA reduction to propionyl-CoA ($E^{0\prime}$ = + 70 mV) with NADH, in contrast to crotonyl-CoA reduction to butyryl-CoA ($E^{0\prime}$ = − 10 mV), does not appear to be coupled to ferredoxin reduction by electron bifurcation, probably because acryloyl-CoA is highly toxic, and, therefore, its concentration needs to be kept very low.

8.5.3 Citrate-Consuming Fermentations

Citrate can serve as external organic electron acceptor in two reductive pathways, in particular, to facilitate glucose fermentation (Fig. 8.25). The name of these fermentations, "citrate" and "citrolactate" fermentation do not conform to the earlier definition for fermentation and simply indicates the use of citrate as electron acceptor (and lactate as endproduct in "citrolactate") while the electron donor in that catabolism is often a carbohydrate.

In **citrolactate fermentation** by *Leuconostoc mesenteroides* with glucose as oxidizable substrate, citrate is converted to pyruvate, which is the ultimate electron acceptor and reduced to lactate (Fig. 8.25, citrolactate):

$$\text{Citrate}^{3-} + 2\,(e^- + H^+) + H^+ \rightarrow \text{Lactate}^- + CO_2 + \text{Acetate}^-$$

The enzymes of citrolactate fermentation include the prototypic reactions introduced as anaplerotic reactions above and are *oxaloacetate decarboxylase, lactate dehydrogenase*, and *citrate lyase*. In this reductive pathway, energy is conserved in the oxaloacetate decarboxylation reaction and by a citrate/lactate antiporter generating a $\Delta\mu H^+$ based on the concentration gradients and the different negative charges of citrate (3−) and lactate (1−).

In fermentation of citrate by *E. coli* and *Klebsiella pneumoniae* (Fig. 8.25 citrate fermentation), with glucose or glycerol as oxidizable substrate, citrate is reduced to succinate:

$$\text{Citrate}^{3-} + 4\,(e^- + H^+) \rightarrow \text{Succinate}^{2-} + CO_2 + \text{Acetate}^-$$

The enzymes of this reductive citrate fermentation pathway are again *citrate lyase* and the succinate ↔ oxaloacetate module. *Fumarate reductase* is membrane-associated and couples fumarate reduction ($E^{0\prime}$ = + 30 mV) to generation of a $\Delta\mu H^+$.

Reductive propionate and reductive citrate fermentation illustrate again the modularity of pathways, as well as molecular trade-offs in energy conservation and end product formation.

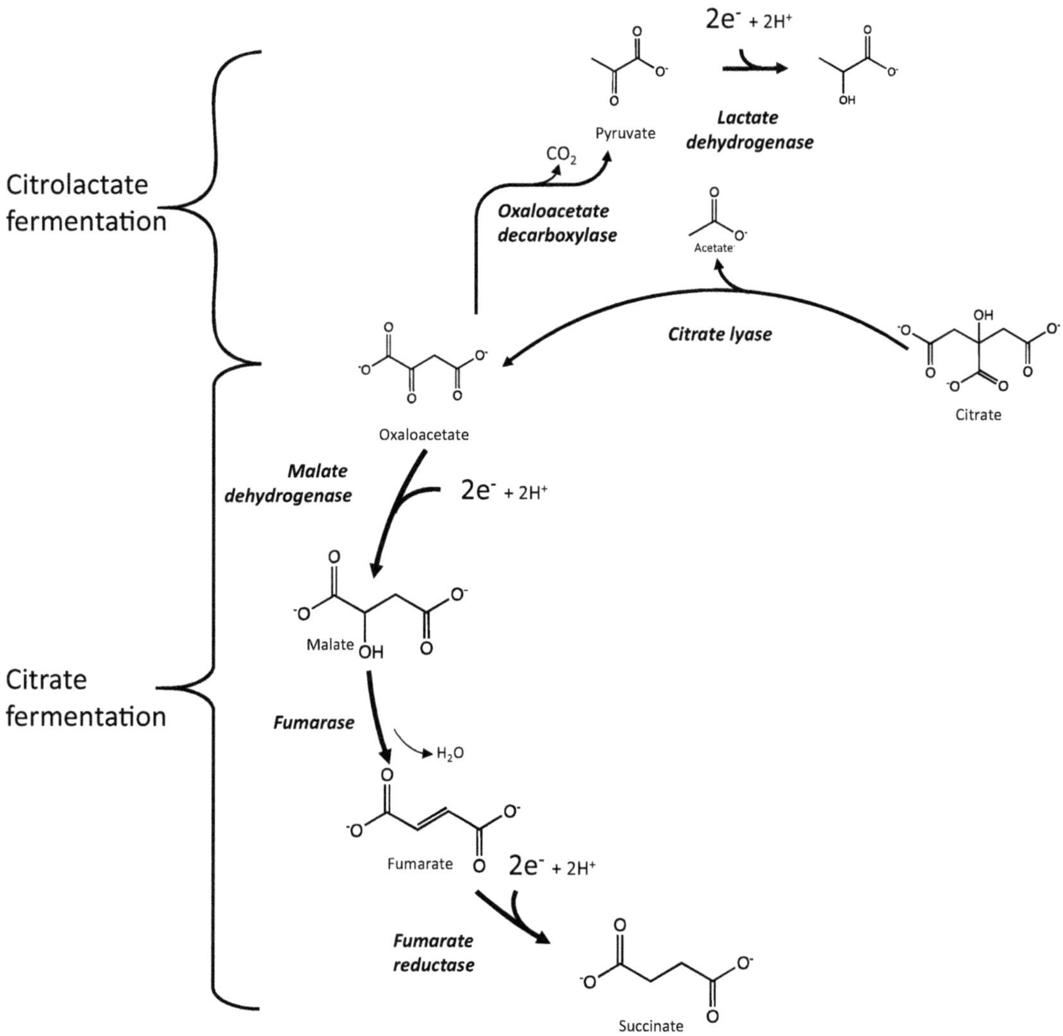

Fig. 8.25 Reductive fermentative pathways of citrate. In citrolactate fermentation (upper part) citrate is cleaved into acetate and oxaloacetate which is decarboxylated to pyruvate. A *lactate dehydrogenase* catalyzes the electron-accepting reaction. In citrate fermentation (lower part), citrate is also cleaved into acetate and oxaloacetate, which is reduced in TCA cycle steps to succinate. The electron-accepting reactions are catalyzed by *malate dehydrogenase* and by membrane-associated, energy-conserving *fumarate reductase*. These pathways are drawn to emphasize the recurrent enzymes and modules of the TCA cycle and anaplerotic reactions. See text for details

8.5.4 1,3-Propanediol-Forming Fermentation

Glycerol from hydrolysis of triacylglycerols can serve in some carbohydrate (glucose) fermentations as co-substrate for and electron sink in the reductive pathway:

$$\text{Glycerol} + 2\,(\text{e}^- + \text{H}^+) \rightarrow 1,3\text{-Propanediol}$$

In a two-reaction reductive pathway (Fig. 8.26), glycerol is first dehydrated in a prototypic reaction by ***glycerol dehydratase*** to the electron acceptor 3-hydroxypropionaldehyde, which is subsequently reduced by the alcohol

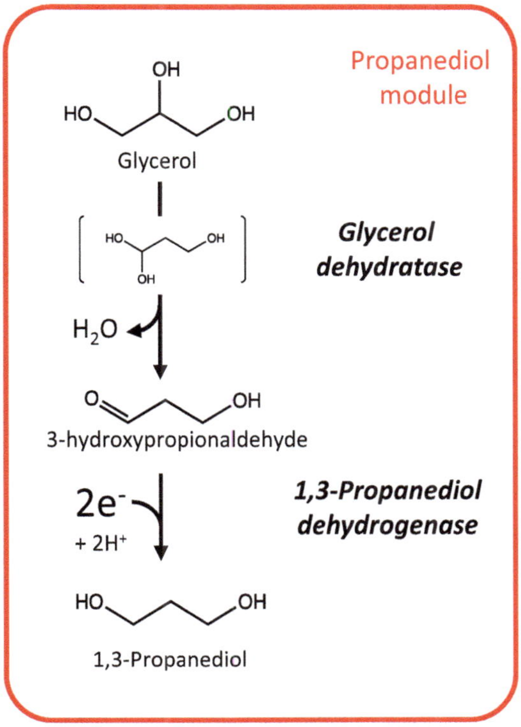

Fig. 8.26 Reductive pathway of 1,3-propanediol fermentation involving *glycerol dehydratase* and *1,3-propanediol dehydrogenase*. See text for details

dehydrogenase ***1,3-propanediol dehydrogenase*** to 1,3-propanediol using NADH as electron donor. 1,3-propanediol-forming fermentation is found in *Klebsiella pneumoniae*, *Citrobacter freundii*, *Enterobacter agglomerans*, and *Clostridium butyricum*. *Glycerol dehydratase* is a radical enzyme, and there are two types of this enzyme depending on how the radical is generated (see Sect. 7.4.4, Fig. 7.25: a coenzyme B_{12}-dependent *glycerol dehydratase* as in *Klebsiella pneumoniae* and a S-adenosyl methionine-dependent *glycerol dehydratase* as in *Clostridium butyricum*.

Glycerol as sole substrate (both electron donor and electron acceptor) can also be fermented to acetate and 1,3-propanediol using this reductive pathway module and the GAP → Pyruvate module for glycerol oxidation to acetate (4 glycerol → 1 acetate +3 1,3-propanediol).

8.5.5 Wood–Ljungdahl Pathway

Next to CO_2 being involved in carboxylation reactions of these reductive pathways, CO_2 can function as fermentative electron sink (i.e., no coupling to energy conservation) or respiratory electron acceptor (i.e., coupling to energy conservation) in reductive pathways of acetogenic and methanogenic Prokaryotes. CO_2 is ubiquitously present in most microbial ecosystems in the form of bicarbonate (pH dependent, ~0.1–1 mM in surface water; 50–100 mM HCO_3^- in anoxic lake sediments, 200–400 mM HCO_3^- in anaerobic digestors). The metabolic logic and biochemical strategies of CO_2-reducing catabolic pathways are discussed extensively in the C1 metabolism in Chap. 10.

The main recurrent modules discussed here are summarized in Table A.4 in the Appendix. Recurring modules in oxidative pathways are summarized in Table A.5 in the Appendix and reductive pathways in Table A.6 in the Appendix.

Further Readings

Lehninger AL (1982) Principles of biochemistry. Worth Publishers

Ahmed H, Ettema TJG, Tjaden B, Geerling ACM, van der Oost J, Siebers B (2005) The semi-phosphorylative Entner–Doudoroff pathway in hyperthermophilic archaea: a re-evaluation. Biochem J 390(2):529–540

Siebers B, Schoenheit P (2005) Unusual pathways and enzymes of central carbohydrate metabolism in Archaea. Curr Opin Microbiol 8:695–706

Fermentative Metabolism

9

Contents

A. M. Spormann, *Principles of Microbial Metabolism and Metabolic Ecology*,
https://doi.org/10.1007/978-3-031-28218-8_9

9.1 Introduction to Fermentation

Catabolic pathways as shown here are snapshots of a microbe's metabolism in a specific environment and at a specific point in time. The displayed pathway maps are not to suggest that a microbe has a "fixed" metabolism. As previous chapters showed, a catabolic phenotype is dynamic and greatly depends on substrate and energy flux (es) in the environment, on an environment's physical-chemical composition, on microbial growth rate, and on competition with other microbes. Depending on the environment, multiple substrates, particularly at low concentrations, can be metabolized by a single microbe either simultaneously (mixotrophy) or sequentially via more complex metabolic patterns involved. The here presented schemes show experimentally-supported metabolism (unless otherwise stated) to emphasize the modularity and mosaic structure of microbial metabolism. The relationship between pathway architecture and the consequence for a microbe's physiology and ecophysiology is elaborated here primarily for fermenting microbes, because the relationship is most visible and intuitive in anaerobic chemoorganotrophic microbes.

Metabolic pathways discussed in this and the following chapter are simplified in that not every single reaction step is shown; a focus is on the modularity of pathways. For example, addition or elimination of H_2O is often omitted as they are discussed previously in the context of modules. Also, cell-internal recycling of cofactors and coenzymes is shown only where needed. Some modules, such as respiratory chains or *ATPase* are abbreviated to indicate their physiological context rather than the detailed chemistry which is described in the previous chapters. Common modules as introduced in Chap. 8 are marked as such only in the metabolic schemes that appear first in this chapter and omitted for clarity in all further schemes. When discussing stoichiometries, the reference to a "mol substrate" is often omitted and simplified as "per substrate" with the understanding that in chemical sciences stoichiometries are described and compared on a mol basis.

Section 9.2 starts with the discussion of fundamentals of fermentative metabolism. In Sects. 9.3 and 9.4, metabolism of fermenting and of syntrophic microbes are discussed. Fermenting microbes and syntrophic microbes differ in that fermentations of the former are exergonic under standard conditions, whereas fermentations of the latter are exergonic only when syntrophic microbes grow via obligate metabolic interaction with respiring anaerobes that keep the concentrations of H_2, formate, and/or acetate low.

9.2 Some Fundamentals of Fermentative vs. Respiratory Metabolism

Catabolic pathways consist of coupled **oxidative** and **reductive pathways**. In **fermentation**, organic compounds (e.g., glucose) are converted via oxidative pathways to intermediates such as pyruvate and/or acetyl-CoA, which are at the end of an oxidative pathway. One or several of these intermediates are then used as electron acceptors in reductive pathways (Fig. 9.1a). Protons or CO_2, that are ubiquitously present, can additionally serve as electron acceptors, and are reduced to H_2, formate, or acetate, respectively. Not covered by this definition of fermentation is (i) Stickland fermentation of two amino acids, where one amino acid is oxidized and the other is reduced, (ii) disproportionation of thiosulfate or sulfite to sulfate and H_2S (inorganic fermentation), (iii) arginine fermentation that does not involve a redox reaction (see Sect. 6.3.3), and (iv) succinate fermentation to propionate and CO_2 (see Sect. 9.4). In **respiration,** the substrate being oxidized can be either organic (e.g., glucose) or inorganic (e.g., H_2), and the electron acceptor is generally inorganic (e.g., O_2, nitrate, sulfate, Fe^{3+}, or CO_2) (Fig. 9.1b). An exception within this definition is fumarate respiration, in which the electron acceptor is organic (see also Fig. 9.1).

The following section discuss: predicting ATP gains from thermodynamics and mechanisms of ATP synthesis (Sect. 9.2.1), metabolic options in fermenting microbes for oxidation of the reduced electron carriers NADH, reduced ferredoxin, and reduced menaquinone (Sect. 9.2.2), the pyruvate—acetyl-CoA junction (Sect. 9.2.3);

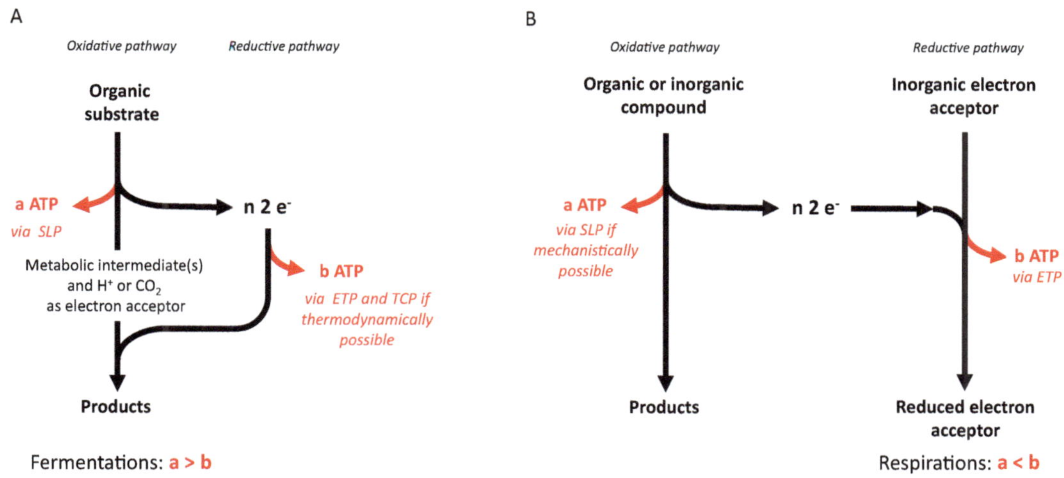

Fig. 9.1 Overview of redox processes and ATP synthesis (**a**) in fermentation and (**b**) in respiration. The electron carriers in fermentations are generally ferredoxins and flavodoxins ($E^{o\prime} = -500$ mV), NAD^+ ($E^{o\prime} = -320$ mV), and/or menaquinone ($E^{o\prime} = -70$ mV). The electron carriers in respirations are NAD^+, menaquinone, ubiquinone ($E^{o\prime} = +90$ mV), and/or cytochromes ($E^{o\prime} = +200$ to +400 mV) (see Sect. 6.2). Note, that fumarate respiration is an exception of (**b**) where the "inorganic electron acceptor" is actually the organic fumarate. *SLP* substrate level phosphorylation, *ETP* electron transport phosphorylation, *TCP* (ion) translocation-coupled phosphorylation. See text for details

and ecological explanations for the diversity of fermentation end products (Sect. 9.2.4).

9.2.1 Predicting n_{ATP} (ATP Gain) from Thermodynamics and Mechanisms of ATP Synthesis

For an overall pathway, the maximal number of ATP (a + b) (Fig. 9.1) per mol substrate used or product formed is $\boldsymbol{n_{ATP}}$ or the "**ATP gain.**" This value can be estimated from (i) the free energy change associated with fermentations or respirations at the concentration and temperature conditions prevailing in the natural habitats of microbes and (ii) the free energy change associated with ATP formation from ADP and phosphate. At intracellular concentrations of ATP, ADP, and phosphate in the mM rather than M range, the free energy change associated with ADP phosphorylation is near +50 ± 5 kJ/mol under equilibrium conditions depending on the pH and the energy charge (ATP + 0.5 ADP/ATP + ADP + AMP) (see Sect. 4.2).

Under equilibrium conditions, the rate of ATP synthesis is zero. Therefore, for driving synthesis of ATP the free energy change of an entire catabolic pathway, including ATP synthesis, needs to be less than zero (see Sect. 5.3). Because in syntrophs the free energy change of their overall fermentation is small and results in low growth rates, the energy required for ATP synthesis by an entire catabolic pathway is estimated to be less than −60 kJ/mol. In all other microbes, dependent on the growth rate, between −70 and −80 kJ/mol ATP are required, where −80 kJ/mol is approached in fast-growing microbes. Thus, syntrophic oxidation of butyrate to 2 acetate and 2 H_2, for example, with a ΔG near −15 kJ/mol butyrate is predicted to be coupled with the synthesis of maximally 0.2 (15/60) ATP, while for aerobic respiration of glucose with a ΔG near −2800 kJ/mol glucose, the estimate is maximally 35 (2800/80) ATP/mol glucose.

Substrate level phosphorylation (SLP) is coupled to oxidation of a carbonyl to a carboxyl group (Sect. 6.3), for example in the oxidation of glyceraldehyde-3-phosphate via 1,3-bisphosphoglycerate to 3-phosphoglycerate$^-$ + H, or of pyruvate$^-$ to acetate$^-$ + CO_2 with acetyl-CoA as intermediate. Therefore, it can be predicted that **per proton or CO_2 net generated** in fermentation or respiration of organic compounds, **one mol of ATP is generated via SLP**, with the exception when CO_2 or glycine is the electron acceptor (see below, Chap. 10). For example, fermentation of glucose to 2 lactic acid is associated with net formation of 2 protons and 2 ATP via SLP; oxidation of glucose with O_2 to 6 CO_2 is associated with formation of 6 ATP via SLP. However, reduction of 2 CO_2 + 4 H_2 to acetate$^-$ and H^+ is not associated with formation of 1 ATP via SLP since in this respiration CO_2 is the electron acceptor.

Since the maximal total number (a + b, Fig. 9.1) of ATP and the number of ATP formed by SLP (a, Fig. 9.1) can be predicted as shown, the number b of ATP formed by ETP and/or TCP can be calculated (Fig. 9.1). For example, the total energy of lactic acid formation from glucose with a ΔG of ~ -200 kJ/mol can drive the phosphorylation of 2.5 (200/80) ATP, of which two are formed by SLP. This leaves -40 kJ/mol glucose for transport-coupled phosphorylation (see below, Sect. 6.4.4, Fig. 9.4). Aerobic oxidation of glucose with a ΔG of ~ -2800 kJ/mol is predicted to be associated with phosphorylation of maximally 35 ATP, of which 6 are formed via SLP. That leaves about -2300 kJ/mol glucose for the synthesis of maximally 29 ATP via electron transport phosphorylation.

The advantage of this thermodynamic approach is that growth yields of microbes, which when under energy-limiting conditions are proportional to n_{ATP}, can be estimated without knowing biochemical details of the metabolic pathways involved, and the number of ATP formed via SLP and TCP can be predicted.

The limitation of the above approach is that only the **maximal** number of ATP per mol substrate can be predicted. As shown repeatedly in the previous chapters, not all microbes pursue a metabolic strategy of maximal n_{ATP} and, thus, of a maximum growth yield. As introduced in detail in Sect. 5.3, the intrinsic rate-yield trade-off predicts that a low ATP gain can be off-set by a higher rate of ATP synthesis and, thus, result in a higher growth rate. Depending on substrate flux and competition in natural environments, either a high growth yield or a high growth rate can be advantageous. For example, *Saccharomyces cerevisiae* ferments glucose to 2 ethanol and 2 CO_2 via the Embden–Meyerhof pathway with a $n_{ATP} = 2$, whereas the bacterium *Zymomonas mobilis* ferments glucose to the same endproducts via the Entner– Doudoroff pathway and a $n_{ATP} = 1$. At high glucose concentration, which is characteristic of the habitat of *Z. mobilis*, this microbe outcompetes *S. cerevisiae*. At low concentration, it is the opposite. Also, in *Pseudomonas aeruginosa* growing aerobically on glucose, oxidation of NADH with 1/2O_2 is coupled to formation of 3 ATP/NADH. In *E. coli*, n_{ATP} for the same reaction results only in 2 ATP per mol NADH (see Sect. 3.5.2). At low O_2, concentration, the facultative *E. coli* outcompetes the strict aerobe *P. aeruginosa*.

Respiratory energy conservation occurs with and without cytochromes and quinones (Sect. 6.4.4.1). Cytochrome- and quinone-independent respiratory energy conservation as in some fermenting-bacteria and -archaea proceeds via membrane-associated *Fd^{2-}: NAD^+ oxidoreductase* (Rnf complex) and membrane-associated [NiFe]-*hydrogenase* (Ech complex) (see Sect. 6.4.4.1.1). Thus, the absence of cytochromes and quinones is not an indication of the absence of energy conservation via transport-coupled phosphorylation! Conversely, the presence of cytochromes and quinones in anaerobes is not an indication of respiration because both electron carriers can be involved in reverse electron transport that allows many syntrophic anaerobes to grow (see Sect. 9.4).

In the metabolic schemes shown in this and Chap. 10, the stoichiometries of ions translocated in energetic coupling reactions, such as in *ATPase*, in ion-translocating *oxidoreductases*, or in antiporters, have not been determined experimentally for many microbes. Thus, assumptions were made to best fit the available metabolic data that exist for a specific pathway and microbe. Furthermore, assumptions were also made based on examples in the literature for other relevant microbes.

What distinguishes fermentative from respiratory metabolism is the **constraint for balancing the intracellular pool of the reduced electron carriers, NADH, reduced ferredoxin (Fd^{2-}), and reduced quinones.** In respiration, re-oxidation of cellular electron carriers depends on the amount of environmentally available inorganic electron acceptor, but is essentially independent of the electron donor substrate. In contrast in fermentations, the electron acceptor available is stoichiometric to the substrate oxidized: internal electron acceptors are derived via oxidation of the substrate molecule in a defined stoichiometry, so there is a stoichiometry between the number of reduced electron carriers formed in the oxidative pathway and the number of available reduction reactions in a reductive fermentative pathway. In other words, a given electron acceptor intermediate from an oxidative pathway has only a limited stoichiometric (and mechanistic) capacity to take up electrons from reduced electron carriers to keep a balanced electron flow for continuous catabolism. Chapters 7 and 8 outlined these prototypic redox reactions and the capacities of modules. In most fermentations, reductive pathways start from the intermediates pyruvate and/or acetyl-CoA. Protons and/or CO_2 are only reduced, if more electrons are released in the oxidative pathway than can be consumed via reduction of these intermediates.

While oxidative reactions and pathways leading to the reduction of NAD^+, ferredoxin, and menaquinone have been covered in detail in Chaps. 7 and 8, the following three sections introduce principles of reductive pathways in fermentations.

9.2.2 Metabolic Options in Fermenting Microbes for Re-oxidation of NADH, Reduced Ferredoxin, and Reduced Menaquinone

Re-oxidation of the NADH via reductive catabolic pathways is necessary to replenish the cellular NAD^+ pool so further catabolic oxidations can occur. NADH is re-oxidized in fermenting microbes via reduction of (i) carbonyl groups, (ii) thioester ester group, (iii) enoyl-CoA (α, β unsaturated fatty acids), (iv) fumarate, and (v) protons and CO_2 (Fig. 9.2).

(i) Exergonic reduction of carbonyl groups (Fig. 9.2a) (aldehydes and ketones) to the respective alcohol groups ($E^{0\prime}$ near −200 mV) with NADH ($E^{0\prime} = -320$ mV) is catalyzed by enzymes of the superfamily of *alcohol dehydrogenases*, including *lactate dehydrogenase* (Fig. 9.2a), *ethanol dehydrogenase* (Fig. 9.2b) and *β-hydroxybutyryl-CoA dehydrogenase* (see Sect. 8.3.2). Such carbonyl-to-alcohol reductions are not coupled to energy conservation.

(ii) Thioester group reduction (Fig. 9.2b), e.g., acetyl-CoA reduction to acetaldehyde ($E' = -410$ mV) with NADH, is exergonic only when steady-state aldehyde concentration is in the μM range (see Sect. 8.2.3). This can be achieved in two ways: by acetyl-CoA reduction to acetaldehyde coupled to a subsequent exergonic reduction of acetaldehyde with NADH to ethanol within one enzyme (AdhE) catalyzing both reactions, or by enclosing both *acetaldehyde dehydrogenase* (CoA-acetylating) and *ethanol dehydrogenase* in a microcompartment (see Sect. 2.2.4, Fig. 2.2). Reduction of acetyl-CoA is associated with the "loss" of an energy-rich bond, equivalent to 1 ATP, and, therefore, is used in fermentations only if n_{ATP} is high relative to the free energy change of the fermentation.

(iii) Reduction of enoyl-CoA to acyl-CoA (Fig. 9.2c, d, e) (e.g., crotonyl-CoA/butyryl-CoA, $E^{0\prime} = -10$ mV) with NADH ($E^{0\prime} = -320$ mV) is highly exergonic (see Sect. 8.2.4) and coupled with endergonic reduction of ferredoxin ($E' = -500$ mV) catalyzed by a cytoplasmic, *electron bifurcating butyryl-CoA dehydrogenase* (Fig. 9.2c, see Sect. 6.5.1). Here, the difference in redox energy between NADH and crotonyl-CoA is partially conserved in the negative redox potential of Fd^{2-}. This

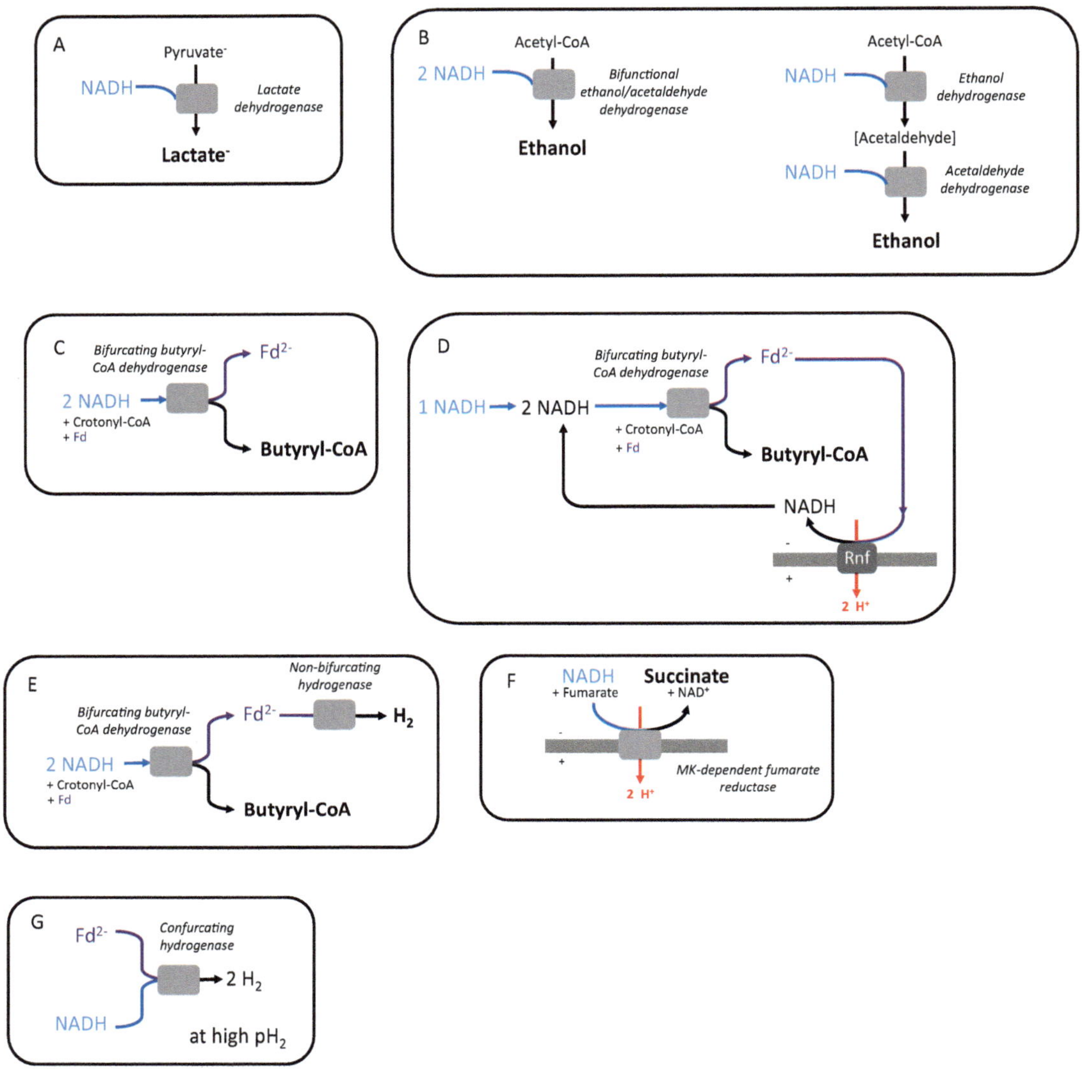

Fig. 9.2 Major NADH oxidation reactions in reductive fermentative pathways with and without coupling to energy conservation. (**a**) cytoplasmic *lactate dehydrogenase*, (**b**) reduction of the acetyl-CoA thioester catalyzed by either a bifunctional *ethanol/acetaldehyde dehydrogenase* or a microcompartment-enclosed *acetaldehyde dehydrogenase* plus *ethanol dehydrogenase.* (**c**) *electron bifurcating butyryl-CoA dehydrogenase* coupling the exergonic reduction of crotonyl-CoA to butyryl-CoA ($E^{0\prime} = -10$ mV) by 2 NADH with the endergonic reduction of Fd ($E' = -500$ mV). (**d**) first variation of (**c**) with energy-conserving Fd^{2-} oxidation to NADH by Rnf. Only one NADH is net consumed, and the other NADH cycles between *bifurcating butyryl-CoA dehydrogenase* and *Rnf* to conserve the redox energy difference between NADH and crotonyl-CoA. A variant of (**d**) involving *Ech* instead of *Rnf* releases H_2 at low pH_2 (not shown here). (**e**) second variation of (**d**) where Fd^{2-} is oxidized by H^+ to H_2 without energy conservation. (**f**) membrane-associated *fumarate reductase* mediating the exergonic reduction of fumarate ($E^{0\prime} = +30$ mV) with NADH ($E^{0\prime} = -320$ mV) coupled to translocation of 2 electrogenic protons involving menaquinone (MK). (**g**) *electron confurcating [FeFe]-hydrogenase* coupling to H_2 ($E^{0\prime} = -420$ mV, high pH_2) formation the endergonic oxidation of NADH with exergonic oxidation of Fd^{2-}. See text for details

conserved energy is converted via TCP to ATP by Rnf, catalyzing Fd^2 oxidation with NAD^+ (Fig. 9.2d) (not shown).

At low H_2 partial pressures ($pH_2 < 10^3$ Pa), Fd^{2-} can also be re-oxidized by protons coupled with energy conservation, if a microbe contains an energy-conserving, membrane-associated *Ech hydrogenase* (not shown). At high pH_2 (>10^3 Pa) Fd^{2-} can only be re-oxidized by protons without energy conservation by a cytoplasmic, non-electron-bifurcating *hydrogenase* (Fig. 9.2e). Interestingly, *propionyl-CoA dehydrogenase* appears to be none-bifurcating although the $E^{0\prime}$ of the acryloyl-CoA/ propionyl-CoA couple is +70 mV. One explanation is that because of the high reactivity of acryloyl-CoA, a very low intracellular concentration of would be required, which is only possible when exergonic acryloyl-CoA reduction with NADH is not coupled to endergonic Fd reduction with NADH (see Sect. 3.2, *E. coli* respiration at low pO_2). There is also evidence that in some anaerobes *butyryl-CoA dehydrogenase* involved in butyrate formation might not be electron-bifurcating (not shown).

(iv) Very exergonic is also fumarate reduction to succinate ($E^{0\prime}$ = + 30 mV) (Fig. 9.2f). The reaction is catalyzed by a membrane-associated, energy-conserving *fumarate reductase*; two protons per fumarate are electrogenically translocated via menaquinone (MK) from the inside to the outside of the cytoplasmic membrane involving cytochromes *b*.

(v) At high pH_2 (10^5 Pa), proton reduction to H_2 ($E^{0\prime}$ = −414 mV) with NADH ($E^{0\prime}$ = −320 mV) is endergonic and can only proceed when coupled to the exergonic oxidation of reduced ferredoxin (E' = −500 mV) via *electron confurcating hydrogenase* (Fig. 9.2g, see also Sect. 6.5.1). At low pH_2 (<10 Pa) (E' = −300 mV) H_2 formation from NADH is exergonic and is catalyzed by *non-confurcating hydrogenases.* At high formate concentration, CO_2 reduction to formate ($E^{0\prime}$ = −430 mV) with NADH ($E^{0\prime}$ = −320 mV) is endergonic and can only proceed when reduction of CO_2 to formate is coupled with exergonic Fd^{2-} oxidation, a reaction catalyzed by *an electron confurcating formate dehydrogenase.* At μM concentration of formate, direct reduction of CO_2 to formate with NADH is thermodynamically possible. H_2 and/or formate are formed by syntrophs as the solely reduced end product(s). The role of the different types of *hydrogenases* and *formate dehydrogenases* in energy conservation becomes most evident when their metabolism is discussed below (see Sect. 9.4).

Re-oxidation of reduced ferredoxin (Fd^{2-}) generated in the oxidative part of the fermentation has much less thermodynamic constraints than NADH oxidation. As described in Sect. 6.2.1, ferredoxins are Fe-S cluster proteins that in strictly anaerobic microbes contain often two redox-active [4Fe-4S] clusters. Ferredoxins do not have the uniformity in terms of structure and redox potential as the non-proteinaceous electron-carrying coenzymes $NAD(P)^+$, quinones, or cytochromes, and the in vivo redox potential of Fd cannot be described accurately by the relative ratio of reduced to oxidized forms (Nernst equation). Although conserved, their amino acid sequence is variable which affects their redox potential, and so does the extent of Fd binding to cognate oxidoreductases during electron transfer. From data of characterized ferredoxins, their redox potential can vary between −300 to −500 mV. For many microbes described below, the relevant electrochemical parameters of the involved ferredoxins have not been determined experimentally. Nevertheless, their oxidized form is abbreviated in this and the following chapter as Fd and their reduced form as Fd^{2-} to indicate that two electrons are transferred; this is not meant to imply that ferredoxins and their exact potentials are identical between the microbes discussed as might be inferred by the unified abbreviation. Therefore, in the following metabolic schemes, assumptions are made that are based primarily on the data known for ferredoxin from *Clostridium pasterianum*. The absence of specific and accurate information should be kept in mind, although the general conclusions and metabolic pathways as outlined are correct.

Reduction with Fd^{2-} (E' = −500 mV) of protons to H_2 ($E^{0\prime}$ = −420 mV), of CO_2 to formate ($E^{0\prime}$ = −430 mV), of CO_2 to CO ($E^{0\prime}$ = −520 mV), of acetyl-CoA + CO_2 to pyruvate ($E^{0\prime}$ = −500 mV) and of NAD^+ to NADH ($E^{0\prime}$ = −320 mV) are exergonic. Oxidation of Fd^{2-} with protons or CO_2 (to formate) are even sufficiently exergonic to be coupled with endergonic reactions such as NADH oxidation to H_2 and formate, respectively.

The following enzymes catalyzing oxidation of Fd^{2-} are of importance in fermentations. These are the **cytoplasmic**

- *ferredoxin-dependent hydrogenase*,
- *ferredoxin-dependent formate dehydrogenase*,
- *ferredoxin-dependent CO dehydrogenase*,
- *pyruvate:ferredoxin oxidoreductase*,
- *reduced ferredoxin- and NAD^+-dependent electron confurcating hydrogenase*, and
- *reduced ferredoxin and NAD^+-dependent electron confurcating formate dehydrogenase.*
- *Nfn*

A special case is the oxidation of Fd^{2-} via direct reduction of acetate to acetaldehyde ($E^{0\prime}$ = −580 mV) catalyzed by cytoplasmic, reversible *acetaldehyde:ferredoxin oxidoreductase* (AOR). This reduction is only possible at cytoplasmic pH values below 6 for thermodynamic reasons. There is also a mechanistic reason for this, because only the protonated form of acetate, acetic acid (pKa = 4.7), can be reduced.

The **membrane-associated** enzymes for Fd^{2-} oxidation are the

- *H^+- or Na^+-translocating hydrogenase* complex Ech catalyzing reduction of protons with Fd^2, and
- H^+- or Na^+- translocating Rnf complex, catalyzing reduction of NAD^+ with reduced ferredoxin.

Rnf and Ech are frequently found in strictly anaerobic fermenting microbes except in those not requiring iron for growth, such as homofermentative lactic acid bacteria and mycoplasma. The reason for the latter is that Rnf and Ech are iron-sulfur proteins.

Re-oxidation of reduced menaquinone ($E^{0\prime}$ = + 70 mV) is, for thermodynamic reasons, difficult with electron acceptors available in fermentations. All metabolite electron acceptors have redox potentials more negative than −200 mV. Oxidation of reduced menaquinone is, therefore, only possible, when coupled to a highly exergonic reaction. The mechanism employed is membrane-associated reversed electron transport (RET) that is driven by the electrochemical proton potential (see Sect. 6.5.2). Four electrogenic protons are most likely required for the reduction of one NAD^+ with reduced menaquinone, in analogy to complex I of the respiratory chain that couples quinone reduction with NADH with the translocation of 4 protons.

9.2.3 The Pyruvate: Acetyl-CoA Junction

For fermentations of carbohydrates, many amino acids, and fatty acids, pyruvate and/or acetyl-CoA + CO_2 or + formate are the major end products of the oxidative pathways. Thus, pyruvate and acetyl-CoA are the starting intermediates for accepting electrons in reductive fermentative pathways. In anaerobes, pyruvate and acetyl-CoA are metabolically connected via reversible *pyruvate: ferredoxin oxidoreductase*, reversible *pyruvate-formate lyase*, and irreversible *NAD^+-dependent pyruvate dehydrogenase* (only in propionic acid bacteria). Figure 9.3 provides an overview of selected **reductive pathways originating from pyruvate and acetyl-CoA** that are frequently found in fermentations. These reductive pathways are agnostic of the chemical nature of the fermentable substrate and the number of reduced electron carriers produced per substrate oxidized.

Pyruvate can be reduced to lactate, propionate, 2,3-butanediol and ethanol with 1 or 2 NADH in exergonic reactions (Fig. 9.3). These reductive pathways generally do not involve energy conservation. There are, however, two known exceptions. Reduction of pyruvate with 2 NADH to propionate via the methylmalonyl-CoA pathway is coupled to the synthesis of 0.5 ATP via electron transport phosphorylation (see below, Fig. 9.9). The site of

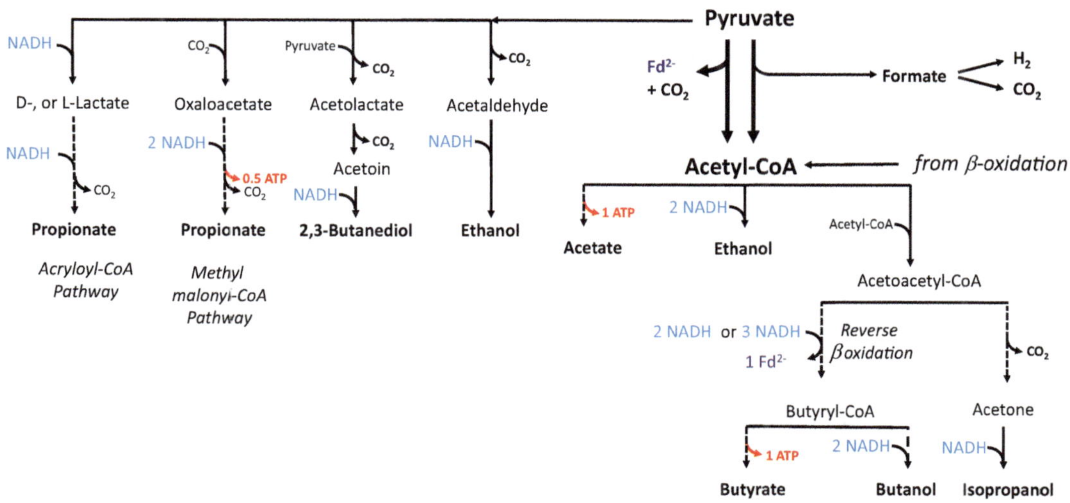

Fig. 9.3 The pyruvate-acetyl-CoA junction: overview of major reductive pathways in fermenting microbes starting at pyruvate or acetyl-CoA. Dashed lines indicate multiple pathway steps. Each pathway re-oxidizes at least 1 NADH. Pyruvate oxidation to acetyl-CoA (see upper right) in fermenting microbes is typically mediated by a *pyruvate: ferredoxin oxidoreductase*, or *pyruvate:formate lyase*. Formate can be disproportionated to H_2 and CO_2 by a *format:hydrogen lyase*. Some reductive pathways are involved in energy conservation, others are not. Details of the individual pathways were omitted here for clarity and are discussed in examples with selected microorganisms below

energy conservation is fumarate reduction to succinate with NADH (see Fig. 9.2f). The second exception is reduction of pyruvate (to lactate) and Fd (to Fd^{2-}) by 2 NADH catalyzed via *electron confurcating lactate dehydrogenase*.

Pyruvate is not only reduced in fermentations but also oxidized to **acetyl-CoA** which is either converted to acetate, generating 1 ATP/acetate via SLP, or reduced with NADH to ethanol, butyrate, butanol, and isopropanol. Butyrate formation from butyryl-CoA is coupled to ATP synthesis via the Acylate ↔ acyl-CoA module. However, since butyryl-CoA is formed from two 2 acetyl-CoA, one energy-rich bond is net “lost.” The metabolic “decision” of whether acetyl-CoA is used for energy conservation or consumed in reductive pathways to re-oxidize NADH determines a pathway's n_{ATP} (see above). Because n_{ATP} values in fermentations are generally numerically low, differences have significant physiological implications and can determine whether a microbe pursues a rate or a yield strategy. Another, equally important reason for microbes to change fermentation strategies from producing organic acids (higher n_{ATP}) to producing alcohols (lower n_{ATP}) is to escape growth inhibition: When organic acids as fermentative end products accumulate in the extracellular environment to high concentration, the environmental pH decreases strongly, resulting in growth inhibition. To avoid such inhibition, fermenting microbes can switch to producing more reduced products at the trade-off of a lower ATP gain (see below *C. acetobutylicum*, Fig. 9.17).

Oxidation of pyruvate to acetyl-CoA and CO_2 in fermenting microbes is not associated with NAD^+ reduction; exceptions being propionic acid bacteria that are essentially secondary anaerobes. Metabolically, it is advantageous in pyruvate oxidation to avoid using NAD^+ as electron acceptor. However, reduced ferredoxin formed by *pyruvate:ferredoxin oxidoreductase* can easily be re-oxidized with protons in the presence of *hydrogenases* (see Fig. 9.2). In case of *pyruvate:formate lyase* the electrons remain with the carbon of formate which elegantly leaves the cell! Thus, the electrons of pyruvate oxidation to acetyl-CoA by these two enzymes are effectively “invisible” in the budget of reduced cellular electron carriers. Oxidizing pyruvate involving either of these enzymes avoids that these electrons are thermodynamically "trapped" as NADH.

In anaerobes fermenting ethanol and CO_2 to propanol, or growing on syngas (H_2, CO_2, and CO), acetyl-CoA and CO_2 are reduced to pyruvate ($E^{0\prime} = -500$ mV) with Fd^{2-} ($E^{0\prime} = -500$ mV) as part of their energy metabolism. The pyruvate-acetyl-CoA junction, thus, functions in both ways.

In fermentations, acetyl-CoA is not only formed from pyruvate as might be suggested by Fig. 9.3, but also **via β-oxidation** of fatty acids (e.g., in syntrophs) (see Sect. 9.4) and in pentose fermentations involving ***phosphoketolase*** (see below and Figs. 9.5, 9.6, 9.7, and 9.8). In the *phosphoketolase* pathway, the pentose substrate is activated by phosphorylation and ketolytically cleaved to acetyl-phosphate and glyceraldehyde-3-phosphate involving an intramolecular redox reaction (see Sect. 8.4.2). Thus, acetyl-phosphate (i.e., ATP) is formed without a concomitant release of a reduced cellular electron carrier! From acetyl-phosphate, either acetate and ATP are formed involving *acetate kinase* or acetyl-CoA involving *phosphotransacetylase*.

9.2.4 Ecophysiological Explanations of Fermentation Product Diversity

There are several reasons for formation of different products from pyruvate and/or acetyl-CoA by different fermenters. First, there is a diversity of fermentation substrates, whose conversion to pyruvate and/or acetyl-CoA is associated with formation of different stoichiometries of NADH relative to pyruvate and acetyl-CoA formed. There are, however, also other and more complex reasons as shown in the three examples below (Fig. 9.3).

(i) Some fermenting bacteria are adapted to grow in the absence of iron, which is only possible if their catabolic enzymes are iron-independent. This excludes *pyruvate:ferredoxin oxidoreductase*, *ferredoxin-dependent formate dehydrogenases,* and *hydrogenases* from participating in these fermentations, because these iron-sulfur enzymes are abundantly present in the catabolic proteome. Thus, homofermenting lactic acid bacteria, that do not require iron, can grow by fermenting lactose to lactic acid in milk whereas microbes requiring iron for its catabolic enzymes cannot. Although milk contains iron, this is bound in lactoferrin and is not available for microbes. When sufficient lactic acid has been formed and the pH in milk decreased to below 5, iron is released from the lactoferrin and iron-requiring microbes can grow if they are acid tolerant.

(ii) Fermenting bacteria which form acetate and/or butyrate are less tolerant to organic acids than lactic acid-forming microbes. The reason for this is that acetic acid has a pK of nearly 4.7 and lactic acid has a pK of 3.7. Whereas the undissociated acids can freely diffuse across the cytoplasmic membrane, the dissociated forms cannot. Therefore, it is much easier for lactic acid-forming bacteria to maintain the internal pH between 6 and 7 than for acetate- and butyrate-forming bacteria. n_{ATP} is considerably higher when forming acetate and butyrate than when only forming lactate. However, the trade-off is that the higher yield mode leaves these microbes to be less acid tolerant. Acetate- and butyrate-forming bacteria can escape this low pH growth inhibition by switching fermentation patterns, reducing the acids with 2 NADH to ethanol and butanol, respectively (see below Fig. 9.17); this is, however, at the trade-off that 1 ATP is conserved less per acid reduced.

(iii) Some microbes ferment hexoses to 2 lactic acid and others to 2 ethanol and 2 CO_2 although in both fermentations 2 ATP are formed per hexose. Bacteria forming ethanol can grow to ethanol concentration of ca. 5% (v/v) (yeast can tolerate up to 10%). Lactic acid bacteria are growth-inhibited by lactic acid at much lower concentration. Ethanol-forming bacteria can, therefore, grow to higher cell concentration than lactic acid-forming bacteria when growing as monocultures on high concentration of hexoses. On the other hand, when growing in syntrophic association with ethanol- and lactic acid-consuming microbes, which keep ethanol and lactate concentration low, lactate-forming microbes grow to a higher cell density because lactate rather than ethanol export out of a cell can be coupled to energy conservation via Ion Translocation Coupled Phosphorylation, where 0.25 ATP are formed per lactate (see Sect. 6.4.4.4.3; Fig. 9.4). Ethanol diffuses through the cytoplasmic membrane almost as easily as water; therefore, its diffusional exit out of the cell is uncoupled from energy conservation.

(iv) Finally, there is a thermodynamic reason for producing numerous different end products from a single substrate is advantageous at low substrate and product concentrations,

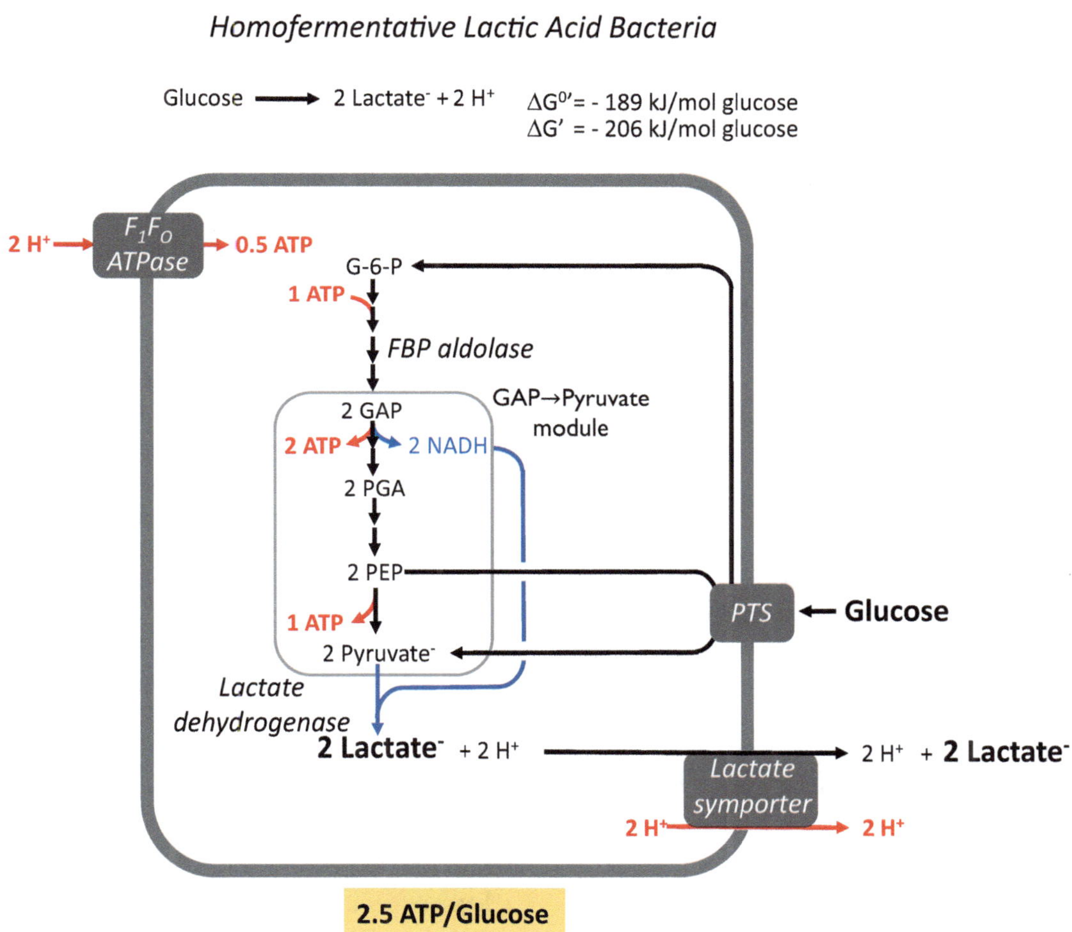

Fig. 9.4 Metabolic pathway of glucose fermentation by the Embden–Meyerhof pathway to two lactate⁻ plus 2 H^+ by a homofermentative lactic acid bacterium. Glucose is taken up by the PEP transferase system, driving simultaneous uptake and phosphorylation of glucose to glucose-6-phosphate with phosphoenolpyruvate hydrolysis. Glucose-6-phosphate is oxidized via the GAP→Pyruvate module to 2 pyruvate, 2 NADH, and 3 ATP. Two pyruvate are the intracellular electron acceptors for re-oxidation of the 2 NADH. The concentration gradient between high internal and low external (in the beginning of the fermentation) lactate drives the export of electro-neutral lactic acid plus an electrogenic H^+ via TCP. An *F_1F_OATPase* conserves the energy of the proton motive force as ATP. $\Delta G'$ is calculated for 1 mM concentration of glucose and lactate. Abbreviations: *G-6-P* glucose-6-phosphate, *2-PGA* 2-phosphoglycerate, *PEP* phosphoenolpyruvate, GAP glyceraldehyde-3-phosphate, *PTS* phosphotransferase system, *FBP aldolase fructose-1,6-bisphosphate aldolase*. See text for details

which is typical of anaerobic ecosystems. As introduced in Sect. 3.3, the Gibbs free energy of a catabolic reaction resulting in multiple endproducts from a single substrate, such as mixed acid fermentation of glucose, becomes more exergonic at mM concentration compared to a reaction substrate/product stoichiometry of ~1: If the stoichiometry is 1, the free energy change of the catabolic reaction is concentration-independent. The more products are formed from one substrate the more exergonic the reaction becomes at low substrate and product concentrations because of a gain in entropy. This might be advantageous for *E. coli* forming from glucose lactate, acetate, ethanol, succinate, CO_2, H_2, and formate, for example.

9.3 Lactic Acid-Forming Fermentations

9.3.1 Homofermentative Lactic Acid Bacteria

Formation of lactic acid as endproduct in the reductive pathway is one of the simplest fermentations of carbohydrates. Fermentation of hexoses and pentoses to lactate occurs by two major pathways that **have in common the *GAP dehydrogenase* module and D-or L-lactate-specific *lactate dehydrogenases***. They differ in the funneling pathways converting diverse hexoses or pentoses to glyceraldehyde-3-phosphate as well as in how NADH is re-oxidized in reductive fermentative pathways.

9.3.1.1 Hexose Oxidation to Two Pyruvate via the Embden–Meyerhof Pathway

Figure 9.4 shows the metabolic scheme of homofermentative glucose fermentation to 2 lactic acid via the Embden–Meyerhof pathway. Specific to this pathway is the key enzyme ***fructose-1,6-bisphosphate aldolase*** (see Sect. 8.4.2) which cleaves the C_6 fructose-1,6-bisphosphate (FBP) into two C_3 carbohydrates, glyceraldehyde-3-phosphate (GAP) and dihydroxyacetone phosphate that are in enzyme-catalyzed isomeric equilibrium. Fructose-1,6-bisphosphate is formed from glucose-6-phosphate (G-6-P) via fructose-6-phosphate. Two GAP are subsequently oxidized via the *GAP dehydrogenase* module to 2 pyruvate, 2 NADH, and 3 ATP (Fig. 9.4); one of the two phosphoenolpyruvate (PEP) formed by this module is used for glucose uptake and activation to glucose-6-phosphate (G-6-P). A second ATP is consumed in the conversion of G-6-P to fructose-1,6-bisphosphate via fructose-6-phosphate. Thus, 2 ATP are net formed by SLP per glucose oxidized to 2 pyruvate and 2 NADH (Fig. 9.4).

The reductive pathway of this lactate-forming fermentation consists of a single enzyme, *D-* or *L*-specific *lactate dehydrogenase*, consuming stoichiometrically the two NADH formed in the oxidative pathway. Some lactic acid-forming microbes couple the efflux of protonated lactate (lactic acid) with a simultaneous net export of one proton (Fig. 6.26). This additional H^+ is electrogenic and its energy is conserved via a $F_1F_0ATPase$ (Fig. 9.4). Prerequisite for such non-redox Ion Translocation Coupled Phosphorylation is a thermodynamic driving force, consisting here of a high intracellular and a low extracellular lactate concentration. Once lactic acid and lactate accumulate outside a cell, the concentration gradient decreases and the available energy approaches zero.

Uptake of Carbohydrates

Because of the low permeability coefficient of ~10^{-10} cm s^{-1} (Table 2.1), import of hexoses through the cytoplasmic membrane must be catalyzed. Hexoses are often taken up by strictly fermenting microbes by the **Phosphoenolpyruvate-dependent Transferase System** (PTS) (Fig. 9.4), a multi-component integral membrane enzyme complex that couples uptake of a hexose into the cell to simultaneous phosphorylation to a hexose-6-phosphate. The energy

(continued)

for both uptake and the "*hexokinase*" step is derived from the enolester of phosphoenolpyruvate. The energy of hydrolysis of the enolester is −52 kJ/bond while the energy required to phosphorylate the C6 hydroxy group of the glucose is only +14 kJ/bond under standard conditions. Thus, −38 kJ/mol are available for and consumed in hexose uptake. The ATP gain per 2 pyruvate formed for the entire pathway involving the PTS for uptake is 2 ($n_{ATP} = 2$).

An alternate carbohydrate uptake system is a **symport** of the uncharged carbohydrate with a proton. The energy for transporting the carbohydrate is derived from an electrogenic proton, energetically equivalent to 0.25 ATP. Once inside, glucose is phosphorylated to G-6-P with ATP in the *hexokinase* reaction to form hexose-6-phosphate. Thus, formation of intracellular G-6-P from external glucose requires 1.25 ATP when glucose is transported via proton symport; the ATP gain per 2 mol pyruvate formed is only 1.75 ($n_{ATP} = 1.75$). A third mode for glucose take-up is via **ABC transporter**, where ATP is hydrolyzed for uptake but uncoupled from substrate phosphorylation. A fourth mechanism is **facilitated diffusion**, which is present in *Zymomonas mobilis* that ferments glucose via the Entner–Doudoroff pathway or in human erythrocytes that ferment glucose to 2 lactic acids. Facilitated diffusion does not require energy input. Not considering glucose transport, the ATP gain in *Z. mobilis* is only 1 ($n_{ATP} = 1$) per 2 mol pyruvate. Which transport system is used depends on the glucose concentration. At very low environmental glucose concentration, the PTS system is most effective and at very high glucose concentrations facilitated diffusion via a *permease* is sufficient. In nature, microbes compete for glucose not only via their glucose threshold concentration but also via their rates of ATP synthesis (growth rate), which, in turn, is dependent on the ATP gain.

9.3.2 Heterofermentative Lactic Acid Bacteria

9.3.2.1 Pentose Fermentation by the *Phosphoketolase* Pathway

Figure 9.5 shows the pathway for pentose (ribose) fermentation to acetic acid and lactic acid via the *phosphoketolase* pathway. Because chemically different endproducts are formed, this fermentation is also referred to as **heterofermentative lactic acid fermentation**. Specific to this pathway is ***phosphoketolase*** that cleaves xylulose-5-phosphate phosphorylytically into acetyl-phosphate and the C_3 glyceraldehyde-3-phosphate. Acetyl-phosphate is converted by the Acylate ↔ acyl-CoA Module to acetate and ATP, while glyceraldehyde-3-phosphate is converted via the GAP → Pyruvate module to pyruvate and 1 NADH (Fig. 9.5). In the *phosphoketolase* reaction, acetyl-phosphate contains an energy-rich bond which is formed by an intramolecular redox reaction and no reducing equivalents are released into the cellular NADH pool.

9.3.2.2 Hexose Fermentation by the *Phosphoketolase* Pathway

The *phosphoketolase* pathway is a pathway specialized for pentose utilization, while the Embden–Meyerhof pathway is restricted to hexoses as substrate. This is due to the substrate specificities of *fructose-1,6-bisphosphate aldolase* and *phosphoketolase* to the C6- and C5-phoshorylated ketoses, respectively. Specificity of *fructose-6-phosphate phosphoketolase* in Bifidobacteria can vary by species between fructose-6-phosphate-specific to fructose-6-phosphate- and xylulose-5-phosphate-specific. The *phosphoketolase* pathway can be extended to metabolize hexoses via a short upstream pathway extension, however, at a metabolic cost. Figure 9.6 shows the modification of the *phosphoketolase* pathway when used by

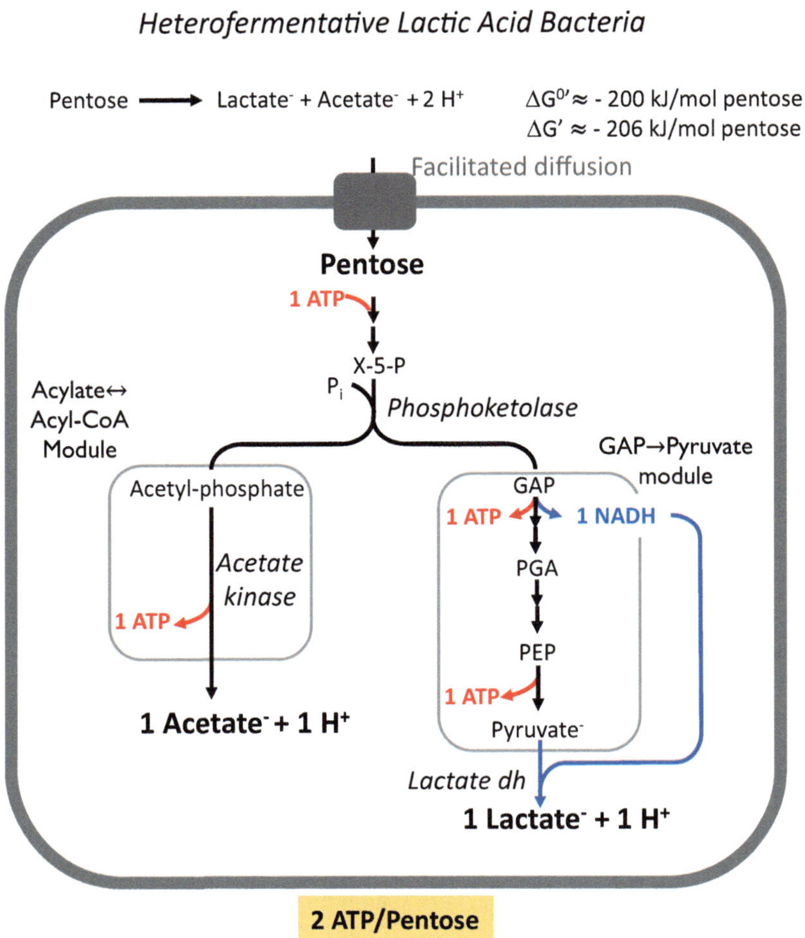

Fig. 9.5 Metabolic pathway of ribose (pentose) fermentation to lactic acid and acetic acid by a heterofermentative lactic acid bacterium via the *phosphoketolase* pathway. Key enzyme is *phosphoketolase*, catalyzing combined ketolytic, phosphorylytic cleavage of xylulose-5-phosphate to acetyl-phosphate and glyceraldehyde-3-phosphate. Pyruvate formed by the *GAP dehydrogenase* module functions as substrate for the reductive pathway to oxidize the one NADH produced in the oxidative pathway. Because *phosphoketolase* catalyzes the redox-neutral formation of an energy-rich bond, this acetyl phosphate is free to be converted to acetate with energy conservation without accruing a pair of electrons. Xylose uptake in *Lactobacillus pentosus* is mediated by a facilitated diffusion via components of the PTS systems but independent of metabolic energy. Δ*G*′ is calculated for 1 mM concentration of pentose, lactate, and acetate, respectively. Abrreviations: *lactate-dh lactate dehydrogenase*, *GAP* glyceraldehyde-3-phosphate, *X-5-P* xylulose-5-phosphate, *2-PGA* 2-phosphoglycerate, *PEP* phosphoenolpyruvate. See text for details

heterofermentative lactic acid bacteria for fermenting glucose. Hexose substrates are first converted to pentoses by two oxidation and a decarboxylation reaction. Glucose 6-phosphate is oxidized to 6-phosphogluconate by NADH-forming *glucose-6-phosphate dehydrogenase,* and 6-phosphogluconate is oxidized to xylulose-5-phosphate by *6-phosphogluconate dehydrogenase*. In the first reaction, the C1 carbon of glucose is oxidized to a carboxyl carbon, and in the second reaction, the C3 carbon is oxidized to a ketone, which results in spontaneous decarboxylation of the C1 carboxyl group via a β-keto acid type decarboxylation reaction. Because 2 NADH accrue in these two oxidative funneling reactions, it is necessary that the end

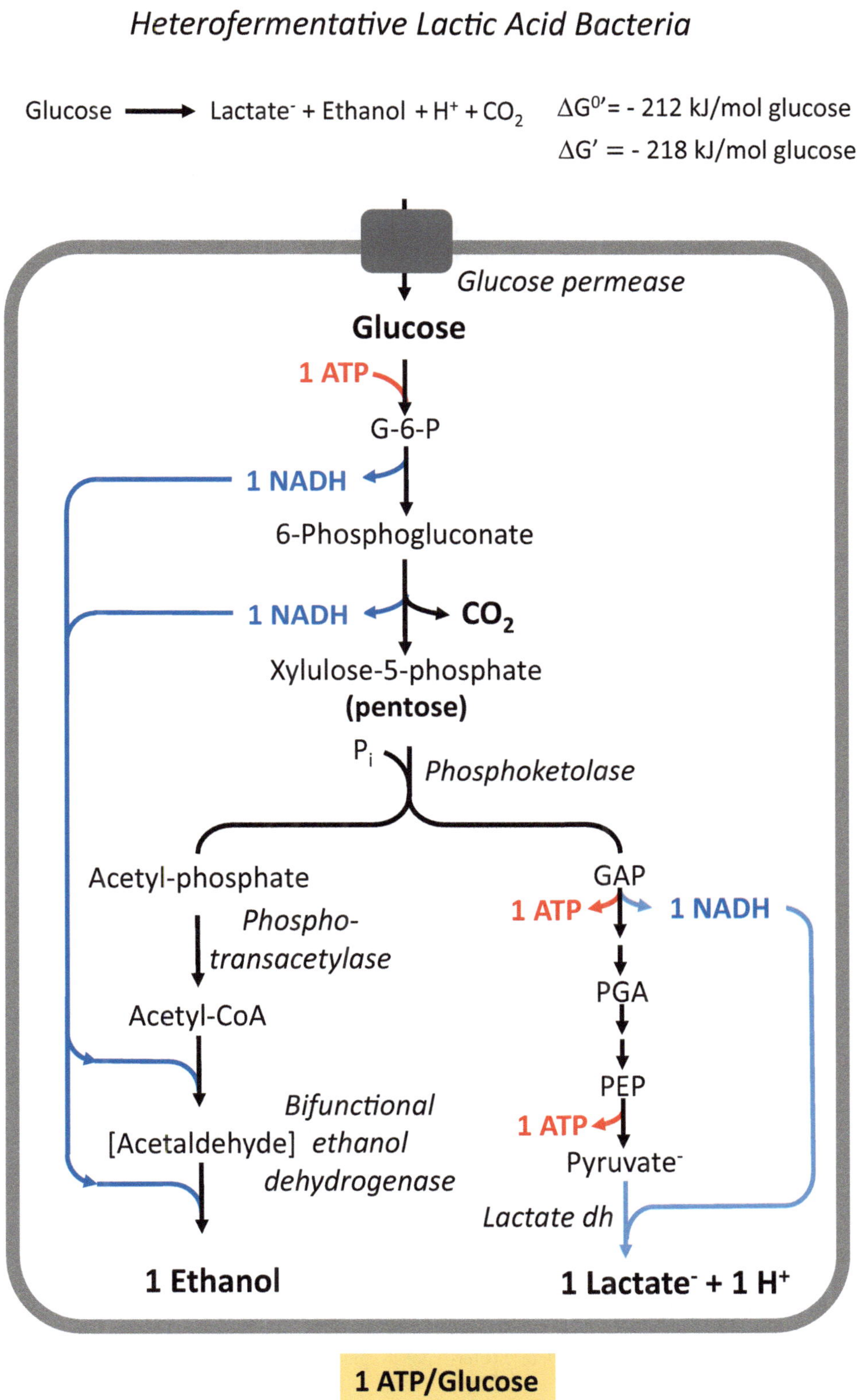

Fig. 9.6 Glucose fermentation by heterofermentative lactic acid bacteria via the *phosphoketolase* pathway. Note, the different fate of acetyl-phosphate in this hexose fermentation relative to pentose fermentation (Fig. 9.5).

intermediates of the oxidative pathway, acetyl-phosphate, and pyruvate, accept the 2 NADH in addition to 1 NADH from the GAP → Pyruvate module. Because two ATP are formed net per pyruvate via the GAP → Pyruvate module, pyruvate is reduced to lactate to re-oxidize an NADH. Acetyl-phosphate is not converted to acetate and ATP, as in pentose fermentation, but reduced to ethanol. This occurs at the trade-off that no ATP is formed via the Acylate ↔ acyl-CoA Module. Consequently, only 1 ATP is formed net in glucose fermentation to 1 ethanol, 1 lactate$^-$, 1 CO_2, and 1 H^+, compared to 2 ATP when glucose is fermented via the Embden–Meyerhof pathway (see Figs. 9.4 and 9.6).

9.3.3 Rate-Yield Trade-Offs in Lactic Acid Bacteria

Lactic acid bacteria can express either a rate or a yield phenotype depending on the flux of the carbohydrate substrate, which is determined by the concentration of the transportable substrate. Carbohydrates are environmentally available mostly as oligo- or polysaccharides. Some oligomeric carbohydrates, such as lactose, sucrose, or maltose, can be taken up directly into a cell. For example, fermentation of soluble lactose (β1 → 4 linked D-galactose and D-glucose) in the unstructured, aqueous environment of milk provides a competitive advantage for fast but low-yield lactic acid-forming microbes for outcompeting others. Lactose at high environmental concentration (up to 8% of milk by weight) can be taken up directly. On the other hand, more complex carbohydrates, such as starch, mucins, or cellulose, cannot be taken up as such, and the effective, transportable concentration of monomers derived from complex carbohydrates is low although the polymer is abundantly present. These complex carbohydrates need to be hydrolyzed first by microbial *glycohydrolases* before they can be accessed as a resource.

Thus, the rate of complex polysaccharide degradation is often limited by the rate of extracellular release of transportable monomers or oligomers (see Fig. 5.9): For example, *α-amylases* rapidly hydrolyze starch or glycogen into simple, transportable oligosaccharides. On the other hand, *cellulases* are organized into extracellular cellulosomes, which include *endo-*, *exo-cellulases*, and *cellobiases*, that collectively hydrolyze cellulose into transportable glucose monomers, however, only at slow rate. Such differences in hydrolysis rates of complex carbohydrates have kinetic and ecological implication for digestion of complex carbohydrates in the mammalian gut or the rumen of the cow, and influences a microbe's "decision" to either pursue a rate or a yield strategy.

Lactococcus lactis, for example, can switch between a rate and a yield mode of metabolism when fermenting carbohydrates. In the presence of high substrate concentration, the microbe metabolizes glucose rapidly via homofermentative lactic acid fermentation with the net formation of 2 ATP per mol glucose ($n_{ATP} = 2$). This pathway is indicated in orange pathway arrows (Fig. 9.7). In the slower but more efficient mode of growth (green arrows in Fig. 9.7), the two pyruvate of the GAP → Pyruvate module are unused as electron acceptor but converted by *pyruvate:formate lyase* into 2 acetyl-CoA and 2 formate. This enzyme, while forming an energy-rich bond via SLP, does not release electrons into the general cellular pool of reduced electron carriers; these

Fig. 9.6 (continued) Acetyl-phosphate is used as electron acceptor and reduced to ethanol in two NADH-consuming reactions. Note also the upstream extension of the *phosphoketolase* pathway to funnel glucose into this fermentative pathway via *glucose-6-phosphate dehydrogenase* and the *6-phosphogluconate dehydrogenase*. Glucose uptake in heterofermentative lactic acid bacteria is generally not assumed to involve a PTS system, but rather a *hexose permease* as in *Leuconostoc mesenteroides*. $\Delta G'$ is calculated for 1 mM concentration of glucose, lactate, and ethanol, respectively. Abbreviations: *FBP aldolase fructose-1,6-bisphosphate aldolase*, *GAP* glyceraldehyde-3-phosphate, *2-PGA* 2-phosphoglycerate. See text for details

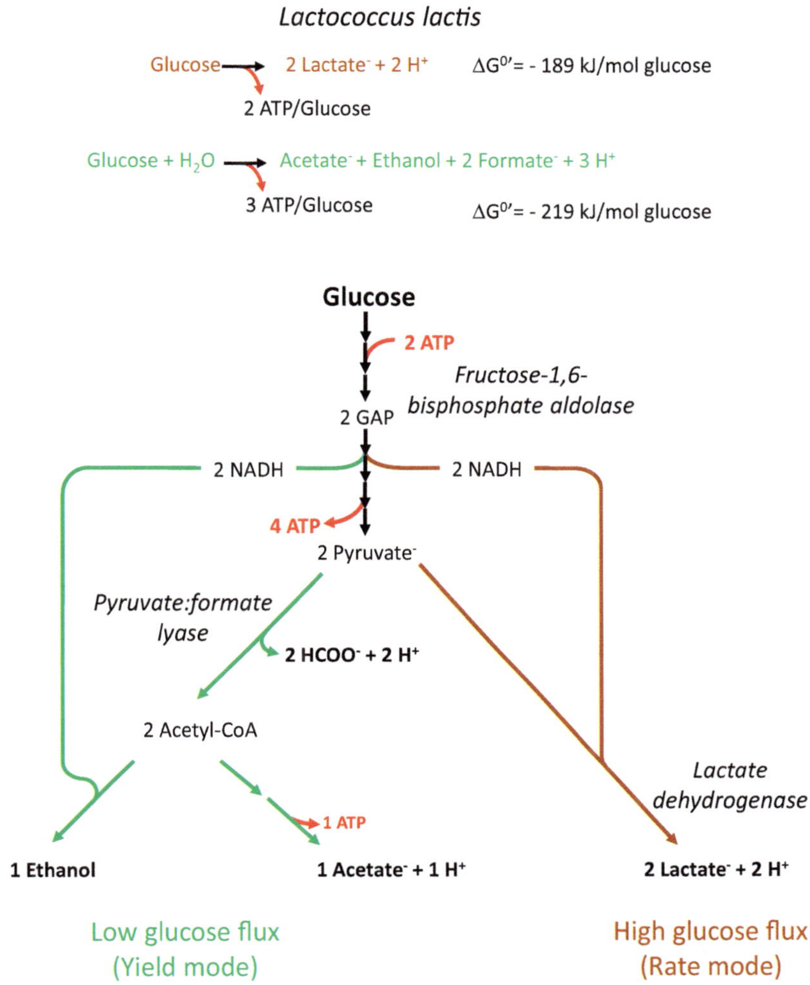

Fig. 9.7 Two modes of glucose fermentation by *Lactococcus lactis*. At high glucose flux (orange lines), glucose is fermented rapidly to 2 lactate⁻ plus 2 H^+, whereas under low flux conditions, the same substrate is fermented to 1 ethanol, 1 acetate⁻, 2 formate⁻, and 3 H^+. The two pathways for the consumption of the same substrate differ in their respective n_{ATP} and pathway length. While glucose fermentation to mixed endproducts increases yield, the pathway involves more steps (from pyruvate: 4 steps in ethanol/acetate formation as opposed to 1 step in lactate formation). Key for the ethanol/acetate-forming pathway is *pyruvate:formate lyase*, which enables acetyl-CoA formation without accruing reduced cellular electron carriers. *GAP* glyceraldehyde-3-phosphate. See text for details

electrons are present in formate! This frees up one acetyl-CoA to be converted to acetate with concomitant formation of one additional ATP, while the second acetyl-CoA is "sacrificial" and reduced to ethanol by *bifunctional alcohol/aldehyde dehydrogenase* (Fig. 9.7). If the microbe would employ a NADH-forming *pyruvate dehydrogenase* instead, both acetyl-CoA would need to be used for accepting those extra electrons and the yield would be the same as in lactate-forming fermentation, however, ethanol would be formed instead. Thus, when fermenting glucose via *pyruvate:formate lyase* and *aldehyde/alcohol dehydrogenase*, 3 ATP are formed per mol glucose ($n_{ATP} = 3$), a 50% increase in yield over homolactate fermentation!

Many lactate-forming microbes can ferment excreted and accumulated lactate once the carbohydrate substrate is consumed. Thus, this "transient" formation of lactate, can be viewed as the result of a rate strategy for carbohydrate utilization at high substrate flux and under competition in an unstructured environment. This is similar to the observation of the Julian Adams experiment, where acetate is first excreted by the Small Colony Variant as a consequence of pursuing a rate strategy to outcompete the parental wild type; the excreted acetate is a new metabolic niche, which is then occupied by the Large Colony Variant. Also, when *E. coli* grows aerobically in high glucose concentration in batch culture, glucose is incompletely used and oxidized to acetate as an "acetate overflow metabolism"; once glucose is consumed, excreted acetate is further metabolized and oxidized to CO_2.

9.3.4 Hexose Metabolism by *Bifidobacterium*

Bifidobacterium sp. ferments 2 glucose to 3 acetate$^-$, 2 lactate$^-$, and 5 H^+, and conserves 2.5 ATP/mol glucose (Fig. 9.8). In contrast to clostridial glucose fermentation to acetate (see below), all carbon atoms of the substrate are conserved in the organic end products and no CO_2 is formed! The *Bifidobacterium* fermentative pathway is an interesting mosaic pathway and contains diverse modules: GAP → Pyruvate module, *lactate dehydrogenase*, pentose phosphate cycle (some reactions), and two types of *phosphoketolases*. Next to xylulose-5-phosphate-specific *phosphoketolase* is a *fructose-6-phosphate phosphoketolase* (Fig. 9.8). In the *Bifidobacterium* pathway, 2 glucose are first phosphorylated to 2 glucose-6-phosphate, which are then isomerized by *glucose-6-phosphate isomerase* to fructose-6-phosphate; these two reactions are also present in the Embden–Meyerhof pathway. One fructose-6-phosphate is cleaved phosphorolytically by *fructose-6-phosphate phosphoketolase* to acetyl-phosphate and erythrose-4-phosphate. In a series of *transketolase* and *transaldolase* reactions of the pentose phosphate pathway, erythrose-4-phosphate plus the second fructose-6-phosphate are converted to sedoheptulose-7-phosphate and glyceraldehyde-3-phosphate, which are then converted to 2 xylulose-5-phosphate. From the xylulose-5-phosphate, subsequent transformation follows the above described *phosphoketolase* pathway of heterofermentative lactic acid bacteria (see Fig. 9.5).

Bifidobacteria inhabit the human intestine and play an important role in human health. They are among the first colonizers of an infant's gut, and are specialized to utilize human milk oligosaccharides that are not absorbed and degraded by the infant. Different Bifidobacteria have different carbohydrate uptake systems: either complex milk oligosaccharides are transported and hydrolyzed inside a cell, or specific *glycohydrolases* are excreted, and resulting mono- or oligo-saccharides are taken up.

9.4 Propionate-Forming Fermentations

While lactic acid-forming fermentations explore the C2 carbonyl carbon of pyruvate as electron acceptor in the reductive pathway, a further reduction of this carbon to a methylene carbon is possible and realized in propionic acid-forming fermentations. Propionate-forming fermenters are obligate anaerobic microbes and have in common that either **pyruvate** or **lactate** function as **electron sink in a core reductive fermentative pathway**.

The two **reductive pathways** for propionate formation originating from pyruvate or lactate are the **Methylmalonyl-CoA** and the **Acryloyl-CoA pathway** (see Sect. 8.5.2). The difference between both pathways reflects a rate-yield trade-off, with the Methylmalonyl-CoA pathway being the more efficient pathway that is present in slower growing propionic acid bacteria, while the thermodynamically less efficient **Acryloyl**-CoA pathway is present in faster propionate-forming microbes.

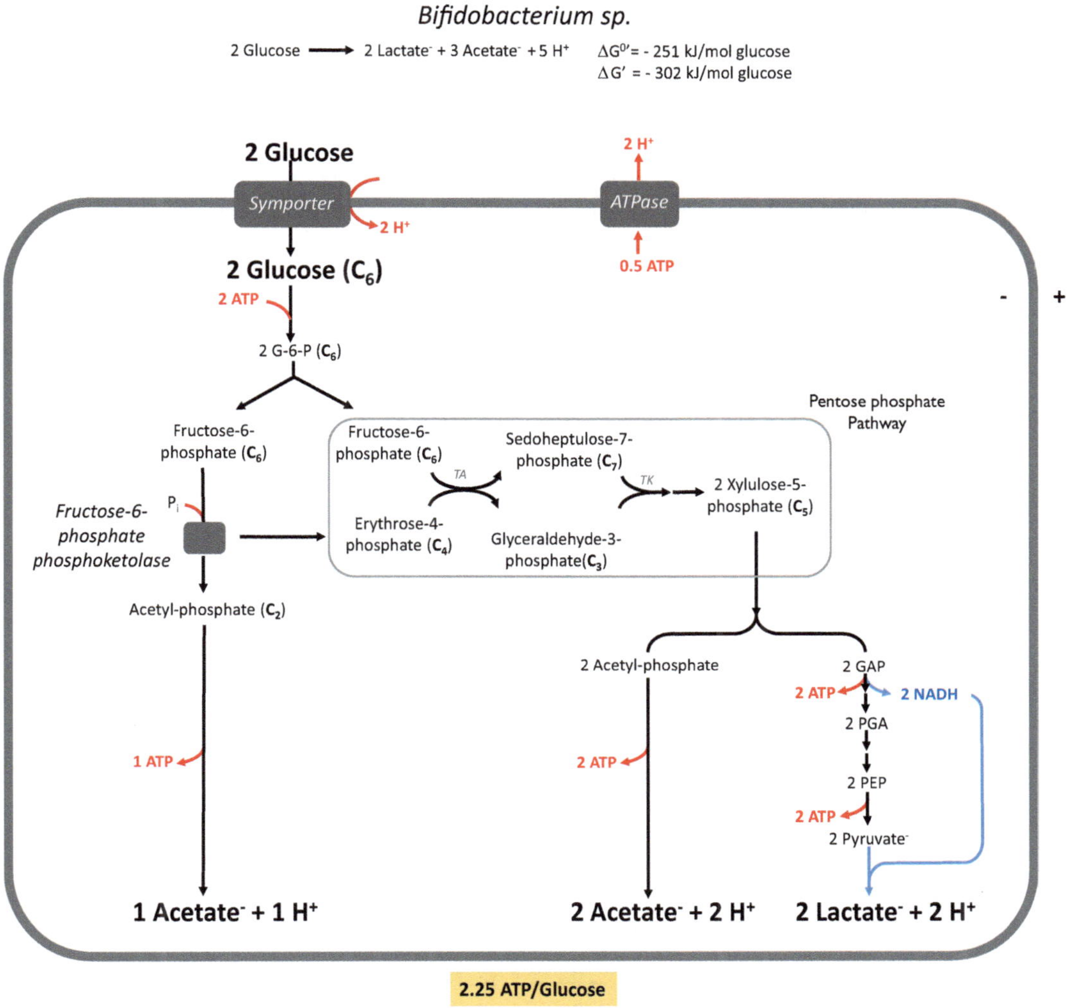

Fig. 9.8 Fermentation of 2 glucose to 3 acetate⁻, 2 lactate⁻, and 5 H^+ by *Bifidobacterium* sp. The pathway is a mosaic fermentation pathway, containing major modules of hexose and pentose fermentations. This can be understood as an overall strategy to maximize the involvement of *phosphoketolase* reactions in hexose degradation. Three of the five reactions forming energy-rich bonds are *phosphoketolase* reactions. Consequently, only 2 NADH are released while 2 glucose are metabolized! This strategy avoids releasing NADH to the cellular pool of educed electron carriers, and, thus, eliminates a re-routing of acetyl-CoA or pyruvate into an electron acceptor *in lieu* of ATP formation. This is different from the strategy of heterofermentative glucose metabolism, where most microbes with that pathway are primarily specialized to ferment pentoses. The *Bifidobacterium* pathway is an optimized pathway for hexose utilization at a high flux of glucose. Glucose is taken up by *Bifidobacterium longum* by a glucose-specific proton symporter; this microbe also contains and expresses a glucose-specific PTS uptake system. Note the increase of ΔG when substrate and products are at mM concentration. $\Delta G'$ is calculated for 1 mM concentration of glucose, lactate, and acetate, respectively. Abbreviations: *2-PGA* 2-phosphoglycerate, *PEP* phosphoenolpyruvate, *GAP* glyceraldehyde-3-phosphate, *TA transaldolase*, *TK transketolase*. See text for details

9.4.1 Propionate-Forming Fermentation via the Methylmalonyl-CoA Pathway

The Methylmalonyl-CoA pathway is found in phylogenetically diverse fermenting microbes such as in *Propionibacteria* sp. (Actinomycetales), *Selenomonas ruminantium* and *Veillonella* sp. (Firmicutes), as well as gram-negative *Bacteroides fragilis* and *B. thetaiotaomicron* (Bacteriodetes). Some propionate-forming fermenters, such as *Propionibacterium freudenreichii*, *P. acidipropionici*, or *Bacteroides fragilis*, can also ferment glucose directly to propionate without lactate as a free, external intermediate (Fig. 9.10).

Lactate is a common substrate to all propionic acid bacteria, and can be both electron donor AND electron acceptor. Overall, two modes of energy conservation are involved in propionic acid bacteria using the Methylmalonyl-CoA pathway: (i) one ATP is formed in the oxidative pathway via SLP from acetyl-CoA, and (ii) 0.5 ATP per succinate is formed in the reductive pathway by fumarate respiration.

The pathway in *P. freudenreichii*, metabolizing 3 lactate$^-$ → 1 acetate$^-$ + 2 propionate$^-$ + CO_2, $\Delta G^{0\prime} = -170$ kJ/mol, is shown in Fig. 9.9. In the oxidative pathway, lactate is oxidized via a *menaquinone-dependent lactate dehydrogenase*. Because of the redox potential of the pyruvate/lactate couple ($E^{0\prime}$ = –190 mV), prokaryotic *lactate dehydrogenases*

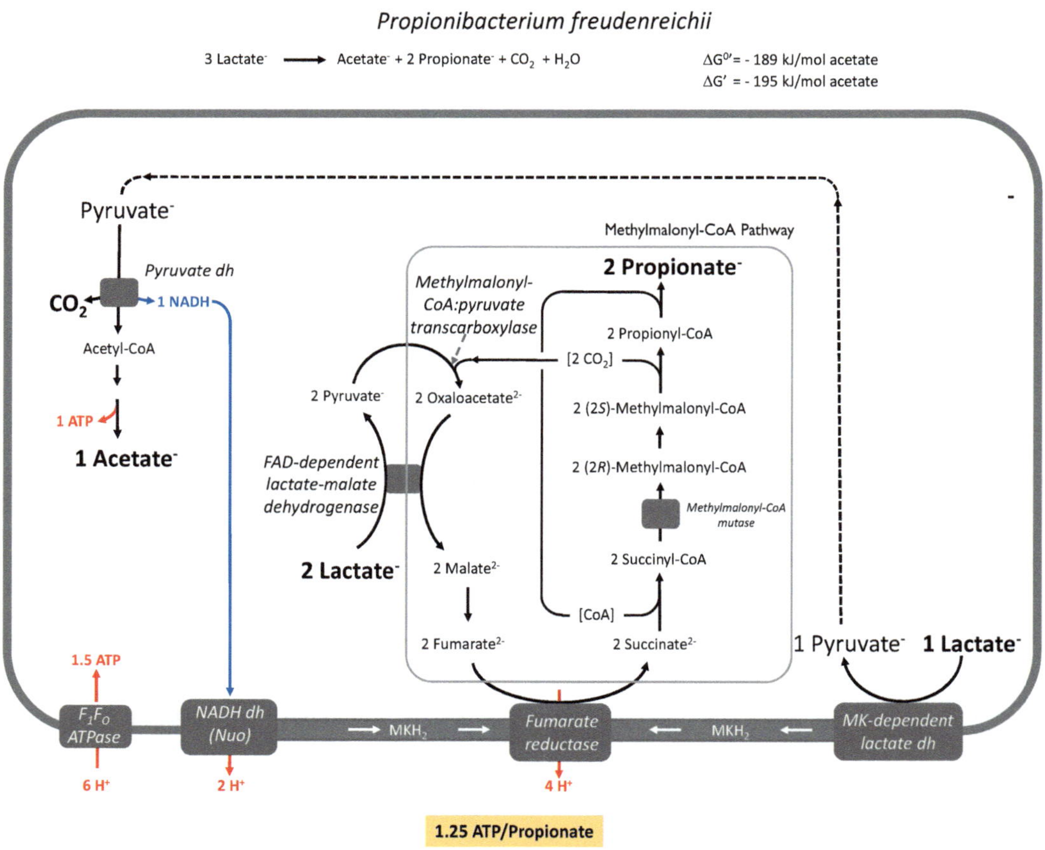

Fig. 9.9 Propionate-forming fermentation of lactate via the Methylmalonyl-CoA pathway in *Propionibacterium freudenreichii*. Abstraction of water in the step from malate to fumarate is not shown. $\Delta G'$ is calculated for 1 mM concentration of lactate, acetate, and propionate, and $pCO_2 = 0.1$ bar, respectively. Abbreviation: *pyruvate dh pyruvate dehydrogenase*. See text for details

couple lactate oxidation with the reduction of more positive (mena)quinone ($E^{0\prime} = -70$ mV), when in a rate mode of growth. Another mode of lactate oxidation in strict anaerobes (see below) is via electron confurcation with Fd^{2-} oxidation to 2 NADH (see Fig. 9.12). While the *menaquinone-reducing lactate dehydrogenase* is not energy-conserving, reduction of menaquinone by *NADH dehydrogenase* (Nuo) in *P. freudenreichii* translocates 2 H^+ and so does the *menaquinol-oxidizing fumarate reductase*. Because 2 fumarate are reduced and 1 NADH is oxidized, a total of 6 electrogenic H^+ are translocated per 3 lactate, which is equivalent to 1.5 ATP.

In the reductive pathway, lactate is oxidized first by a FAD-dependent *lactate-malate dehydrogenase* to pyruvate with concomitant reduction of oxaloacetate to malate. Oxaloacetate is derived from the previously formed pyruvate via carboxylation by a *methylmalonyl-CoA:pyruvate transcarboxylase*. Malate is an intermediate in the Methylmalonyl-CoA pathway and is dehydrated to fumarate for subsequent reduction by reduced menaquinone and energy conservation in *menaquinol-oxidizing fumarate reductase*. A *propionyl-CoA:succinate CoA transferase* couples activation of succinate to succinyl-CoA with conversion of propionyl-CoA to propionate in an equilibrium reaction. Thus, in lactate fermentation to 1 acetate and 2 propionate, 1.25 ATP are conserved net per propionate.

When *P. freudenreichii* ferments glucose to acetate and propionate, energy is conserved via SLP and fumarate respiration, except that electrons for oxaloacetate reduction are from NADH oxidation (Fig. 9.10). There is also no need for a *menaquinone-dependent lactate dehydrogenase.* During glucose fermentation to propionate, 2.75 ATP are formed per propionate in the longer oxidative pathway, conferring a slower yield phenotype (Fig. 9.10).

This trait of some propionic acid bacteria containing the Methylmalonyl-CoA pathway to switch between fermentation of either glucose or lactate to propionate via common pathway elements allows these microbes to escape competition with lactic acid bacteria for rapid glucose utilization. When fermenting glucose to propionate, these propionic acid bacteria perform the combined lactate- and propionate-forming fermentation within a single microbe rather than this overall metabolism being speciated into a two-member consortium (Fig. 9.11). In a sense, this ecology reflects the phenotype of the Large Colony Variant (LCV) of the Julian Adams experiment: LCV can still metabolize glucose in the absence of competition, but in the presence of competition with inefficient but fast Small Colony Variants, LCV adapted to utilizing the excreted end product acetate.

9.4.2 Propionate-Forming Fermentation via the Acryloyl-CoA Pathway

The Acryloyl-CoA pathway has been shown in the Firmicutes *Clostridium propionicum, Coprococcus catus,* and *Megasphaera elsendii* as well as in the Bacteriodetes *Prevotella ruminicola*. Also in these microbes, 3 $lactate^-$ are fermented to 1 $acetate^-$, 2 $propionate^-$ and 1 CO_2, and lactate is both the electron donor and electron acceptor in this pathway (Fig. 9.12). In the oxidative pathway of *Clostridium homopropionicum*, 1 lactate is oxidized by *confurcating lactate dehydrogenase* using Fd^{2-} from the subsequent *pyruvate:ferredoxin oxidoreductase* reaction to form acetyl-CoA and 2 NADH. The Acylate ↔ acyl-CoA module converts 1 acetyl-CoA to 1 acetate and 1 ATP.

Two lactate are reduced in the reductive pathway to two propionate. Key enzymes of the Acryloyl-CoA pathway are ***lactoyl-CoA dehydratase*** and ***propionyl-CoA dehydrogenase*** (*acryloyl-CoA reductase*). In contrast to *electron bifurcating butyryl-CoA dehydrogenase* from *Acidaminococcus fermentans, Clostridium kluyveri,* or *C. difficile* (see below), for the reasons described above, *propionyl-CoA dehydrogenase* in *C. propionicum* is non-bifurcating despite the relatively positive redox potential of $E^{0\prime} = +70$ mV of the acryloyl-CoA/propionyl-CoA couple.

In contrast to the Methylmalonyl-CoA pathway, only 0.5 ATP per propionate are conserved in microbes with the Acryloyl-CoA pathway and only in the oxidative pathway so that the

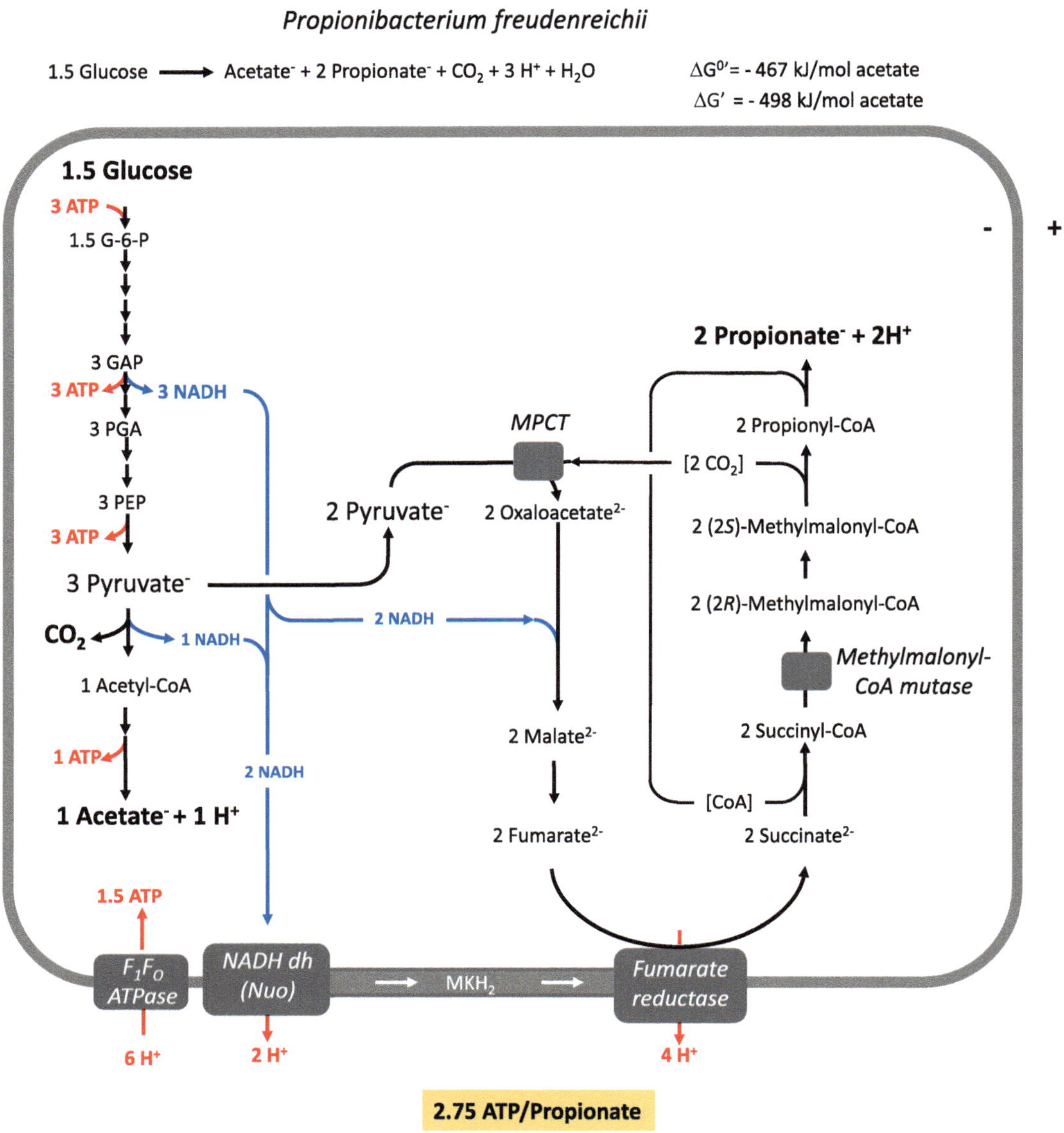

Fig. 9.10 Propionate-forming fermentation of glucose via the Methylmalonyl-CoA pathway in *Propionibacterium freudenreichii*. A switch from a slow glucose to a fast lactate utilization in *P. freudenreichii* or *P. acidipropionici* reflects a switch between expressing a rate or a yield phenotype, and enables propionic acid bacteria to excape competition with fast lactic acid bacteria. Abstraction of water in the step from malate to fumarate is not shown. *MPCT methylmalonyl-CoA: pyruvate transcarboxylase, pyruvate dh pyruvate dehydrogenase*. $\Delta G'$ is calculated for 1 mM concentration of glucose, acetate, and propionate, and $pCO_2 = 0.1$ bar, respectively. Abbreviations: MPCT *methylmalonyl-CoA: pyruvate transcarboxylase, pyruvate dh pyruvate dehydrogenase*. See text for details

Acryloyl-CoA pathway functions only as electron sink and not in energy conservation. Consequently, the growth yield of these microbes is less than ½ the growth yield compared to the former pathway. This enables propionic acid bacteria with the Acryloyl-CoA pathway to grow faster; another illustrative example of a rate-yield trade-off.

Megasphera elsdenii ferments glucose to butyrate and acetate, but lactate to propionate and acetate via the acryloyl-CoA pathway.

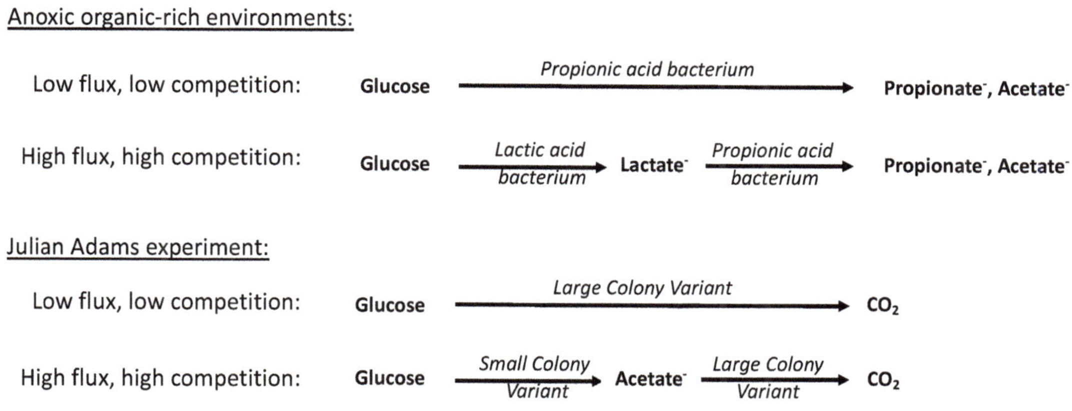

Fig. 9.11 Metabolic speciation and pathway segregation resulting in a rate or yield phenotype in propionate-forming fermentations as in *Propionibacterium freudenreichii*, *P. acidipropionici* or *Bacteroides fragilis*, compared to the *E. coli* strains obtained in the Julian Adams experiment. Equations are not stoichiometrically balanced. The short, niche-specialized pathway of fermentation of lactate to acetate and propionate involves a food chain where a lactic acid bacterium produces the lactate substrate for the propionic acid-forming microbes. This mode of metabolism is expressed in a high-flux, competitive environment. In a low flux environment, these propionic acid-forming microbes switch to fermenting glucose directly and intracellularly to acetate and propionate. In this mode, the microbe is expressing the entire, longer pathway, while at high flux, the shorter pathway is expressed. This speciation of a pathway is reminiscent of the Small and Large Colony Variant phenotype observed in the Julian Adams experiment. See text for details

Other Anaerobic Pathways of Propionate Formation

Fermentation of **1,2-propanediol to propionate and propanol** as in *Salmonella typhimurium*, *Ruminococcus obeum*, *R. inulinivorans*, or *Listeria innocua* represents a third pathway of fermentative propionate formation. 1,2-propanediol is formed as fermentation end product by some microbes, such as *Lactobacillus buchneri*, in fermentation of lactate or of the deoxy-carbohydrates fucose or rhamnose. Abstraction of H_2O from 1,2-propanediol by a *glycerol dehydratase*-type *dehydratase* forms the reactive propanal, which is disproportionated to propionate and propanol. One ATP is formed here via aldehyde (propionaldehyde) oxidation coupled to SLP; the resulting electrons are transferred to a second aldehyde substrate, which is reduced to the corresponding alcohol (propanol). This is essentially a variation of a fermentation of 2 pyruvate to 1 acetate and 1 lactate or of 2 ethanolamine to 1 acetate and 1 ethanol.

A truncated version of the methylmalonyl-CoA Methylmalonyl-CoA pathway is catabolic fermentation of **succinate to propionate**. In this pathway of *Propiogenium modestum*, which is one of the shortest known catabolic pathways, the intermediate methylmalonyl-CoA is decarboxylated to propionyl-CoA. Energy is conserved in this non-redox decarboxylation of methylmalonyl-CoA via membrane-bound *methylmalonyl-CoA decarboxylase* by translocating 2 electrogenic Na^+ across the cytoplasmic membrane.

In contrast to the two examples above, in **fermentation of threonine, methionine, or isoleucine,** propionate is not formed as end product of the reductive but the **oxidative pathway**. Threonine and methionine fermentation converge at 2-ketobutyrate as intermediate that is cleaved to propionyl-CoA and formate by a *2-ketobutyrate:formate lyase* reminiscent of *pyruvate:formate lyase*. Fermentation of isoleucine to propionate in the oxidative pathway of Stickland fermentation proceeds via oxidative decarboxylation and β-oxidation reactions.

Clostridium homopropionicum

3 Lactate$^-$ ⟶ Acetate$^-$ + 2 Propionate$^-$ + CO_2 + H_2O $\Delta G^{0\prime}$ = - 189 kJ/mol acetate
$\Delta G'$ = - 195 kJ/mol acetate

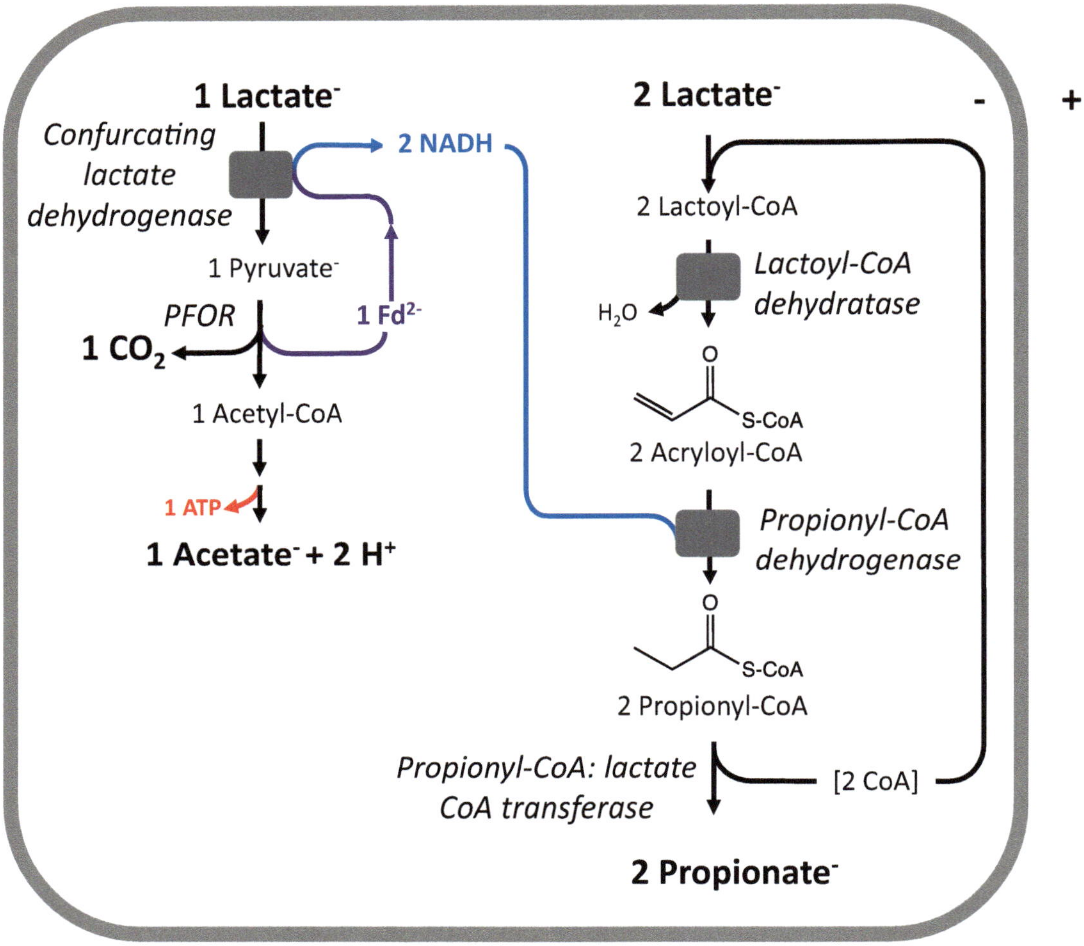

0.5 ATP/Propionate

Fig. 9.12 Lactate fermentation to propionate via the Acryloyl-CoA pathway as in *Clostridium propionicum*. An *electron confurcating lactate dehydrogenase* drives lactate oxidation with Fd^{2-} consumption to 2 NADH (see also Fig. 10.13). In the reductive Acryloyl-CoA pathway, lactate is activated by a *propionyl-CoA:lactate CoA transferase* to lactoyl-CoA, which is dehydrated by a radical enzyme *lactoyl-CoA dehydratase* to acryloyl-CoA. Acryloyl-CoA is reduced by *NAD$^+$-dependent propionyl-CoA dehydrogenase*. $\Delta G'$ is calculated for 1 mM concentration of lactate, acetate, and propionate, and $pCO_2 = 0.1$ bar, respectively. *Abbreviation*: *PFOR* pyruvate:ferredoxin oxidoreductase. See text for details

Propionic acid-forming fermentations play industrially an important role in conversion of fermentatively produced lactate (pK = 3.7) to propionate (pK = 4.7) (Swiss-, and Leerdamer cheese, e.g.).

9.5 Clostridial Fermentations

Clostridia are gram-positive, endospore-forming Firmicutes and have a strictly anaerobic, fermentative metabolism. They contain neither cytochromes nor menaquinone. *Clostridia*, which include *Ruminococcus* species, have in common that **acetyl-CoA, H^+**, and/or **CO_2** function as electron acceptor in their reductive catabolism. Clostridia are the major butyrate-forming microbes in the human gut and industrially important solvent-producing bacteria. Depending on the major substrate specialized in, they can be categorized into **saccharolytic, proteolytic, purinolytic, or acetogenic *Clostridia*.**

9.5.1 Fermentations of Carbohydrates by *Ruminococcus albus, Clostridium pasteurianum, C. acetobutylicum* and for Comparison by *Pyrococcus furiosus*

Core metabolic modules of saccharolytic Clostridia are the **GAP → Pyruvate module** and ***pyruvate:ferredoxin oxidoreductase*** producing acetyl-CoA, CO_2, NADH, and Fd^{2-} when fermenting hexoses or pentoses. Depending on the specific microbe, hexoses are oxidized via the Embden–Meyerhof pathway, while pentoses are metabolized via reactions of the pentose phosphate pathway. The pentose phosphate pathway converts a pentose to a hexose for oxidation by the Embden–Meyerhof pathway. *C. acetobutylicum* ferments xylose at high concentration in parallel also via the *phosphoketolase* pathway.

Glucose and other carbohydrates are taken up via the PTS system although some carbohydrates can be transported in symport with protons. The common core of the reductive pathway of carbohydrate-fermenting Clostridia consists of a **reductive β-oxidation module** and **H_2 formation via *hydrogenases*.**

Carbohydrate- and, in particular, hexose-fermenting Clostridia have three characteristic fermentation pathway patterns: fermentations when (i) environmental pH_2 is high, (ii) environmental pH_2 is low, and (iii) environmental pH < 6. These patterns are shown as overview in Fig. 9.13. Transitions between these patterns are continuous with mixed fermentation patterns being expressed under conditions between these end points.

At low pH_2 (i.e., 10 Pa ⍰ $E'_{H+/H2}$ = −300 mV), ***electron confurcating hydrogenases*** mediate Fd^{2-}-dependent oxidation of NADH from cellular electron pool as H_2; thereby freeing up the energy-rich thioester of acetyl-CoA for ATP formation (Fig. 9.13). At high pH_2 (i.e., 101 kPa ⍰ $E'_{H+/H2}$ = −414 mV), only oxidation of Fd^{2-}, but not of NADH, to H_2 is thermodynamically feasible. Thus, 2 NADH are consumed by ***electron bifurcating butyryl-CoA dehydrogenase*** where 2 electrons of one NADH are eventually transferred via Fd^{2-} as intermediate to 2 H^+, thereby removing these electrons as H_2 from the cellular electron pool (see also Fig. 9.2). For formation of crotonyl-CoA as the electron sink for bifurcation, however, one of the two thioesters of acetyl-CoA is consumed and not available for ATP formation.

As fermentations proceed under batch conditions, acetic and butyric acid accumulate, and the external pH is decreasing. As a response, metabolism of those *Clostridia* shifts to prevent further acidification and to remove acidity by reduction of previously formed acetic acid and butyric acid to the corresponding alcohols ethanol and butanol. Electrons for these reductions are from NADH and Fd^{2-} (Fig. 9.13). The underlying principles and shifts are discussed in the examples below in more detail.

Fermentative metabolism of glucose by ***Ruminococcus albus***, which belongs phylogenetically to *Clostridia*, shifts as a function of pH_2 (Fig. 9.14). At low pH_2, glucose is

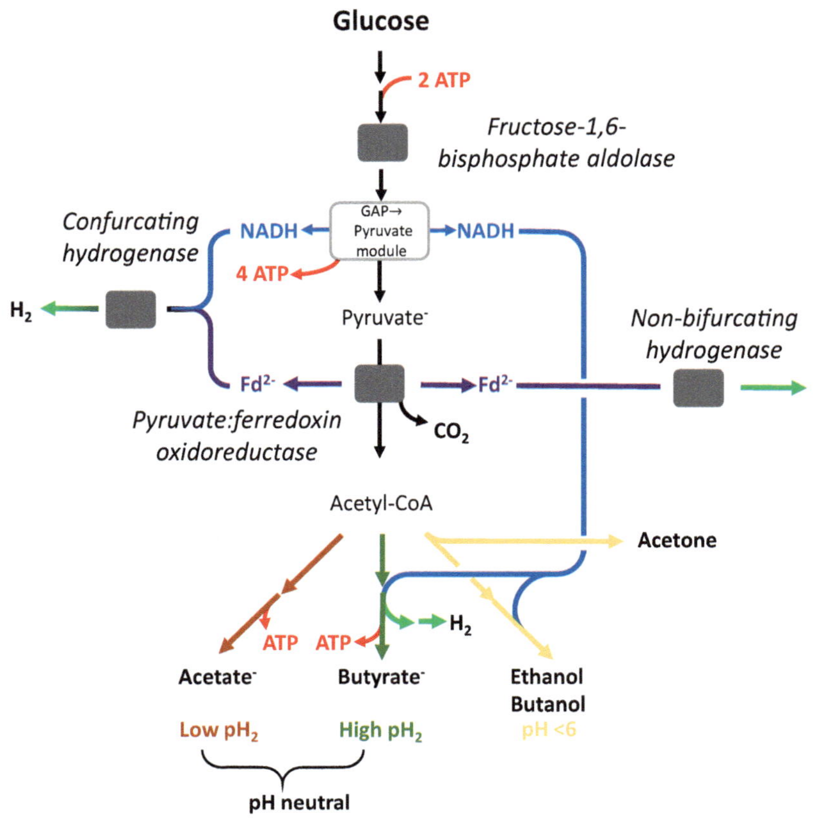

Fig. 9.13 Overview of major fermentation patterns of saccharolytic Clostridia as a function of environmental pH_2 and pH. Emphasized is the carbon and electron flow as well as energy conservation. For emphasizing the pattern, not all metabolic steps are shown, and electron flux and end product formation is shown only qualitative and not stoichiometrically balanced. Blue arrows indicate flux of NADH, magenta of Fd^2, light green of H_2. Brown: reductive pathway at low pH_2; green: reductive pathway at high pH_2; yellow: reductive pathway at low pH (<pH6). See text for details

fermented to 2 acetate$^-$, 4 H_2, 2 CO_2, and 2 H^+, while at high pH_2, glucose is fermented to 1 acetate$^-$, 1 ethanol, 2 H_2, 2 CO_2 and 1 H^+. In vivo, low pH_2 indicates conditions where hydrogen is consumed by a methanogenic or sulfate-reducing partner microbe in close spatial proximity.

In *R. albus*, glucose is taken up in a symport with one H^+ and metabolized via *fructose-1,6-bisphosphate aldolase*, the GAP → Pyruvate module, and *pyruvate:ferredoxin oxidoreductase* to 2 acetyl-CoA and 2 CO_2 (Fig. 9.14). *R. albus* does not use a reductive β-oxidation module as electron sink but at **low pH_2 a *confurcating hydrogenase*** to form 2 H_2 from 1 NADH and 1 Fd^{2-} (Fig. 9.14a). At **high pH_2**, a bifunctional ***acetaldehyde/ethanol dehydrogenase*** oxidizes 2 NADH from the GAP → Pyruvate module and reduces one acetyl-CoA to ethanol (Fig. 9.14b); the electrons from 2 Fd^{2-} from *pyruvate:ferredoxin oxidoreductase* reaction are released as 2 H_2. At low pH_2, there is no stoichiometric need for using the second acetyl-CoA to balance cellular electrons, and this second acetyl-CoA is free to be converted to 1 acetate and 1 ATP. At low pH_2, 3.75 ATP are formed per glucose (55% thermodynamic efficiency) while at high pH_2 only 2.75 ATP (53% thermodynamic efficiency). Thus, commensal H_2 consumption with, e.g., hydrogenotrophic methanogens, increases the fitness (n_{ATP}) of *R. albus*. This H_2 consumption and the beneficial shift of fermentation to a higher n_{ATP} at low pH_2 is NOT a

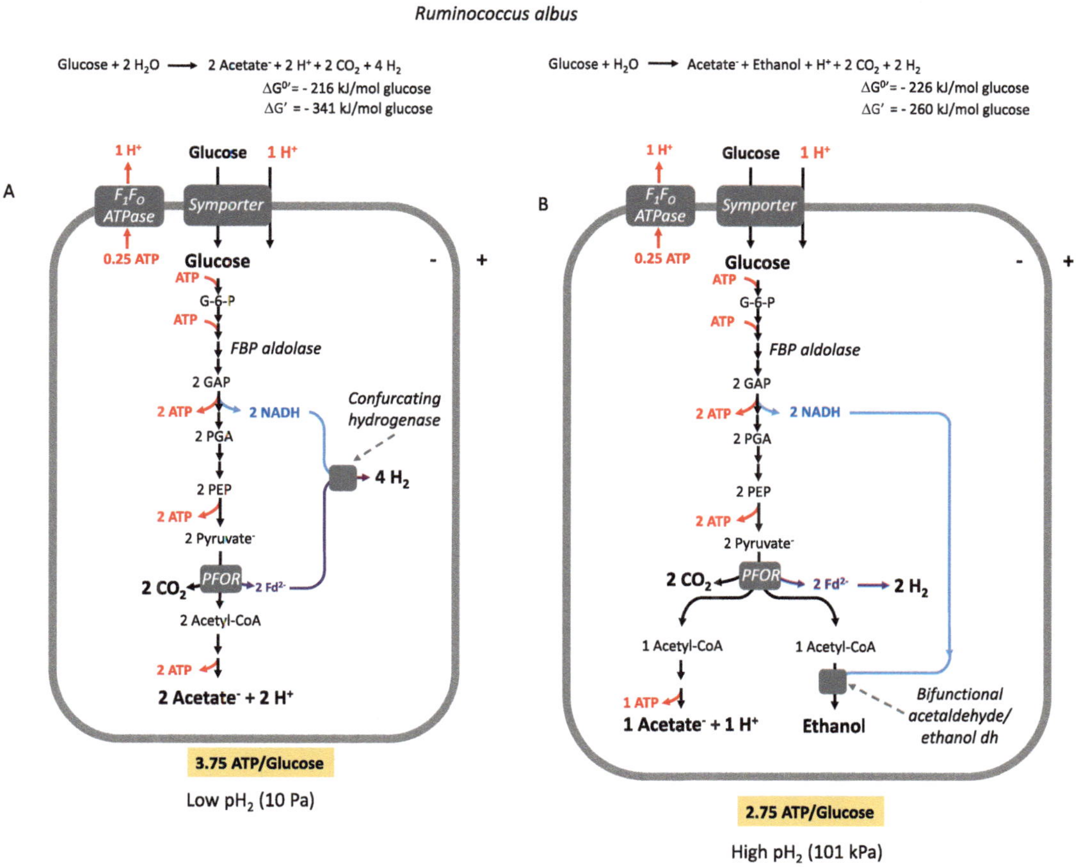

Fig. 9.14 Glucose fermentation by *Ruminococcus albus* as a function of environmental pH_2. (**a**) Low pH_2 conditions are not syntrophic conditions although H_2 is consumed by another microbe. (**B**) High pH_2 conditions. $\Delta G'$ is calculated for 1 mM glucose, 0.1 mM acetate, 1 mM ethanol, and $pCO_2 = 0.1$ bar; low pH_2 is 10 Pa, respectively. Note, the large decrease in ΔG at low pH_2 is primarily driven by low pH_2. Abbreviations: FBP aldolase, *fructose-1,6-bisphosphate aldolase*; PFOR, *pyruvate:ferredoxin oxidoreductase*. See text for details

syntrophic interaction as the change in free energy of the overall catabolic reaction under standard conditions is negative in both cases (see Sect. 9.4).

While not a Clostridium, an interesting variation of glucose fermentation to acetate and H_2 is present in the hyperthermophilic archaeon ***Pyrococcus furiosus*** (Fig. 9.15). Glucose is taken up by an ABC transporter consuming 1 ATP. Instead of an *NAD⁺-dependent, phosphorylating glyceraldehyde-3-phosphate dehydrogenase* as in *Ruminococcus albus*, this microbe contains a ***Fd-dependent, non-phosphorylating glyceraldehyde-3-phosphate dehydrogenase*** (see Chap. 8.4, Fig. 8.14), where in the associated GAP → Pyruvate module only 1 ATP is formed net per pyruvate. Pyruvate is oxidatively decarboxylated by a *pyruvate:ferredoxin oxidoreductase* to Fd^{2-} and acetyl-CoA, which is converted by the Acylate ↔ acyl-CoA Module to acetate and ATP. Thus, per glucose consumed 4 Fd^{2-} are formed that are converted to 4 H_2 by a membrane-associated *hydrogenase* abbreviated ***Mbh***, which is an energy-conserving Ech type *hydrogenase*, leading to the formation of 2 ATP per glucose by ETP. Overall, 3 ATP are formed net from glucose.

Fermentation of glucose to acetate, butyrate, and H_2 in ***Clostridium pasteurianum*** is shown in Fig. 9.16. While the oxidative fermentative pathway includes ***fructose-1,6-bisphosphate aldolase*, the GAP → Pyruvate module, *pyruvate:***

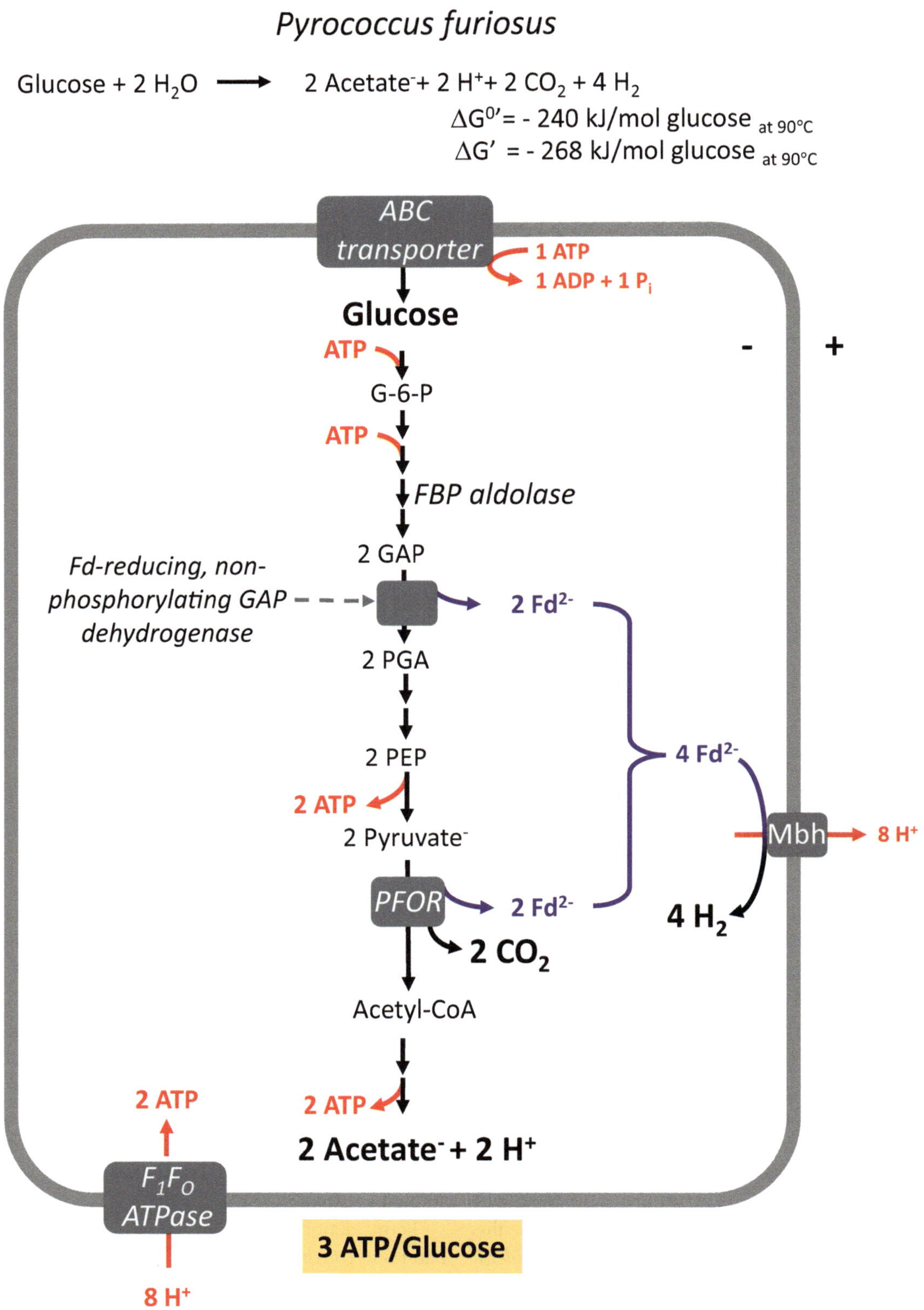

Fig. 9.15 Glucose fermentation to 2 acetate$^-$, 2 H^+, and 2 CO_2 by the archaeon *Pyrococcus furiosus*. Because *Fd-dependent glyceraldehyde-3-phosphate dehydrogenase* is non-phosphorylating, ATP is formed in the

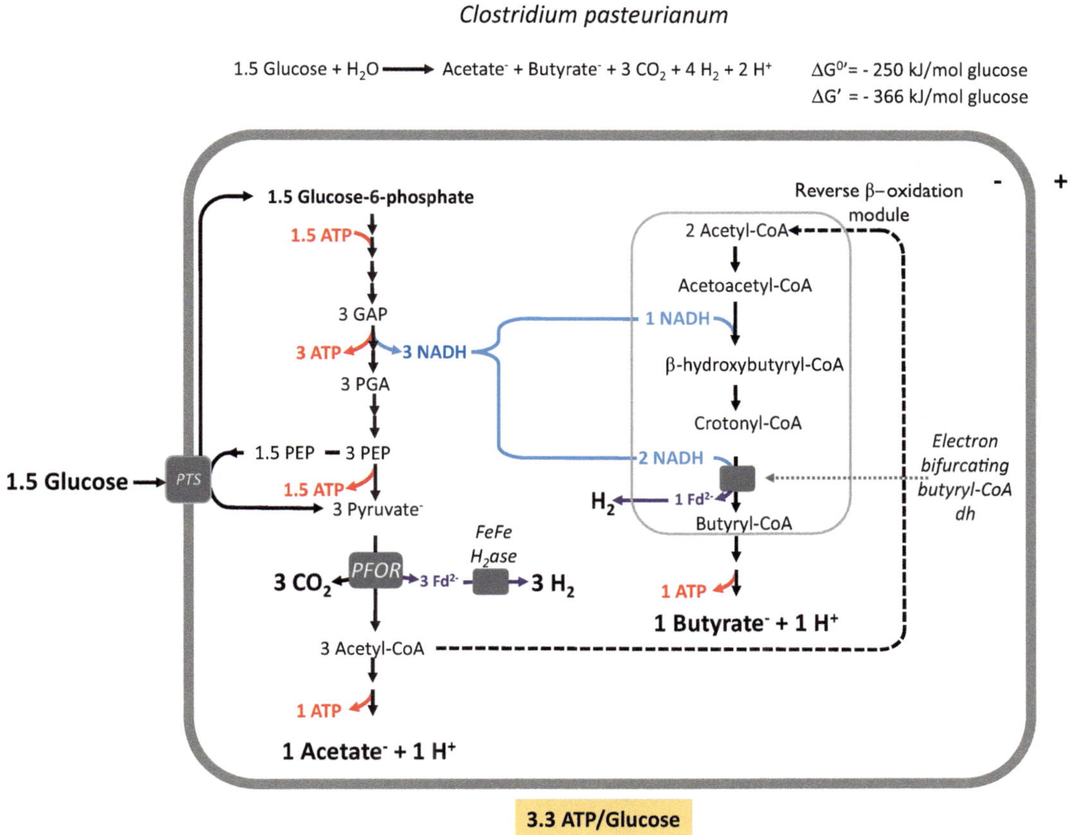

Fig. 9.16 Glucose fermentation to butyrate$^-$, acetate$^-$, $2H^+$, and H_2 in *Clostridium pasteurianum*. *Electron bifurcating butyryl-CoA dehydrogenase* consumes 2 NADH and forms butyryl-CoA and Fd^{2-}, which is oxidized by a *non-bifurcating hydrogenase* to form H_2. Steps abstracting water are not shown. $\Delta G'$ is calculated for 1 mM concentration of glucose, acetate, butyrate, and $pCO_2 = 0.1$ bar; low pH_2 is 10 Pa, respectively. Note the large decrease in ΔG at low pH_2, primarily driven by low pH_2. Abbreviations: *PFOR pyruvate:ferredoxin oxidoreductase*, *[FeFe] H_2ase* non-bifurcating *[FeFe] hydrogenase*. See text for details

ferredoxin oxidoreductase, and **the Acylate ↔ acetyl-CoA Module,** the reductive pathway includes the **reverse β-oxidation module with *electron bifurcating butyryl-CoA-dehydrogenase*** and ***[FeFe] hydrogenase***. In the oxidative pathway, 1.5 glucose are net oxidized to 3 acetyl-CoA, 3 NADH, 3 Fd^{2-}, and 4.5 ATP. One of the three acetyl CoA is converted to acetate, leading to formation of a further ATP via the Acylate ↔ acyl-CoA Module.

The other two acetyl-CoA are metabolized in the reductive pathway to form two electron sink reactions catalyzed by *β-hydroxybutyryl-CoA dehydrogenase* and ***electron bifurcating butyryl-CoA dehydrogenase***. The latter couples oxidation of 2 NADH with crotonyl-CoA

Fig. 9.15 (continued) *pyruvate kinase* reaction of the GAP→Pyruvate module (i.e., 2 ATP per glucose), in the *acetate kinase* reaction (2 ATP per glucose), and in Mbh (2 ATP per glucose). Mbh is an energy-conserving Ech *hydrogenase*. $\Delta G'$ is calculated for 1 mM concentration of glucose, acetate, and $pCO_2 = 0.1$ bar, respectively. *PFOR pyruvate:ferredoxin oxidoreductase*. See text for details

reduction and reduction of Fd. Fd^{2-} is subsequently re-oxidized by non-bifurcating *[FeFe] hydrogenase* to release from cellular metabolism NADH electrons as H_2 (Fig. 9.16, see also Fig. 9.2). The net total ATP balance is 3.3 ATP/glucose.

Butyrate in the Human Gut

Butyrate is a key microbial metabolite in the human gut and is a catabolic substrate for colonocytes of the epithelial gut layer. Colonocytes respire microbially-formed butyrate with host-supplied O_2, resulting in oxygen depletion; thereby, generating a hypoxic mucosal environment which favors anaerobic microbial fermentations in the gut lumen and prevents bacterial escape through the intestinal mucosal barrier. A reduction in fermentative butyrate production (e.g., via antibiotic treatment) deprives colonocytes of the electron donor butyrate and associated O_2 consumption resulting in an increased local O_2 level. The increase of O_2 in an otherwise anoxic gut opens novel niches for facultative respiring and rapidly growing bacteria, such as γ-proteobacteria, that outcompete other microbes for similar electron donor substrates, resulting in an ecological shift and dysbiosis of the gut microbial community.

Clostridium acetobutylicum is a major industrial microbe for production of the solvents acetone and butanol (Fig. 9.17), based on its pH-dependent fermentation shift (see Fig. 9.13). During acidogenic growth at circumneutral pH, fermentation of glucose is similar to that of *C. pasteurianum* involving an *electron bifurcating butyryl-CoA dehydrogenase.* When cells transition into stationary phase, glucose fermentation changes to solventogenesis by shifting from acetate/butyrate production to acetone/butanol/ethanol formation (Fig. 9.17). Characteristic

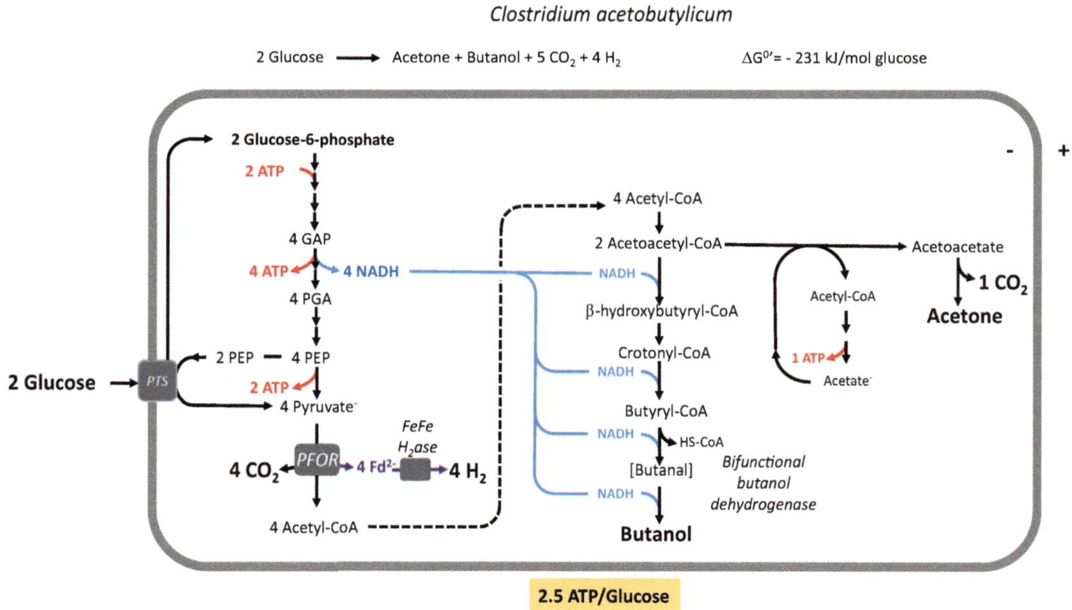

Fig. 9.17 Glucose fermentation to butanol, acetone, H_2, and CO_2 in solventogenic *Clostridium acetobutylicum.* The pathway of glucose fermentation is similar to that of *C. pasteurianum*, except that electron flux is shifted from H_2 formation to butyryl-CoA reduction to butanol. Formation of acetone drives NADH formation via the GAP→Pyruvate module. Abbreviation: *PFOR pyruvate: ferredoxin oxidoreductase*. See text for details

to this shift is an **alternative fate of acetyl-CoA**. Acetyl-CoA and butyryl-CoA are reduced to the corresponding alcohols, thereby, avoiding a further acidification of the extracellular environment. For each acetyl-CoA formed from glucose, one NADH is formed in the GAP → pyruvate module. Because no electron sink is generated in the 2 acetyl-CoA → acetone reaction sequence, the electrons from the two NADH are freed up for the two-step butyryl-CoA reduction to butanol. In this fermentation, only 2.5 ATP are formed net per glucose.

In the solventogenesis phase, butyrate and acetate from the environment are also partially re-used and reduced to butanol and ethanol, respectively. H_2, released from cells in the acidogenesis phase via an Fe-Fe *hydrogenase* (HydA), appears to be taken up in the solventogenesis phase by a membrane-associated Ni-Fe *hydrogenase* (HupS) for butyrate and acetate reduction via NADH as electron carrier (not shown).

The shift in fermentation of *C. acetobutylicum* appears to depend on a combination of (i) a decrease in extracellular pH, (ii) extracellular accumulation of acetic and butyric acid, and (iii) cell density. In particular, accumulation of these weak acids at pH <6 induces a change of intercellular pH, which the microbe cannot maintain at decreasing external pH. This loss of internal pH homeostasis is probably due to the production of the weak acids acetic and butyric acid with pKa of ~4.8. At external pH <6, a significant fraction (~6% or more) of the weak acids are protonated and membrane permeable (see Fig. 6.12). After diffusing into a cell, these weak acids dissociate at the higher internal pH and release a proton that cumulatively acidifies the cytoplasm as the concentration of both acids increases during batch fermentations. This effectively collapses the internal pH.

9.5.2 Ethanol-Acetate Fermentation of *Clostridium kluyveri*

Metabolism of ***Clostridium kluyveri*** is another example where energy conservation is driven by electron bifurcation associated with enoyl-CoA reduction in a reductive fermentative pathway (Fig. 9.18). This microbe ferments 6 ethanol + 4 acetate$^-$ to 5 butyrate$^-$ + 4 H_2O + 2 H_2 + 1 H^+. The C_6 fatty acid caproate is also formed in this fermentation when butyrate substitutes for acetate as extracellular substrate entering the reverse β-oxidation module. For the sake of clarity, this congruent aspect of this metabolism is omitted here.

The core elements in this fermentation are the **reverse β-oxidation pathway with *electron bifurcating butyryl-CoA dehydrogenase/Etf*, the Acylate ↔ acetyl-CoA Module, the *alcohol/aldehyde oxidoreductase* sequence, *[FeFe] hydrogenase,*** and **Rnf.** In contrast to glucose fermentation, ethanol is the electron donor and oxidized to acetyl-CoA. One of the 6 acetyl-CoA is converted to 1 acetate$^-$ and 1 ATP via the Acylate ↔ acetyl-CoA Module (Fig. 9.18). The reverse β-oxidation pathway functions as electron sink for NADH oxidation to drive Fd^{2-} formation via *electron-bifurcating butyryl-CoA dehydrogenase*. Another 1.5 ATP is formed via ferredoxin-dependent Rnf mediating ferredoxin-dependent Ion Translocation Coupled Phosphorylation to reduce NAD^+.

Oxidation of 6 ethanol to 6 acetyl-CoA yields 12 NADH (Fig. 9.18). These reactions occur in a microcompartment (see Sect. 8.3.5), presumably to mitigate the cytotoxic effects of acetaldehyde as an intermediate. Five of the six acetyl-CoA are funneled into the reverse β oxidation pathway where they condense with 5 acetyl-CoA that are derived from the external substrate acetate upon activation by a *CoA transferase*. Subsequent formation of 5 butyryl-CoA consumes 5 NADH in the *β-hydroxybutyryl-CoA dehydrogenase* reaction. Since *C. klyuveri* contains an *electron bifurcating butyryl-CoA dehydrogenase*, 10 NADH are required for producing 5 butyryl-CoA from 5 crotonyl-CoA. Of these 10 NADH, 7 NADH are from ethanol → acetyl-CoA oxidation, while the other 3 NADH are from oxidation of 3 Fd^{2-} to 3 NADH by Rnf. The 3 Fd^{2-} consumed by ***Rnf*** result in conservation of the energy of six electrogenic protons via *F_1F_OATP synthase*. Two of the five Fd^{2-} are oxidized

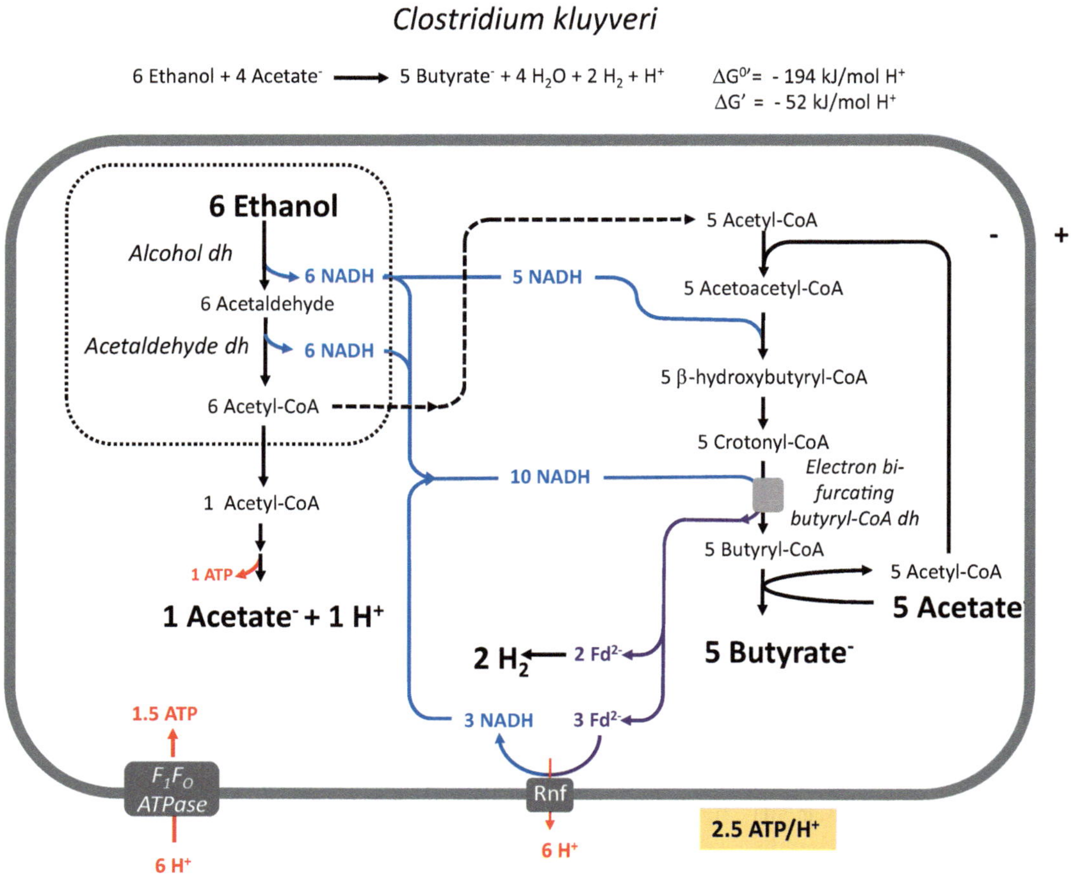

Fig. 9.18 Fermentation of ethanol plus acetate to butyrate and H_2 by chain elongation by *Clostridium kluyveri*. For simplification, it is assumed that only butyrate and not caproate is formed. Chain elongation of fatty acids (acetate, butyrate) via the reverse β-oxidation module is an electron sink for ethanol oxidation; energy conservation occurs via Rnf through NADH-consuming, *bifurcating butyryl-CoA/Etf* to form Fd^{2-}. Enzyme reactions occurring enclosed in the microcompartment are indicated in the dashed rectangle. Note, how the available energy (ΔG) decreases when substrate and product concentrations are close to 10 μM because of the entropic effect (converting 10 molecules into 7). Steps abstracting water are not shown. See text for details

with protons to 2 H_2 (Fig. 9.18). This fermentation effectively represents a **chain elongation** of a substrate fatty acid (acetate) by a C2 unit to the C4 compound butyrate.

The ATP gain (n_{ATP}) of this chain elongation fermentation strongly depends on the in situ concentrations of substrates, as evident from the ratio of organic substrates to end products of 10:5. Thus, at an in situ concentrations of 10 μM, for example, the $\Delta G'$ changes from −184 kJ/mol H^+ to −41.5 kJ/mol H^+!

Most fermentations of Clostridia and other microbes discussed in this chapter so far lead to H_2 formation. Fermentation patterns vary with in situ pH_2, resulting in substantially different n_{ATP}. In most anoxic environments, pH_2 is kept low by H_2-consuming microbes. While some respiring microbes couple catabolic H_2 oxidation with, e.g., O_2, NO_3^-/NO_2^-, Fe(III), or SO_4^{2-} reduction, many methanogens and acetogens reduce CO_2 to methane and acetic acid, respectively. Therefore, their metabolism controls via their H_2-pull the fitness of other community members. Figure 9.19 captures schematically such metabolic interaction based primarily on H_2 and on acetate as intermediates. The thickness of arrows

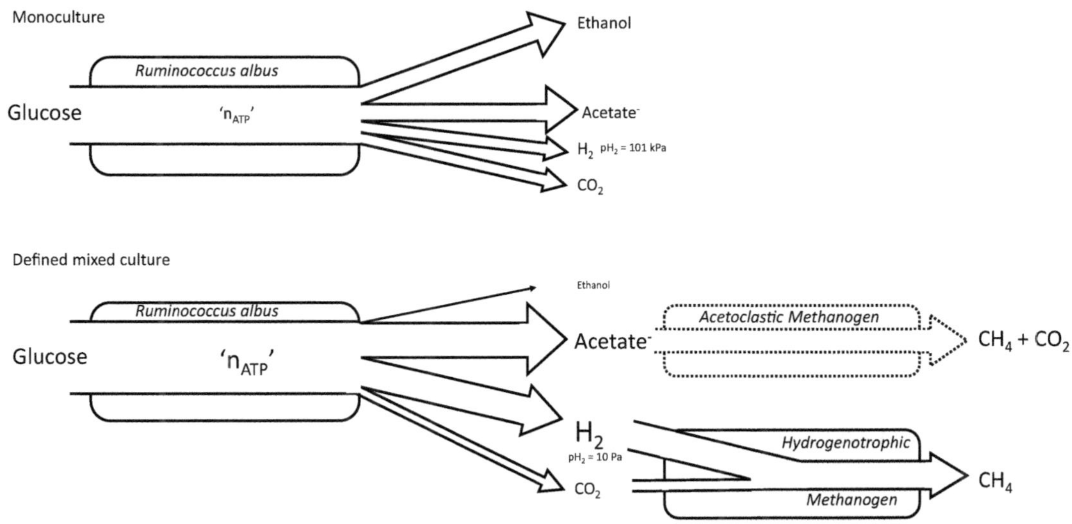

Fig. 9.19 Metabolic interactions between a H_2-forming (e.g., *Ruminococcus albus*, Fig. 9.14) and H_2-consuming microbe (e.g., hydrogenotrophic methanogen, Fig 10.19) and its effect on the fitness of both partners. The example here shows how H_2 consumption by the methanogen shifts ethanol formation at high pH_2 to more acetate and H_2 formation at low pH_2. The higher stoichiometry of H_2 formation at low pH_2 effectively increases the flux of electrons through a H_2 pool. This metabolic coupling increases the ATP gain of both microbes. Acetate consumption by an acetoclastic methanogen (Fig. 10.21) further increases the flux of carbon and electrons, and n_{ATP} of each of the community members. The width of the arrows indicates relative flux sizes under monoculture (upper Figure) and defined mixed culture (lower Figure) conditions. Stoichiometries were omitted for clarity. Modified after F. Widdel. See text for details

indicates relative metabolic fluxes. For example, active metabolism of H_2 by a hydrogenotrophic methanogen re-routes glucose fermentation from ethanol and H_2 to acetate and increased H_2 formation in *Ruminococcus albus* (Fig. 9.19). A higher fraction of acetate in a fermentation balance represents an increased intracellular flux of acetyl-CoA through the Acylate $\leftrightarrow$ acyl-CoA Module and, with that, results in a higher n_{ATP}. Thus, the presence of a methanogen increases n_{ATP} of the fermenting microbe by a thermodynamically-enabled re-routing of NADH oxidation to proton reduction, which increases H_2 flux to the methanogen and of acetate to the acetoclastic methanogen (see Chap. 10.4). This example also shows that a high flux of electrons through a particular intermediate in the environment, here of H_2, does not necessitate a high in situ concentration; on the contrary, the high flux of H_2 and the increased ATP gain for both microbes occurs only at low in situ concentration of H_2! This insight is often expressed anecdotally as: "The most important metabolites in a microbial ecosystem often are hardly detected, because they are rapidly consumed."

A similar effect of maintaining a low pH_2 is essential in syntrophic oxidation of end products of primary fermenters (see below).

9.5.3 Proteolytic Clostridial Fermentations

The spectrum of end products of amino acid fermentations is more diverse relative to carbohydrate fermentations, because the 20 different proteinaceous amino acids are metabolized to different end products in the oxidative and reductive pathways. Despite this diversity, amino acid fermentation proceeds according to a few principles.

9.5.3.1 Stickland Fermentations

Amino acids are fermented in pairs by "**Stickland Fermentation**," where both electron donor and electron acceptor are amino acids; either the same

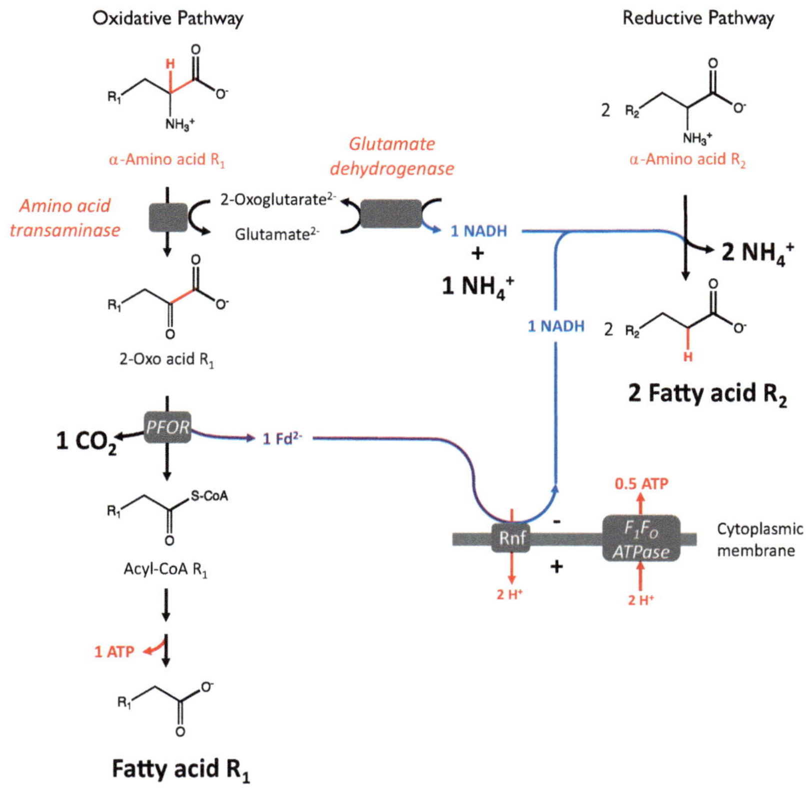

Fig. 9.20 Overview of clostridial strategies to ferment pairs of amino acids via Stickland fermentation. Core reactions in the oxidative pathway are the oxidative deamination (*amino acid deaminase*, *glutamine dehydrogenase*), *2-oxo acid:ferredoxin oxidoreductase,* and the acylate↔acyl-CoA module. The bonding electron pairs that are oxidatively removed from the electron donor amino acid are indicated by red bonding lines and hydrogen atom. There is more metabolic diversity in how the C2 carbon of the electron-accepting amino acid is reduced (red lines and hydrogen atom), however, abstractions of H_2O by *dehydratases* after deamination are common. Energy conservation occurs via the Acylate↔acyl-CoA module and via Rnf and Ion Translocation Coupled Phosphorylation. Abbreviation: *2-oxo acid:FOR 2-oxo acid: ferredoxin oxidoreductase*. See text for details

amino acid or two different ones. The common metabolic strategy of such amino acid fermentation is outlined in Fig. 9.20: in the oxidative pathway, one amino acid (with side chain R_1) is **oxidatively deaminated** to the corresponding 2-oxo acid. This deamination and oxidation of the α carbon occurs via an amino acid-specific ***deaminase,*** transferring the amino group to 2-oxoglutarate to form glutamate, and ***glutamate dehydrogenase***, releasing NADH plus 1 NH_4^+ (Fig. 9.20). The 2-oxo acid formed, similar as the 2-oxo acids pyruvate and 2-oxo-glutarate (see Sect. 8.2.3), is then oxidatively decarboxylated via a ***2-oxo acid:ferredoxin oxidoreductase*** to an acyl-CoA that is shorter by a C_1 (CO_2). The Acylate ↔ acyl-CoA Module converts acyl-CoA to the corresponding fatty acid (R_1) plus 1 ATP. Both NADH and Fd^{2-} are formed in the oxidative pathway. Rnf couples energy conservation of Fd^{2-} oxidation to NAD^+ reduction and translocation of two electrogenic H^+ or Na^+ per Fd^2.

Glycine Reductase

in Stickland fermentation. *Glycine reductase* catalyzes exergonic reductive elimination of the α amino group of glycine

(continued)

coupled to formation of acetyl-phosphate + NH_4^+:

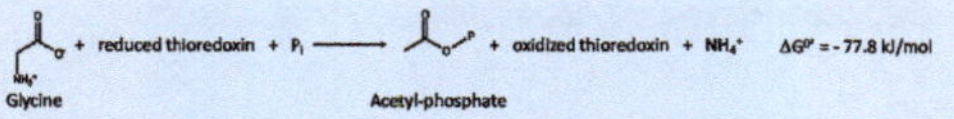

Electron donor of the three protein-component selenocysteine enzyme is thioredoxin that is reduced by NADPH in *Clostridia spp*. Glycine reduction not only serves as electron sink in the reductive pathway of amino acid fermentation by re-oxidizing NADPH, but also forms an energy-rich bond as acetyl-phosphate, which is converted to ATP. Thus, energy is conserved here by SLP in a reductive pathway! *Glycine reductase* has been shown to be active in *C. sticklandii*, *C. difficile*, *C. litorale*, and *Eubacterium acidaminophilum* as well as in the Gram-negative *Treponema dendricola*.

In the **reductive pathway** either the same amino acid as the electron donor or a different amino acid functions as electron sink where the oxidized C2 carbon (α carbon) is reduced to another fatty acid of the same carbon length (red bond and H in Fig. 9.20).

9.5.3.2 Alanine Fermentation via the Acryloyl-CoA Pathway

Alanine, being both electron donor and acceptor, is fermented to acetate and propionate via a pathway that closely resembles the above-discussed pathway of lactate fermentation to propionate (see also Fig. 9.12). Alanine and lactate are structurally similar C_3 compounds, where the corresponding C2 carbon atoms in alanine and lactate are at the same oxidation state. Fermentation of 3 alanine to 1 acetate$^-$ + 2 propionate$^-$ + 1 CO_2 + 3 NH_4^+ by *Clostridium propionicum* is shown in Fig. 9.21. There is an important difference, however, between fermentative oxidation of lactate and alanine. Alanine oxidation to pyruvate is an oxidative deamination, consisting of an *alanine deaminase* and a *glutamate dehydrogenase* (Figs. 9.20 and 9.21). NADH is formed and no Fd^{2-} is required as in the lactate-oxidizing *confurcating lactate dehydrogenase* (see Fig. 9.12). This frees up the Fd^{2-} of the *pyruvate:ferredoxin oxidoreductase* in alanine fermentation for energy conservation via Rnf to form 1 NADH plus 0.5 ATP (Fig. 9.21). Of the total of 4 NADH formed from oxidation of 3 alanine (3 in oxidative deamination, and one via Rnf), two NADH are used by *lactate dehydrogenase* to form 2 lactate from 2 pyruvate, while the other 2 NADH are consumed in non-bifurcating *propionyl-CoA dehydrogenase*. Thus, alanine fermentation to propionate conserves 0.75 ATP/propionate vs 0.5 ATP/propionate in lactate fermentation although the oxidation state of the corresponding carbon is equal!

9.5.3.3 Glutamate Fermentations

The currently known pathways for glutamate fermentation to acetate and propionate proceed either via **hydroxyglutarate**, **methylaspartate**, or **2-oxo-glutarate** as intermediates. Pathway variants with 2-oxo-glutarate as intermediate generally contain reaction sequences of the TCA cycle or the full Methylmalonyl-CoAmodule and are not discussed here further. Below are the pathways via hydroxyglutarate and methylaspartate, respectively.

9.5.3.3.1 Glutamate Fermentation via Hydroxyglutarate in *Acidaminococcus fermentans* and *Clostridium symbiosum*

In fermentations by *Acidaminococcus fermentans*, which belong phylogenetically to the *Clostridia*, and *Clostridium symbiosum*, glutamate is both electron donor and electron acceptor (Fig. 9.22). 5 Glutamate are fermented to 6 acetate$^-$ + 2 butyrate$^-$ + H_2 + 5 CO_2 + 5 NH_4^+. The first three metabolic steps are common to both the oxidative and reductive pathways. Five glutamates are deaminated to 5 hydroxyglutarate and activated via an *acetyl-CoA:CoA transferase* to 5 hydroxyglutaryl-CoA. The radical enzyme *hydroxyglutaryl-CoA dehydratase* abstracts H_2O from hydroxyglutaryl-CoA forming 5 glutaconyl-

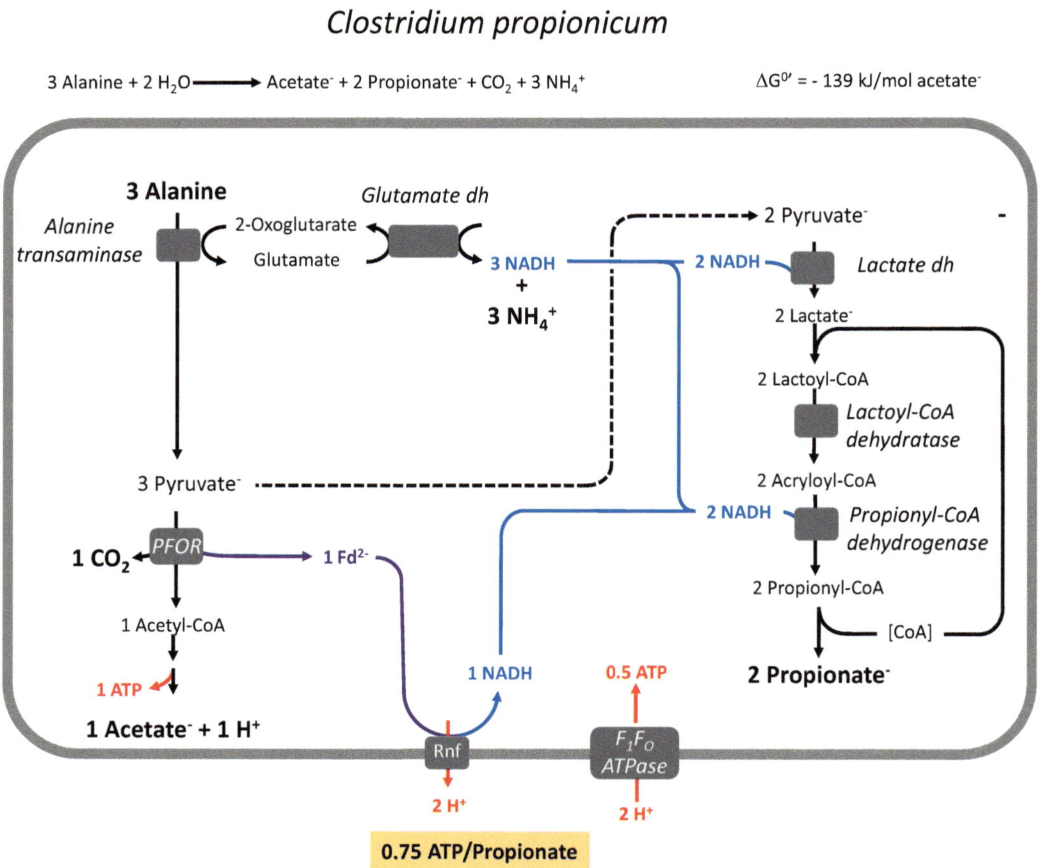

Fig. 9.21 Fermentation of alanine to acetate and propionate by *Clostridium propionicum*. Alanine is both the electron donor (left side pathway) and electron acceptor (right side pathway) via pyruvate from the oxidative pathway. In the reductive pathway, pyruvate is first reduced to lactate. Reduction of lactate to propionate is via the acryloyl-CoA pathway that is also used by this microbe for lactate fermentation (see Fig. 9.12). Metabolic energy is conserved via Rnf from Fd^{2-} formed via *pyruvate: ferredxin oxidoreductase*. NADH formed by either Rnf or *glutamate dehydrogenase* is oxidized by reduction of the C2 carbon of lactate to form propionate. Alanine fermentation to propionate is very similar to lactate fermentation to propionate, except for the redox potential in the first oxidation reaction in each pathway (here via *glutamate dehydrogenase* reducing NAD^+, or via *confurcating lactate dehydrogenase*). Steps abstracting water are not shown. Abbreviation: *PFOR pyruvate:ferredoxin oxidoreductase*. See text for details

CoA, which are decarboxylated by a biotin-containing, membrane-associated, Na^+-translocating *glutaconyl-CoA decarboxylase* to 5 crotonyl-CoA. In these common steps neither NADH nor Fd^{2-} are formed. Crotonyl-CoA is a funneling intermediate, and both pathways diverge at crotonyl-CoA involving reactions of the β-oxidation module either in the forward or reverse direction, respectively.

In the oxidative branch, 3 crotonyl-CoA are oxidized to 6 acetyl-CoA via the **oxidative β-oxidation module** forming 3 NADH. Five acetyl-CoA are consumed in the *acetyl-CoA: CoA transferase* so that only the energy of the sixth acetyl-CoA is conserved as 1 ATP. In the reductive pathway, 3 NADH from the oxidative pathway plus a fourth NADH formed via Rnf are consumed by *bifurcating butyl-CoA dehydrogenase* of the **reductive β-oxidation module.** Two butyryl-CoA and 2 Fd^2 are formed. One Fd^{2-} is oxidized by *non-bifurcating hydrogenase* to H_2, while the second Fd^{2-} is oxidized to NADH by

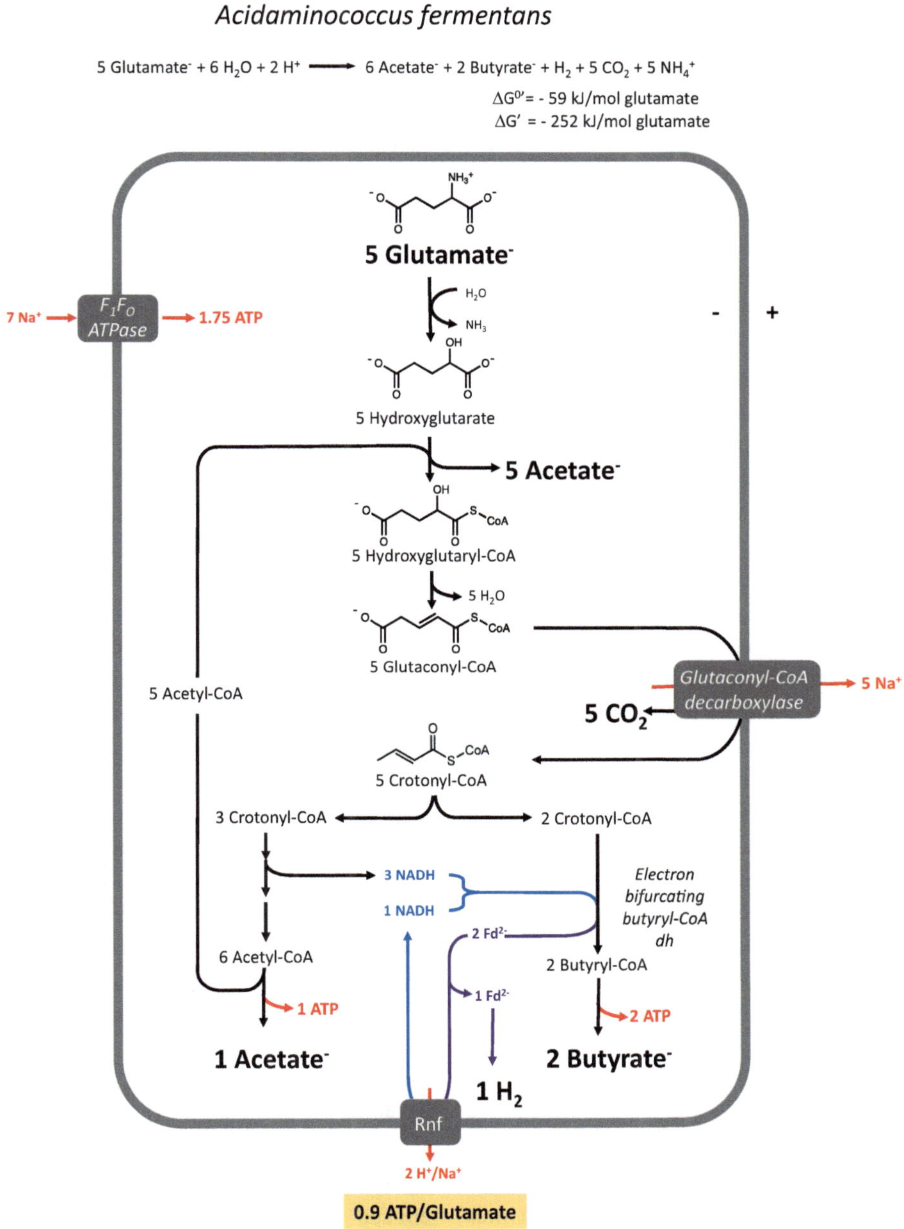

Fig. 9.22 Glutamate fermentation via hydroxyglutarate in *Acidaminococcus fermentans* and *Clostridium symbiosum*. Glutamate is common for both the oxidative and reductive pathways. The switch between both pathways is at the level of crotonyl-CoA, and in both pathways, the β-oxidation module is operating; however, in the opposite directions. Central to this metabolism is Fd^{2-}-formation by *electron bifurcating butyryl-CoA dehydrogenase* to drive energy-conserving electron flux through Rnf. The intermediate glutaconyl-CoA allows for energy conservation by membrane-bound decarboxylation to crotonyl-CoA and concomitant electrogenic translocation of 1 Na^+/crotonyl-CoA The latter is an example of energy conservation in a net non-redox reaction. ΔG calculated for 1 mM concentration of glutamate, acetate, butyrate, and NH_4^+, and $pCO_2 = 0.1$ and $pH_2 = 10$ Pa. Note the large decrease in ΔG, primarily driven by an entropic effect at low concentrations. See text for details

Rnf to form the fourth NADH for *bifurcating butyryl-CoA dehydrogenase*. Theoretically, Fd^{2-} "earmarked" for *non-bifurcating hydrogenase* could also be funneled into Rnf, thereby increasing n_{ATP}; however, the resulting NADH cannot be re-oxidized because of the lack of a sufficient number of available reduction steps per substrate. This is an example that pathway yields, n_{ATP}, are also determined by underlying chemistries and not solely by ecological rate-yield trade-offs.

9.5.3.3.2 Glutamate Fermentation via Methylaspartate in *Clostridium tetanomorphum*

In the first reactions of glutamate fermentation via **methylaspartate**, glutamate is funneled into the pathway of pyruvate fermentation as found in *Clostridium pasteurianum*, except that also Rnf is involved in Fd^{2-} oxidation (Fig. 9.23). This "upstream pathway extension" of glutamate-to-pyruvate conversion, is, again, common to both the oxidative and reductive pathway. 5 Glutamate are converted by coenzyme B_{12}-dependent *glutamate mutase* to 5 methylasparate, which are deaminated by *methylasparate deaminase* to 5 mesaconate. Five mesaconate are subsequently hydrated to 5 citramalate, which are substrates of an aldol-type cleavage to yield 5 acetate and 5 pyruvate. Five pyruvate are oxidized via *pyruvate:ferredoxin oxidoreductase* to 5 acetyl-CoA and 5 Fd^{2-}. Four acetyl-CoA are the carbon substrate of the reductive pathway to form electron sinks for four of the five Fd^{2-} via the reverse β-oxidation module. The fifth acetyl-CoA is converted to acetate with concomitant ATP formation. Five plus one Fd^{2-} are oxidized via energy conservation by Rnf to 6 NADH, which are used to reduce 2 acetoacetyl-CoA to 2 butyrate; 2 NADH are consumed in the *β-hydroxybutyryl-CoA dehydrogenase* and 4 NADH in the *electron bifurcating butyryl-CoA dehydrogenase* reaction. The latter enzyme forms, next to butyryl-CoA, 2 Fd^{2-} of which one is oxidized by *non-bifurcating hydrogenase*. The second Fd^{2-} is oxidized by Rnf to form the sixth NADH consumed in the reductive pathway. Two ATP are formed from 2 butyryl-CoA. In this glutamate fermentation, about half of the ATP are formed via Rnf (Fig. 9.23). While n_{ATP} in the hydroxyglutarate pathway is 0.95, the methylaspartate pathway is more efficient with $n_{ATP} = 1.2$, a ~ 26% increase in efficiency, which is ecologically significant.

9.5.3.4 Metabolism of *Clostridium difficile*

Several Clostridia, such as *C. sticklandii*, *C. difficile*, *C. sporogenes*, or *C. diphteriae*, are medically relevant for causing intestinal diseases. These are proteolytic Clostridia performing Stickland fermentations. *C. difficile* is metabolically similar to *C. sticklandii*; however, *C. difficile* can also ferment carbohydrates, which *C. sticklandii* cannot. *C. difficile* has complex nutritional requirements and metabolic patterns which reflect its adaptation to the intestine environment. Figure 9.24 shows an overview of major oxidative and reductive pathways, of substrate fluxes and end products, and of NADH and Fd^{2-} forming and consuming steps in these microbes. The fluxes are not balanced and not all oxidative and reductive pathways are expressed at the same time. The purpose of this Figure is to show the mosaic and recurrent use of core metabolic modules (i.e., the microbe's metabolic "business" plan) as well as the resulting capacity of this microbe for accessing chemically diverse electron donors and electron acceptors.

The kinetically preferred metabolism of *C. difficile* is amino acid fermentation by Stickland fermentation using as electron donor the amino acids leucine, isoleucine, valine, and alanine. The preferred electron-accepting amino acids are proline, glycine, and leucine that are reduced to 5-aminovalerate, acetate, and isocaproate, respectively. For many oxidative pathways, the reverse β-oxidation module with an *electron bifurcating butyryl-CoA dehydrogenase* as well as the Wood–Ljungdahl pathway (see Fig. 10.11) function as electron sinks for catabolic NADH oxidation. Similar as in *C. tetanomorphum*, 2-oxo acids are oxidized via *2-oxo acid:ferredoxin oxidoreductase* to acyl-CoA and Fd^{2-}; however, *C. difficile* contains also a *pyruvate:formate lyase*. In addition to Stickland fermentation of amino acids,

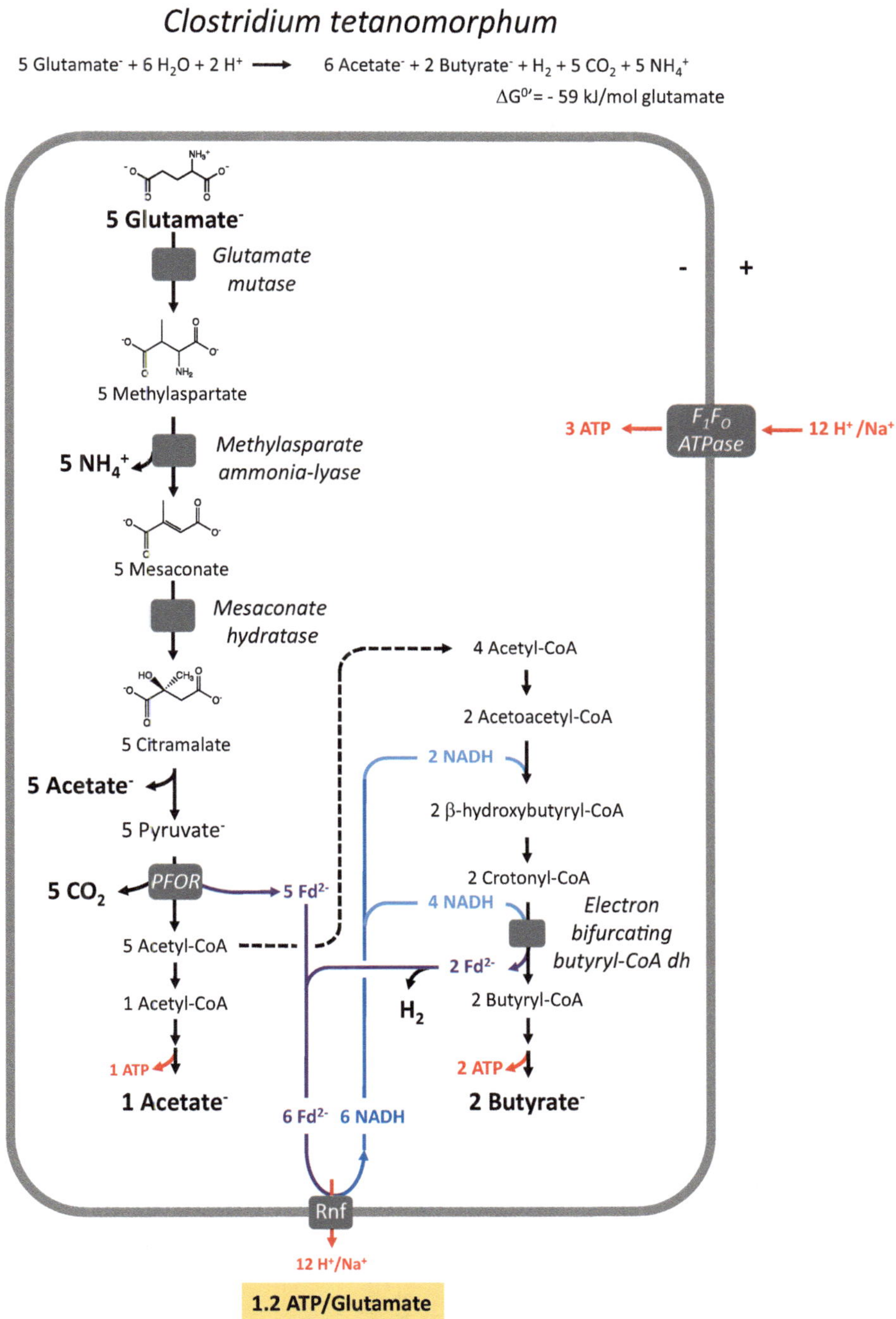

Fig. 9.23 Glutamate fermentation via Methylaspartate pathway in *Clostridium tetanomorphum*. Glutamate is first converted via non-redox reactions to pyruvate, which is oxidized by *pyruvate:ferredoxin oxidoreductase* to acetyl-CoA. Acetyl-CoA is the switch between the oxidative and reductive pathway as in many other fermentations. Involvement of *electron bifurcating butyryl-CoA dehydrogenase* in the reductive β-oxidation module is similar to clostridial carbohydrate fermentations. Also here, the reductive pathway drives, as electron sink, an energy-conserving electron flux through Rnf. ΔG calculated for 1 mM concentration of glutamate, acetate, butyrate, and NH_4^+, and pCO_2 = 0.1 bar and pH_2 = 10 Pa. Note the large decrease in ΔG, primarily driven by an entropic effect at low concentrations. Steps abstracting water are not shown. Abbreviation: *PFOR pyruvate:ferredoxin oxidoreductase*. See text for details

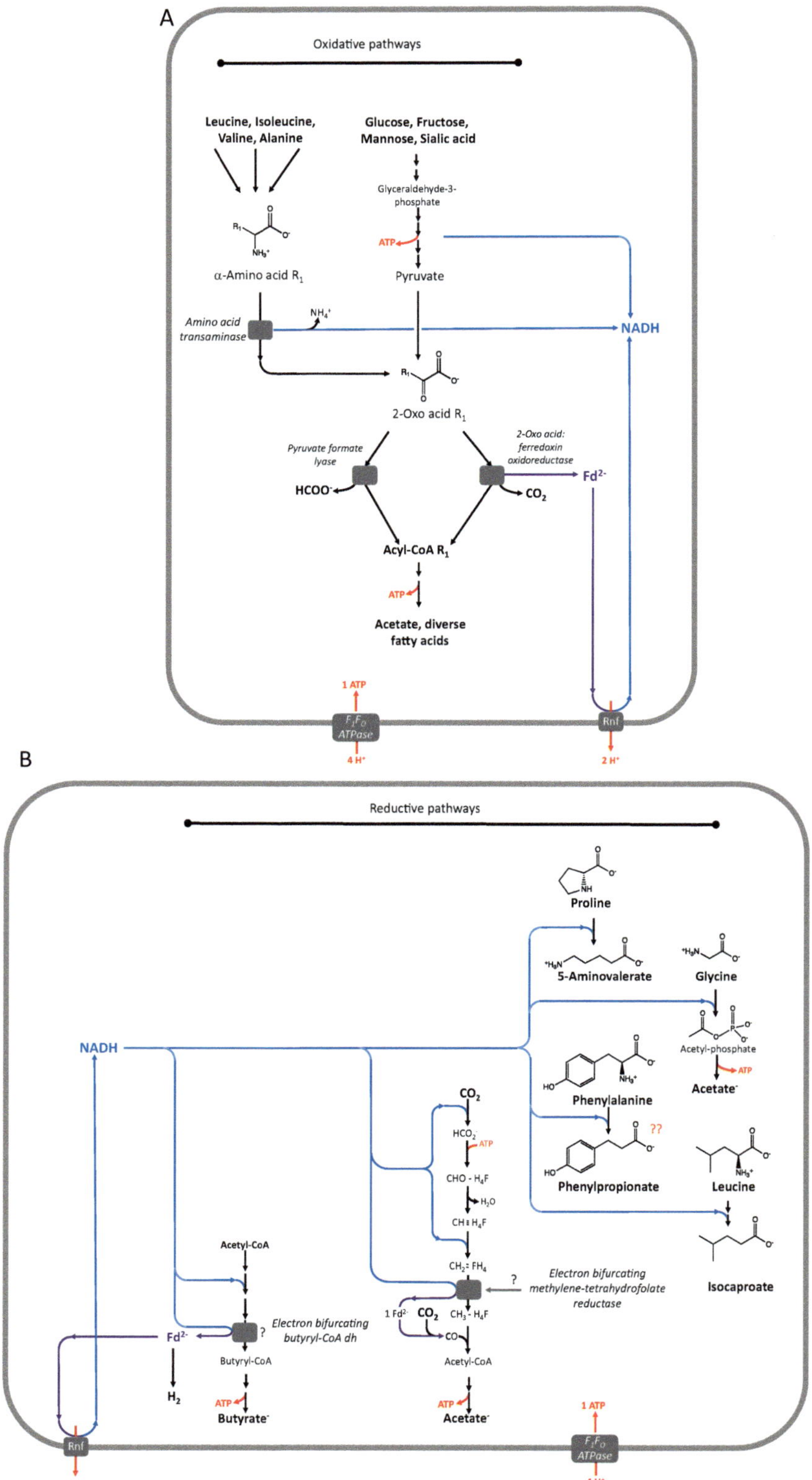

Fig. 9.24 Overview of fermentative metabolism of diverse amino acids and carbohydrates in *Clostridium difficile*. (**a**) Summary of postulated oxidative pathways; (**b**) summary of postulated reductive pathways. For the purpose of this illustration, NADH couples oxidative and reductive pathways. Electrons, substrate, and products are not stoichiometrically balanced to emphasize the overall fermentation capacity and the mosaic of metabolic

C. difficile ferments the carbohydrates glucose, fructose, and mannose via the Emden–Meyerhof pathway and the GAP → Pyruvate module similarly as in *C. pasteurianum*.

The primary elements for energy conservation in *C. difficile* consist of SLP in the oxidative GAP → pyruvate module when using carbohydrates. *2-Oxo-acid:ferredoxin oxidoreductase,* as well as *electron bifurcating butyryl-CoA dehydrogenase* provide Fd^{2-} for Ion Translocation Coupled Phosphorylation by Rnf. ATP formation also occurs by *glycine reductase.*

Many Clostridia fermenting carbohydrates, amino acids, or nucleic acids can reduce 2 CO_2 as electron acceptor in the reductive pathway to 1 acetate$^-$ via the Wood–Ljungdahl pathway. These acetogenic Clostridia, such as *C. autoethanogenum*, are discussed in Chap. 10.3.

9.6 Metabolism of Syntrophs Involved in Interspecies H_2 or Formate Transfer

"Microbial syntrophy" is a concept that describes chemical interactions between microbes with ecological and evolutionary significance. Syntrophic associations are metabolite-specific, mutually beneficial, and augmented interactions; in other words, microbial syntrophy is a substrate-and reaction-specific lifestyle. Syntrophic associations are ubiquitous in anaerobic microbiomes and have been shaped by co-evolution of both partner microbes. Syntrophic interactions between fermenting and methanogenic microbes via interspecies H_2 or formate transfer is well known, and the cycling of H_2 or formate is pivotal for these anaerobic microbiomes to function. Some of these interactions are obligate for thermodynamic reasons. Only these are addressed here.

$$\text{Ethanol} + H_2O \rightarrow \text{Acetate}^- + H^+ + 2H_2(g) \qquad \Delta G^{0\prime} = +9.6\ \text{kJ/mol} \tag{9.1}$$

$$\text{Lactate}^- + H_2O \rightarrow \text{Acetate}^- + CO_2 + 2H_2(g) \qquad \Delta G^{0\prime} = -8.8\ \text{kJ/mol} \tag{9.2}$$

$$\text{Butyrate}^- + 2H_2O \rightarrow 2\text{Acetate}^- + H^+ + 2H_2(g) \qquad \Delta G^{0\prime} = +48\ \text{kJ/mol} \tag{9.3}$$

$$\text{Propionate}^- + 2H_2O \rightarrow \text{Acetate}^- + CO_2(g) + 3H_2\ (g) \qquad \Delta G^{0\prime} = +71\ \text{kJ/mol} \tag{9.4}$$

$$\text{Acetate}^- + H^+ + 2H_2O \rightarrow 2\ CO_2(g) + 4H_2(g) \qquad \Delta G^{0\prime} = +95\ \text{kJ/mol} \tag{9.5}$$

$$\text{Hexadecane}(C_{16}H_{34})(l) + 16H_2O \rightarrow 8\ \text{Acetate}^- + 8H^+ + 17\ H_2\ (g) \qquad \Delta G^{0\prime} = +470\ \text{kJ/mol} \tag{9.6}$$

$$\text{Methane}(g) + 2H_2O \rightarrow CO_2(g) + 4\ H_2(g) \qquad \Delta G^{0\prime} = +131\ \text{kJ/mol} \tag{9.7}$$

where (g) indicates that for $\Delta G^{0\prime}$ calculation the substrate or product was considered to be in the gaseous state and (l) when in the liquid state. When CO_2 rather than protons are used as electron acceptor for oxidation of the substrates, then $\Delta G^{0\prime}$ values of the fermentations 1–7 are more positive by 3.5 kJ per mol formate formed: $H_2(g) + CO_2(g) \rightarrow \text{formate}^- + H^+$; $\Delta G^{0\prime} = +3.5$ kJ/mol.

The fermentations discussed previously have in common that they are sufficiently exergonic under physiological standard conditions of solutes (1 M), gases (10^5 Pa), and pH = 7 to

Fig. 9.24 (continued) modules/reactions in this microbe. The core oxidative enzymes (**a**) are *2-oxo acid:ferredoxin oxidoreductases* including *pyruvate:ferredoxin oxidoreductase, amino acid deaminase/glutamate dehydrogenase,* and the GAP→pyruvate module. Core metabolic modules of the reductive pathways (**b**) are reverse β-oxidation, the reductive Wood–Ljungdahl pathway (see Chap. 10.3), *glycine reductase*, and *2-hydroxy acid dehydratases*. Core for energy conservation, next to the acylate↔acyl-CoA module, is postulated electron flux through energy-conserving Rnf, and Fd^{2-}-forming *bifurcating butyryl-CoA dehydrogenase* and *2-oxo acid:ferredoxin oxidoreductases*. See text for details

drive the phosphorylation of (at least 1) ADP. Fermentations discussed in the following section (reactions 9.1–7) are, in contrast, endergonic or only slightly exergonic (reaction 9.2) under physiological standard conditions, and exergonic only, when the concentrations of the products H_2 and/or formate are very low. Anaerobic microbes capable of such fermentations are referred to as syntrophs when, for thermodynamic reasons, they can grow using an organic substrate only in association with a H_2-or formate-consuming partner microbe, such as methanogens or sulfate reducers. Growth of syntrophic microbes is inherently slow (doubling times in the order of days and weeks) as predicted from the positive free energy changes under physiological standard conditions ($\Delta G^{0\prime}$) (see also Haldane relationship Sect. 5.4.2). The slowest step is probably interspecies H_2 or formate transfer via diffusion because of the thermodynamically-required low H_2 or formate concentration. In some cases, as in anaerobic oxidation of methane (AOM) with sulfate (reaction 9.8), a diffusional H_2 or formate transfer is probably too slow to sustain sufficient growth rates. Instead, electrons are transferred directly between these species via polyheme cytochromes that function as conducting wires.

Many syntrophic microbes are not obligate syntrophs because they also have alternative, non-syntrophic lifestyles when fermenting other organic substrates. For example, *Pelobacter acetylenicus* ferments ethanol to acetate and H_2 (fermentation 9.1) only in the presence of H_2-consuming methanogens or H_2-consuming sulfate reducers but ferments acetylene to ethanol and acetate ($\Delta G^{0\prime} = -129$ kJ/mol acetylene) in the absence of a syntrophic partner. Similarly, *Syntrophomonas wolfeii* can grow syntrophically using butyrate (fermentation 9.3) and non-syntrophically using crotonate (2 crotonate$^-$ + 2 $H_2O \rightarrow$ 1 butyrate$^-$ + 2 Acetate$^-$ + H^+; $\Delta G^{0\prime} = -102$ kJ/mol butyrate). Some syntrophs can also use electron acceptors other than H^+ or CO_2, such as sulfate or fumarate if present. Some lactate-fermenting syntrophs (fermentation 9.2) shift to sulfate reduction when sulfate becomes available (2 lactate$^-$ + SO_4^{2-} + 2 $H^+ \rightarrow$ 2 acetate$^-$ + 2 CO_2 + 2 H_2O + H_2S; $\Delta G^0 = -175$ kJ/mol SO_4^{2-}). Such metabolic flexibility enables these syntrophs to switch metabolism between electron donors and acceptors, as one or the other is depleted in an environment.

Syntrophs have in common that they oxidize end products of primary fermenters (alcohols, short-chain fatty acids), as well as long-chain alkanes, etc., and possibly methane to acetate or CO_2 and H_2 (fermentation 9.1–7) or formate. In syntrophic metabolism, they exclusively use reduction of protons or CO_2 to H_2 and formate, respectively, as electron sink in their reductive catabolism. Many syntrophs belong phylogenetically to the δ-proteobacteria, but also other Gram-negative and some Gram-positive bacteria have a syntrophic lifestyle. Anaerobic methane oxidation (AOM) to CO_2 and H_2 (fermentation 9.7) has been proposed to be catalyzed by thermophilic archaea belonging to the Euryarchaeota.

The free energy changes associated with syntrophic fermentations decrease considerably (i.e., the fermentation becomes more exergonic) with increasing temperature, mainly driven by an increase in entropy. This decrease is most pronounced in the case of fermentation 9.6, where one substrate ($C_{16}H_{34}$) is converted to twenty-five products (8 acetate and 17 H_2).

Fermentation of benzoate to 3 acetate, 1 CO_2, and 3 H_2 ($\Delta G^{0\prime} = +60$ kJ/mol) was first believed to be catalyzed by a syntroph, *Syntrophus aciditrophicus*, because this microbe was discovered in co-cultures with hydrogenotrophic methanogens. However, *S. acidotrophicus* can also grow by the exergonic fermentation of benzoate to 0.5 mol cyclohexanecarboxylate and 1.5 acetate in the absence of an H_2-consuming microbe. Therefore, *sensu stricto*, *S. acidotrophicus* is not an obligate syntroph just as *Ruminococcus albus*, for example, is not one either (see Fig. 9.14). In case of *R. albus*, a H_2-consuming microbe enhances n_{ATP} by shifting metabolism toward acetate-forming fermentation. This metabolic interaction, however, is not obligated for glucose fermentation by *R. albus*. As shown above, this rumen microbe ferments glucose in the presence of hydrogenotrophic methanogens to 2 acetate, 2 CO_2, and 4 H_2 and in their absence to 1 ethanol, 1 acetate, 2 CO_2, and 2 H_2 (Fig. 9.14).

In syntrophic microbes catalyzing fermentations 9.1–6, NAD^+ ($E^{0\prime} = -320$ mV), ferredoxin ($E^{0\prime} = -420$ mV), and/or menaquinone (MK) ($E^{0\prime} = -74$ mV) have been identified as electron carriers. In these microbes, intracellular NADH/NAD^+ ratio is probably near 1/5 ($E' \approx -300$ mV), the Fd^{2-}/Fd ratio near 100/1 ($E' \approx -480$ mV), and the MK_{red}/MK_{ox} ratio near 10/1 ($E' \approx -100$ mV). For syntrophic oxidative pathways to operate, electrons from these reduced electron carriers have to be transferred to protons or CO_2. E' of the $2H^+/H_2$ couple is −420 mV and that of the CO_2/formate couple is −430 mV. When, as in anaerobic environments, H_2 partial pressure is lower than 10 Pa, $E'(2H^+/H_2)$ is more positive than −294 mV (see Sect. 3.4); when, as in anaerobic environments, CO_2 partial pressure is near 10^4 Pa and formate concentration near 2.3 μM, $E'(CO_2/\text{formate}) \approx -294$ mV. Thus, H^+ and CO_2 can thermodynamically accept electrons directly from NADH (−300 mV); however, because of the very small difference in in situ redox potential between the NAD^+/NADH and $2H^+/H_2$ couple, fermentations operate near thermodynamic equilibrium and, thus, at very low rates. Proton and CO_2 reduction with NADH could also be coupled to the exergonic reduction of NAD^+ with Fd^{2-} via NADH-dependent *electron confurcating hydrogenase or confurcating formate dehydrogenases*, rendering H_2 and formate formation much more exergonic. However, there is no experimental evidence for the function of these enzymes in most of the syntrophs. In fact, in *Syntrophomonas wolfeii*, *hydABC* predicted to encode an *electron confurcating hydrogenase* was shown to encode a *non-confurcating hydrogenase* that catalyzes ferredoxin-independent reduction of protons with NADH to H_2. Because of the large redox potential difference, reduction of protons and CO_2 with reduced menaquinone ($E' \approx -100$ mV) is thermodynamically only possible when coupled to a strongly exergonic reaction. The coupling mechanism employed is reversed electron transport (see Sect. 6.5.2). Conversely, reduction of protons or CO_2 ($E' = -294$ mV) with reduced ferredoxin ($E' = -480$ mV) is highly exergonic.

Energy-conserving metabolism under syntrophic conditions (fermentations 9.1–7) requires that steady-state concentration of H_2 or formate is maintained at low level by H_2- or formate-consuming partner microbes. In anoxic sediments and in the rumen of cows, H_2 partial pressure lower than 10 Pa and formate concentration lower than 1 μM have been determined. In order to maintain these low steady-state concentrations, a flux J of H_2 or formate from the syntroph to the methanogens or sulfate reducers is required. According to Fick's Law, $J = -D\frac{dC}{dX}$, where dC is the solution concentration difference between source and sink, dX the distance between source and sink, and D the diffusion coefficient [cm^2/s] (see Table 2.1). Thus, the highest flux J_S, which is proportional to J_{ATP} and, thus, to the growth rate (see Sect. 5.3) occurs under conditions of a large dC, small dX, and high diffusion coefficients. dX is smallest when the cells are in direct contact forming aggregates as in biofilms.

The concentration difference dC is proportional to in situ concentration, which, in anoxic environments, is higher for formate than for H_2. When both are at the same redox potential (e.g., $= -294$ mV at $pH_2 = 10$ Pa (0.07 μM in solution), pH = 7, [formate] = 2.5 μM, and $pCO_2 = 10^4$ Pa), concentration of formate is 36 (= 2.3/0.07) times higher than the H_2 concentration at the same redox potential. The diffusion coefficient of H_2 ($D_{H2} = 5.1 \times 10^{-5}$ cm^2/s) in water at 25 °C is 3.4 times higher than that of formate ($D_{formate} = 1.5 \times 10^{-5}$ cm^2/s). Therefore, at the same redox potential and over the same distance, the effective flux J_S of electrons (as H_2 or formate) between syntrophic partners is 10.6 times (36/3.4) higher via formate than via H_2. In other words, an increase of the distance between the syntrophic partners by a factor of 10.6 can be compensated by microbes switching from interspecies H_2 to interspecies formate transfer to maintain the same electron transfer rate. This is probably why most planktonically growing syntrophs contain both *hydrogenases* and *formate dehydrogenases*.

For the following examples of individual syntrophic fermentations, four thermodynamic and structural features are important to consider:

(i) The free energy required to drive synthesis of ATP at intracellular mM concentrations of ADP and inorganic phosphate is near −50 +/− 5 kJ/mol (see Sect. 3.4). Under equilibrium conditions, the rate of ATP synthesis is zero. Therefore, a fermentation reaction must be exergonic by more than −50 +/− 5 kJ/mol to kinetically allow for the synthesis of 1 ATP during finite growth (see Sect. 5.3). How much more exergonic is unknown. Empirically, the more exergonic an overall catabolic reaction is, the closer a value of near −80 kJ/mol per mol ATP is approached. In the very slowly growing syntrophic microbes a value of −55 to −60 kJ/mol ATP most likely seems sufficient.

(ii) Membrane-associated *F_1F_o ATPase* can have varying numbers of subunits in the rotating C-ring, as low as 8 in the *ATPase* in mitochondria and as high as 15 in the respective enzyme in cyanobacteria (see Sect. 6.4.2). Rotation is driven via proton flux by the electrochemical proton potential; one vectorial proton per C subunit being required for one full rotation that is associated with the synthesis of 3 ATP. Thus in the case of mitochondria, 2.666 (8/3) vectorial protons are required for the synthesis of 1 ATP, whereas in cyanobacteria the number of vectorial protons required is 5 (15/3). The number of C-subunits in *ATPase* expressed under syntrophic conditions is unknown. It is most likely 15 or higher, in order to allow mechanistically ATP synthesis to occur at the low ΔG associated with the syntrophic fermentations. For the same reason, the proton motive force ($\Delta\mu H^+/F$) is expected to be lower in syntrophs (near −12 kJ/mol H^+) than in non-syntrophs (near 20 kJ/mol H^+) (see Sect. 6.4.2).

(iii) Exergonic ubiquinone reduction with NADH in mitochondria is associated with electrogenic translocation of 4 protons. It is, therefore, assumed that reduction of protons or CO_2 with reduced menaquinone via ***reverse electron transport*** (see Sect. 6.5.2) is "pushed" by translocation of four protons. This is highlighted here because in the literature 2 vectorial protons are assumed generally to be sufficient.

(iv) In anaerobic environments, steady-state H_2 concentration depends on the H_2-threshold concentration of the microbe catalyzing H_2 oxidation (see Sect. 3.5, Fig. 3.9). When protons are the only available electron acceptor and only acetogenic bacteria are present, steady-state pH_2 is close to and below 10^3 Pa. In the presence of most methanogens, it is close to and below 10 Pa, and close to 1 Pa in the presence of *Methanospirillum hungatei* that has a specially low H_2-threshold. When sulfate is available, pH_2 can approach 0.1 Pa in the presence of most H_2-consuming sulfate reducers. When CO_2 is the electron acceptor, the redox potential of the CO_2/formate couple is also dependent on the CO_2 concentration, which in most anaerobic habitats is estimated to be in the order of 10^4 Pa (10% in the gas phase). CO_2 concentration is relatively high because CO_2 is the end product of anaerobic mineralization. In all thermodynamic calculations that follow, a $pCO_2 = 10^4$ Pa was used.

9.6.1 Syntrophic Oxidation of Ethanol to Acetate and H_2

The discovery of obligate syntrophy was made when investigating metabolism of *Methanobacillus omelianskii* that grows on 2 ethanol and CO_2, forming methane. Subsequent studies revealed that *M. omelianskii* was a mixed culture of two microbes, the S-organism, and *Methanobacterium bryantii*. The S-organism grew on ethanol, forming 1 acetate and 2 H_2 per ethanol (fermentation 9.1) and *M. bryantii* using 4 H_2 and CO_2 forming methane. Whereas the S-organism was dependent on the methanogen for growth in ethanol, the methanogen grew well on H_2 and CO_2 in the absence of the S-organism. Thermodynamic calculations revealed that ethanol conversion to acetate and 2 H_2 is only exergonic when acetate and H_2 concentrations are kept low by the methanogenic partner microbe. Notably, the exceptional *Methanogenium organophilum* and *Methanofolis ethanolicus* convert ethanol to acetate plus methane non-syntrophically in a single methanogenic microorganism.

The S-organism was lost, but some properties have been reported such as the activity in cell extracts of *acetaldehyde:ferredoxin oxidoreductase, hydrogenase, NAD⁺-specific alcohol dehydrogenase, NAD⁺-specific formate dehydrogenase, phosphotransacetylase* and *acetate kinase*. These enzyme activities are also present in *Pelobacter acetylenicus* and *P. carbinolicus* that can also grow using ethanol, forming acetate and H_2 when in co-culture with a methanogen or a sulfate reducer. The genomes of the two *Pelobacter* species contain genes encoding the enzymes mentioned above. Additionally, genes for a membrane-associated, energy-conserving Rnf complex were present (Fig. 9.25).

The free energy change associated with ethanol fermentation to acetate and 2 H_2 strongly depends on the pH_2 being adjusted by the syntrophic partner microbe. In the presence of a methanogen, at $pH_2 \approx 10$ Pa and ≈ 1 mM concentrations of ethanol and acetate, $\Delta G'$ is near −30 kJ/mol, which is sufficient to drive synthesis of 0.4 ATP (Fig. 9.25a). If the partner microbe is a sulfate reducer and pH_2 is ≈ 0.1 Pa, $\Delta G'$ is near −60 kJ/mol, which is sufficient to drive the synthesis of 1 ATP (Fig. 9.25b). Interestingly, the two *Pelobacter* species discussed above can regulate their metabolism such that the thermodynamic efficiency of ATP synthesis remains nearly constant. The metabolic changes required to increase n_{ATP} from 0.4 at $pH_2 = 10$ Pa to near 1 at $pH_2 = 0.1$ Pa are shown in Fig. 9.25, respectively.

9.6.2 Syntrophic Oxidation of Lactate to Acetate, CO_2, and H_2

Desulfovibrio vulgaris ferments lactate to acetate, CO_2, and 2 H_2 (fermentation 9.2) in the absence of sulfate and when in syntrophic association with H_2-consuming methanogens. The cells harbor a membrane-associated *lactate dehydrogenase*, cytoplasmic *pyruvate:ferredoxin oxidoreductase*, periplasmic *cytochome c3-dependent hydrogenase*, membrane-associated *energy converting hydrogenase* (Ech), a cytoplasmic *hydrogenase, phosphotransacetylase,* and *acetate kinase*, as well as menaquinone and ferredoxin. In cell suspensions, H_2 formation from lactate rather than from pyruvate is inhibited by protonophores.

The free energy change associated with lactate fermentation to acetate, CO_2, and 2 H_2 is also strongly dependent on the pH_2 adjusted by the syntrophic partner microbe. In case of a hydrogenotrophic acetogen, steady pH_2 is in the range of below 10^3 Pa. At such pH_2, a $pCO_2 = 10^4$ Pa, and [lactate] ≈ 1 mM and [acetate] ≈ 1 mM, $\Delta G'$ is near −25 kJ/mol and ATP gain is near 0.2 ATP/mol lactate. In case of a hydrogenophilic methanogen as syntrophic partner, steady-state H_2 concentration is near 10 Pa. At this pH_2, a $pCO_2 \approx 10^4$ Pa, [lactate] ≈ 1 mM and [acetate] ≈ 1 mM, $\Delta G'$ is near - 60 kJ/mol, and ATP gain is 0.6 ATP/lactate. As in the above-described ethanol-fermenting syntroph, the lactate-fermenting syntroph can regulate its metabolism such that the thermodynamic efficiency of ATP synthesis remains nearly constant. The metabolic shift required to increase ATP gain from 0.2 at $pH_2 = 10^3$ Pa to 0.6 at $pH_2 = 10$ Pa is shown in Fig. 9.26, respectively.

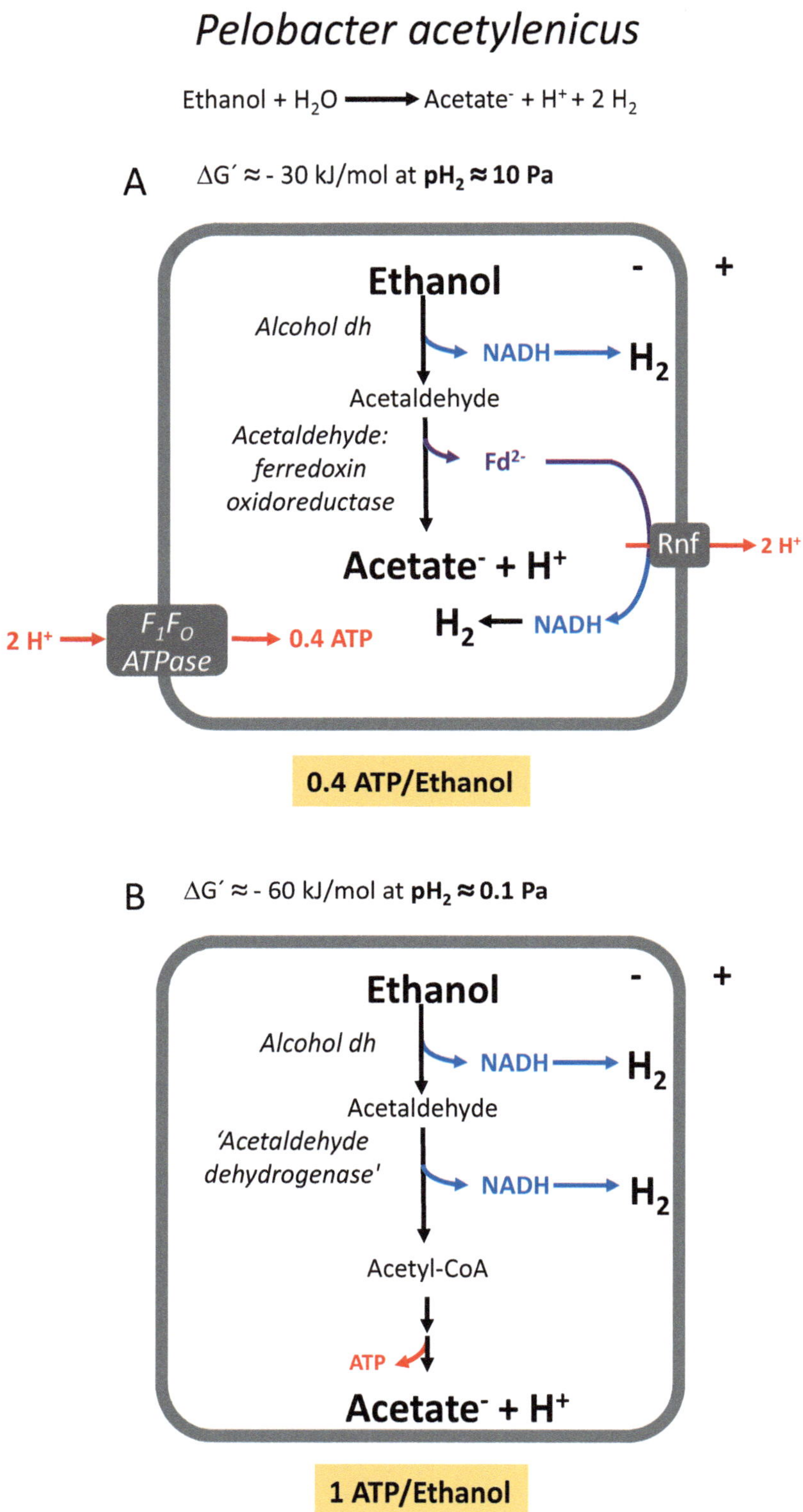

Fig. 9.25 Proposed ethanol fermentation to acetate and 2 H_2 by *Pelobacter acetylenicus*. (**a**) Fermentation in the presence of a H_2-consuming methanogens, keeping pH_2 near 10 Pa; (**b**) Fermentation in the presence of a H_2-consuming sulfate reducer, keeping pH_2 near 0.1 Pa. See text for details

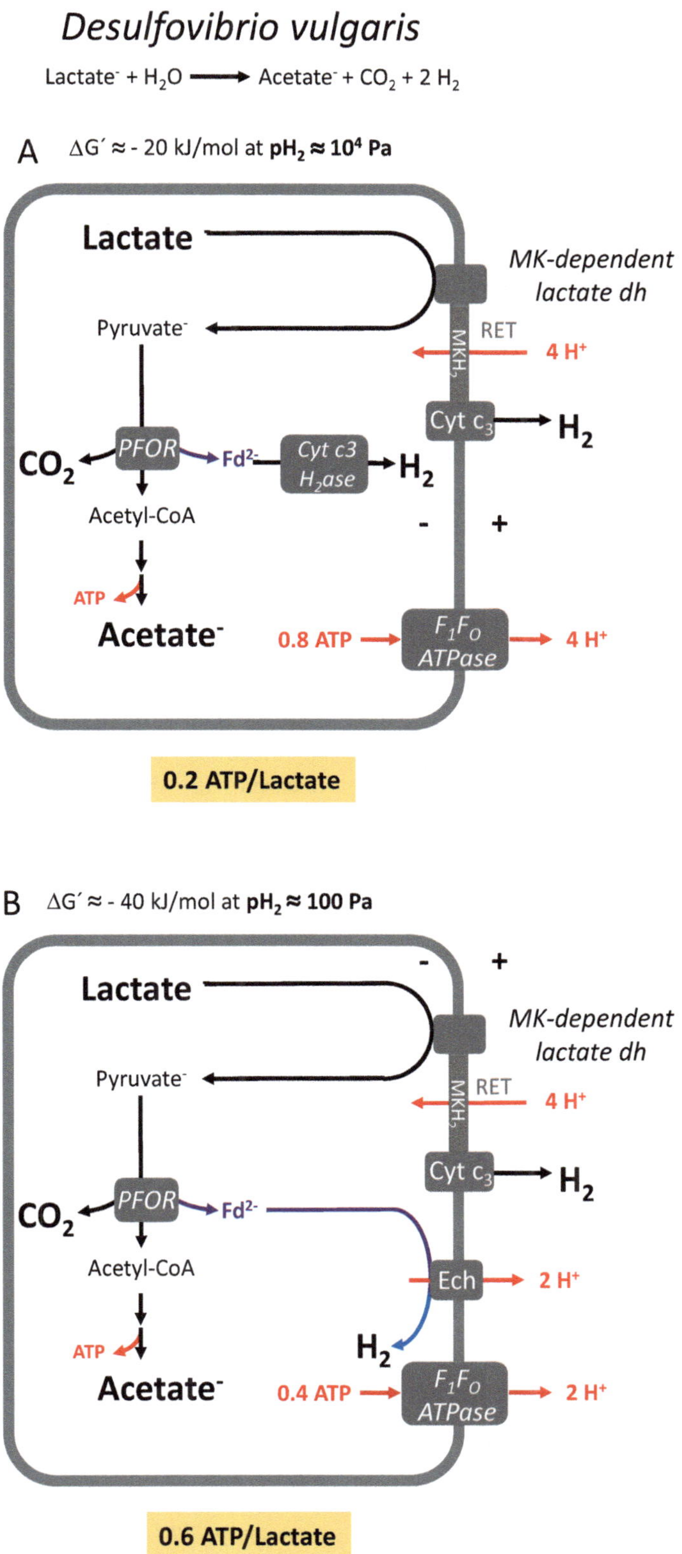

Fig. 9.26 Proposed lactate fermentation to acetate, CO_2, and 2 H_2 by a *Desulfovibrio vulgaris* strain. (**a**) Fermentation in the presence of a H_2-consuming acetogen keeping pH_2 near 10^4 Pa; (**b**) Fermentation in the presence

9.6.3 Syntrophic Oxidation of Butyrate to Acetate and H_2

Syntrophomonas wolfei is the best-studied microbe of the many syntrophs that mediate fermentation of butyrate to 2 acetate and 2 H_2 (fermentation 9.3) in the presence of a hydrogenotrophic/formate-utilizing methanogen. This γ-proteobacterium also ferments hexanoate but not longer-chain fatty acids. *S. wolfei* contains all enzymes required for oxidation via the β-oxidation module, menaquinone, and cytochromes *b*, as well as electron-transferring flavoproteins, EtfAB, but no Rnf complex. It contains a cytoplasmic *NAD⁺-specific hydrogenase* HydABC and a *hydrogenase* oriented toward the periplasm (Fig. 9.27). Cytoplasmic *NAD⁺-specific formate dehydrogenase* and periplasm-oriented *formate dehydrogenase* are also present.

Syntrophic oxidation of butyryl-CoA to crotonyl-CoA ($E^{0\prime} = -10$ mV) is of special interest. In butyric acid-forming microbes, crotonyl-CoA reduction with 2 NADH to butyryl-CoA and Fd^{2-} is a highly exergonic electron bifurcating reaction that, for thermodynamic reasons, is essentially irreversible. In *S. wolfeii*, oxidation of butyryl-CoA to crotonyl-CoA, therefore, most likely proceeds as in aerobic fatty acid oxidation via electron transferring flavoproteins EtfAB as electron acceptor that transfers 2 electrons to menaquinone (MK) in a membrane-associated, enzyme-catalyzed reaction. Electrons of reduced menaquinone are thermodynamically activated by reversed electron transport (see Sect. 6.5.2), consuming the energy of 4 electrogenic protons. These activated electrons have a redox potential that is sufficiently negative to reduce two protons to H_2 via the *hydrogenase* oriented toward the cytoplasm (Fig. 9.27).

In the proposed fermentative catabolism (Fig. 9.27), 1 ATP is generated from acetyl-CoA and 0.8 ATP is consumed via *F_1F_OATPase* to generate 4 electrogenic protons for reversed electron transport. Thus, net only 0.2 ATP is formed per butyrate. When *S. wolfei* grows in syntrophic association with methanogens that maintain a steady-state pH_2 near 10 Pa, the free energy change of the fermentation is close to −15 kJ/mol butyrate, assuming that butyrate- and acetate-concentrations are ≈ 1 mM. Therefore, in *S. wolfei* growing syntrophically using butyrate, about -60 kJ may be sufficient to drive the synthesis of 1 mol ATP.

9.6.4 Syntrophic Oxidation of Propionate to Acetate, CO_2, and H_2

Two types of syntrophic bacteria oxidize propionate in association with methanogens. The first type utilizes the Methylmalonyl-CoA module (see Sects. 8.3.2 and 9.4.1) in which three pairs of electrons from propionate oxidation are released in the form of hydrogen (H_2) (Fig. 9.28) and/or formate. Microbes of this metabolic type include *Syntrophobacter* spp. and *Pelotomaculum* spp. The second type oxidizes propionate via a dismutating pathway, in which two propionate molecules are first joined to form a C_6 compound, which is then dismutated to acetate and butyrate before the latter is being oxidized via β-oxidation. The main microbe of this type includes *Smithella* spp. Only the Methylmalonyl-CoA pathway is discussed here that has recently been studied extensively in *Pelotomaculum thermopropionicum* growing in syntrophic association with *Methanocella conradii*.

Key to the understanding of the energy metabolism of *P. thermopropionicum* is the presence of membrane-associated *succinate dehydrogenase* in complex with *[NiFe]-hydrogenase* and the

Fig. 9.26 (continued) of a H_2- consuming methanogen keeping pH_2 near 100 Pa. *Cyt c3 H_2ase* cytochrome c3-dependent periplasmic hydrogenase, *MK* menaquinone, *RET* reverse electron transfer. See text for details

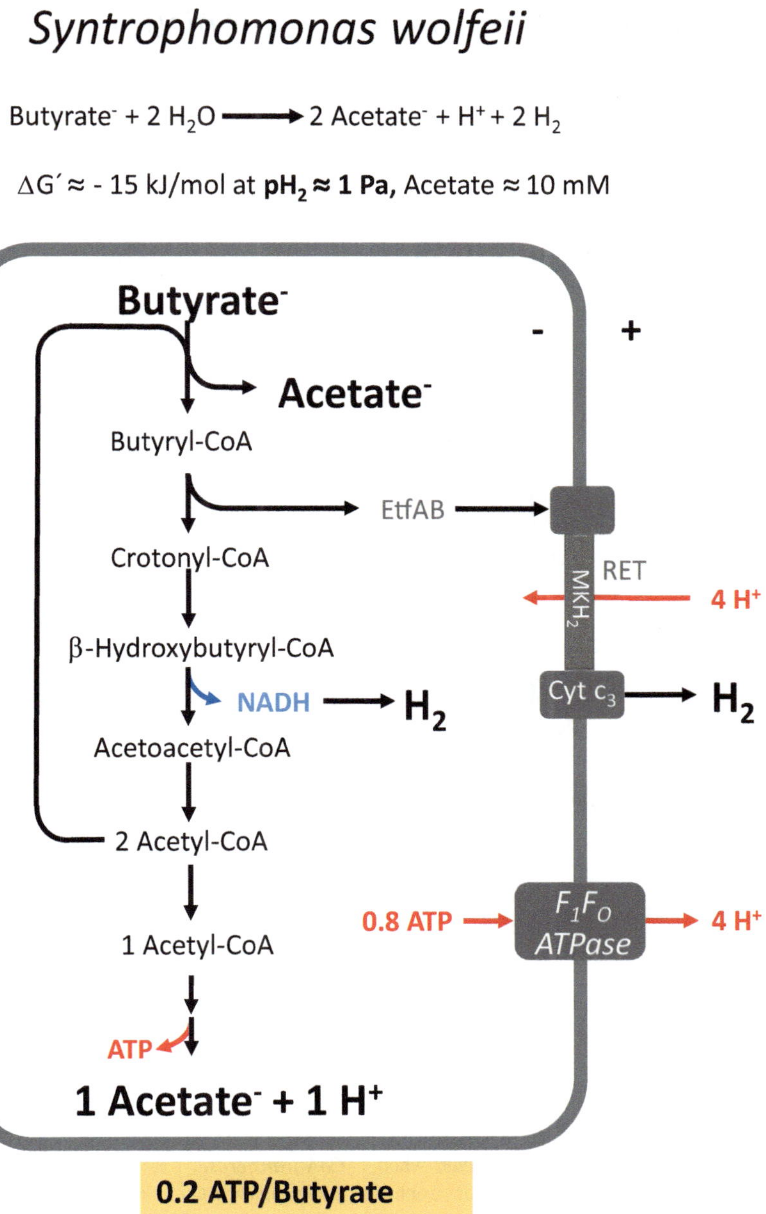

Fig. 9.27 Proposed butyrate oxidation to 2 acetate, 2 H^+, and 2 H_2 by *Syntrophomonas wolfeii* when growing in the presence of H_2-consuming *Methanospirillum hungatei*. Steps abstracting water are not shown. Abbreviations: *Etf* electron transfer flavoproteins, *MK* menaquinone, *RET* reverse electron transport. For details, see the text

presence of cytoplasmic *NADH-* and *reduced ferredoxin dependent confurcating [FeFe]-hydrogenase*. The metabolic scheme shown in Fig. 9.28 predicts that per mol propionate 0.2 mol ATP is synthesized, for which about −15 k is available under conditions of syntrophic growth with *M. conradii*. Thus, also in this syntroph, only about −60 kJ is most likely required to drive synthesis of 1 ATP.

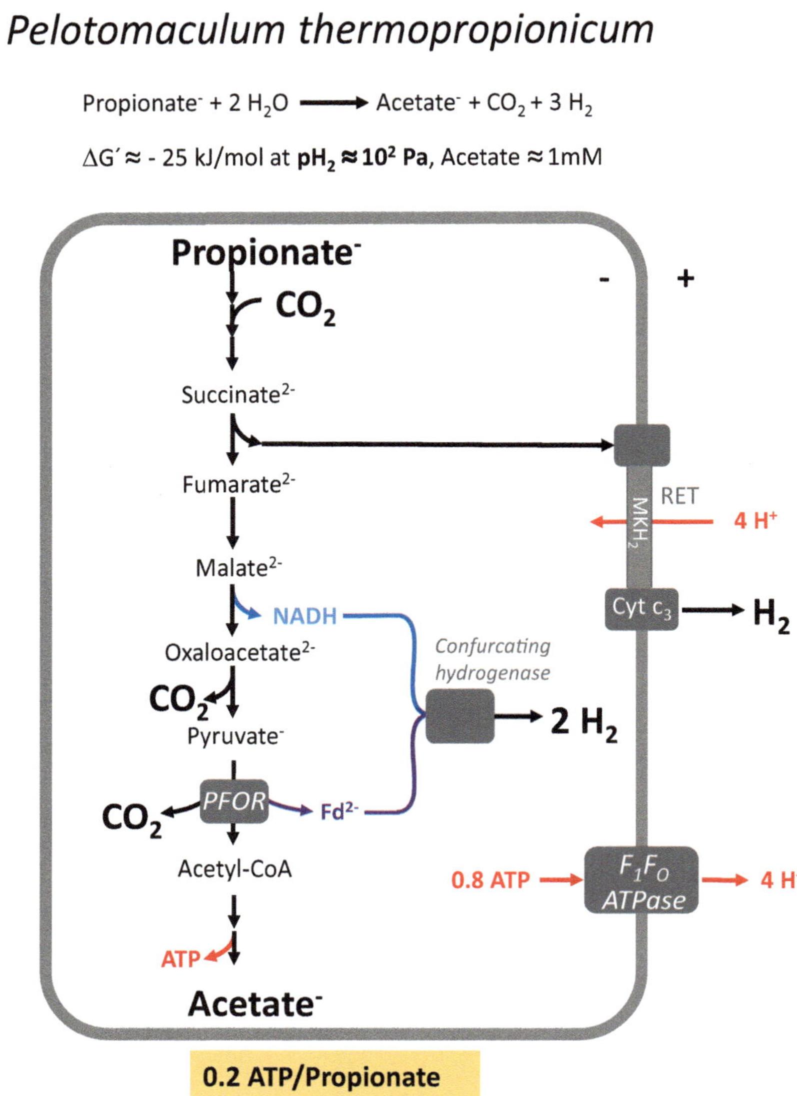

Fig. 9.28 Proposed propionate oxidation to acetate, CO_2, and 3 H_2 by *Pelotomaculum thermopropionicum* when growing in the presence of hydrogenotrophic *Methanocella conradii.* The five enzymes of the Methylmalonyl-CoA module involved in propionate conversion to succinate are: *succinyl-CoA:propionate CoA-transferase*, *propionyl-CoA carboxylase*, *methylmalonyl-CoA mutase,* and *methylmalonyl-CoA racemase*. Steps abstracting water are not shown. Abbreviation: *RET* reverse electron transport. See text for details

9.6.5 Syntrophic Acetate Oxidation to H_2 and CO_2

Relatively little is known about the energy metabolism of microbes catalyzing syntrophic oxidation of acetate to 2 CO_2 and 4 H_2 (fermentation 9.5) when growing in the presence of H_2-consuming methanogens. Acetate oxidation to CO_2 and H_2 is rarely observed in mesophilic methanogenic environments, where acetate is fermented to methane and CO_2 primarily by acetoclastic methanogens. In thermophilic methanogenic environments, both fermentations operate in parallel, and in hyperthermophilic environments, syntrophic acetate oxidation to CO_2 and H_2 predominates. A thermodynamic reason for that is that the free energy change associated with this fermentation at 25 °C is only about −15 kJ/mol, when [acetate] ≈ 1 mM, pH $_2$ near 1 Pa, and pCO_2 near 10^{-4} Pa. At increasing temperature, however, the free energy change is considerably more negative because of an increase in entropy: 6 mol products are formed (2 mol CO_2 and 4 mol H_2) from one mol substrate (acetate). It is estimated that $\Delta G'$ is more negative by −5 kJ/mol when the temperature increases by 10 °C. In addition, the free energy change of syntrophic acetate fermentation is more exergonic by −5.7 kJ/mol when the proton concentration increases by a factor of 10.

Only a few isolated microbes have been described so far to syntrophically ferment acetate: mesophilic *Clostridium ultunense*, *Syntrophaceticus schinkii*, Candidatus Syntrophonatronum acetioxidans, thermotolerant *Tepidanaerobacter acetatoxydans*, and thermophilic *Pseudothermotoga lettingae* and *Thermacetogenium phaeum*. All these species can also grow non-syntrophically using organic compounds producing acetate from 2 CO_2. *T. phaeum* can even grow on CO_2 and H_2 forming acetate.

As far as known, these syntrophic bacteria oxidize acetate to CO_2 and H_2 by a reversed Wood– Ljungdahl pathway (see Chap. 10.3). With respect to energy conservation, however, there are differences, which are not completely understood yet. The metabolic scheme proposed in Fig. 9.29 is taking into account the genome and biochemical data of various acetogens including that of *T. phaeum*.

9.6.6 Syntrophic Oxidation of Hexadecane to Acetate and H_2

Saturated long-chain hydrocarbons such as hexadecane ($C_{16}H_{14}$) are fermented to acetate and H_2 (fermentation 9.6) by anaerobic microbes when growing in syntrophic association with H_2-consuming methanogens. The metabolic pathway of hydrocarbon oxidation is a variation of the *benzyl/alkylsuccinate synthase* pathway of anaerobic alkane and aromatic hydrocarbon degradation in sulfate-reducing or denitrifying microbes (see Sect. 8.3.4). The first reaction is the abstraction of a terminal H radical from the alkane via a glycyl radical enzyme generating a terminal alkane radical that reacts with fumarate to an 2-alkylsuccinate radical. The functionalized alkane is then subjected to β-oxidations involving the steps outlined for syntrophic butyrate oxidation in Fig. 9.27. Fumarate is re-generated by funneling acetyl CoA from β-oxidation into reactions of the TCA cycle.

9.6.7 Syntrophic Methane Oxidation to CO_2 and H_2

In early 2000, data was provided that anaerobic oxidation of methane (AOM) with sulfate as electron acceptor in marine sediments is catalyzed by a syntrophic association of archaea (ANME) (related to methanogenic archaea) and sulfate-reducing bacteria. Early on it was speculated that the archaea oxidized methane to CO_2 and H_2, where the latter is consumed by sulfate reducers. This hypothesis was based on the finding that in some microscopic images, ANME microbes and sulfate reducers were spatially separated. Thermodynamic calculations showed that at very low H_2 concentrations of <0.1 Pa, interspecies H_2 transfer is feasible (Fig. 9.30). Inconsistent with this hypothesis was, however, that in metagenomes of the mesophilic archaea involved in AOM, no genes for *hydrogenases* were found. Also, when H_2 concentration in the sediments was artificially increased, no inhibition of AOM was observed. However, genes for *hydrogenases* were found in metagenomes of thermophilic methanotrophic archaea. Furthermore, the first thermophilic H_2-consuming sulfate reducers were isolated from a thermophilic AOM enrichment culture. Notably also in methane oxidation to CO_2 and 4 H_2, the free energy change becomes more negative at elevated temperatures by approximately −5 kJ/mol per 10 °C (5 products, CO_2 and 4 H_2, are formed from one methane). However, the measured rates of H_2 formation were inconsistent

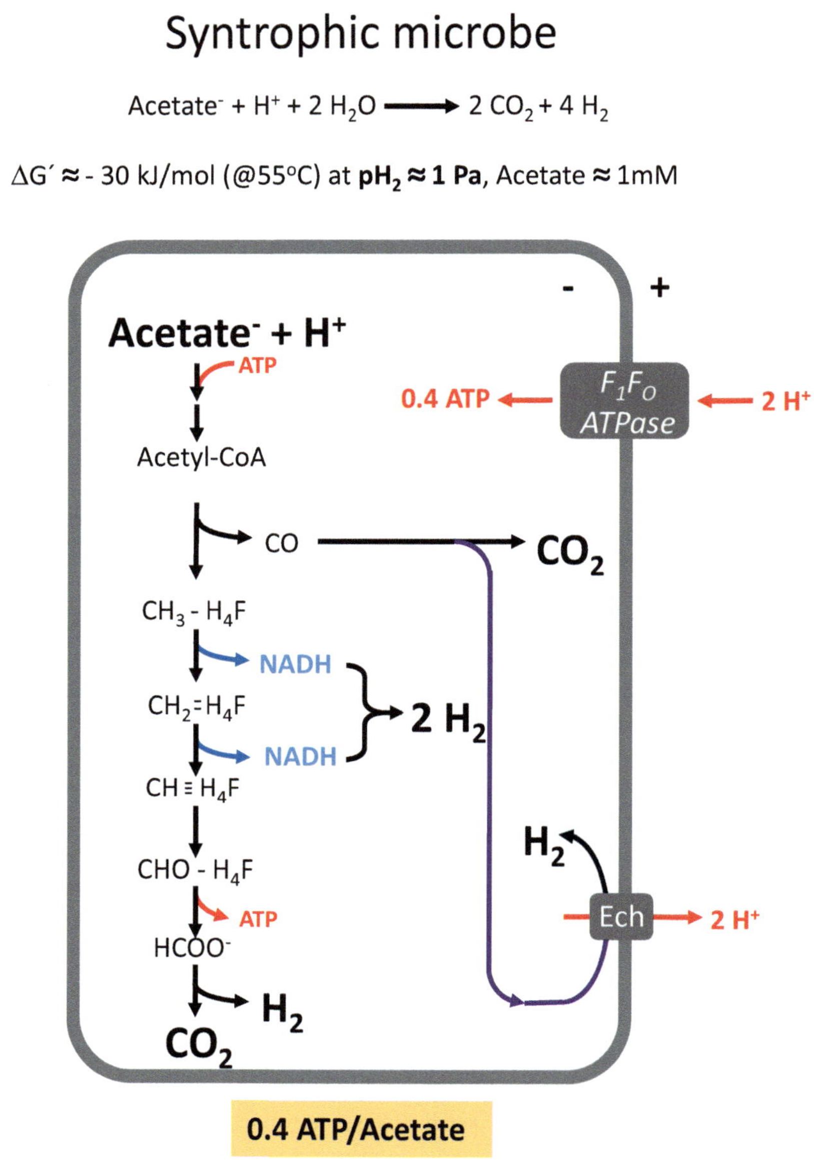

Fig. 9.29 Proposed acetate oxidation to 2 CO_2 and 4 H_2 by syntrophic bacteria growing in the presence of H_2-consuming methanogens. In syntrophic bacteria, F_1F_o *ATPase* is assumed to have a H^+/ATP stoichiometry of 5. Steps abstracting water are not shown. Abbreviation: *H_4F* tetrahydrofolate. See text for details

with the rates of sulfate reduction with methane. The question of thermophilic AOM, therefore, is still unanswered. The general opinion today is that in most biotopes, in which methane is oxidized with sulfate, AOM involves a direct interspecies electron transfer (Fig. 9.31).

As far as is known, ANME microbes involved in anaerobic methane oxidation with sulfate oxidize methane to CO_2 in reverse of the pathway used by methanogens to reduce CO_2 to methane (see Chap. 10.4). The rates of anaerobic methane oxidation are very low with doubling times in the order of weeks to months. With respect to energy conservation, there must, however, be differences, which are not yet completely understood. In Fig. 9.31 a metabolic scheme is

$$CH_4 + 2\,H_2O \longrightarrow CO_2 + 4\,H_2$$

$$\Delta G \approx -20\ \text{kJ/mol}, @\ 55°C, \mathbf{pH_2 < 0.1\ Pa}$$

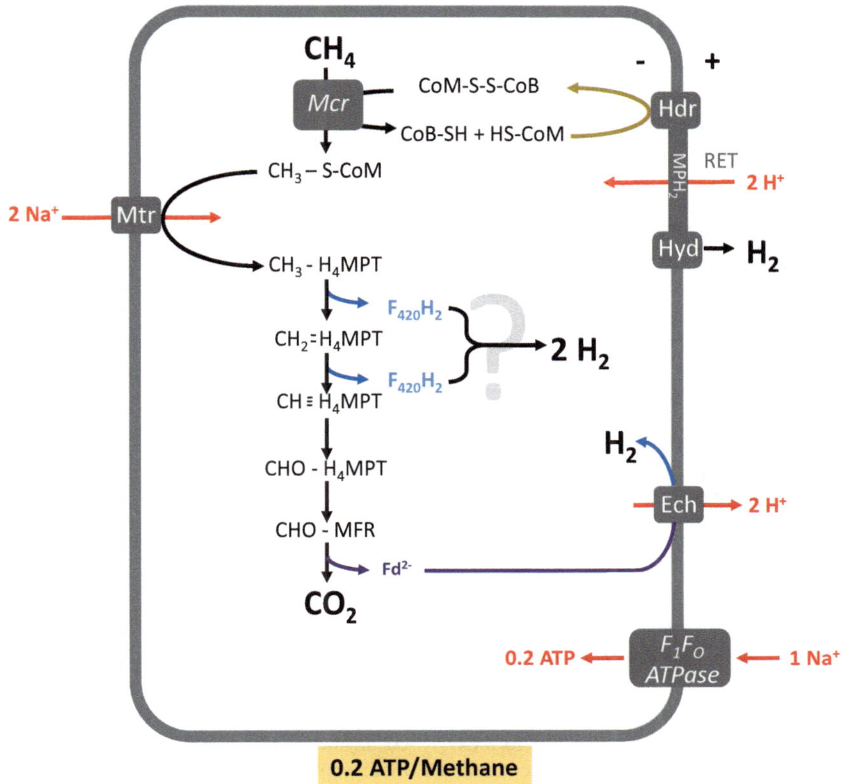

Fig. 9.30 Proposed metabolism of the still hypothetical methane oxidation to 2 CO_2 and 4 H_2. Energy-conserving steps are unclear and missing. Steps abstracting water are not shown. Abbreviations: *H_4MPT* tetrahydromethanopterin, *Fd* ferredoxin, *F_{420}* coenzyme F_{420}, *HS-CoM* coenzyme M, *HS-CoB* coenzyme B, *MP* methanophenazine, *Hdr* cytochrome *b*- and membrane-associated *heterodisulfide reductase*, *Ech* energy-converting *[NiFe]-hydrogenase*, *RET* reversed electron transport, *Hyd* cytochrome *b* and membrane-associated *[NiFe]-hydrogenase*, *Mtr* membrane-associated energy-conserving *methyltransferase*, *Mcr methyl-coenzyme M reductase*. See text for details

proposed taking into account genomic and biochemical data of various ANME microorganisms and of the cytochrome-containing order *Methanosarcinales*. The hypothetical scheme shows how methane oxidation to CO_2 and 4 H_2 could proceed.

9.7 Metabolism of Syntrophs Involved in Direct Interspecies Electron Transfer

As indicated above, AOM with sulfate (fermentation 9.8) involves a methanotrophic archaeon and sulfate-reducing bacteria.

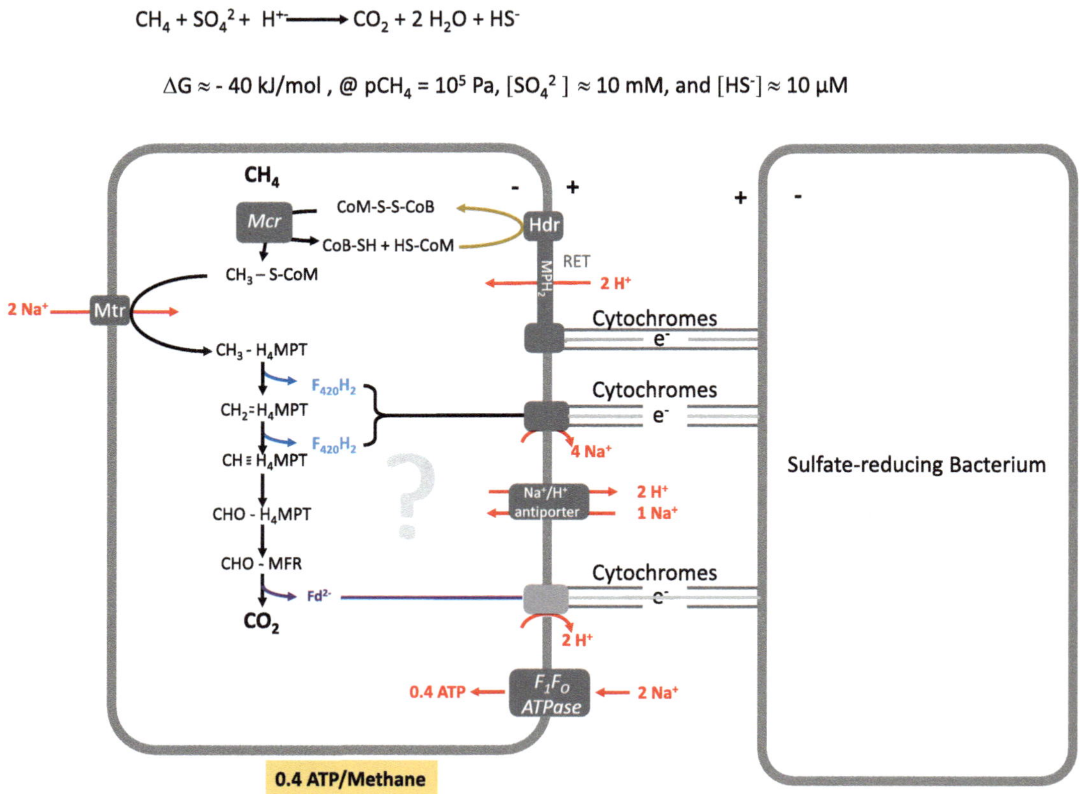

Fig. 9.31 Proposed metabolic scheme of anaerobic methane oxidation with sulfate as electron acceptors to 2 CO_2 and H_2S by a two-member consortium. Consistent with a HS^- concentration in marine sediments of ≈ 10 µM or less, the free energy change of methane oxidation with sulfate is ≈ −40 kJ/mol or less. H_2S is continuously removed by diffusion into oxic zones or precipitation as metal sulfides. Abbreviations: *H_4MPT* tetrahydromethanopterin, *Fd* ferredoxin, *F_{420}* coenzyme F_{420}, *HS-CoM* coenzyme M, *HS-CoB* coenzyme B, *MP* methanophenazine, *Hdr* cytochrome *b*, and membrane-associated *heterodisulfide reductase*, *RET* reversed electron transport, *Mtr* membrane-associated energy-conserving *methyltransferase*, *Mcr* methyl-coenzyme M reductase. For comparison, see Fig. 9.30, and text for details

$$CH_4(g) + SO_4^{2-} + H^+ \rightarrow CO_2(g) + 2H_2O + HS^- \quad \Delta G^{0\prime} = -21\,\text{kJ/mol} \tag{9.8}$$

If, as initially assumed, electron transfer between the two species is via H_2, this metabolism would need to occur at $pH_2 < 0.1$ Pa ≙ <0.7 nM in order for anaerobic methane oxidation to CO_2 and 4 H_2 to be sufficiently exergonic to support ATP synthesis (Fig. 9.30). At nM concentrations, the rate of diffusional electron transport via H_2 is probably too low to sustain growth of the methanotrophic archaeon and the sulfate-reducing bacterium (see above). One microbial strategy to compensate for the requirement of a low H_2 concentration is to reduce the distance between the two syntrophically-interacting microbes and to switch from H_2 to formate for interspecies electron transfer (see above). If this is kinetically insufficient, an electron acceptor with a redox potential more positive than that of the $2H^+/H_2$ couple ($pH_2 = 0.1$ Pa; $E' \approx -230$ mV) or of the CO_2/formate couple ($pCO_2 \approx 10^4$ Pa; [formate] = 25 nM; $E' = -230$ mV) needs to be employed. In the case of the methanotrophic archaea growing in association with sulfate-reducing bacteria, the electron acceptor appears to be extracellular poly-heme cytochromes that forms nano-wires and mediates transfer of these

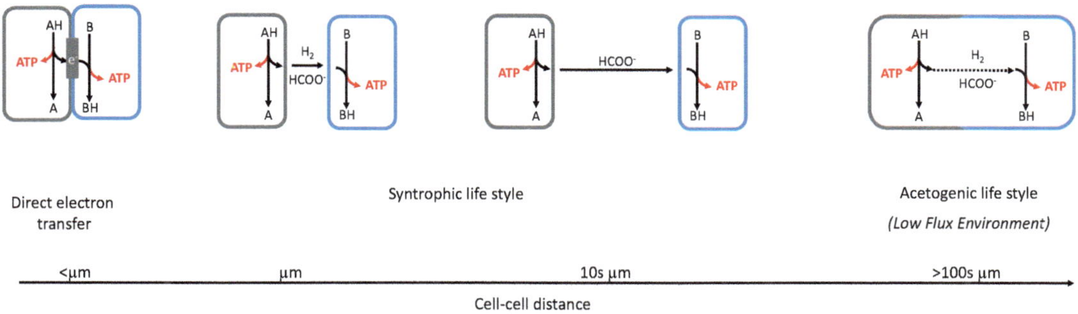

Fig. 9.32 Modes of Interspecies electron transfer as a function of cell-cell distance in communities of primary and secondary fermenters, methanotrophs (grey cells), and H_2 consumers (blue cells). Distances shown are to emphasize the order of magnitude rather than precise numbers. In low flux, small ΔG environments, when cells are in physical contact ($d < 1$ μm), direct extracellular electron transfer can be a predominant mode. In flocs of cells, such as in anaerobic digesters or the intestine with stable cell-cell distances in the range of 1 μm, H_2- and formate-based interspecies electron transfer is beneficial, while at larger distances (10s of μm), formate-based electron transfer is advantageous. At even larger distances, selection favors both, H_2-forming and H_2-consuming metabolism to be in one microorganism, here acetogenic metabolism for H_2 consumption (see Chap. 10). See text for details

electrons to the sulfate reducer (Fig. 9.31). Direct extracellular electron transfer via extracellular cytochromes is also observed in *Geobacter, Shewanella*, and *Rhodoferax sp.* that can transfer catabolic electrons to and from redox-active surfaces such as iron minerals or electrodes.

Interspecies electron transfer (i) via direct electron transfer, (ii) via H_2, or (iii) via formate have in common that they are different solutions of coupling an oxidative pathway in one microbe to a reductive pathway in a different microbe, allowing both to conserve metabolic energy (Fig. 9.32). When cells are in close (<1 μm) proximity to each other, direct extracellular electron transfer, as postulated for methanotrophic archaea and sulfate-reducing bacteria, takes place. Syntrophy based on interspecies H_2 or formate transfer is beneficial when microbes are apart from each other on a scale of μm (Fig. 9.32). For the reasons discussed above on the relationship between diffusion, redox potential, and concentration, formate is a better interspecies electron transfer compound than H_2, in particular in environments with increasing cell-cell distances. When the distance between microbes is large (>100s of μm) as in low flux environments such as deep-sea sediments, the large distance between microbes is a kinetic barrier for interspecies, diffusion-based electron transfer, and selection favors microbes with the presence of both "hydrogen/formate"-forming and - consuming pathways in single microbe, such as an acetogen (see Chap. 10.3). This observation is reminiscent of the outcome of the Julian Adams experiment.

Further Reading

Boll M et al (2016) Fermentative cyclohexane carboxylate formation in *Syntrophus aciditrophicus*. J Mol Microbiol Biotechnol 26:165–179

Buckel W, Thauer RK (2013) Energy conservation via electron bifurcating ferredoxin reduction and proton/ $Na^{(+)}$ translocating ferredoxin oxidation. Biochim Biophys Acta 1827(2):94–113

Chadwick GL, Skennerton CT, Laso-Perez R, Leu AO, Speth DR, Yu H et al (2022) Comparative genomics reveals electron transfer and syntrophic mechanisms differentiating methanotrophic and methanogenic archaea. PLoS Biol 20(1):e3001508

Mock J et al (2015) Energy conservation associated with ethanol formation from H_2 and CO_2 in *Clostridium autoethanogenum* involving electron bifurcation. J Bacteriol 197(18):2965–2980

Westerholm M, Calusinska M, Dolfing J (2022) Syntrophic propionate-oxidizing bacteria in methanogenic systems. FEMS Microbiol Rev 46:1–26

Wiechmann A et al (2020) It does not always take two to tango: "Syntrophy" via hydrogen cycling in one bacterial cell. ISME 14:1562–1570

Prototypic Reactions, Modules, and Pathways of C_1 catabolism

10

Contents

A general perspective of C_1 metabolism is presented first, followed by modules, prototypic reactions, and key enzymes, before major C_1 catabolic pathways are introduced.

10.1 Introduction to C_1 Transformations

Catabolic C_1 transformations cover transitions in oxidation state of the carbon atom of C_1 compounds between −4 (CH_4) to +4 (CO_2) (Fig. 10.1). Methane, methanol, methylamines, methanethiol, carbon monoxide, or formate can be oxidized to CO_2 by aerobic and anaerobic prokaryotes. On the other hand, some anaerobic prokaryotes conserve energy associated with reductive C_1 catabolism, e.g., of CO_2, formate, or methyl compounds, as well as with acetate disproportionation. While the individual steps of microbial C_1 transformations are simple, the involvement of diverse and complex, C_1-carrying coenzymes can complicate their understanding. These coenzymes will be introduced in the context of modules and pathways. Moreover, some C_1 intermediates are either unbound or bound to one or more C_1-carrying coenzymes, which can add to the complexity. To anchor C_1 metabolism, it is useful

The original version of the chapter has been revised. A correction to this chapter can be found at https://doi.org/10.1007/978-3-031-28218-8_11

A. M. Spormann, *Principles of Microbial Metabolism and Metabolic Ecology*, https://doi.org/10.1007/978-3-031-28218-8_10

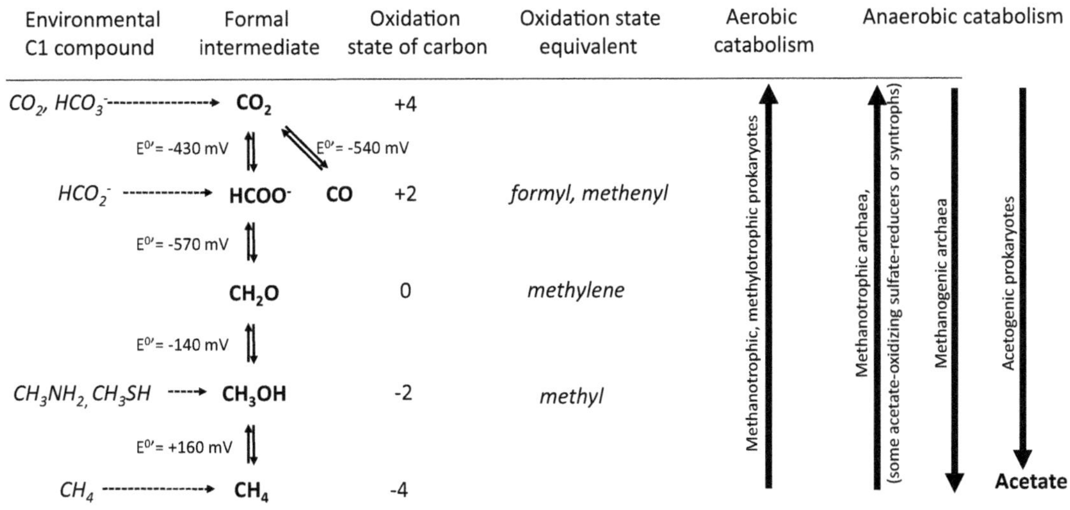

Fig. 10.1 Oxidation state-centric view of microbial C_1 metabolism. A "formal intermediate" refers to the oxidation state of a C_1 carbon and not its chemical context, e.g., whether it is bound to a C_1-carrying coenzyme. A "formal intermediate" may not be an actual, free pathway intermediate. "Oxidation state equivalent" includes metabolic C_1 compounds in coenzymes, whose carbon is at an equivalent oxidation state as the "formal intermediate." Redox potentials indicated are those for redox transitions of CO_2, formate, formaldehyde, methanol, and methane. The redox potential of CH_2O to formate is an approximation because of the inaccuracy in experimentally determining the enthalpy of formation of formaldehyde. The main C_1 catabolic pathways, their directions, and microorganisms involved are indicated on the right: aerobic Methano/methylotrophic Pathway, Methanogenesis Pathway, and Wood–Ljungdahl pathway. See text for details

to first focus on the **oxidation state of the carbon atom of a C_1 compound,** regardless of whether or not it is bound to a coenzyme. Fig. 10.1 introduces the concept of a **"formal intermediate"** to emphasize the oxidation state of the carbon atom independent of a binding to a coenzyme. The chemical name of the most simple, unbound C_1 compound equivalent in oxidation state (therefore, "formal") is assigned to a C_1 intermediate with the understanding that such "formal intermediate" may or may not be an actual pathway intermediate (Fig. 10.1). For example, the formal intermediate "formaldehyde" indicates a C_1 carbon at the oxidation state of zero, such as in formaldehyde or in a methylene carbon when bound to C_1-carrying coenzyme tetrahydrofolate or tetrahydromethanopterin (see below). Conversely, a "methenyl" or "formyl" carbon, when bound to a C_1-carrying coenzyme, is at the same oxidation state as a "formate" carbon. Abbreviations such as "methyl" will be used to describe any C_1 carbon (free or coenzyme bound) that is at the oxidation state of a methyl group, i.e., oxidation state - 2, regardless of the actual chemical context (Fig. 10.1).

10.2 C_1 Modules

There are currently three known C_1-transforming pathways; two operating in anaerobic microbes, and one methane/methyl-oxidizing pathway in aerobes (Fig. 10.2). Notably, the two former pathways can operate in either the **reductive** or **oxidative** direction, depending on a microbe's physiology. One of the pathways is the **Wood-Ljungdahl Pathway**, which functions as a catabolic or anabolic CO_2-reductive pathway in many strictly anaerobic microbes as well as, in the reverse, in anaerobic (and also syntrophic) acetate oxidation to CO_2; the second **Methanogenesis Pathway** functions both in anaerobic formation and oxidation of methane in Archaea. The third aerobic methane/methyl-oxidizing pathway is related to the anaerobic methane oxidation pathway but contains some important variations. This aerobic C_1 pathway operates only in the oxidative direction (Fig. 10.1).

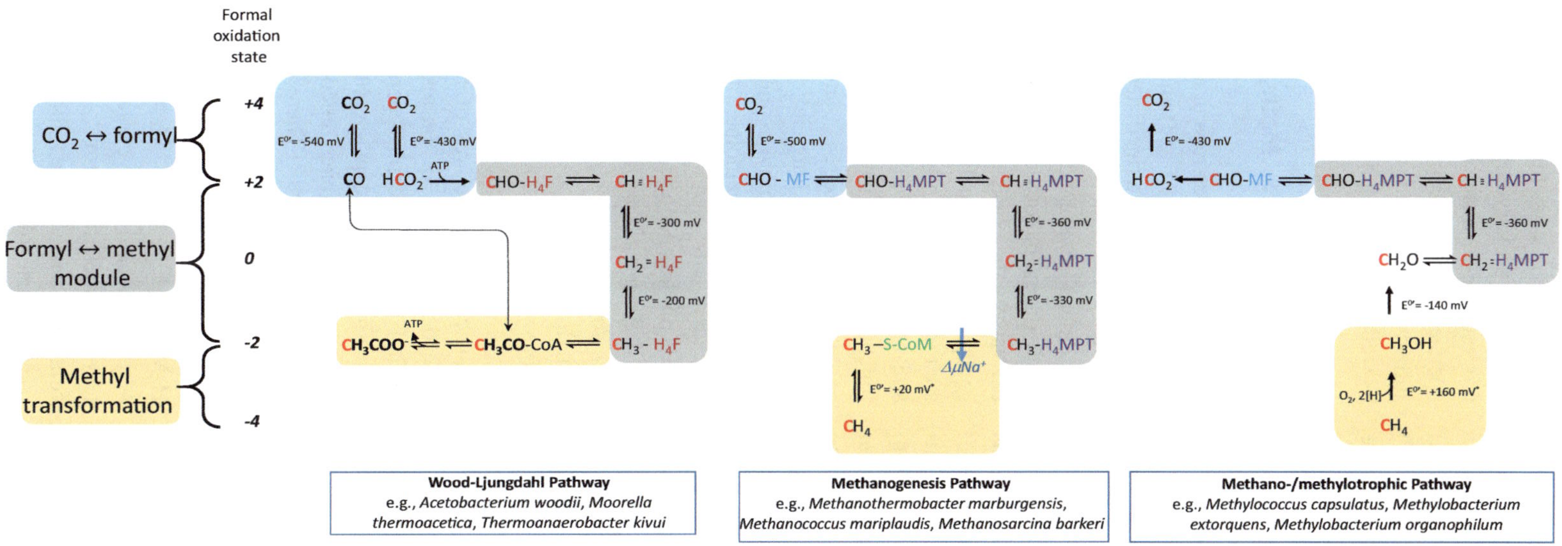

Fig. 10.2 C_1-centric overview of reductive and oxidative microbial C_1 catabolism. The overview is organized based on the oxidation state of the C_1 carbon, ranging from +4 (CO_2) to −4 (CH_4). There are three major types of C_1 pathways: (i) Wood–Ljungdahl pathway, (ii) Methanogenesis Pathway, and (iii) aerobic Methano/methylotrophic Pathway. Common to all pathways is the formyl ↔ methyl module, indicated by the grey-shaded area between the carbon oxidation states +2 and −2. There are pathway-specific differences on either end of each pathway: the CO_2 ↔ formyl reactions and the methyl transformations differ between acetogenic, methanogenic, and aerobic methylotrophic metabolism. The vertical transitions in oxidation state of the C_1 carbon are indicated with the respective redox potentials. Reaction-specific electron-carrying coenzymes were omitted as they may vary for a given reaction between microbes as discussed in this chapter. C_1-carrying coenzymes tetrahydrofolate (H_4F) are indicated in dark red, tetrahydromethanopterin (H_4MTP) in magenta, methanofuran (MF) in light blue, and co-enzyme M (CoM-SH) in green. With the exception of the *methyl transferase* catalyzing transfer of the methyl group of CH_3-H_3MPT to CoM-SH in the Methanogenesis Pathway, no net energy conservation is associated with any specific step in any C_1 pathway; thus, these pathways are solely C_1 oxidation or reduction pathways. Energy conservation in aerobic methanotrophic bacteria occurs via respiration, while energy conservation in the CO_2-reductive Wood–Ljungdahl pathway is via electron flux through Rnf or Ech, see below. Asterisk indicates calculated standard redox potential. See text for details

The oxidation state-centric view of C_1 compounds of Fig. 10.1 is applied to these three major pathways in Fig. 10.2. The oxidation state of a C_1 compound (in red) in each of the four rows is equal and ranges between the most oxidized (+4, top row) and most reduced (−4, bottom row) state. The C_1 intermediates shown in this figure are actual intermediates in the respective pathways and in chemical context of specific C_1-carrying coenzymes (see below). As this figure shows, in a single pathway a C_1 compound can occur at the same oxidation state in three different chemical forms. For example in the Wood-Ljungdahl pathway, a 'formyl' carbon is present in formate, in formyl-tetrahydrofolate, and in methenyl-tetrahydrofolate. Specific physiological electron donor(s)/acceptor(s) of each vertical oxidation/reduction transition are omitted as they vary between prokaryotes and the metabolic/ecophysiological context as outlined below. Consequently, reversibility of each redox reaction, and with that the direction of a C_1 pathway can vary depending on the redox potential of the electron-transferring coenzyme involved, similarly as in fermentative pathways (see Chap. 9). With the understanding of the significance of electron-carrying coenzymes involved for determining the direction of a pathway, reactions are introduced as reversible in Fig. 10.2 unless a reaction has been consistently observed to be irreversible for mechanistic and overall thermodynamic reasons.

C_1 metabolism as shown in Fig. 10.2 can be divided into a **common** and **distinct pathway elements**. The **common** pathway element is the **formyl↔methyl module** (Fig. 10.2, grey shaded reactions). Pathway elements **distinctive** and consequently characteristic to each major pathway are the **CO_2↔formyl** (blue) and the **methyl transformations** (yellow) (Fig. 10.2).

The common formyl↔methyl transition includes reversible interconversions of C_1 compounds/intermediates between the oxidation state of formyl and methyl. In anaerobic microbes, the formyl↔methyl module can function either in the oxidative or reductive direction depending on enzymes and electron carriers involved (see below). For illustrative purposes, the formyl↔methyl module is introduced below in the context of anaerobic, CO_2-reductive catabolic pathways as found in strictly anaerobic acetogenic prokaryotes. There are variations of this module in anaerobic acetate oxidation that operates in many complete-oxidizing anaerobic sulfate-reducing and some syntrophic bacteria, as well as in autotrophic CO_2 fixation.

10.2.1 Formyl ↔ Methyl Module

All redox interconversions in anaerobic C_1 catabolism between formyl (+2) and methyl (−2) carbon occur via C_1 intermediates covalently bound to coenzymes. The two main C_1 carriers are the pteridines tetrahydrofolate (H_4F) and tetrahydromethanopterin (H_4MPT) as well as derivatives thereof (Fig. 10.3). Important for C_1 group binding and activation by both coenzymes are the N^5 and N^{10} nitrogen atoms of the pteridine and arylamine moiety (Fig. 10.3 blue N). H_4F functions as C_1-carrier in catabolic CO_2 reduction pathways of acetogens and in acetyl-CoA oxidizing Wood–Ljungdahl pathway, whereas methanogens employ catabolically H_4MTP C_1-carriers. Genes annotated to encode H_4MTP-dependent enzymes have been found broadly across phylogenetic groups including but not limited to *Archaeglobus fulgidus* and other Archaea, *Methylobacterium extorquens* (and other methylotrophic and methane-oxidizing bacteria), *Planctomycetes*, *Actinobacteria*, *Chloroflexi*, and α-, β-, γ-proteobacteria. Predicted amino acid sequences and structures of H_4F- and H_4MPT-dependent enzymes of the respective formyl ↔ methyl modules are phylogenetically unrelated, indicating independent evolution, and so are the biosynthetic genes encoding biosynthesis of each pterin cofactor. There are also significant differences in reactivity between the N^5 and N^{10} nitrogen within each and between the two coenzymes (see Box "H_4F versus H_4MPT", Fig. 10.3). *Methanococcus sp.* and *Methanosarcina sp.* contain sarcinapterin, a tetrahydromethanopterin derivative which contains an amide bond between the α-carboxylate of the H_4MPT glutamate and an L-glutamate.

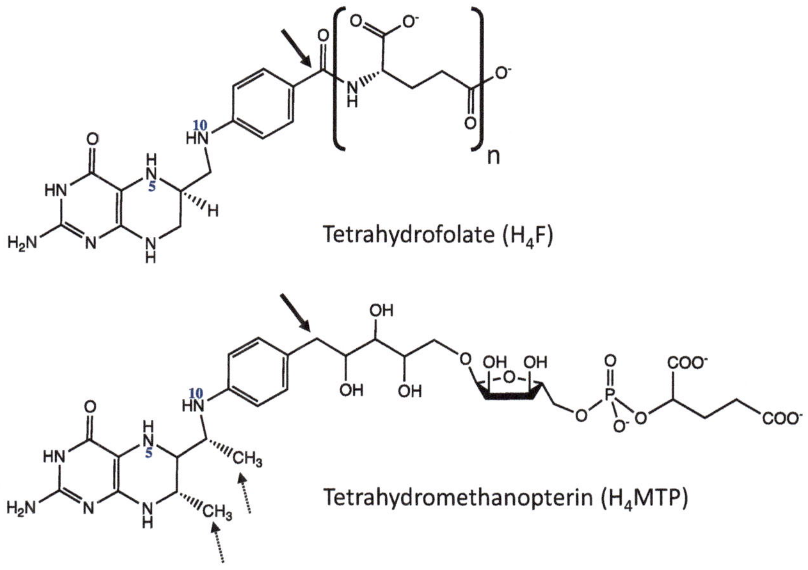

Fig. 10.3 Structure and relevant functional groups of C_1-carrying tetrahydrofolate and tetrahydromethanopterin of the Formyl ↔ Methyl module. Tetrahydrofolate consists of a pteridine-, a *p*-aminobenzoate-, and a polyglutamate moiety, where the glutamate chain length can vary between 1–8. Tetrahydromethanopterin consists of a pteridine linked to an arylamine that is covalently bound to a ribitol residue that is linked to a ribose-5-phosphate and one (shown here) or more hydroxyglutamate residues. The phospho-hydroxyglutamate moiety of H_4MPT can be substituted differently in prokaryotes. Methylation pattern (dotted arrows) of H_4MPT can also vary. The functionally relevant N atoms in both pterins are N^5 and N^{10} (indicated in blue). Reactivity of N^{10} in H_4F is controlled by the electron-withdrawing carbonyl group conjugated to the aromatic ring, which is substituted in H_4MTP by a slightly electron-donating methylene group at the equivalent position (solid arrows, see Box for details)

H_4F versus H_4MPT

Why do acetogenic microbes use in their formyl ↔ methyl module H_4F as C_1 carrier while methanogenic Archaea and aerobic methylotrophs use H_4MPT? The standard redox potentials of free formate to free formaldehyde is $E^{0\prime} = -570$ mV and of formaldehyde to methanol is $E^{0\prime} = -140$ mV. A reduction of formate to unbound formaldehyde by NADH or NADPH ($E^{0\prime} = -320$ mV) is endergonic and, thus, unsuitable as a catabolic pathway step. Moreover, the end product of formate reduction, free formaldehyde, is an effective electrophile, and, therefore, toxic to a cell. These thermodynamic and cytotoxicity issues are resolved when the C_1 intermediate is covalently bound to an N atom in H_4F or H_4MPT. Relevant N atoms are the N^5 and N^{10} nitrogen of the pteridines, where each of those two N atoms affects the reactivity of an N-bound C_1 group differently. Reactivity of the N^{10} nitrogen between the two coenzymes is also different. In tetrahydrofolate, reactivity of N^{10} is largely determined by the electron-withdrawing carbonyl group conjugated via the aromatic ring to N^{10} (arrow in Fig. 10.3). This electron-withdrawing carbonyl group is absent in H_4MPT and substituted at the equivalent position by a slightly electron-donating methylene group with the effect that N^{10} of H_4MPT is a better nucleophile than that of H_4F. Therefore, redox potentials of C_1-intermediates bound to H_4MPT are more negative than those of equivalent C_1-H_4F intermediates. For example, redox potential of the methenyl/

(continued)

methylene couple in H_4MTP is $E^{0\prime}$ = –360 mV while in H_4F it is $E^{0\prime}$ = –300 mV; similarly, the redox potential of the methylene/methyl couple in H_4MTP is $E^{0\prime} = -330$ mV while in H_4F is $E^{0\prime}$ = –200 mV, respectively. In addition, cyclization of N^{10}-HCO-H_4F to N^5,N^{10}-HC ≡ H_4F is endergonic, whereas cyclization of N^5-HCO-H_4MTP is exergonic ($\Delta G^{0\prime}$ = + 6 kJ/mol vs – 4 kJ/mol at pH7). If free formate were to be converted to N^5,N^{10}-HC ≡ H_4F, this reaction would be endergonic by +28 kJ/mol (thus, the ATP requirement in the H_4F-dependent pathway, see below) while the equivalent conversion to N^5,N^{10}-HC ≡ H_4MPT is endergonic only by +5 kJ/mol. H_4MPT also contains two methyl groups (dotted arrows in Fig. 10.3) that are absent in H_4F. These can potentially constrain the rotation of the bound C_1 group and, thereby, affect its activation.

Collectively, molecular differences between the two C_1 carriers H_4F and H_4MPT result in reduction reactions of C_1-H_4F intermediates to occur in acetogenic prokaryotes at more positive redox potentials, which allows NADH and NADPH to function as physiological electron donor, while the more negative redox potentials of C_1-H_4MPT intermediates require the deazaflavin $F_{420}H_2$ at $E^{0\prime}_{F420/F420H2} = -350$ mV as electron carrier in methanogens.

Every microbe, regardless of its catabolism, has an anabolic C_1 metabolism for synthesis of purines, pyrimidines, amino acids, or coenzymes, that includes the formyl ↔ methyl module which is H_4F-dependent except in Archaea. Therefore, identifying genomically in an uncharacterized microbe genes as encoding a H_4F-dependent formyl ↔ methyl module cannot be taken as evidence *per se* for the presence of a H_4F-dependent catabolic pathway.

In the reductive direction, the first intermediate of the formyl ↔ methyl module is formyl-H_4F and formyl-H_4MPT (Fig. 10.4). In H_4F, the formyl group is covalently bound to the N^{10} nitrogen, whereas in H_4MPT it is bound to N^5. For simplification, the formyl ↔ methyl module is discussed further in context of H_4F-dependent C_1 conversions; however, it is comparatively displayed in Fig. 10.4 for H_4MPT-dependent reactions. In the first step, H_2O is abstracted from the formyl group of N^{10}-HCO-H_4F by *methenyltetrahydrofolate cyclohydrolase* to form N^5,N^{10}-HC ≡ H_4F (methenyl-H_4F). Reduction of methenyl-H_4F to methylene-H_4F is catalyzed by *methylenetetrahydrofolate dehydrogenase* with either NADH or NADPH as hydride donor depending on the microbe. Reduction of methylene-H_4F to methyl-H_4F at $E^{0\prime} = -200$ mV is mediated by a *methylenetetrahydrofolate reductase.* There are two variants of *methylenetetrahydrofolate reductase*, discussed below, with implications for energy conservation and for expressing a rate-yield phenotype. In a pathway functioning in the direction of methyl group oxidation, there appears to be also a membrane-associated *methylene-H_4F reductase* that catalyzes the oxidation of CH_3-H_4F to methylene-H_4F with menaquinone as electron acceptor. CH_3-H_4F is the end point intermediate on the methyl side of the formyl ↔ methyl module.

Because of the lower redox potential of C_1-H_4MPT intermediates, the deazaflavin $F_{420}H_2$ with a redox potential ($E^{0\prime} = -350$ mV) is the hydride carrier in respective reductive reactions and functions analogously to a pyridine nucleotide in hydride transfer (see Box "H_4F versus H_4MPT", Fig. 10.5). With some modifications, this formyl ↔ methyl module also operates in the oxidative direction in oxidative C_1 catabolism (e.g., of CH_4, CH_3OH, CO, $HCOO^-$, etc.) in aerobic and anaerobic microorganisms (see below).

10.2.2 CO_2 ↔ Formyl Conversions

In contrast to the common formyl ↔ methyl module, steps of CO_2 reduction to formyl are different in the three pathways (Fig. 10.2, blue background).

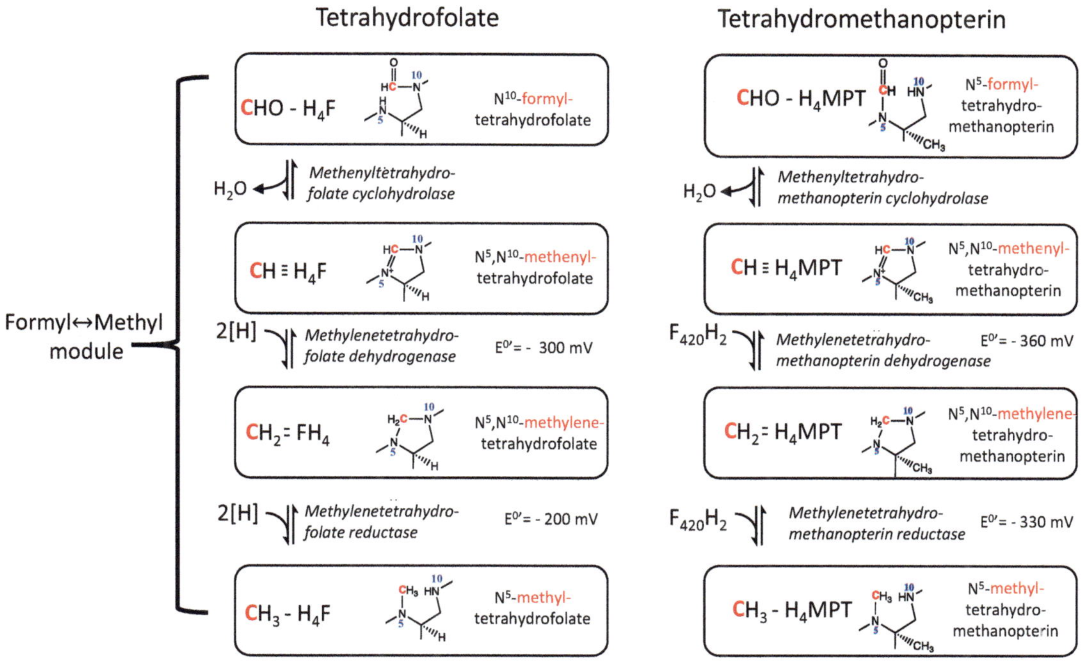

Fig. 10.4 Reactions of the Formyl ↔ Methyl module involving H_4F or H_4MPT as C_1-carrying coenzymes. Reactions are written in the reductive direction and are reversible. Intermediates, their relevant structure, redox potentials, and transforming enzymes are discussed in the text. The electron donor in the H_4F-dependent formyl ↔ methyl module can be NADH or NADPH, whereas $F_{420}H_2$ is the electron donor in the H_4MPT-dependent module. Note the difference between "methylene" and "methenyl." See text for details

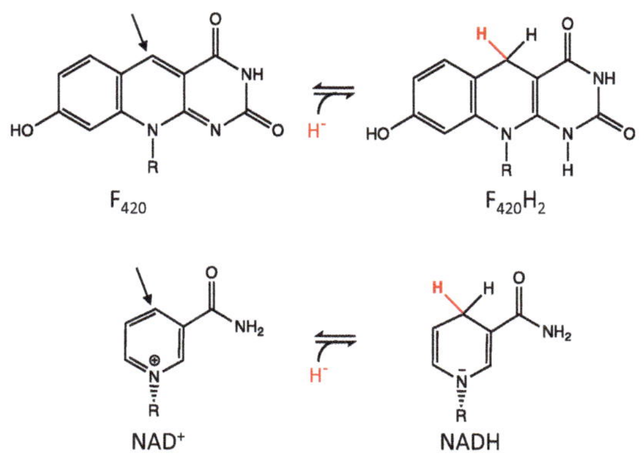

Fig. 10.5 Structure and function of the deazaflavin F_{420} as a hydride carrier and similarity to NAD^+. Upper row indicates the reversible reduction of F_{420} by a hydride ($2\ e^- + 1\ H^+$) at the ring carbon which is at equivalent position of the carbon being reduced in NAD^+. Note migration of double bonds during reduction. R denotes a ribityl-, lactyl-, and polyglutamate-containing moiety

In acetogenic prokaryotes using the Wood–Ljungdahl pathway, CO_2 reduction to formate is catalyzed by an NADPH-specific *formate dehydrogenase* or by a *H_2-dependent CO_2 reductase* (see below). Formate is a free intracellular intermediate which is subsequently covalently bound to N^{10} of the C1-carrier H_4F by *formyl-H_4F synthetase* under consumption of 1 ATP (Fig. 10.2). When the Wood–Ljungdahl pathway operates in the oxidative direction, as in some acetate-oxidizing, sulfate-reducing prokaryotes, NAD^+ may act as electron acceptor in the thermodynamically downhill *formate dehydrogenase* reaction because NADH is re-oxidized in these microbes via an external electron acceptor with a more positive redox potential (e.g., adenosine phosphosulfate or sulfite in the sulfate-reduction pathway). Reduction of the second CO_2 to "formyl" in the Wood–Ljungdahl pathway is the reduction of CO_2 to carbon monoxide, CO (Fig. 10.2). CO is formed reversibly from CO_2 and reduced ferredoxin^{2-} (Fd^{2-}) by *carbon monoxide dehydrogenase* which mostly occurs in a complex with *acetyl-CoA synthase* (see below).

In methanogenic and methanotrophic Archaea CO_2 reduction to formyl and ligation to the C_1-carrying coenzyme methanofuran is catalyzed by a single enzyme *formylmethanofuran dehydrogenase* and requires Fd^{2-} as electron donor (Fig. 10.2, and see below). In this *formylmethanofuran dehydrogenase* reaction, CO_2 is reduced to formate, which is not a free intermediate but trapped within the enzyme before reacting with methanofuran at the active site. The formyl group of formyl-MF is subsequently transferred to N^5-H_4MPT by *formylmethanofuran:H_4MPT formyltransferase* to form N^5-HCO-H_4MPT (Fig. 10.2, and see below). These reactions are reversible and postulated to operate also in anaerobic methane oxidation.

Why is $CO_2 \leftrightarrow$ formyl conversion in hydrogenotrophic methanogens independent of ATP while acetogens spend 1 ATP for this conversion?

Forming a covalent bond between free formate and the C_1-carriers MF, H_4MPT, and H_4F are endergonic reactions. Because the formamide bond of N^{10}-formyl-H_4F is stronger than the same bond of N^5-formyl-H_4MPT, formation of formyl-H_4F requires the hydrolysis of 1 ATP, while formation of formyl-H_4MPT via formyl-MF requires only the equivalent of 0.5 ATP (i.e., the difference in redox potential between Fd^{2-} and NADH, see below). In CO_2-reducing methanogens, this formylation reaction is coupled to reduction of CO_2 with electrons from a reduced ferredoxin at a redox potential of ~−500 mV, which is far more negative than that required for CO_2 reduction with H_2 to formate ($E' = -300$ mV at $pH_2 = 10$ Pa). Thus, the coupling of exergonic CO_2 reduction with endergonic formylation at the enzyme drives the overall *formylmethanofuran dehydrogenase* reaction. Reduced ferredoxin at −500 mV is formed by *electron bifurcating heterodisulfide reductase* driven by the reduction of the heterodisulfide CoM-S-S-CoB at a redox potential of $E^{0\prime} = -140$ mV with H_2 as electron donor ($E' = -300$ mV) (see below). In acetogens, electron donor is either H_2 at $E' = -300$ mV or NADPH at $E^{0\prime} = -320$ mV; which is sufficient for CO_2 reduction to formate; however, the absence of a low potential electron donor (such as Fd^{2-}) as the driving force for covalent bonding of formate to H_4F necessitates the ATP-dependent step in the acetogenic Wood–Ljungdahl pathway. Acetogens generate one low redox potential Fd^{2-} for CO_2 reduction to CO ($E^{0\prime} = -520$ mV) by electron bifurcation coupled to the two remaining reduction reactions in this pathway (*electron bifurcating $CH_2 = H_4F$ reductase* or *electron bifurcating hydrogenase* coupled with NADH consumption in the *$CH_2 = H_4F$ reductase* or *dehydrogenase*, see below). Because the redox potential of $CH_2 = H_4F/CH_3$-H_4F at −200 mV is too positive to directly drive Fd reduction by electron bifurcation, there is in the Wood–Ljungdahl pathway no further reaction available as a "pull" for reduction of a second Fd by electron bifurcation. Consequently, no Fd^{2-} is available for a direct CO_2 reduction and formylation.

In methano/methylotrophic bacteria, formate oxidation to CO_2 proceeds via an *NAD^+-dependent formate dehydrogenase* and is, thus, irreversible.

10.2.3 Prototypic Reactions in Methyl Conversions

The fate of the methyl carbon of CH_3-H_4F and CH_3-H_4MPT (Fig. 10.2 orange background) is resolved differently in the anaerobic acetogenesis and methanogenesis pathways. Formation of the methyl intermediate in the aerobic methane-oxidizing pathway is also different from the corresponding steps in the two anaerobic pathways. Therefore, the three methyl-transforming enzymes are idiosyncratic to and define each pathway:

(i) ***carbon monoxide dehydrogenase/acetyl-CoA synthase*** defines the **Wood–Ljungdahl Pathway** in both the reductive and oxidative direction (Sect. 10.2.3.1);
(ii) ***methyl-coenzyme M reductase*** (Sect. 10.2.3.3) and ***heterodisulfide reductase*** (Sect. 10.2.3.4) define **methanogenesis** and **anaerobic methanotrophy**;
(iii) ***methane monooxygenase*** defines O_2-dependent methane oxidation (see Sect. 7.4.1).

10.2.3.1 Carbon Monoxide Dehydrogenase/Acetyl-CoA Synthase

Carbon monoxide dehydrogenase/acetyl-CoA synthase (CO-DH/ACS) in combination with *CH_3-H_4F: corrinoid iron sulfur protein methyltransferase* catalyzes both reduction of CO_2 to CO AND CO-dependent carbonylation of CH_3-H_4F to acetyl-CoA with HS-CoA (Fig. 10.6). CO-DH/ACS is the key enzyme of the Wood–Ljungdahl pathway in catabolic acetate formation in acetogenic prokaryotes and in catabolic acetate oxidation in some strictly anaerobic sulfate-reducing and syntrophic prokaryotes. CO-DH/ACS is also found in acetoclastic methanogens in the pathway of

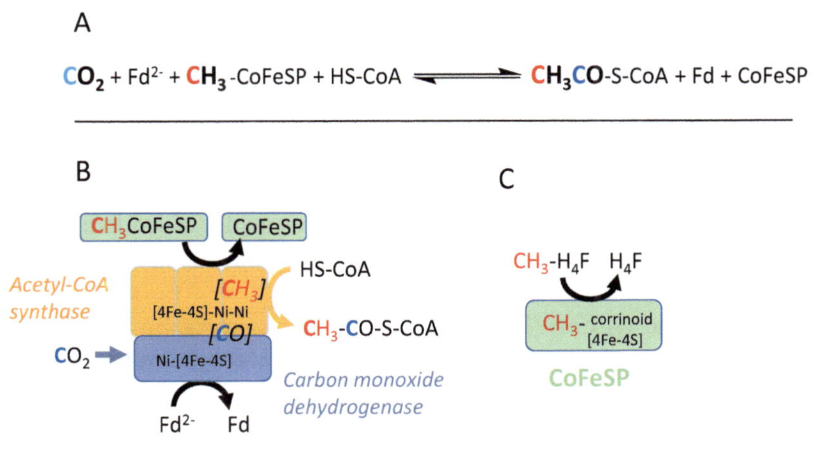

Fig. 10.6 The *carbon monoxide dehydrogenase/acetyl-CoA synthase* complex. (**a**) Overall reversible *carbon monoxide dehydrogenase/acetyl-CoA synthase* reaction. (**b**) *Carbon monoxide dehydrogenase* subunit (blue), and *acetyl-CoA synthase* (orange). For simplicity, only one tetramer is shown. CO-DH catalyzes the reduction of CO_2 to CO (blue), which is trapped in the enzyme and diffuses via a channel to the active site of ACS, where it binds to one of the Ni atoms. The corrinoid FeS cluster-containing *methyl transferase* transfers the methyl group (red), derived from CH_3-H_4F, to the active site where the carbonylation reaction occurs. HS-CoA thiolytically releases the enzyme-bound acetyl group as acetyl-CoA. (**c**) Corrinoid Fe-S cluster-containing *methyl transferase* (CoFeSP) demethylates CH_3-H_4F and is the donor of the methyl group to the ACS subunit in (B). See text for more details

acetate disproportionation to CH_4 and CO_2. Acetoclastic methanogens carry a CO-DH/ACS which catalyzes cleavage of acetyl-CoA into CH_3-H_4MPT and CO as well as CO oxidation, in combination with a H_4MPT-specific *H_4MPT: corrinoid iron-sulfur protein methyltransferase*.

The Wood–Ljungdahl pathway functions also anabolically as a pathway for CO_2 fixation in hydrogenotrophic methanogens. Thus, finding genes in a prokaryotic genome encoding CO-DH/ACS and other enzymes of the Wood–Ljungdahl pathway does not indicate by itself whether the microbe has the capacity to operate a catabolic Wood–Ljungdahl pathway.

CO-DH/ACS contains combinations of several interesting metal cofactors. Bifunctional heterotetrameric CO-DH/ACS is a Ni-containing iron-sulfur enzyme with a unique Ni-[FeS] cluster in *CO dehydrogenase* and a Ni-Ni-[FeS] cluster in *acetyl-CoA synthase* (Fig. 10.6). All reactions catalyzed by the subunits as well as the overall reaction are reversible under physiological conditions. *CO dehydrogenase* subunit, which is mostly found in a complex with *acetyl-CoA synthase*, catalyzes reversible reduction of CO_2 to CO with Fd^{2-} as electron donor. In *CO dehydrogenase*, carbon monoxide is an enzyme-trapped intermediate which is transferred to the active Ni-Ni site of *acetyl-CoA synthase*. A corrinoid and Fe-S cluster-containing *methyl transferase* (CoFeSP), which is not part of the complex, transfers the methyl group of CH_3-H_4F to the *acetyl-CoA synthase* subunit, where the methyl group is carbonylated with CO to form an acetyl-*ACS* adduct. The latter is thiolytically released with HS-CoA as acetyl-CoA (Fig. 10.6). The presence of reduced Fe-S clusters with a redox potential near −500 mV makes *carbon monoxide dehydrogenase/acetyl-CoA synthase* oxygen-sensitive and restricts its operation to strictly anaerobic prokaryotes.

Mechanistically and structurally unrelated to *carbon monoxide dehydrogenase/acetyl-CoA synthase* are *decarbonylases* found in anabolic decarbonylations of long-chain aldehydes to (n − 1)alkanes. Next to a C_{-1}-alkane end product, the aldehyde carbon is released by *decarbonylases* (i) in plants and green algae as carbon monoxide, (ii) in insects oxidatively as CO_2, and (iii) in *Prochlorococcus marinus* as formate. These enzymes are of interest for microbial production of alkanes.

Carbon monoxide oxidation in catabolism of aerobic microbes proceeds by *CO dehydrogenases* that have a different structure and active site. *CO dehydrogenase* of the aerobic, carboxidotrophic *Oligotropha carboxidovorans* is a molybdopterine-containing, two [2Fe-2S] center flavoprotein that is related in overall composition to other Mo-dependent *hydroxylases* such as *xanthine dehydrogenase*. The enzyme oxidizes CO to CO_2 and funnels the electrons into an aerobic respiratory chain for energy conservation.

10.2.3.2 Methyl-H_4MPT: Coenzyme M Methyltransferase

In the Methanogenesis Pathway, CH_3-H_4MPT of the formyl ↔ methyl module is connected to methane formation via cytoplasmic membrane-associated, corrinoid-containing *methyl-H_4MPT:coenzyme M methyltransferase* catalyzing the transfer of the methyl group from CH_3-H_4MPT to coenzyme HS-CoM (2-mercaptoethanesulfonate, Figs. 10.2 and 10.7) to form CH_3-S-CoM. This methyl transfer is exergonic by −30 kJ/mol and coupled to the translocation across the cytoplasmic membrane of 2 electrogenic Na^+ ions per methyl transferred. Ion-pumping by this *methyl transferase* is one of the few examples of ion translocation-dependent energy conservation in the absence of a net redox reaction.

10.2.3.3 *Methyl-Coenzyme M Reductase*

Methyl-coenzyme M reductase (MCR) is one of the most abundant catabolic enzymes in Nature. MCR is a heterohexameric protein complex catalyzing reduction of the methyl group of CH_3-CoM to CH_4 with CoB-SH (7-mercaptoheptanoylthreonine phosphate) as electron donor (Figs. 10.7 and 10.8). MCR contains the Ni-porphinoid Factor$_{430}$ (F_{430}) at its active site with Ni in the highly reactive +1 state (Fig. 10.8b). MCR has a thiyl radical-based reaction mechanism, where methane is formed by the

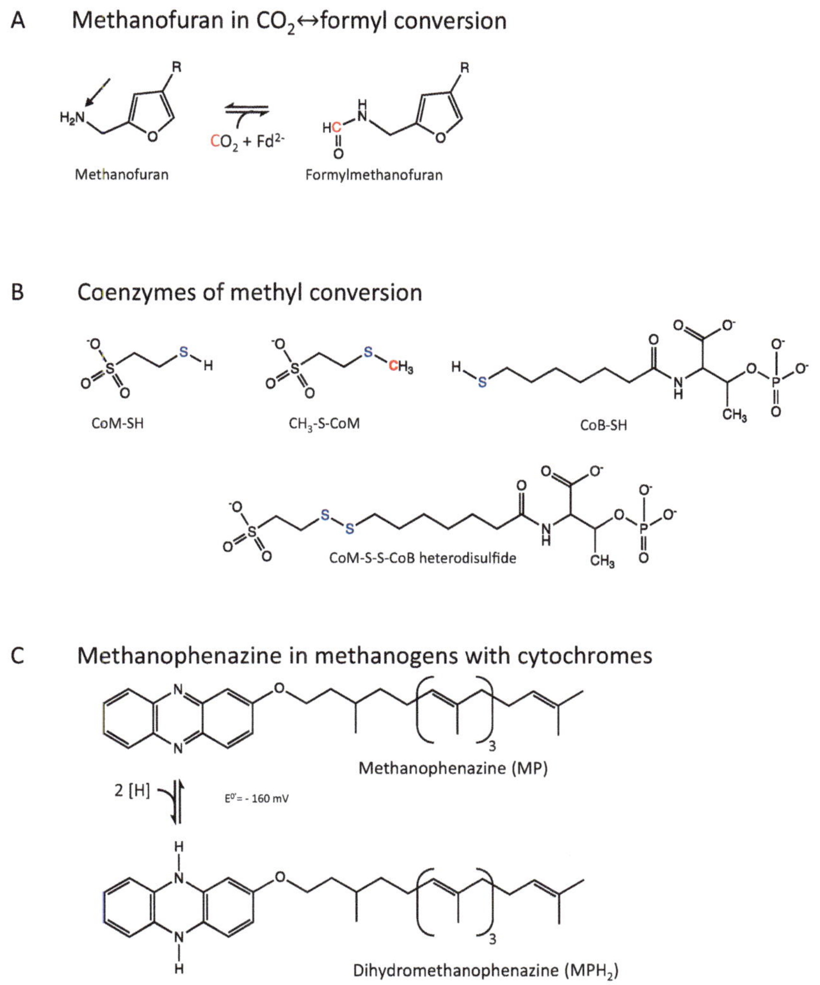

Fig. 10.7 Coenzymes involved in methanogens in CO_2 ↔ formyl conversion, in methyl transformations, and methanophenazine. (**a**) CO_2 ↔ formyl conversion: methanofuran (only the furan ring of the coenzyme shown here) is the coenzyme for CO_2 reduction and binding of formate to the amino group (arrow) of methanofuran as formylmethanofuran. (**b**) Coenzymes for the methyl conversion are coenzyme M (2-mercaptoethanesulfonate) as methyl group carrier (red), CH_3-CoM, and coenzyme B (7-mercaptoheptanoylthreonine phosphate, CoB-SH) as electron carrier, where the sulfhydryl is the reductant of CH_3-S-CoM in *methyl-coenzyme M reductase* to form methane and the CoM-S-S-CoB heterodisulfide. (**c**) Oxidized and reduced form of methanophenazine, a menaquinone analog in electron transport of hydrogenotrophic methanogens containing cytochromes/ methanophenazine. See text for details

reaction of an intermediate methyl radical reacting with the hydrogen atom derived from HS-CoB. Because of the presence of a Ni^{1+} in F_{430} ($E^{0\prime} = < -600$ mV), MCR is one of the most oxygen-sensitive enzymes known and found only in strictly anaerobic Archaea. Although reduction of CH_3-S-CoM to CH_4 is exergonic by −30 kJ/mol methane, MCR catalyzes also the reverse reaction in anaerobic archaeal CH_4 oxidation and probably also of anaerobic ethane, propane, and butane oxidation with CoM-S-S-CoB as cellular electron acceptor. *Methyl-coenzyme M reductase* functions physiologically in conjunction with a *heterodisulfide reductase* that catalyzes reduction of the heterodisulfide

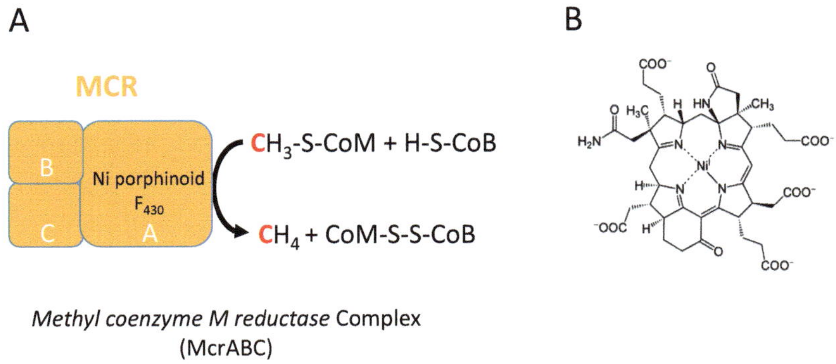

Fig. 10.8 *Methyl-coenzyme M reductase* (MCR) in methanogenic archaea. (**a**) MCR is a heterohexameric enzyme and consists of α, β, and γ subunits. For simplification, only one heteromonomer is shown. The enzyme is highly modified posttranslationally and contains the Ni-porphinoid F_{430}. (**b**) Structure of F_{430}, the Ni-porphinoid prosthetic group, of *methyl-coenzyme M reductase* as found in methanogens and anaerobic methane-oxidizing ANME 2 and 3. The lactam ring contains modifications in anaerobic methane-oxidizing ANME1. F_{430} has an extensively reduced tetrapyrrole ring system

CoM-S-S-CoB to HS-CoM and HS-CoB (see below).

Among the many aspects that make this reaction unusual is the use of the thiol of CoB-SH as electron donor to reduce CH_3-CoM to CH_4 plus CoM-S-S-CoB. CoB-SH is the functional equivalent of a cellular electron carrier except that the oxidized form is a covalent disulfide with HS-CoM. This is in contrast to NAD^+/NADH, F_{420}/F_{420}, Fd/Fd^{2-}, or quinone/quinol that are the common cellular electron carriers in all other oxidation/reduction modules discussed so far. The CoM-S-S-CoB/(CoB-SH + HS-CoM) couple is at a standard redox potential of $E^{0\prime} = -140$ mV. *Methyl-coenzyme M reductase* is the so far only known catabolic enzyme using CoB-SH as electron donor. Regeneration of CoB-SH (and of CoM-SH) from CoM-S-S-CoB in methanogens is mediated by *heterodisulfide reductase* (Hdr) (see below). As the physiological function of *methyl-coenzyme M reductase* is intrinsically linked to activity of a *heterodisulfide reductase*, these two enzymes function as a physiological module in methanogens, although they have not been observed to form a mega complex.

10.2.3.4 *Heterodisulfide Reductase*

There are two types of mechanistically dissimilar *heterodisulfide reductase*, (i) *bifurcating heterodisulfide reductases* HdrABC as found in methanogens without methanophenazine/cytochromes and (ii) *methanophenazine-dependent heterodisulfide reductases* HdrDE as found in methanogens with cytochromes (Fig. 10.9). Both enzyme types only have in common the overall reaction of CoM-S-S-CoB reduction and differ in electron transfer components and in managing available energy of the same overall reaction. This difference in energetic coupling has significant ecophysiological consequences (see below). Both enzymes evolved separately, are structurally different, and differ in the cofactor content.

10.2.3.4.1 Cytoplasmic *Electron-Bifurcating Heterodisulfide Reductase*

HdrABC is a cytoplasmic *electron bifurcating heterodisulfide reductase* in complex with Mvh *hydrogenase* (in *Methanothermobacter marburgensis*) and couples flavin-based electron bifurcation of 2 H_2 to CoM-S-S-CoB and Fd reduction. This overall bifurcating reaction is exergonic by $\Delta G^{0\prime} = -39$ kJ/mol, so a significant fraction of the available energy is unused (Fig. 10.9). Fd^{2-} is stoichiometrically consumed in CO_2 reduction to formyl-MF in hydrogenotrophic methanogens. HdrABC is found in methanogens without cytochromes, such as *Methanothermobacter marburgensis* and

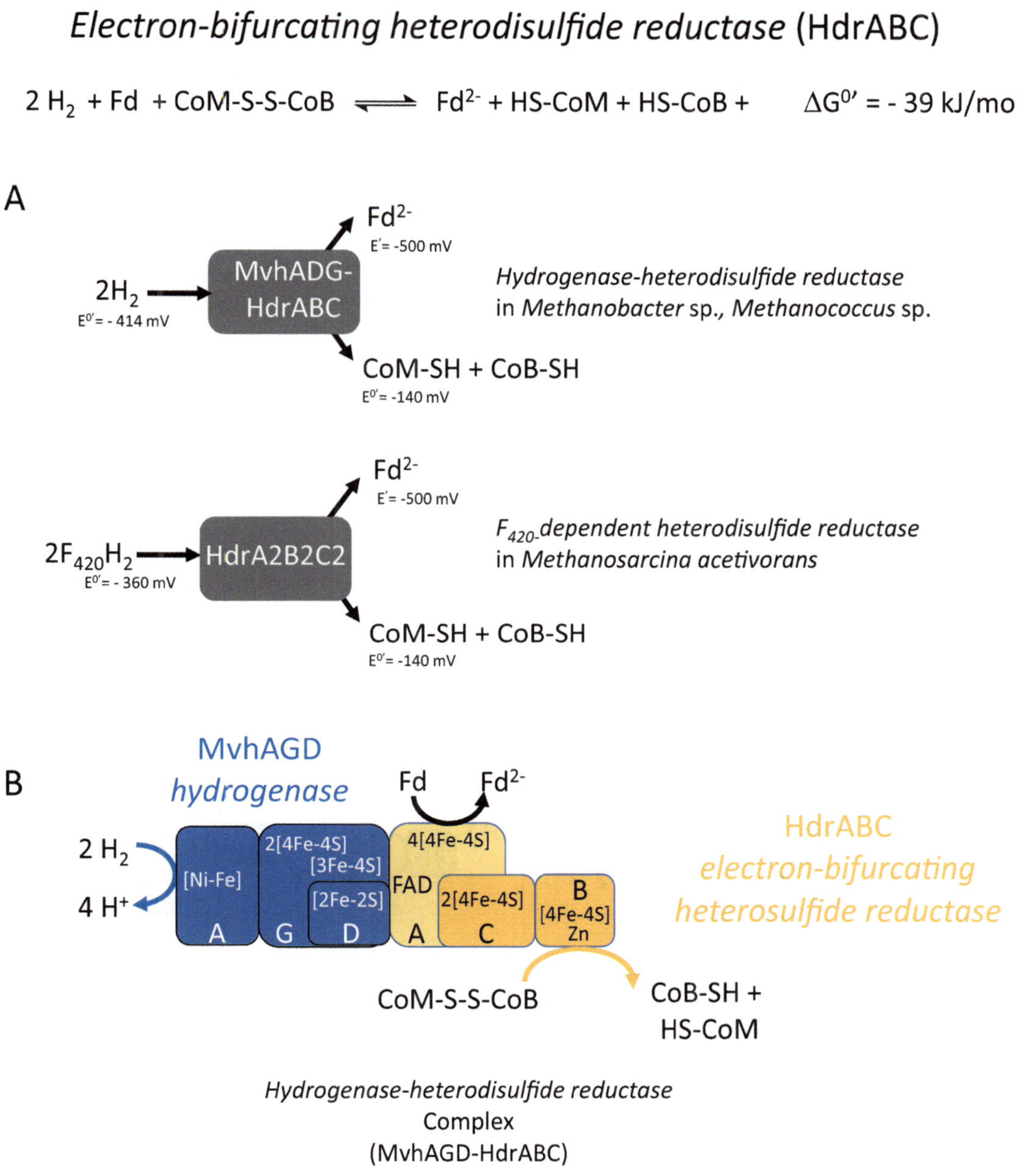

Fig. 10.9 Flavin(FAD)-based *electron bifurcating heterodisulfide reductase* HdrABC in complex with MvhAGD *hydrogenase* in methanogens without cytochromes (here *Methanothermobacter marburgensis*). (**a**) Schematic representation of electron bifurcation using either 2 H_2 or 2 $F_{420}H_2$ as electron donor, driven by CoM-S-S-CoB reduction. (**b**) Molecular composition of the MvhAGD-HdrABC complex. Active HdrABC is a heterodimer; for simplification, only one heteromonomer is shown. See text for details

Methanococcus maripaludis (Fig. 10.9a). HdrABC can also form a complex with $F_{420}H_2$ *dehydrogenase* or with Fdh *formate dehydrogenase* (in *Methanococcus maripaludis*), replacing MvhA, when using formate instead of H_2 as environmental electron donor (not shown). HdrABC is highly oxygen-sensitive because of the involvement of Fe-S clusters and Fd.

10.2.3.4.2 Membrane-Associated *Methanophenazine-Dependent Heterodisulfide Reductase*

HdrDE is a membrane-bound *methanophenazine-dependent heterodisulfide reductase* and functions in respiratory energy conservation with formation of $\Delta\mu H^+$ as intermediate coupled to CoM-S-S-CoB reduction (Fig. 10.10). Physiological electron donor is reduced methanophenazine (MPH_2). In contrast to HdrABC, CoM-S-S-CoB reduction with MPH_2 is coupled to a net translocation of 2 electrogenic H^+. The overall reaction is close to thermodynamic equilibrium, and most of the available energy is conserved. When H_2 is electron donor, as in *Methanosarcina barkeri*, H_2 is oxidized at the outside of the cytoplasmic membrane by Vht *hydrogenase* and electrons are transferred via membrane-integrated methanophenazine to HdrDE (Fig. 10.10). Two additional electrogenic H^+ are translocated in this H_2 oxidation with methanophenazine by HdrDE (Fig. 10.10a). Methanophenazine ($E^{0\prime} = -165$ mV) functions as electron carrier in analogy to menaquinone (Fig. 10.7c). Thus, a total of 4 electrogenic H^+ ($\Delta\mu H^+$), equivalent to 1 ATP, are conserved during CoM-S-S-CoB reduction with H_2 by HdrDE. HdrDE can also be present in a membrane complex with an *$F_{420}H_2$ dehydrogenase* (FpoA-O complex) that couples $F_{420}H_2$ oxidation to CoM-S-S-CoB reduction. Since *$F_{420}H_2$ dehydrogenase* is located at the inner surface of the

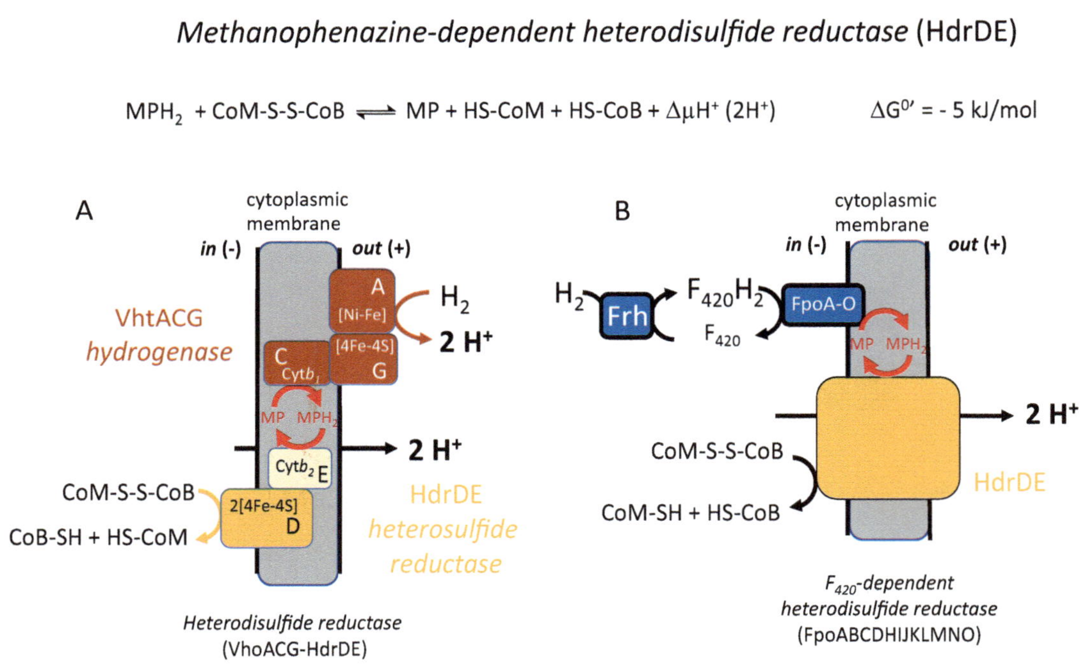

Fig. 10.10 *Membrane-bound methanophenazine-dependent heterodisulfide reductase* (HdrDE) in methanogens with cytochromes (here *Methanosarcina barkeri*). HdrDE is the terminal reductase of a cytochrome- and methanophenazine-based electron transport chain, reminiscent of quinone-based electron transport phosphorylation. Methanophenazine at $E^{0\prime} = -165$ mV is a menaquinone equivalent. HdrBC and HdrD are homologues; HdrA and HdrE are structurally and functionally unrelated. (**a**) In H_2-dependent *methanophenazine-dependent heterodisulfide reductase*, H_2 is oxidized at the outer surface of the cytoplasmic membrane. A total of four electrogenic protons are net translocated. (**b**) In $F_{420}H_2$-dependent *methanophenazine-dependent heterodisulfide reductase*, because of the more positive redox potential of F_{420}/$F_{420}H_2$ ($E^{0\prime} = -360$ mV) and topology of the *$F_{420}H_2$ dehydrogenase* subunit, only 2 electrogenic protons are translocated. Abbreviations: *Frh F_{420}-reducing hydrogenase*, *Fpo $F_{420}H_2$ dehydrogenase*, *MP* methanophenazine, *MPH_2* dihydromethanophenazine. See text for details

cytoplasmic membrane and the redox potential of $F_{420}/F_{420}H_2$ ($E^{0\prime} = -360$ mV) is more positive than H_2, no vectorial electron transport is possible and only 2 electrogenic H^+ are translocated per methanophenazineH_2 oxidized. HdrDE does not contain a flavin cofactor and is found in methanogens containing methanophenazine and *b*-type cytochrome.

While HdrBC and HdrD are homologues and catalyze reduction of the CoM-S-S-CoB heterodisulfide, HdrA and the cytochrome-containing HdrE subunits are unrelated and confer the distinctive energetic coupling (Figs. 10.9 and 10.10).

HdrABC Genes in Genomes

Based on amino acid sequence similarity HdrABC type *heterodisulfide reductases* are frequently reported to be encoded in genomes of non-methanogenic prokaryotes. Their metabolic role is often unclear. Despite the sequence similarity, making an assignment that these genes encode a functional *"heterodisulfide reductase"* as first discovered for methanogens and as experimentally shown to catalyze the electron-bifurcating reduction of CoM-S-S-CoB is problematic. In the absence of experimental verification, in particular, in the absence of evidence for a cellular metabolism of CoM-S-S-CoB, CoM-SH, or CoB-SH, such annotation may be inaccurate. The annotated genes may encode for an enzyme that might catalyze an electron bifurcation, just not based on CoM-S-S-CoB reduction.

10.3 CO_2-Reducing Microbes Involving the Reductive Wood–Ljungdahl Pathway

The Wood-Ljungdahl Pathway (WLP, Fig. 10.11) is the catabolic CO_2-reductive pathway of acetyl-CoA formation in anaerobic acetogenic bacteria and Lokiarchaeota, and, in reverse, in catabolic acetyl-CoA oxidation in some syntrophs and in most strictly anaerobic, acetate-oxidizing sulfate-reducing bacteria. This pathway uniquely couples catabolic C_1 and C_2 metabolism and functions also in autotrophic CO_2 fixation as an anabolic pathway in methanogens, acetogens, and in some sulfate-reducing bacteria. The catabolic Wood-Ljungdahl Pathway is often, but not always, directly coupled to energy conservation. CO_2 is ubiquitously present in microbial ecosystems as bicarbonate (pH dependent, ~0.1–1 mM in surface water; 50–100 mM HCO_3^- in anoxic lake sediments, 200–400 mM HCO_3^- in anaerobic digestors). Anaerobic microbes that use the Wood-Ljungdahl pathway in their reductive catabolism are referred to as acetogens, sometimes as homoacetogens where the term 'homoacetogen' originated from on early observation of fermentation of 1 glucose to 3 acetate (and 3 H^+) as sole and exclusive end product as found in *Moorella thermoacetica* (see below).

10.3.1 Energetics and Reversibility of the Wood-Ljungdahl Pathway

The complete Wood-Ljungdahl pathway consists of idiosyncratic CO_2↔formate conversion, the formyl↔methyl module, idiosyncratic methyl transformation, and formation of acetyl-CoA via CO-DH/ACS. When functioning in catabolism, acetyl-CoA is coupled to ATP and acetate formation via the acylate↔acyl-CoA module (Fig. 10.2). Figure 10.11 summarizes the pathway steps including the four oxidoreductases plus variations in enzymes, electron carriers, and redox potentials of the C1 compounds involved.

As this pathway overview shows, no net ATP formation is directly coupled to the catabolic Wood-Ljungdahl pathway: one ATP is formed from acetyl-CoA and one is consumed in the *formyl-H_4F synthase* reaction (Fig. 10.11). The first reaction is CO_2 reduction to formate. When H_2 is electron donor, the reaction is close to

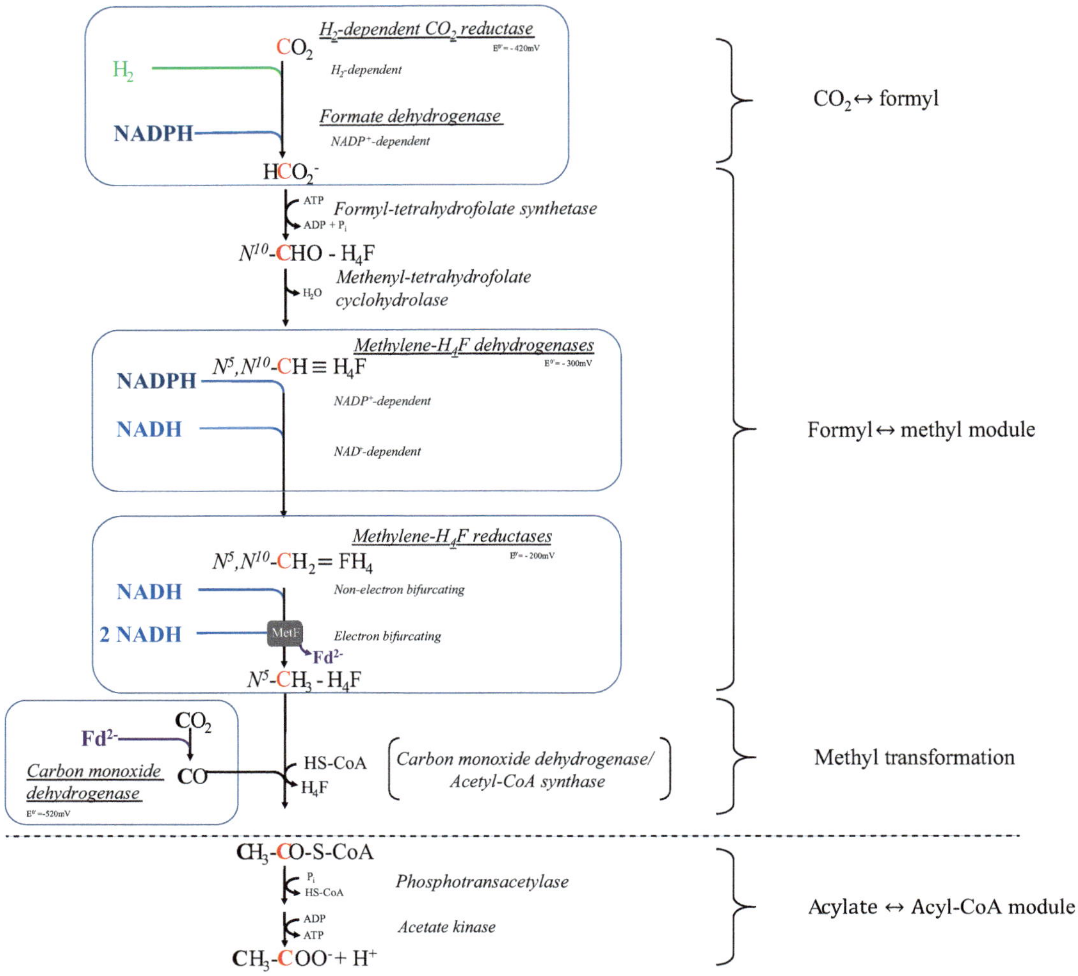

Fig. 10.11 CO_2-reducing Wood-Ljungdahl pathway of acetyl-CoA formation. When functioning as a catabolic pathway, as shown here, acetyl-CoA formed is converted to acetate and ATP via *phosphotransacetylase* and *acetate kinase*. The four oxidoreductase reactions are boxed, and diversity in electron carriers as well as possible enzymes is indicated for each oxidoreductase reaction. Use of alternative enzymes depends on the metabolic and ecophysiological context of the microbe. These molecular differences determine electron balancing as well as whether to pursue a rate or yield strategy. The two CO_2 carbon atoms are colored red (forming the methyl group carbon) and black (forming the carboxyl group carbon), respectively. Light blue is electron flow from NADH, dark blue from NADPH, magenta from Fd^{2-}. Redox potentials for C_1 compounds, free or coenzyme-bound, are given below the enzyme name. If H_2 is electron donor, this catabolic pathway catalyzes the overall reaction: $4\ H_2 + 2\ CO_2 \rightarrow$ acetate + H^+ + $2\ H_2O$, and $\Delta G^{0'}$ = -95 kJ/mol. [*Carbon monoxide dehydrogenase/Acetyl-CoA synthase*] indicates the enzyme complex. See text for details

equilibrium, while NADPH as electron donor renders this reaction at +22 kJ/mol thermodynamically uphill under standard conditions. The standard redox potential of methenyl-H_4F/methylene-H_4F is at −300 mV close to the *in situ* redox potential of NAD^+/NADH. Therefore, this reaction is reversible and not associated with energy conservation. *Carbon monoxide dehydrogenase* requires Fd^{2-} and catalyzes CO_2 reduction to CO close to equilibrium. The non-redox reactions catalyzed by *formyl-H_4F synthetase*, *methenyl-H_4F cyclohydrolase*, and *phosphotransacetylase/acetate kinase* are close to thermodynamic equilibrium (Fig. 10.11), and, therefore, reversible.

The reductive WLP is an electron sink pathway to re-oxidize reduced electron carriers from diverse oxidative pathways: H_2 or NADPH is oxidized in CO_2 reduction to formate; NADH or NADPH is oxidized in methenyl-H_4F reduction to methylene-H_4F; NADH is oxidized in methylene-H_4F reduction to methyl-H_4F; and Fd^{2+} in CO_2 reduction to CO (Fig. 10.11). Fd^{2-}, NADH, or H_2 are the reduced electron carriers that are formed in oxidative pathways; e.g., Fd^{2-} from *pyruvate:ferredoxin oxidoreductase*, *electron bifurcating methylene-H₄F reductase*, or *electron bifurcating hydrogenase*; NADH from *dehydrogenases*, including *alcohol dehydrogenases*; and H_2 from H_2-producing primary fermenters or from Ech. If energy conservation is coupled to the Wood-Ljungdahl Pathway, it is via an electron flux through Rnf or Ech and transport-coupled phosphorylation. Energy conservation via Rnf and Ech requires as substrate Fd^{2-} and metabolic steps to re-oxidize NADH (for Rnf). NADH is oxidized only by *methylene-H_4F reductase* and in some acetogens by *methylene-H_4F dehydrogenase*. *Carbon monoxide dehydrogenase* competes with Rnf/Ech for reduced ferredoxin (Figs. 10.11 and 10.12). Thus, the characteristic metabolic feature of an acetogenic microbe reflects its solution for how to couple a stoichiometrically-balanced electron flux from NADH, Fd^{2-}, H_2 to the four reduction reactions. In other words, the electron carriers recovered in oxidative pathway(s), including the products of Rnf and Ech, need to (1) yield Fd^{2-} for energy conservation and (2) be matched to the electron carrier-specificity of the available oxidoreductases (Fig 10.12). H_2 released from Ech can either escape a cell or be oxidized by *H_2-dependent CO_2 reductase* or *electron bifurcating hydrogenase* (see below).

Key for mediating the required switching of electron carriers are flavin-based electron bifurcating *transhydrogenase* (Nfn), *electron-bifurcating hydrogenase* (Hyd), and *methylene-H_4F reductase* (Fig. 10.12). Non-electron bifurcating *methylene-H_4F reductase,* as in *Acetobacterium woodii,* consumes 1 NADH and is exergonic by −22 kJ/mol (Fig. 10.11). Dissipation of this energy as heat results in a lower pathway n_{ATP} (see below) and enables (1) a faster rate of acetate formation and (2) utilization of substrate(s) at low concentration. On the other hand, an *electron bifurcating methylene-H_4F reductase* as in *Clostridium autoethanogenum* or *Moorella thermoacetica* (see below), consumes 2 NADH and is coupled to Fd^{2-} formation, a reaction that is close to equilibrium at +8 kJ/mol (Fig. 10.12d). The involvement of an *electron bifurcating methylene-H_4F reductase* enables microbes to pursue a yield strategy (see below, *Acetobacterium woodii* vs *Moorella thermoacetica*).

If energy conservation is coupled to electron flux between oxidative and reductive pathways, acetogens use either **Rnf or Ech** for energy conservation (Fig. 10.12). NADH formed by **Rnf** is consumed either by *methylene-H_4F reductase* or, in some acetogens, by *methylene-H_4F dehydrogenase.* **Ech**-based energy conservation functions in conjunction with *electron bifurcating hydrogenase* (Hyd) (Fig. 10.12d, f), representing an overall Fd^{2-} oxidation coupled to NADH formation similar as in Rnf. However in case of Ech, H_2 is an intermediate and the pathway n_{ATP} using this combination is higher: 4 electrogenic H^+/Na^+ are translocated. Thus, acetogens, such as *Moorella thermoacetica,* conserving energy via Ech/Hyd are yield strategists while Rnf-employing acetogens, such as *Acetobacterium woodii,* are more rate strategists. One note of caution for this metabolic reasoning is, however, that a common pool and common redox potential of cellular ferredoxin is assumed, which may not be the case.

When the only source of electrons is NADH, *electron bifurcating methylene-H_4F reductase* is the only Fd^{2-}-generating enzyme. Under those metabolic conditions, no energy is conserved by Rnf or Ech from Fd^{2-} (i.e., n_{ATP} = 0 of the reductive pathway), because Fd^{2-} is consumed by *carbon monoxide hydrogenase.*

For the Wood-Ljungdahl pathway operating in the oxidative direction, two cases need to be considered: (1) acetate oxidation with sulfate as electron acceptor as in many complete-oxidizing sulfate reducers such as *Desulfotomaculum*

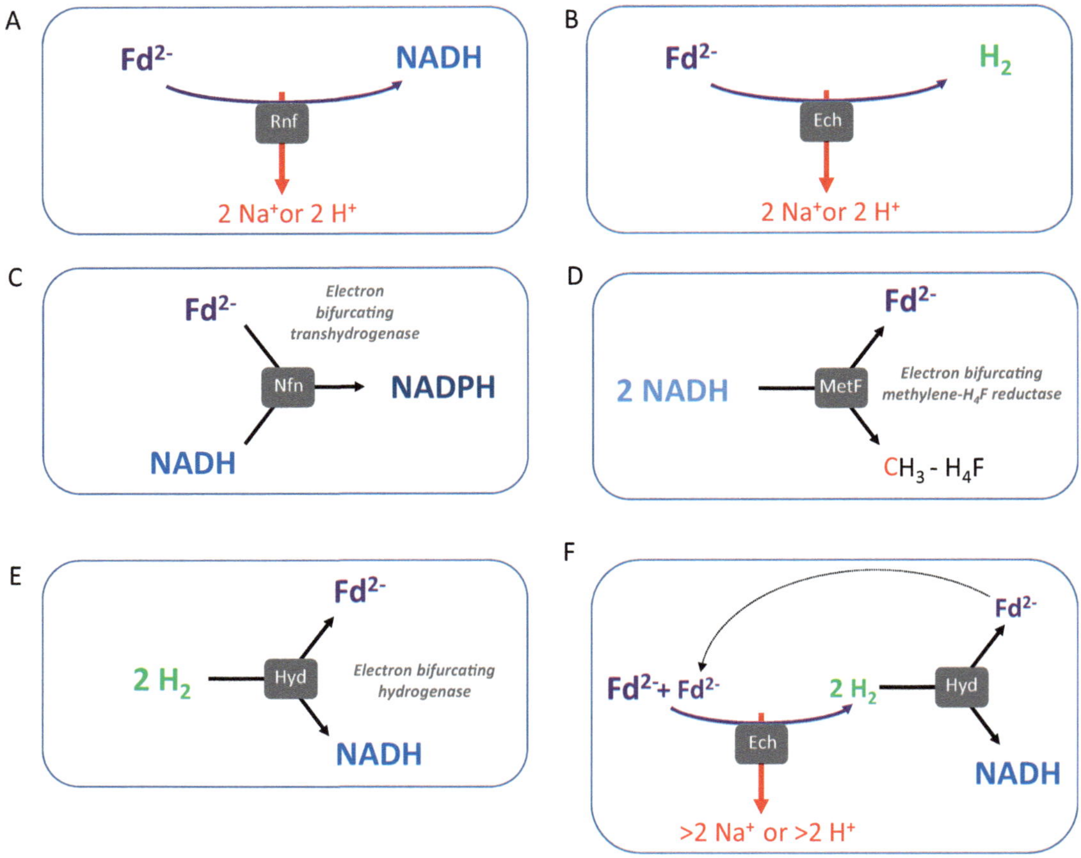

Fig. 10.12 Summary of enzymes in acetogens involved in energy conservation and in switching of electron carriers from oxidative pathways (NADH, Fd^{2-}, H_2) to electron carriers of the reductive Wood–Ljungdahl pathway (H_2, NADH, NADPH, Fd^{2-}). (**a**) Membrane-bound energy-conserving Rnf, coupling Fd^{2-} oxidation and NAD^+ reduction to translocation of $2H^+$ or $2Na^+$ (see Sect. 6.4.4.1.1). (**b**) Membrane-bound energy-conserving Ech, coupling Fd^{2-} oxidation and H^+ reduction to translocation of $2H^+$ or $2Na^+$ (see Sect. 6.4.4.1.1). (**c**) *Electron bifurcating transhydrogenase* Nfn, reversibly interconverting Fd^{2-} plus NADH to 2 NADPH (see Sect. 6.5.1.1). (**d**) *Electron-bifurcating methylene-tetrahydrofolate reductase* MetF, converting 2 NADH to Fd^{2-} plus methyl-H_4F (see Sect. 6.5.1.1). (**e**) *Electron bifurcating hydrogenase* Hyd, reversibly interconverting 2 H_2 to Fd^{2-} plus NADH (see Sect. 6.5. 1.1). (**f**) Metabolic reaction sequence of energy-conservation by Ech and *electron bifurcating hydrogenase* Hyd; the redox energy difference between Fd^{2-} and NADH is conserved as electrogenic H^+ or Na^+. The E′ of the ferredoxin involved is unknown and can be lower than −500 mV. Note, the cycling of one H_2 between both enzymes. See text for details

acetoxidans, *Desulfobacterium autotrophicum*, or *Desulfovibrio baarsii*, and (2) syntrophic acetate oxidation with H^+/CO_2 as electron acceptor as in *Clostridium ultunense* or *Thermoacetogenium phaeum*. Cofactors involved in *methylene-H_4F reductase* are different in the WLP when operating in the oxidative direction. Because of the relatively positive redox potential of $E^{0\prime}$ = −200 mV, oxidation of CH_3–H_4F requires an electron acceptor with a more positive redox potential in order to drive a metabolically significant flux in the oxidative direction. Oxidation of CH_3–H_4F with NAD^+ is too endergonic and would result in a significant kinetic bottleneck despite potentially high *methylene-H_4F reductase* concentration [E_T]. When sulfate is electron acceptor, $E^{0\prime}$ of the HSO_3^-/H_2S couple of −115 mV in the catabolic sulfate reduction pathway is more positive than the CH_2=H_4F/CH_3–H_4F couple. Thus, a coupling of CH_3–H_4F

oxidation via menaquinone ($E^{0'}$ = −70 mV), where menaquinol is subsequently oxidized by *bisulfite reductase*, provides an exergonic flux for acetate oxidation. In syntrophic acetate oxidizers, the electron acceptors H^+ or CO_2 (Chap. 9) are reduced to H_2 or formate at E' of ~−300 mV. Also here, oxidation of CH_3–H_4F is a thermodynamic/kinetic bottleneck. Syntrophic acetate-oxidizing microbes solve this problem by employing a menaquinone-dependent *methylene-H_4F reductase* where menaquinol is subsequently oxidized via reverse electron transport to activate the 'CH_3–H_4F electrons'.

Acetogens and Syngas Fermentation

Acetogens, in particular *Clostridium autoethanogenum*, *C. ljungdahlii*, *C. ragsdalei*, and *Thermoanaerobacter kivui*, are emerging as a new class of biotechnologically important microbes for production of commodity chemicals, such as ethanol or acetate and derivatives, from CO_2. Gaseous CO_2, H_2, and CO are catabolic substrates for numerous acetogens. Depending on the strains with their idiosyncratic fermentation patterns and pH-dependent bioenergetics, CO_2 (or CO) can be converted in catabolic processes at high selectivity and high titer to acetate and ethanol. Such microbial use of CO_2 is different from anabolic CO_2 fixation because of the higher rates and much higher selectivity. CO_2, H_2, and CO are released as industrial 'waste gases' from steel manufacturing, petroleum refining, or as synthesis gas from gasification of biomass, agricultural waste, coal, and steam reforming of methane ($CH_4 + H_2O \rightarrow CO + 3\ H_2$). Thus, in conjunction with CO_2 capture and storage, these combined chemical and microbial processes convert, for example, chemically heterogeneous (plant) biomass into homogeneous carbon-neutral organic compounds, including commodity chemicals and fuels. This intriguing metabolic capacity is due in large part to metabolism of CO, which, at $E^{0'}$ = −520 mV, is a stronger reductant than H_2 ($E^{0'}$ = −420 mV) and can, therefore, be explored by Fd-dependent energy conservation at stoichiometries that H_2 cannot.

10.3.2 Metabolism of Acetogenic Microbes

Three well-studied acetogens, *Acetobacterium woodii, Moorella thermoacetica,* and *Clostridium autoethanogenum,* will be discussed here in detail. All these microbes use the reductive Wood-Ljungdahl pathway in their reductive catabolism. *A. woodii* does not contain quinones or cytochromes and conserves energy via Na^+-dependent Rnf, where 2 Na^+ are translocated per Fd^{2-} oxidized to NADH. The quinone, cytochrome-containing *M. thermoacetica* conserves energy by H^+ translocation via Ech. Ethanol-producing *C. ljungdahlii* does not contain cytochromes but a H^+-dependent Rnf.

10.3.2.1 *Acetobacterium woodii*

The core metabolic features in *A. woodii*, when using the Wood–Ljungdahl pathway in the reductive metabolism, are that.

(i) cytoplasmic *methylene-H_4F dehydrogenase* and *methylene-H_4F reductase* use NADH as electron donor, and
(ii) H_2 is electron donor for CO_2 reduction by *hydrogen-dependent CO_2 reductase* (Fig. 10.9, 10.10, 10.11, 10.12 and 10.13).

The key modules for energy conservation are a Na^+-dependent Rnf in conjunction with a con-/bifurcating Fd^{2-}-, NADH-*dependent hydrogenase* (Hyd).

When *A. woodii* disproportionates 2 lactate$^-$ into 3 acetate$^-$ +1 H^+ (Fig. 10.13a), lactate oxidation is driven by a an *electron confurcating lactate dehydrogenase* which couples lactate oxidation to NAD^+ reduction, driven by Fd^{2-} oxidation: 2 lactate are oxidized to 2 pyruvate with concomitant consumption of 2 Fd^{2-} to form 4 NADH. A

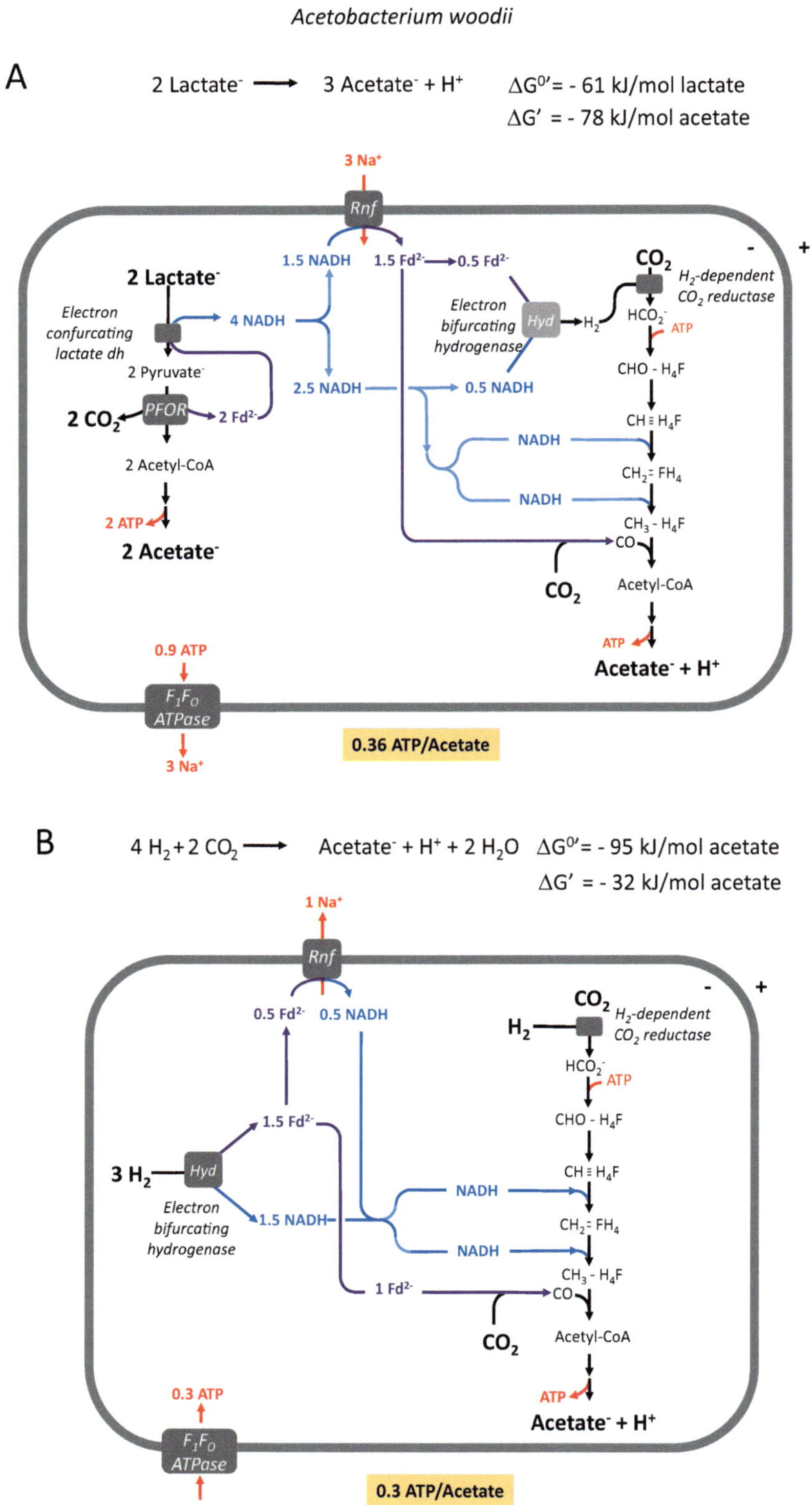

Fig. 10.13 Predicted metabolic map of *A. woodii* metabolizing (**a**) lactate or (**b**) H_2 + CO_2 via the WLP. Magenta lines indicate the fate of reduced ferredoxin, Fd^{2-}, and blue lines indicate the fate of NADH. ΔG′ was calculated with 1 mM acetate or lactate, 100 Pa H_2, and 0.1 bar CO_2. Note that the WLP is non-energy

pyruvate:ferredoxin oxidoreductase catalyzes oxidative decarboxylation of 2 pyruvate to 2 acetyl-CoA and 2 Fd^{2-} + 2 CO_2. In net balance, these two CO_2 build the carbon skeleton of the third acetate formed in this metabolism, although they can exchange with environmental CO_2/bicarbonate as the actual substrate. Two NADH are consumed in each, the *methylene-H$_4$F dehydrogenase* and *methylene-H$_4$F reductase* reaction, and 1.5 NADH of the two remaining NADH are activated by Rnf to 1.5 Fd^{2-} under consumption of 3 electrogenic Na^+. This is equivalent to consuming ~0.9 ATP. One Fd^{2-} is oxidized by *CO dehydrogenase*, while the remaining 0.5 Fd^{2-} are confurcated by Hyd with 0.5 NADH to 1 H_2. Thus, a total of 0.36 ATP is formed per acetate in this lactate fermentation by *A. woodii*. Because lactate oxidation consumes reduced ferredoxin, and one Fd^{2-} is consumed in the *CO dehydrogenase* reaction, no Fd^{2-} is left for energy conservation and the Wood–Ljungdahl pathway is not coupled here to net energy conservation. At 1 mM lactate and acetate concentration, the thermodynamic efficiency is 40% assuming $\Delta G'_{ATP} = -50$ kJ/mol.

When *A. woodii* grows with H_2 as sole electron donor (Fig. 10.13b), Hyd *hydrogenase* acts in the reverse direction as a *bifurcase* and bifurcates 3 H_2 to 1.5 Fd^{2-} and 1.5 NADH. One Fd^{2-} is used in the *CO dehydrogenase* reaction, and 0.5 Fd^{2-} is oxidized by Rnf to form 0.5 NADH coupled to translocation of 1 Na^+. The two NADH collectively formed are consumed in the *methylene-H$_4$F dehydrogenase* and *methylene-H$_4$F reductase* reactions, respectively. Since the Wood–Ljungdahl pathway is net ATP neutral and no substrate level phosphorylation is possible with H_2 as sole electron donor, the only net energy conserved per acetate is one electrogenic Na^+ via a Na^+-dependent *ATPase,* equivalent to ~0.3 ATP/acetate. This corresponds to a thermodynamic efficiency of ~40% assuming $\Delta G'_{ATP} = -50$ kJ/mol at relevant *in situ* concentrations of 1 mM acetate, 100 Pa H_2, and 0.1 bar CO_2. In this H_2/CO_2 metabolism, the Wood–Ljungdahl pathway is energy-conserving. The low n_{ATP} of this catabolism enables *A. woodii* to use H_2 concentration as low as 0.1% for growth (see Fig. 3.4).

The above examples of *A. woodii* show how the Wood–Ljungdahl pathway as a reductive pathway can be coupled or uncoupled to net energy conservation. When it is energy conserving, the pathway functions as an electron sink to initiate in this microbe an Rnf-based electron flux. Also to note is how the enzymes that define the microbe's energy metabolism (Rnf and Hyd) function in either the forward or reverse reaction depending on the overall metabolic context.

10.3.2.2 *Moorella thermoacetica*

Moorella thermoacetica is a thermophilic, endospore-forming, cytochrome- and menaquinone-containing Firmicute. The energy-conserving core of this acetogen consists of

(i) H^+- based Ech,
(ii) *bi/confurcating transhydrogenase* Nfn,
(iii) *NADP$^+$-reducing hydrogenase,*
(iv) *bifurcating NADH-dependent methylene-H$_4$F reductase.*
(v) *bifurcating* Hyd *hydrogenase* (Fig. 10.14, see also Figs. 10.12 and 10.13).

The significant differences in the Wood–Ljungdahl pathway reactions relative to *A. woodii* are:

- *M. thermoacetica* contains an NADPH-specific *formate dehydrogenase,*
- a *NADPH-dependent methylene-H$_4$F dehydrogenase*, and
- a postulated *bifurcating NADH-dependent methylene-H$_4$F reductase* (in complex with

Fig. 10.13 (continued) conserving and functions as electron sink in (**a**) but is associated with energy conservation in (**b**). Abbreviations: *PFOR pyruvate:ferredoxin oxidoreductase, Hyd electron-bifurcating hydrogenase, lactate dh lactate dehydrogenase*. See text for details

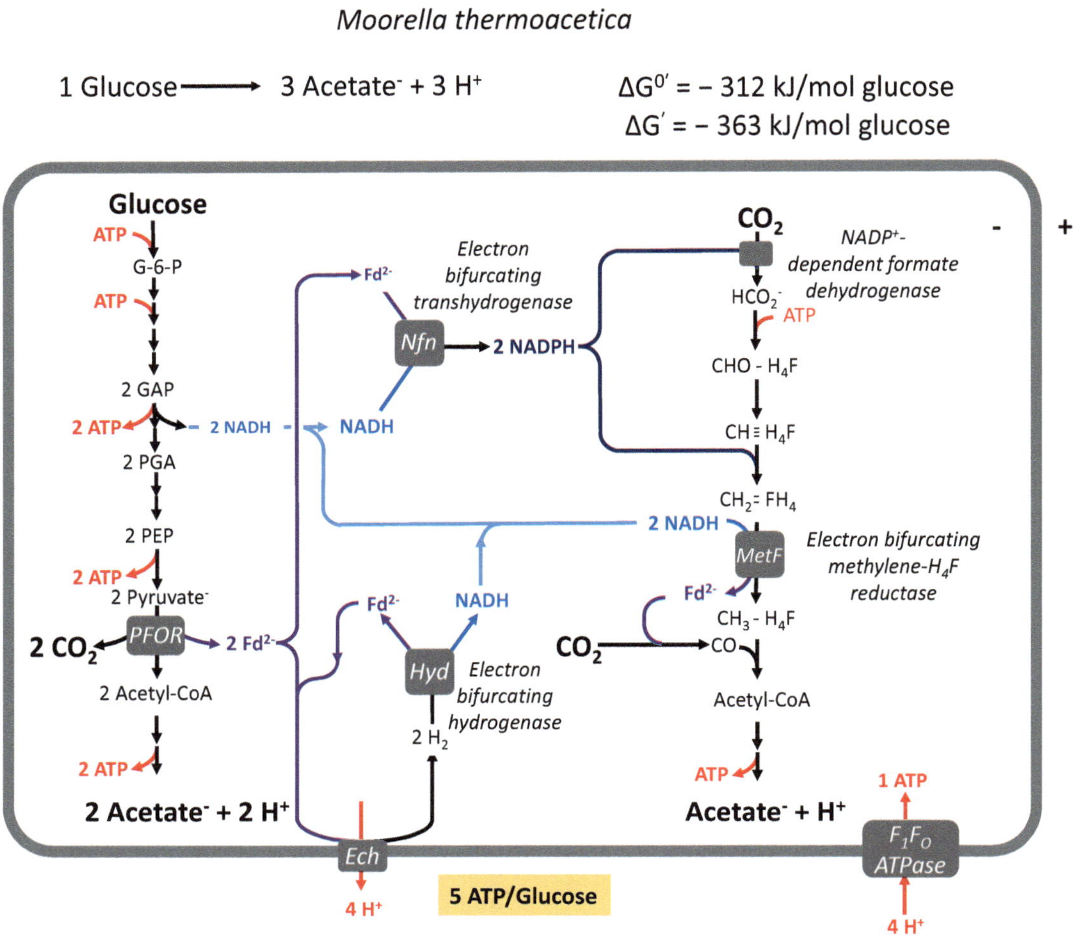

Fig. 10.14 Predicted metabolic map of glucose fermentation to 3 acetate by *Morella thermoacetica*. Two acetate are derived from glucose oxidation via 2 pyruvate, and the third acetate is derived via reduction of 2 CO_2. $\Delta G'$ was calculated with 0.1 mM acetate and 1 mM glucose. Abbreviations *PFOR pyruvate:ferredoxin oxidoreductase*, *Nfn electron bifurcating transhydrogenase, Hyd electron bifurcating hydrogenase*. See text for details

an HdrABC) coupling the exergonic reduction of methylene-H_4F with an Fd-like protein or an Ech-like complex for energy conservation by proton translocation (Fig. 10.14).

As other Clostridia, *M. thermoacetica* ferments glucose to acetate (glucose → 3 acetate$^-$ + 3 H^+). Two acetate are derived from glucose, while the third acetate is derived from 2 CO_2. Glucose is oxidized via the Embden–Meyerhof pathway, the GAP → Pyruvate module, *pyruvate:ferredoxin oxidoreductase*, and the acylate ↔ acyl-CoA module to 2 acetate, 2 CO_2, and 4 ATP net. The electrons accrued in this oxidative pathway are 2 NADH and 2 Fd^{2-}. One NADH and one Fd^{2-} are confurcated by Nfn to 2 NADPH, which are consumed in the *formate dehydrogenase* and *methylene-H_4F dehydrogenase* reaction. The second Fd^{2-} as well as an additional Fd^{2-} are oxidized by Ech to 2 H_2, and four electrogenic H^+ are translocated in energy conservation. This reasoning assumes only a single cellular pool of ferredoxin with the same redox potential, which may not be the case. The two H_2 are subsequently bifurcated by *bifurcating hydrogenase* (Hyd) to form that additional Fd^{2-} and NADH (Figs. 10.12f and 10.14). This reaction sequence of Ech and NADH-forming *bifurcating hydrogenase* effectively represents a hydrogen loop, oxidizing net 1 Fd^{2-} to 1 NADH and conserving the difference in redox energy as electrogenic protons. The NADH formed plus the second NADH of the *GAP dehydrogenase* reaction are consumed by bifurcating *methylene-H₄F reductase* (MetF) to CH_3-H_4F and Fd^{2-}; the latter is electron donor in the reduction of CO_2 to CO. Collectively, a total of 5 ATP are conserved per glucose (2 ATP net in the GAP → Pyruvate module, 2 ATP in the acylate ↔ acyl-CoA module, and 1 ATP by Ech). Also here, one acetate is derived from the two CO_2 that were released in the *pyruvate:ferredoxin oxidoreductase* reaction. These two CO_2 are not necessarily the actual CO_2 molecules that constitute the third acetate; they can be derived from external CO_2/bicarbonate. The thermodynamic efficiency of this pathway at 0.1 mM acetate and 1 mM glucose of ~70% is high and indicates that the *in situ* concentration of acetate need to be very low in order to kinetically drive such high n_{ATP} metabolism.

When *M. thermoacetica* grows using H_2 and CO_2 (Fig. 10.15), 4 H_2 are oxidized by an *NADP⁺-dependent hydrogenase* to 4 NADPH, which are bifurcated by *transhydrogenase* Nfn to 2 Fd^{2-} and 2 NADH. These 2 NADH are consumed in the putative *bifurcating methylene-H₄F reductase* to reduce methylene-H_4F to methyl-H_4F and a reduced Fd-like electron carrier that is most likely oxidized by Ech *hydrogenase* to H_2 while 2 electrogenic protons are translocated. The hydrogen formed is oxidized by *NADP⁺-reducing hydrogenase*. One of the Fd^{2-} formed in the Nfn bifurcation is used by *carbon monoxide dehydrogenase/acetyl-CoA synthase* complex to reduce CO_2 to carbon monoxide, and the second Fd^{2-} is oxidized by Ech to H_2 under energy conservation via proton translocation; H_2 is oxidized by NADPH-forming *hydrogenase*. These two NADPH are consumed by *formate dehydrogenase* and *methylene-H₄F dehydrogenase*. Thus, all oxidation and reduction reactions are balanced, and net energy conservation is 2 × 2 electrogenic H^+, which equates to 1 ATP. A thermodynamic efficiency of this pathway at *in situ* concentrations of 0.1 mM acetate, 100 Pa H_2, and 0.1 bar CO_2 of close to 100% indicates "death in equilibrium." Thus, such yield metabolism of *Moorella thermoacetica* can only operate at significantly higher pH_2, which is the reason why *A. woodii*, at a thermodynamic efficiency of ~40% for the same overall catabolism, can utilize lower H_2 concentration than *M. thermoacetica* (see Fig. 3.9 and discussion for hydrogenotrophic methanogens).

When *M. thermoacetica* grows using carbon monoxide as sole electron donor (4 CO + 2 H_2O → 1 $Acetate^-$ + 1 H^+ + 2 CO_2) (Fig. 10.16), the same pathway elements as above function in conjunction with an additional cytoplasmic, *Fd-reducing carbon monoxide dehydrogenase*. Three CO are oxidized by this *carbon monoxide dehydrogenase* to 3 Fd^{2-} and 3 CO_2. In conjunction with 3 additional Fd^{2-}, a total of 6 Fd^{2-} are oxidized by Ech to form 6 H_2 and translocate 12 H^+. These 6 H_2 are bifurcated to 3 NADH and 3 Fd^{2-}; the latter providing the 3 additional Fd^{2-} for Ech. This reaction sequence again functions as an internal H_2 loop. Of the 3 NADH formed, 2 NADH are oxidized by a putative bifurcating *methylene-H₄F reductase* (MetF) to form CH_3-H_4F and Fd^{2-}; one Fd^{2-} is confurcated with NADH to 2 NADPH. Both NADPH are consumed in the Wood–Ljungdahl core by *formate dehydrogenase* and *methylene-H₄F dehydrogenase*, and 3 ATP are predicted to be formed per acetate.

10.3.2.3 *Clostridium autoethanogenum*

C. autoethanogenum is an acetogenic Firmicute that is industrially relevant for syngas (H_2/CO/CO_2) fermentation, in particular, to ethanol. This capacity is due to the microbe's preferential metabolism of CO, over H_2/CO_2, to ethanol and acetate, butane-2,3-diol, and lactate. Optimal pH for growth is around pH5, which is an important parameter for its distinguishing capacity of ethanol formation (see below). Phylogenetically, *C. autoethanogenum* is closely related and almost identical in genome sequence and

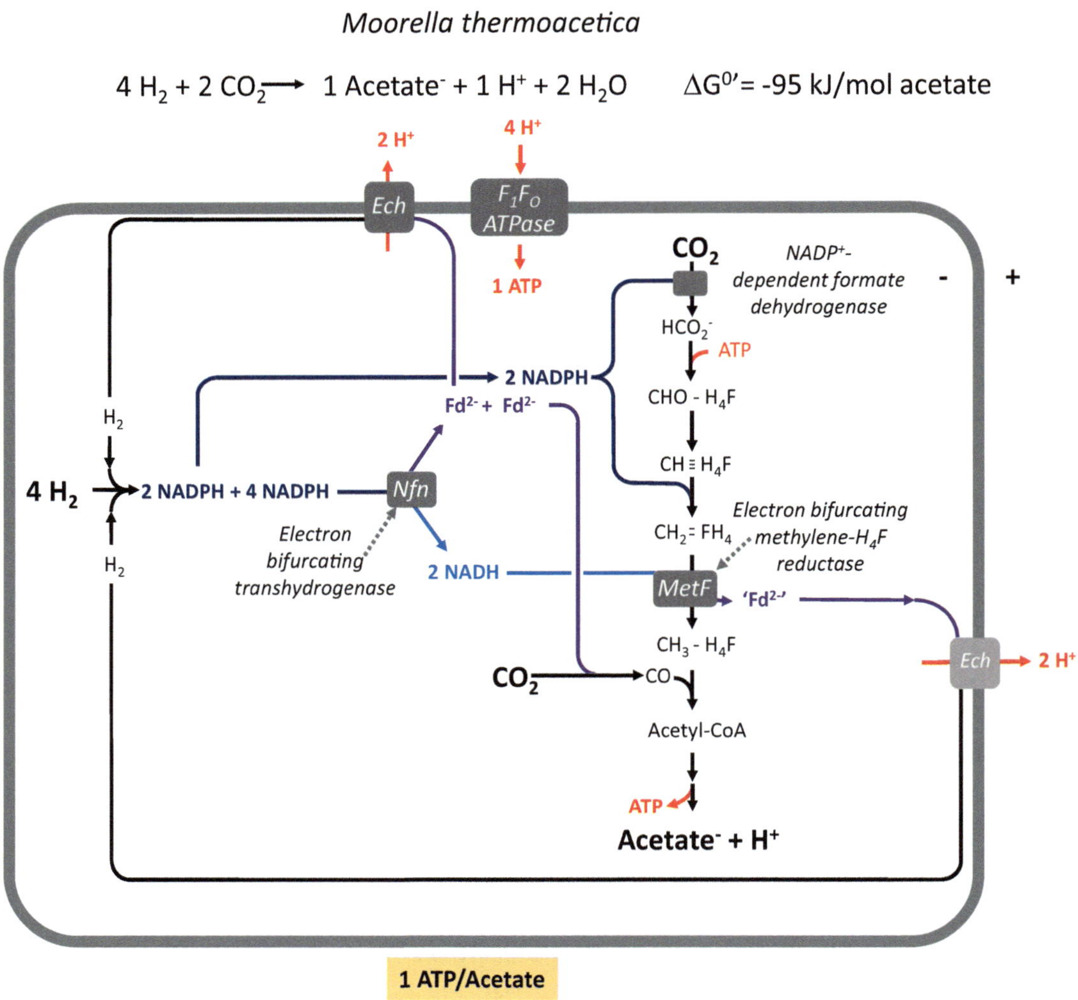

Fig. 10.15 Predicted metabolic map of *Morella thermoacetica* growing with $4H_2 + 2CO_2$. Magenta arrows indicate flux of electrons from Fd^{2-}, light blue from NADH, and dark blue from NADPH. In this metabolism, Nfn operates in the bifurcating direction compared to glucose fermentation (Fig. 10.14). ΔG′ was calculated with 0.1 mM acetate, 100 Pa H_2 and 0.1 bar CO_2. *PFOR pyruvate:ferredoxin oxidoreductase*, *Nfn electron bifurcating transhydrogenase*. See text for details

organization to *C. ljungdahlii* and *C. ragsdalei*. *C. autoethanogenum* lacks cytochromes and menaquinone; however, its energy metabolism is H^+- and not Na^+-dependent, and involves Rnf for transport-coupled phosphorylation. Further key catabolic features are

(i) *NADP⁺-dependent bifurcating hydrogenase* (Hyt),
(ii) *ferredoxin-dependent bifurcating transhydrogenase* (Nfn),
(iii) *NADH-dependent bifurcating methylene-H_4F reductase* (MetFV), and
(iv) *acetaldehyde:ferredoxin oxidoreductase.*

Under standard conditions, fermentation of $H_2 + CO_2$ to ethanol ($6\ H_2 + 2\ CO_2 \rightarrow$ Ethanol $+3\ H_2O$, $\Delta G^{0\prime} = -105$ kJ/mol ethanol, Fig. 10.15) is more exergonic than H_2/CO_2 fermentation to acetate ($\Delta G^{0\prime} = -95$ kJ/mol acetate). However, because of the different H_2

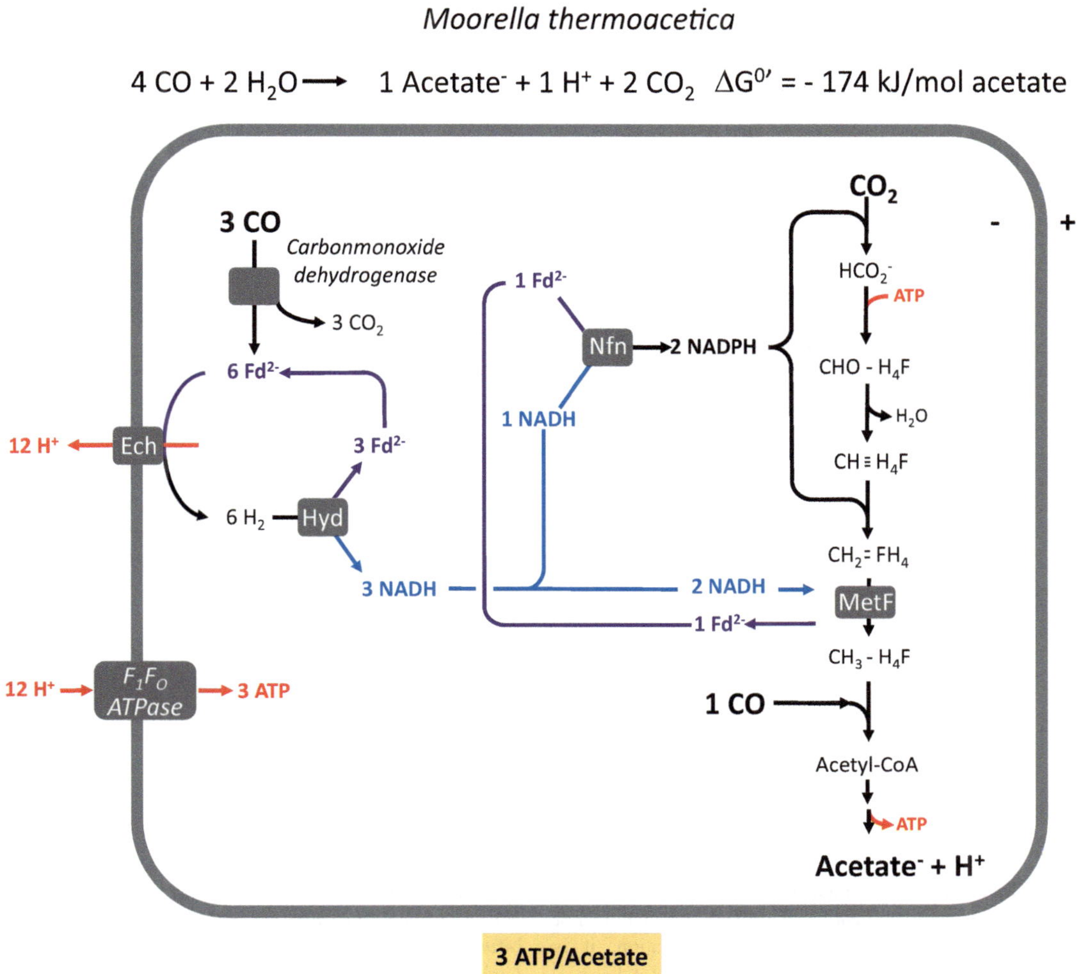

Fig. 10.16 Predicted metabolic map of *Morella thermoacetica* growing with carbon monoxide. Abbreviations: *Nfn electron bifurcating transhydrogenase, Hyd electron bifurcating hydrogenase*. See text for details

stoichiometry, acetogenesis is more exergonic at low *in situ* pH_2 (~10 Pa) because of the entropic effect: 6 H_2 in ethanol-forming vs 4 H_2 in acetate-forming fermentation.

In H_2/CO_2 fermentation to ethanol by *C. autoethanogenum* (Fig. 10.17), 1 H_2 is consumed by *H_2-dependent formate dehydrogenase/CO_2 reductase*. Five H_2 are bifurcated by Hyt *hydrogenase* to 2.5 Fd^{2-} and 2.5 NADPH. 1 NADPH is consumed by *methylene-H_4F dehydrogenase* and the remaining 1.5 NADPH is bifurcated by Nfn *transhydrogenase* to 0.75 Fd^{2-} and 0.75 NADH. Of the 2.5 Fd^{2-} formed in Hyt bifurcation, 2.25 Fd^{2-} are oxidized by Rnf to 2.25 NADH, where 1 NADH is consumed by *ethanol dehydrogenase* and the remaining 1.25 NADH plus the 0.75 NADH from the Nfn *transhydrogenase* reaction are the two NADH, which are bifurcated by *bifurcating methylene-H_4F reductase* to methyl-H_4F and Fd^{2-}. Fd^{2-} is consumed in the *CO dehydrogenase* reaction of the *CO dehydrogenase/acetyl-CoA synthase* reaction. 0.25 Fd^{2-} from Hyt bifurcation and 0.75 Fd^{2-} from Nfn bifurcation add up collectively to 1 Fd^{2-}, which is consumed in the *acetaldehyde: ferredoxin oxidoreductase*. Also here and as in glucose fermentation by *M. thermoacetica*, the Wood–Ljungdahl pathway functions as electron

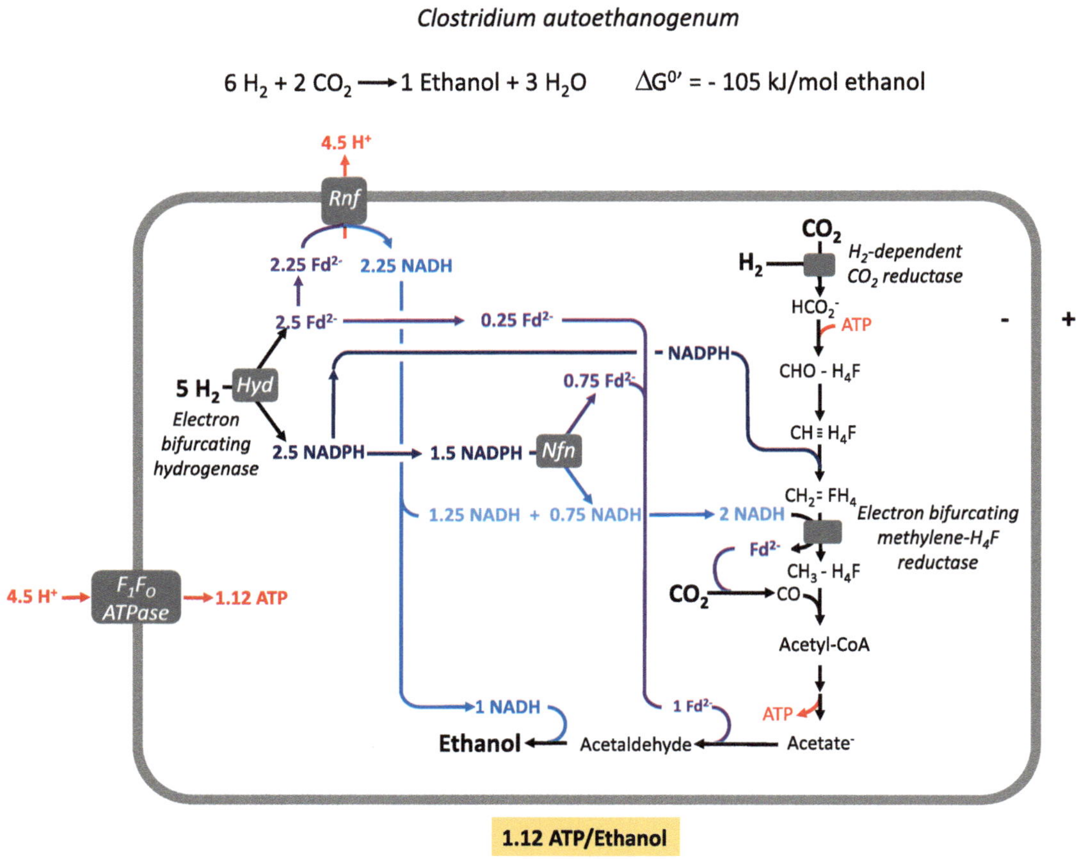

Fig. 10.17 Predicted metabolic map of *Clostridium autoethanogenum* fermenting H_2 + CO_2 to ethanol. For simplification and focus on ethanol formation, it is assumed that only ethanol is formed. A net of 1.12 ATP is formed per ethanol. Only the *acetaldehyde:ferredoxin oxidoreductase* pathway version is shown. Abbreviations: *Nfn electron bifurcating transhydrogenase*, electron-bifurcating butyryl-CoA dh, *electron-bifurcating butyryl-CoA dehydrogenase*. See text for details

sink in the reductive pathway, and the combination of bifurcating Hyt and Nfn in conjunction with Rnf mediates an Fd-dependent electron flux through energy-conserving Rnf. Collectively 1.12 ATP is predicted to be formed per ethanol.

An alternative pathway of acetyl-CoA reduction to ethanol can be used by *C. autoethanogenum* in ethanol-forming fermentation when **acetyl-CoA is not converted to acetate** but reduced by an *NADPH-dependent acetaldehyde dehydrogenase* (CoA-acetylating) followed by the *NAD⁺-dependent ethanol dehydrogenase*. This re-routing of electrons requires that only 0.5 NADPH are bifurcated in Nfn rather than 1.5 NADPH, so that one NADH can be consumed in acetyl-CoA reduction to acetaldehyde. Consequently, 2.75 rather than 2.25 Fd^{2-} are oxidized by Rnf when in conjunction with the acetyl-CoA reduction pathway. Under those conditions, only 0.5 ATP/ethanol net is formed, because acetyl-CoA is unavailable for conversion to 1 ATP and 1 acetate. Thus, for the latter pathway of acetyl-CoA reduction, the overall n_{ATP} is smaller than for direct Fd^{2-}-dependent acetate reduction. Because of that, the less efficient pathway can function at a pH_2 as low as 1%, whereas the direct, Fd^{2-}-dependent acetate reduction operates only at pH_2 of >10%. This example is reminiscent of the different architectures of the respiratory chain of *E. coli* as a function of O_2 availability (Sect. 3.5.2, Fig. 3.8) where a difference in n_{ATP} enables competitive metabolism at low concentrations in a trade-off with efficient pathways.

10.3.2.4 Metabolic Flexibility of Microbes with the Wood–Ljungdahl Pathway

The metabolic strategies of *A. woodii* and *M. thermoacetica* (Figs. 10.13, 10.14, 10.15, and 10.16) revealed that in heterotrophic growth, the reductive Wood–Ljungdahl pathway functions as an effective metabolic "escape" from the stoichiometric constraints of classical primary fermenters (Sect. 9.2). These constraints arise from the requirement that the electron acceptor is derived from an intermdiate of an oxidative fermentation pathway (e.g., acetyl-CoA), often at the trade-off of a high pathway n_{ATP}. The WLP as internal electron sink uses CO_2 rather than a (energy-rich) metabolic intermediate as electron acceptor and provides, thus, an escape from a classical fermentative lifestyle. As exemplified in *A. woodii* vs *M. thermoacetica*, the pathway can be coupled or uncoupled to energy conservation and enable to pursue either a rate or yield strategy. A second consequence of the internal electron sink via the WLP is related to metabolic interactions with other microbes, in particular with those in interspecies H_2 or formate transfer, which is a consequence of metabolic speciation (Chap. 5). Consider, for example, glucose fermentation to 2 acetate, 2 CO_2, and 4 H_2 ($\Delta G' = -341$ kJ/glucose) by *Ruminococcus albus* at low pH (Fig. 9.14) and compare to glucose fermentation to 3 acetate ($\Delta G' = -345$ kJ/glucose) by *M. thermoacetica* at low pH_2. Disregarding the different growth temperatures, in both cases glucose is fermented to acetate as the sole organic end product. The two metabolisms differ in how electrons from the oxidative pathway are managed. Metabolism of *R. albus* at low pH_2 requires the presence of a hydrogenotrophic methanogen to maintain a low pH_2 at ~10 Pa; therefore, the overall pathway of glucose conversion to methane and CO_2 is speciated to metabolically different microbes. The "all-in-one" metabolism of *M. thermoacetica* with an active WLP (Fig. 10.14) makes efficient glucose metabolism independent of a H_2-consuming partner microbe! While *R. albus* can metabolize glucose independently of a H_2-consuming microbe to acetate, ethanol, and H_2, less energy is available in this fermentation ($\Delta G' = -259$ kJ/mol glucose, Fig. 9.14) than in the combined fermentative plus WLP of *M. thermoacetica* ($\Delta G' = -345$ kJ/mol glucose, Fig. 10.15).

So, under which environmental conditions is a metabolic speciation (e.g., *R. albus* plus hydrogenotrophic methanogen, or *A. woodii* plus hydrogenotrophic methanogen) advantageous over a strategy where the two types of metabolism can operate in a single microbe and is biologically controlled (e.g., *M. thermoacetica*)? In other words, when would it be advantageous for a microbe to release H_2 and CO_2 to the environment rather than use an internal electron sink in an overall longer pathway? The *in situ* ΔG of both microbes is approximately the same ($\Delta G' = -341$ kJ/mol glucose for *R. albus*, Fig. 9.14 vs $\Delta G' = -345$ kJ/mol glucose, *M. thermoacetica*, Fig. 10.14); however, the two types of metabolism differ in the thermodynamic efficiency (~55% vs ~70%) and pathway length. As previously discussed, in high substrate and high flux environments it is advantageous to partition and speciate an overall pathway into separate microbes containing each short, low n_{ATP} pathway in order to maintain a high rate of substrate oxidation and ATP synthesis (J_{ATP}) (see also the Julian Adams experiment, Chaps. 1–5). On the other hand, low fluxes result in low cell density and, consequently, in low abundance of potential partner microbes at a distance sufficient for a kinetically significant interspecies H_2 or formate transfer (Fig. 9.32). Moreover, low fluxes select for microbes with more efficient metabolism, where the high thermodynamic efficiency is off-set by an anabolic "pull" to mediate an overall growth metabolism at a finite rate.

There is great metabolic flexibility in acetogenic microbes for substrates in oxidative pathways, including hexoses, pentoses, alcohols, (ethanol, propanol, butanol), diols (1,2-propanediol, ethylene glycol, 2,3-butanediol), carboxylic acid (lactate, pyruvate, citrate, glyoxylate, glycolate), as well as C_1 compounds (CO, methanol, formate). Many acetogens have three modes of growth and can switch between:

(i) that of a primary/syntrophic fermenter producing H_2 in the absence of an expressed WLP,
(ii) that of a H_2-consuming acetogen (with active WLP), and
(iii) that of a combined primary/secondary fermenter AND H_2-consuming acetogen (active WLP) with intracellular H_2 flux.

For example, *A. woodii* can syntrophically ferment glucose to 2 acetate, where hydrogen is released and consumed by a methanogen rather than funneled into the *A. woodii*-specific Wood–Ljungdahl pathway. Glucose metabolism by *M. thermoacetica* reveals the intracellular transfer of electrons via an internal H_2 loop (Ech + Hyd) to the reductive WLP. Thus, many acetogens are generalists and can control whether to operate short-speciated ((i), (ii)) or long non-fractionated pathways (iii). This facultative lifestyle of an acetogen expressing either short single or combined pathways via H_2-scavenging WLP is advantageous in both a high flux or a low flux, low organic matter, and low cell density environment, respectively (Fig. 9.32). This can explain the relatively high abundance of versatile acetogens (e.g., *Chloroflexi* or Lokiarchaeota) in marine sediments at low cell densities. CO_2/bicarbonate is ubiquitous in anoxic environments. The above-outlined options for operating a complete or incomplete metabolic pathway are similar to that found in propionic acid bacteria that form propionate from either lactate (under high flux, competitive conditions) or glucose (under low flux conditions) (Figs. 9.9–9.11). Both metabolic types can switch between a fast-fermentative growth and a slower and more efficient, partner-independent growth.

10.4 CO_2-Reducing Microbes Involving the Methanogenesis Pathway

The **Methanogenesis Pathway** is the catabolic pathway of methane formation by the strictly anaerobic group of methanogenic archaea and a pathway for anaerobic archaeal oxidation of methane with a sulfate-reducing partner microbe. Some non-catabolic mini-methane formation is observed in some acetogenic microbes and related to an active Wood–Ljungdahl pathway. Catabolism of known methanogens is highly specialized and speciated, and restricted to the methanogenic pathway as the sole pathway for energy conservation. Well-known substrates of methanogens are typically H_2/CO_2, C_1 compounds (CO_2, methylamines, betaine, dimethylsulfide, etc.), and acetate.

With a few exceptions, catabolism in methanogens has in common the involvement of

- *methyl-Coenzyme M reductase* (Mcr),
- *Na^+-translocating methyl-H_4MPT:coenzyme M methyltransferase*, and.
- either a cytoplasmic *electron bifurcating heterodisulfide reductase* (HdrABC) or a membrane-bound, energy-conserving *methanophenazine-dependent heterodisulfide reductase* (HdrDE).

10.4.1 Energetics and Reversibility of the Methanogenesis Pathway

For the following sections, the methanogenesis pathway is discussed in the direction of methane formation. Two types of methanogenic archaea are distinguished: methanogens containing methanophenazine/cytochrome and those without. When using H_2/CO_2 as catabolic substrate, both types of methanogens use a common core of C_1 transforming reactions consisting of an idiosyncratic $CO_2 \leftrightarrow$ formyl activation, the formyl $\leftrightarrow$ methyl module, and the idiosyncratic methyl conversion, summarized in Fig. 10.18 (see also Fig. 10.2). Both groups differ in the reduction of the heterodisulfide using either an *electron bifurcating heterodisulfide reductase* (HdrABC, in methanogens without methanophenazine/cytochrome) or a *methanophenazine-dependent heterodisulfide reductase* (HdrDE) (Figs. 10.9 and 10.10).

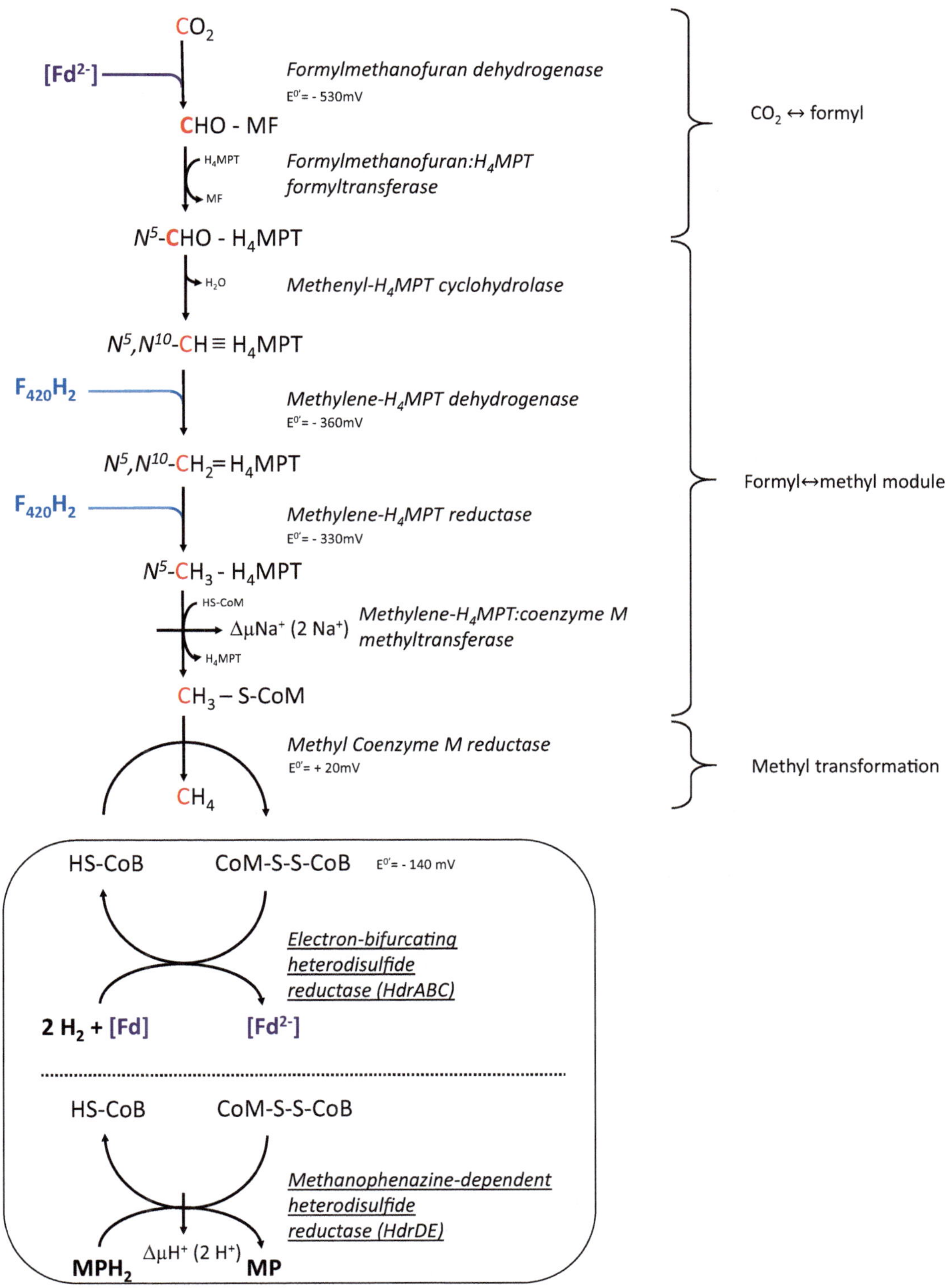

Fig. 10.18 The CO_2-reducing Methanogenesis Pathway in methanogenic archaea. Light blue is electron flow from $F_{420}H_2$, and magenta from Fd^{2-}. Redox potentials for C_1 compounds, free or coenzyme-bound, are given below the enzyme name. If H_2 is electron donor, F_{420} is reduced by *F_{420}-reducing hydrogenases* (not shown), and this catabolic pathway catalyzes the overall reaction: $4\ H_2 + CO_2 \rightarrow CH_4 + 2\ H_2O$, and $\Delta G^{0\prime} = -131$ kJ/mol at 25 °C.

With the exception of *methyl-H_4MPT: coenzyme M methyltransferase*, *methyl-coenzyme M reductase*, and *electron bifurcating heterodisulfide reductase*, all pathway reactions operate close to thermodynamic equilibrium, and are, therefore, reversible. In all known hydrogenotrophic methanogens, exergonic *methyl-H_4MPT: coenzyme M methyltransferase* reaction ($\Delta G^{0\prime} = -30$ kJ/mol) is coupled to energy conservation by translocating 2 electrogenic Na^+. The type of *heterodisulfide reductase* employed distinguishes the two types of methanogens: *electron bifurcating heterodisulfide reductase* in hydrogenotrophic methanogens is exergonic by -39 kJ/mol, while the *methanophenazine-dependent heterodisulfide reductase*, coupling translocation of 2 H^+, is close to thermodynamic equilibrium ($\Delta G^{0\prime} = -5$ kJ/mol).

These two different modes of energetic coupling by HdrABC and HdrDE *heterodisulfide reductases* have ecophysiological consequences and are another example of molecular pathway solutions resulting in either a yield or rate phenotype. In the electron-bifurcating mode of HdrABC, only -15 kJ ($\Delta G^{0\prime}$) of the overall -54 kJ/mol (H_2 + CoM-S-S-CoB $\rightarrow$ CoM-SH + CoB-SH) are used for endergonic reduction of Fd. In the electron transport phosphorylation mode associated with HdrDE, -54 kJ/mol are consumed for releasing four electrogenic H^+ ($\Delta\mu H^+$). Consequently, methanogens without methanophenazine/cytochrome using HdrABC have a lower n_{ATP} and a lower molar growth yield but grow faster and at lower pH_2 than methanogens using cytochrome-dependent HdrDE (see also Sect. 3.5.4, Fig. 3.8). This capacity to access the catabolic substrate H_2 at low concentration at a trade-off of a more efficient energetic coupling is similar to (i) acetate activation by *acetate kinase/phosphotransacetylase* module or by *acetate synthetase* (see Sect. 8.3.1, and see below), (ii) acetate formation from H_2/CO_2 in *A. woodii* vs *M. thermoacetica* (Figs. 10.13 and 10.15), and (iii) expression in *E. coli* of an uncoupled *cytochrome bd oxidase* at low pO_2 relative to a coupled *cytochrome bo* oxidase at higher pO_2 (see Chap. 6).

Methanogens without methanophenazine/cytochrome are more abundant in environments at temperatures of 60 °C and higher, while methanogens containing methanophenazine/cytochromes are predominant at lower temperatures (e.g., 4 °C). This difference in ecophysiology is also related to the relationship between thermodynamics and n_{ATP} in both pathway variants. As the $\Delta G^{0\prime}$ of hydrogenotrophic methanogenesis decreases from -131 kJ/mol at 25 °C to - 100 kJ/mol at 100 °C, the unused energy of the HdrABC reaction (Fig. 10.9) provides these methanogens with a thermodynamic "buffer" to thermodynamically outcompete methanophenazine/cytochrome-containing methanogens: The large n_{ATP} of the pathway of the latter group drives these microbes at higher temperature into "death in thermodynamic equilibrium" at substrate concentrations, which still enable a metabolic flux in methanogens without methanophenazine/cytochrome (see below).

A second ecophysiological implication of the difference between HdrABC and HdrDE is access to diverse catabolic resources. HdrABC couples

Fig. 10.18 (continued) Boxed are the two types of *heterodisulfide reductases*: HdrABC is electron bifurcating and 2 H_2 are consumed for heterodisulfide reduction plus Fd^{2-} formation; no energy conservation is associated. HdrDE couples exergonic reduction of heterodisulfide CoM-S-S-CoB with dihydromethanophenazine oxidation to translocation of 2 H^+. *Electron bifurcating heterodisulfide reductase* is found in methanogens without methanophenazine-cytochrome; *methanophenazine-dependent heterodisulfide reductase* is present in methanogens with methanophenazine-cytochrome. Abbreviations: *MF* methanofuran, *H_4MPT* tetrahydromethanopterin, *CoM-SH* coenzyme M (2-mercaptoethanesulfonate), *HS-CoB-SH* (7-mercaptoheptanoylthreonine phosphate), *CoM-S-S-CoB* heterodisulfide, *MP* methanophenazine, and *MPH_2* dihydromethanophenazine. See text for details

stoichiometrically the first and the last step of hydrogenotrophic methanogenesis from CO_2 (see below). These methanogens without cytochrome are therefore locked into a methanogenesis pathway obligatorily starting with CO_2 and ending with CH_4. This is not the case in HdrDE-containing methanogens, where the source of a C_1 carbon can also be a methyl group from methylamines, betaine, dimethylsulfides, as well as the methyl group (C2 carbon) of acetate (HdrDE requires only CH_3-CoM and low redox potential electrons). Such metabolic alternative to methanogenesis from CO_2 is advantageous in high flux environments by enabling HdrDE-containing methanogens to escape competition for H_2 with HdrABC-containing methanogens via accessing different substrates such as acetate or other methyl-group containing substrates.

Most hydrogenotrophic methanogens, such as *Methanothermobacter marburgensis*, *Methanococcus maripaludis*, or *Methanospirilum hungatei* contain cytoplasmic, *electron bifurcating heterodisulfide reductase*, and do not contain cytochromes or methanophenazine. Only members of the Methanosarcinales, such as *Methanosarcina barkeri* or *Methanosarcina acetivorans,* contain *methanophenazine/cytochrome-dependent heterodisulfide reductases.* Some *Methanosarcina sp.* contain both types of *heterodisulfide reductases*, however.

The common exergonic reaction in the methanogenesis pathway is the *methyl-coenzyme M reductase* reaction ($\Delta G^{0\prime} = -30$ kJ/mol) (Fig. 10.18). This thermodynamic and kinetic "trap" of the *methyl-coenzyme M reductase* reaction has important ecophysiological implications for anaerobic methane oxidation (see below).

10.4.2 Metabolism of Methanogenic Archaea

10.4.2.1 *Methanothermobacter* and *Methanococcus Species*

Figure 10.19 shows the methanogenic pathway in hydrogenotrophic methanogens without methanophenazine/cytochrome as in *Methanothermobacter marburgensis* or *Methanococcus maripaludis.* In this example, H_2 is electron donor but when using formate as electron donor, a similar pathway, where formate substitutes for H_2 and *formate dehydrogenase* substitute for *hydrogenase*, operates in *M. maripaludis.* Two H_2 are consumed in the *bifurcating heterodisulfide reductase* (HydABC) and two H_2 in the *F_{420}-reducing hydrogenase* reaction. Via HdrABC *heterodisulfide reductase*, the first reaction of methanogenesis, catalyzed by *formylmethanofuran dehydrogenase*, and the last step, catalyzed by *methyl-coenzyme-M reductase*, are stoichiometrically coupled: 2 H_2 are bifurcated to reduce CoM-S-S-CoB (to CoM-SH and CoB-SH) and Fd to Fd^{2-} or bound [Fd^{2-}]. Fd^{2-} is electron donor to *formylmethanofuran dehydrogenase* to reductively activate CO_2 to formylmethanofuran. *Electron bifurcating heterodisulfide reductase* may form a super complex with *formylmethanofuran dehydrogenase* where electron transfer involves an enzyme-bound polyferredoxin rather than a free, diffusable Fd^{2-}. After transfer of the formyl group from formylmethanofuran to H_4MPT, *methenyl-H_4MPT cyclohydrolase* abstracts H_2O from methenyl-H_4MPT, which is subsequently reduced by *F_{420}-dependent methylene-H_4MTP dehydrogenase* to methylene-H_4MTP and further by *$F_{420}H_2$-dependent methylene-H_4MTP reductase* to methyl-H_4MTP. Membrane-bound, corrinoid-containing, Na^+-translocating *methyl-H_4MPT:CoM methyltransferase* transfers the methyl group from methyl-H_4MTP to CoM-SH to form CH_3-CoM. This exergonic reaction is coupled to translocation of 2 Na^+ per methyl group transferred ($\Delta\mu Na^+$). CH_3-CoM is reduced by *methyl-CoM reductase* with CoB-SH as electron donor, and CH_4 and heterodisulfide CoB-S-S-CoM are released. CoM-S-S-CoB is reduced by electron-bifurcating HdrABC under simultaneous formation of Fd^{2-} from 2 H_2.

The energy of the electrochemical Na^+ gradient is converted into ATP by *Na^+-dependent A_1A_oATPase*, resulting in net formation of 0.5 ATP per mol CH_4.

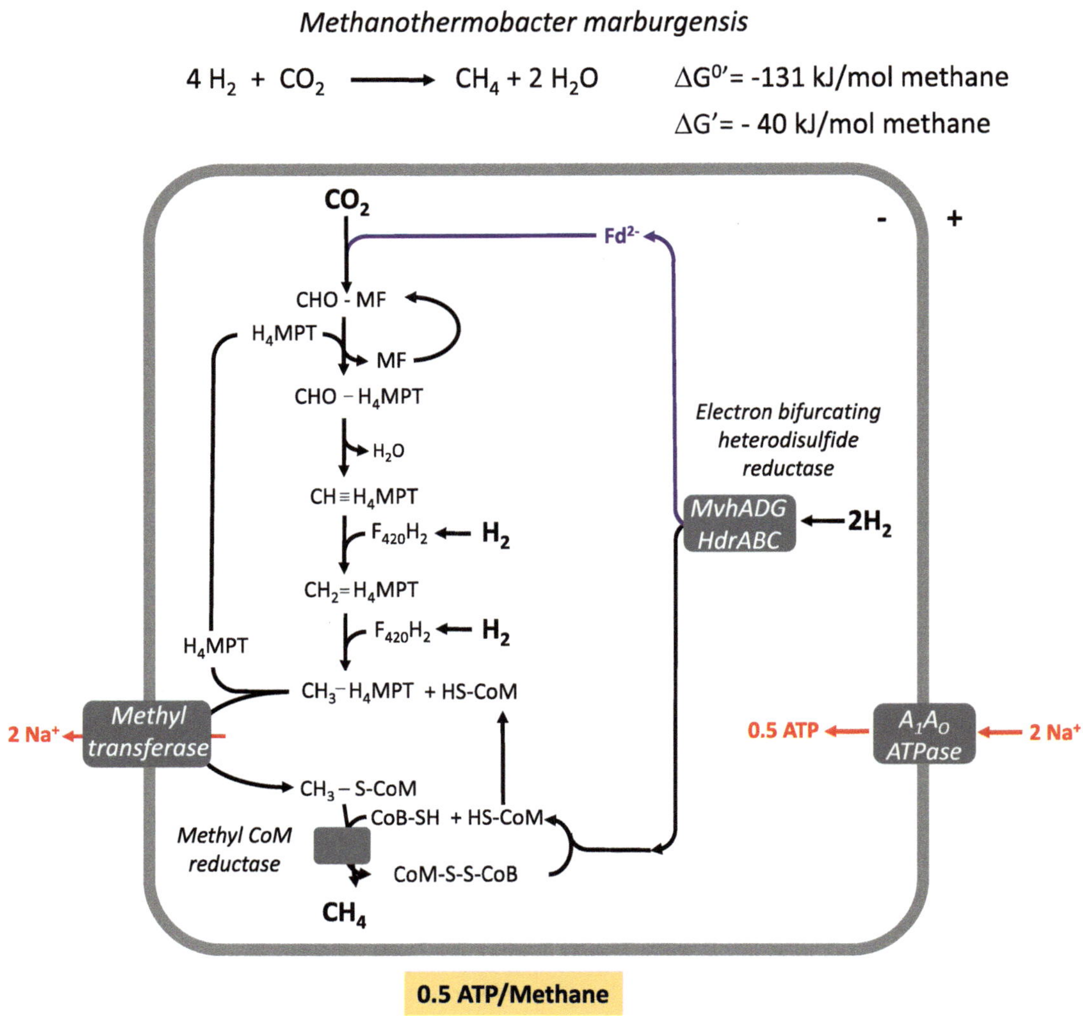

Fig. 10.19 Proposed metabolic map of *Methanothermobacter marburgensis* oxidizing hydrogen with CO_2 to methane. The sole energy-conserving step in this catabolism is the *CH_3-H_4MPT:HS-CoM methyltransferase* reaction, and 0.5 ATP is conserved per CH_4 formed. *Methanococcus maripaludis*, when using formate rather than H_2 as electron donor, has a similar pathway, except a *F_{420}-reducing formate dehydrogenase* and an FdhHdrABC rather than a H_2-consuming VhuHdrABC complex are involved. ΔG′ was calculated for 10 Pa H_2. See text for details

10.4.2.2 *Methanosarcina* and *Methanosaeta* Species

Methanosarcinales, such as *Methanosarcina barkeri*, *Methanosarcina acetivorans*, or *Methanothrix soehngenii*, are methanogens containing methanophenazine/cytochrome and have an energy-conserving, membrane-bound *heterodisulfide reductase* (HdrDE) (Fig. 10.10). The methanogenesis pathway from H_2/CO_2 in *M. barkeri* differs from that of *M. marburgensis* in that the first and last reaction are uncoupled (Fig. 10.20). Because *heterodisulfide reductase* HdrDE is non-bifurcating and does not form Fd^{2-}, *M. barkeri* activates H_2-electrons for CO_2 reduction to formyl via a reverse electron transport by Ech under consumption of 2 electrogenic H^+. Fd^{2-} is the electron donor for CO_2 reduction by *formylmethanofuran dehydrogenase*. Two H_2

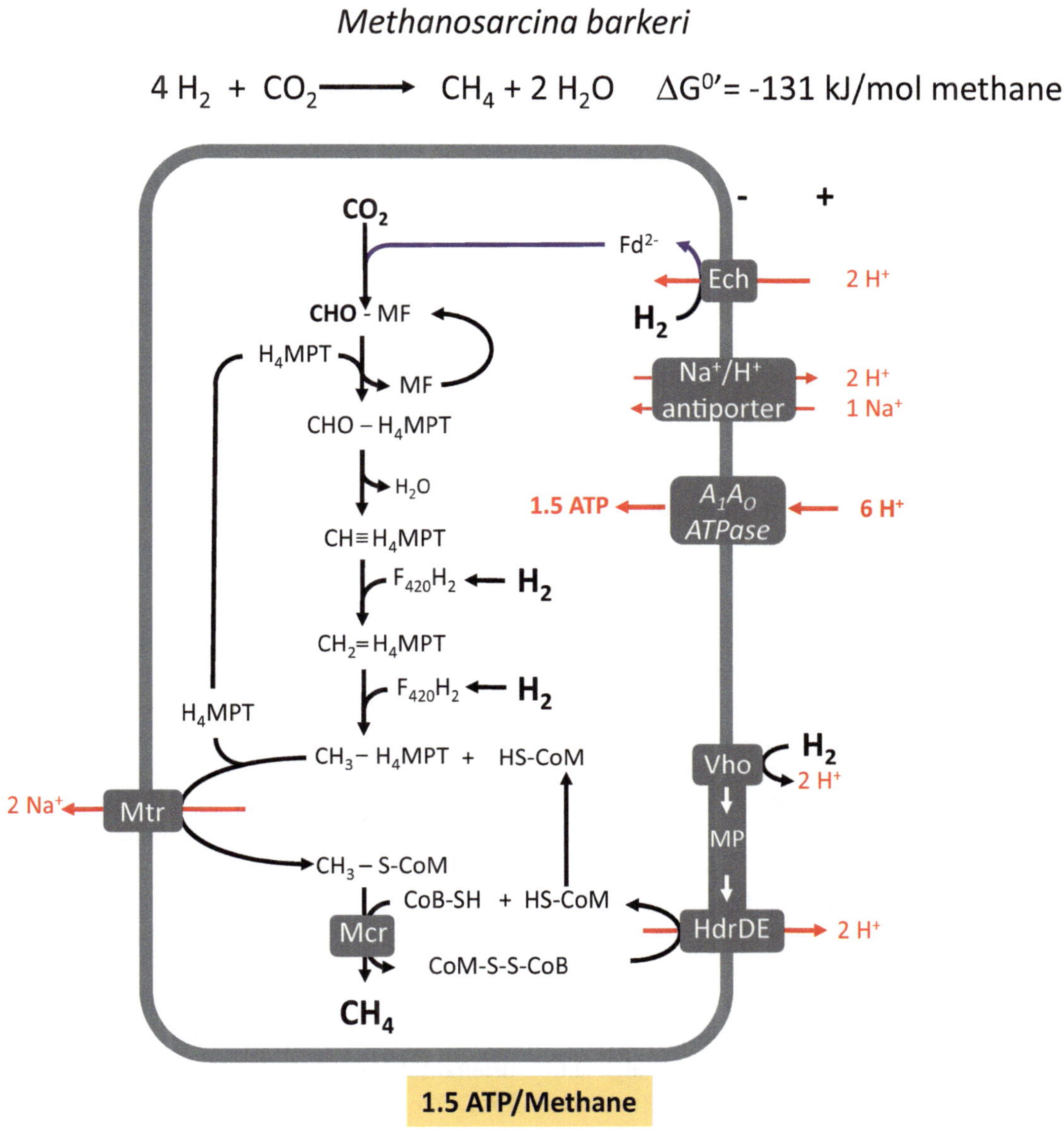

Fig. 10.20 Proposed metabolic map of *Methanosarcina barkeri* oxidizing H_2 with CO_2 to methane. The two energy-conserving steps in this catabolism are the *CH_3-H_4MPT:HS-CoM methyltransferase* reaction and the methanophenazine/cytochrome-containing HdrDE *heterodisulfide reductase*. A total of 1.5 ATP is conserved per CH_4 formed. Abbreviations: *Mtr methyl transferase*, *Vho* hydrogenase, *HdrDE methanophenazine-dependent heterodisulfide reductase*, and *Mcr methyl CoM reductase*. See text for details

are consumed in *F_{420}-reducing hydrogenase* for reduction of methenyl-H_4MPT and methylene-H_4MPT, respectively. The CoB-S-S-CoM heterodisulfide is reduced with H_2 as reductant via a methanophenazine/cytochrome-containing electron transport chain that couples this exergonic reduction with net translocation of 4 electrogenic H^+ (Fig. 10.10). Two Na^+ are translocated in the *methyl-H_4MPT:CoM methyltransferase* reaction as in *M. marburgensis*. When *M. barkeri* grows with H_2/CO_2, four electrogenic H^+ are translocated during CoM-S-S-CoB reduction in addition to 2 electrogenic Na^+, which are converted by a Na^+/H^+ antiporter to 4 electrogenic H^+. Thus, collectively 8 electrogenic H^+ are formed. Two electrogenic H^+ are consumed in the H_2 activation to Fd^{2-}, so that a net of 6 H^+ per CH_4 are available for energy conservation by *ATP synthase*, amounting to 1.5 ATP/mol CH_4.

This comparison between hydrogenotrophic methanogens with and without methanophenazine/cytochrome reveals a genetically fixed (i.e., speciated metabolism) difference in n_{ATP} (0.5 vs 1.5) and a resulting rate-yield trade-off. While the molar growth yields of both methanogens differ by a factor of two (*M. marburgensis* ~ 3 g (dw)/mol CH_4 vs *M. barkeri* ~ 7 g (dw)/mol CH_4), the doubling times are ~1 h (*M. marburgensis*) and >10 h (*M. barkeri*). An important difference between HdrABC- and HdrDE-based hydrogenotrophic methanogenesis is the competitiveness of each group at low pH_2. As outlined and discussed in detail in the Chapter Energetics (Sect. 3.5.4, Fig. 3.8), when the free energy of a catabolic reaction is coupled to ATP synthesis via a pathway with a higher yield (n_{ATP}), the lowest substrate concentration that can be used for growth is thermodynamically limited. Therefore, microbes with a high n_{ATP} pathway will run into "death by thermodynamic equilibrium" at a substrate concentration that is higher than in microbes with a low n_{ATP} pathway. This is particularly relevant if gases are the catabolic substrate and a high substrate stoichiometry is required. In the above case, the threshold of hydrogenotrophic methanogensis in *M. marburgensis* is <10 Pa, whereas >10 Pa in *M. barkeri*. This lack of competitiveness at low pH_2 may be one explanation for why many methanogens with cytochromes lost their capacity to use H_2 and adapted to a different niche based on using methyl group-containing compounds, including acetate, for methanogenesis.

Estimates are that globally 2/3 of the microbially-formed methane is derived from the methyl group of acetate via acetoclastic methanogens or via syntrophic acetate oxidation. Acetate disproportionation to CH_4 and CO_2 is, thus, an ecologically important process in methanogenic ecosystems, and *Methanosarcina* sp. and *Methanosaeta* sp. seem to be the most abundant acetoclastic methanogens. *Methanosarcina* sp. are generalist with a large (4–5.7 Mbp) genome and a broad methanogenic substrate range that includes methanol, methylamines, methylated thiols as well as H_2/CO_2. On the other hand, *Methanosaeta* sp. with a smaller (1.9–3.0 Mbp) genome seems to be more restricted and specialized to the acetate niche.

The proposed pathway of acetoclastic methanogenesis in *M. barkeri* is shown in Fig. 10.21. This pathway has fewer steps than the hydrogenotrophic pathway and is a mosaic of *carbon monoxide dehydrogenase/acetyl-CoA synthase*, *methyl-coenzyme M reductase*, *methyl transferase,* and HdrDE *heterodisulfide reductase*. Key for energy conservation is Ech and a H_2 loop with some resemblance to that in *M. thermoacetica* and many sulfate-reducing microbes. Acetate, after uptake of the acetate$^-$ anion, is activated by consumption of 1 ATP via the acylate ↔ acyl-CoA module to acetyl-CoA, which is cleaved by *CO-DH/ACS* into a methyl group, that is transferred to H_4MPT to form methyl-H_4MPT, and carbon monoxide. The methyl group of CH_3-H_4MPT is transferred in an exergonic reaction to CoM-SH by Na^+-translocating *methyl-H_4MPT:CoM methyltransferase* and 2 Na^+ are translocated. Methyl-CoM is reduced to methane using CoB-SH as reductant, and the resulting heterodisulfide is reduced via energy-conserving HdrDE complex. Electrons for that reduction are derived from CO oxidation via Fd^{2-}; energy in form of 2 H^+ translocated is conserved via Ech. The produced H_2 diffuses to the outside and is oxidized by Vho *hydrogenase*. Vectorial electron transport to the heterodisulfide via methanophenazine leaves 2 H^+ at the outside of the cell. Since a total of 8 electrogenic H^+/Na^+ were formed and synthesis of 1 ATP requires 5 H^+, the equivalent of 1.6 ATP is recovered. Of that, 1 ATP is consumed in acetate activation, which leaves *M. barkeri* with a net energy of 0.6 ATP/mol acetate conserved. For such acetate disproportionation, heterodisulfide reduction only via HdrDE is possible, because no net

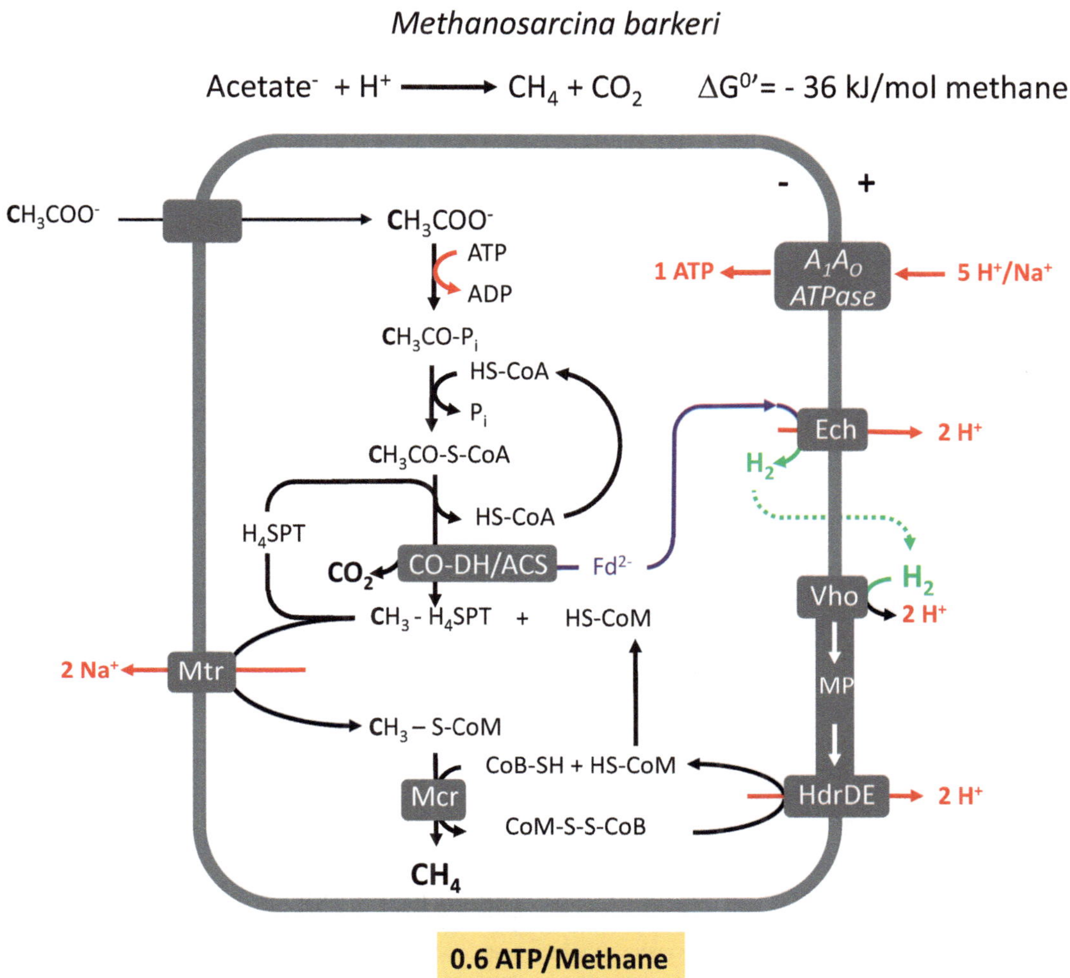

Fig. 10.21 Proposed metabolic map of *Methanosarcina barkeri* disproportionating acetate to methane and CO_2. This short catabolic pathway consists of the *methyl-CoM reductase/heterodisulfide* module, an acylate ↔ acyl-CoA module, and the CO-DH/ACS reaction. Energy is conserved via Ech, in the *methyl transferase* reaction, and via TCP via methanophenazine in CoM-S-S-CoB reduction (HdrDE). The catabolic pathway of *Methanotrix* differs in the acetate activation step. *Methanothrix* sp. contains an *acetyl-CoA synthetase,* which enables to use acetate at lower concentrations at the trade-off of thermodynamic efficiency. Abbreviations: *CO-DH/ACS carbon monoxide dehydrogenase/acetyl-CoA synthase*, *SPT* sarcinopterin, *Mtr methyl transferase*, *Vho* hydrogenase, *HdrDE methanophenazine-dependent heterodisulfide reductase*, *MP* methanophenazine, and *Mcr methyl CoM reductase.* See text for details

energy is conserved by *electron bifurcating heterodisulfide reductase* (HdrABC).

M. barkeri grows as sarcina (name is derived from "marching pack of roman legionaries"), which are symmetrically ordered aggregates of 8 or more cells. This cell physiology has been speculated to be advantageous for cells to consume H_2 that escapes a neighboring cell, thereby, minimizing loss from the population of the H_2 essential for energy conservation.

The two acetoclastic methanogens *M. barkeri* and *Methanosaeta sp.* (previously *Methanotrix* spp.) have different metabolic strategies for acetate activation with ecophysiological consequences. While *M. barkeri* employs a 1 ATP-consuming *acetate kinase-*

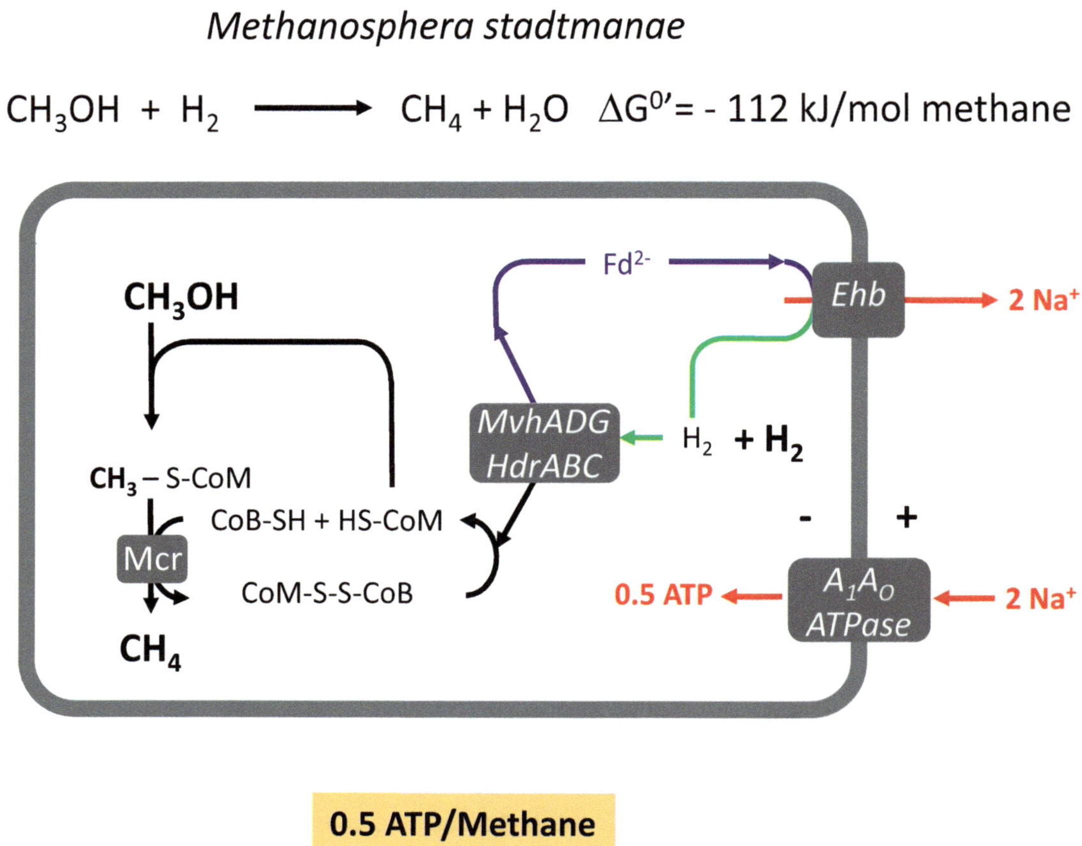

Fig. 10.22 Proposed metabolic pathway of catabolic methanol reduction to methane with H_2 in *Methanosphera stadtmanae*. Only two of the canonical enzymatic steps of the methanogenesis pathway from CO_2, *methyl-coenzyme M reductase,* and *heterodisulfide reductase* are present. In conjunction with a non-energy-conserving cytoplasmic *methyltransferase,* they drive an electron flux through Ehb, an Ech-like, Fd^{2-}-oxidizing, H^+-reducing, Na^+-translocating enzyme. Abbreviations: *MvhADG/HdrABC methylviologen-reducing hydrogenase/electron-bifurcating heterosulfide reductase* complex, and *Mcr methyl CoM reductase*. See text for details

phosphotransacetylase reaction sequence, *Methanosaeta* sp. (*Methanotrix soehngenii)* uses an *acetate synthetase* (see Sect. 3.5.4), which consumes 2 ATP equivalents. This cost of acetate activation reduces n_{ATP} to 0.33 ATP/acetate in *Methanosaeta* sp., which is only half of the yield of *Methanosarcina* sp. ($n_{ATP} = 0.6$). This lower yield, however, enables *Methanosaeta* sp. to utilize acetate at much lower concentration than *Methanosarcina* sp. ($K_M \approx 0.5$ mM and minimum concentration for growth at 7–70 μM, *vs* $K_M > 5$ mM in *M. barkeri*) (see Sect. 3.5.4). This trait, however, comes at a cost of slower growth (*M. barkeri* $t_d \sim 24$ h, vs $t_d \sim 3.5$–9 days in *Methanosaeta* sp.) (see above and Fig. 3.4); another example for a rate/yield trade-off and its ecophysiological consequences.

10.4.2.3 *Methanosphera stadtmanae*

Methane formation from methanol and H_2 by *Methanosphaera stadtmanae* occurs in the absence of the canonical energy-conserving *methyl-H_4MPT:CoM methyltransferase* (Fig. 10.22). The methanol methyl group is transferred to CoM-SH by a cytoplasmic *methyltransferase* in an exergonic (−17 kJ/mol) reaction. This energy remains unused. Methyl-CoM is subsequently reduced with CoB-SH by *methyl-*

coenzyme M reductase to methane, and the heterodisulfide is released. *Electron bifurcating heterodisulfide reductase* (HdrABC) reduces the heterodisulfide (CoM-S-S-CoB) and Fd with 2 H_2 to CoM-SH plus CoB-SH and Fd^{2-}. In contrast to the previously discussed methanogens, an Ech-like membrane-bound Ehb catalyzes Fd^{2-} oxidation with reduction of 2 H^+ to H_2 and concomitant translocation of 2 electrogenic Na^+ ($\Delta\mu Na^+$). This H_2 formed is the second hydrogen of *electron bifurcating heterodisulfide reductase* and enters an internal H_2 loop. Thus, the sole mode of energy conservation in this methanogen is by an electron flux through an Ech-like complex, driven by a rudimentary methanogenesis pathway. More specifically, it is an internal hydrogen loop through an Ech-like complex, similar as in *Moorella thermoacetica* where the Wood-Ljungdahl pathway is the electron sink to drive an electron flux through the Ech-containing H_2 loop (Figs. 10.10 and 10.12).

The above metabolism is another example of the modularity of pathway elements; here, a combination of methanogenesis and Wood-Ljungdahl pathway reactions.

10.5 Anaerobic Methanotrophs

While details of the metabolism of anaerobic methane oxidation, the transfer of electrons to the sulfate-reducing partner, and energy conservation in both microbes are not well understood, oxidation of methane proceeds as a reversal of archaeal methanogenesis (see Figs. 9.31 and 10.18, in reverse, see above): methane is oxidized by CoM-S-S-CoB to CH_3-CoM, followed by a transfer of the methyl group to H_4MPT to form CH_3-H_4MPT, oxidation of CH_3-H_4MPT to formyl-H_4MPT, transfer of the formyl group to methanofuran, and finally oxidation of formyl-methanofuran to CO_2. The first reaction in the oxidative pathway is the highly endergonic *methyl-coenzyme M reductase* reaction:

$$CH_4 + \text{CoM-S-S-CoB} \rightarrow CH_3 - \text{CoM} + \text{CoB} - \text{SH} \quad \Delta G^{0\prime} = +30\,\text{kJ/mol}$$

The flux of CH_4 oxidation is, thus, determined by the thermodynamic constraints of this reaction, and follows the relationship outlined by the Haldane equations (Sect. 5.4.2). Since a ΔG of +30 kJ/mol is equivalent to an equilibrium constant of $K_{eq} = 10^{-5}$ and assuming K_M values for the forward and reverse reaction in the same order of magnitude, the rate of methane oxidation to CH_3-CoM by this enzyme is only 10^{-5} of the rate of methane formation, and, therefore, very slow. Such a slow rate is compensated to some extent by a high cellular level of *methyl-CoM reductase* in anaerobic methane-oxidizing cells; however, there is a limit to the extent that this enzyme can occupy additional proteome space because of general competition for proteome space. Assuming a level of *methyl-CoM reductase* that is ten-fold higher than in methane-forming methanogens, the rate of methane oxidation would still be only 1/10,000 of the rate of methane formation. Indeed, *methyl-CoM reductase* has been observed to be present in consortia of anaerobic methane oxidation with sulfate at very high levels and to catalyze anaerobic methane oxidation with CoM-S-S-CoB. Thus, the Haldane relationship helps to explain why anaerobic methane oxidation via a reversal of the methanogenesis pathway is a slow process intrinsically.

Interestingly, the archaeon "*Ca.* Methanoliparia" converts hexadecane to acetate and methane non-syntrophically as a single microbe, where the alkane is oxidized to alkyl-S-CoM, and CO_2 is reduced to methyl-coenzyme M involving two different *methyl-coenzyme M reductases.*

The apparent anoxic oxidation of methane with nitrite as electron acceptor by "*Candidatus* Methylomirabilis oxyfera" can be explained by the operation of a well-established pathway of aerobic methane oxidation (see below).

10.6 The Aerobic Methanotrophic Pathway

Aerobic methane oxidation, as shown here for *Methylococcus capsulatus* (Bath) (Fig. 10.23), is initiated by *methane monooxygenase*, here by soluble sMMO, consuming one O_2 and one NADH per CH_4 to form methanol and H_2O. Methanol diffuses into the periplasm

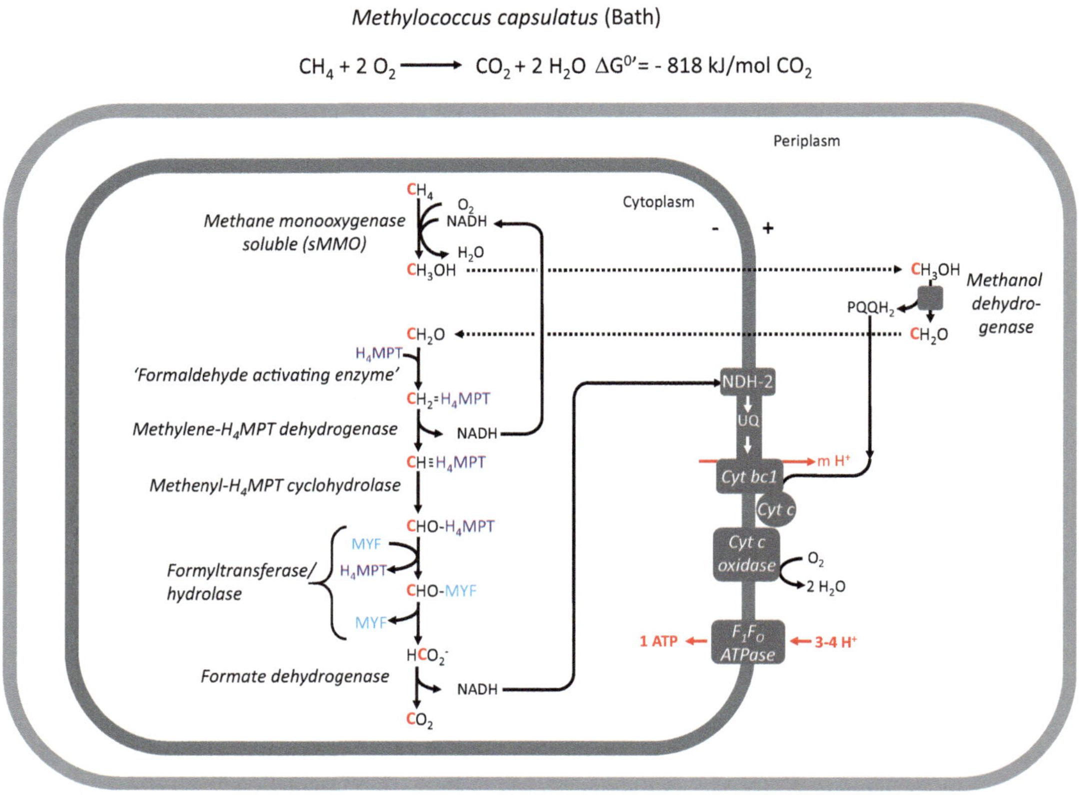

Fig. 10.23 The aerobic methanotrophic pathway of *Methylococcus capsulatus* (Bath). *Monooxygenase* (MMO) shown here is the soluble MMO (sMMO). Only two of the four pairs of electrons of methane are usable for energy conservation. Exact stoichiometry of H^+ translocation associated with the respiratory chain is omitted. Rather than the archaeal methanofuran, in bacteria such as *Methylrubrum extorquens* (formerly *Methylobacterium extorquens*), methylfuran (MYF) has 12 to 24 glutamate residues rather than a tyramine. MYF is a prosthetic group and tightly associated with the bifunctional *formyl transferase/hydrolase*. PQQ is membrane-integrated, and the reduced form reduced the Cyt bc1 complex. *PQQ* pyrroloquinoline quinone ($E^{0\prime}_{PQQ/PQQH2} = +120$ mV), *H_4MPT* tetrahydromethanopterine, *MYF* methylfuran, *NDH NADH dehydrogenase*, *UQ* ubiquinone. See text for details

where it is oxidized by a periplasmic, pyrroloquinoline quinone (PQQ)-dependent *methanol dehydrogenase* to formaldehyde. Electrons from $PQQH_2$ enter the electron transport chain at the level of cytochrome *bc1*. Formaldehyde diffuses back into the cell and is believed to react via a *formaldehyde-activating enzyme* with H_4MPT to form methylene-H_4MPT. *Methylene-H_4MPT dehydrogenase* catalyzes oxidation of methylene-H_4MPT to methenyl-H_4MPT, which becomes hydrolyzed by *methenyl-H_4MPT cyclohydrolase* to formyl-H_4MPT. A *formyltransferase/hydrolase* complex with a tightly-bound methylfuran prosthetic group catalyzes hydrolysis of formyl-H_4MPT to formate via formyl-methylfuran as enzyme-associated intermediate. In contrast to the archaeal methanofuran, bacterial methylfuran of this pathway contains a tyrosine residue instead of the tyramine and has a polyglutamic (12–24) acid side chain. This modification does not significantly alter the relevant chemistry of the furan ring. Formate is oxidized by an *NAD^+-dependent formate dehydrogenase* to CO_2. With the exception of NADH, which is stoichiometrically required for the initial *methane monooxygenase* reaction, the remaining two pairs of electrons released by this pathway are transferred to the aerobic electron transport to reduce O_2 (Fig. 10.23).

"Anaerobic" methane oxidation with nitrite by "*Candidatus* Methylomirabilis oxyfera"

"*Candidatus* Methylomirabilis oxyfera" couples the apparent anaerobic oxidation of methane with nitrite reduction to N_2:

$$3CH_4 + 8NO_2^- + 8H^+ \rightarrow 3CO_2 + 4N_2 + 10H_2O \quad \Delta G^{0'} = -928\,kJ/mol$$

Genomic, transcriptomic, and physiological studies revealed that this apparent anaerobic methane oxidation is, in fact, a canonical aerobic methane oxidation involving all essential enzyme complexes of that pathway (Figs. 10.23). O_2 required for methane activation by *methane monooxygenase* is generated by a modified denitrification pathway (Fig. 10.24): NO_2^- is reduced by *nitrite reductase* to NO, which is not reduced to N_2 via N_2O as in denitrification, but disproportionated by a *nitric oxide dismutase*

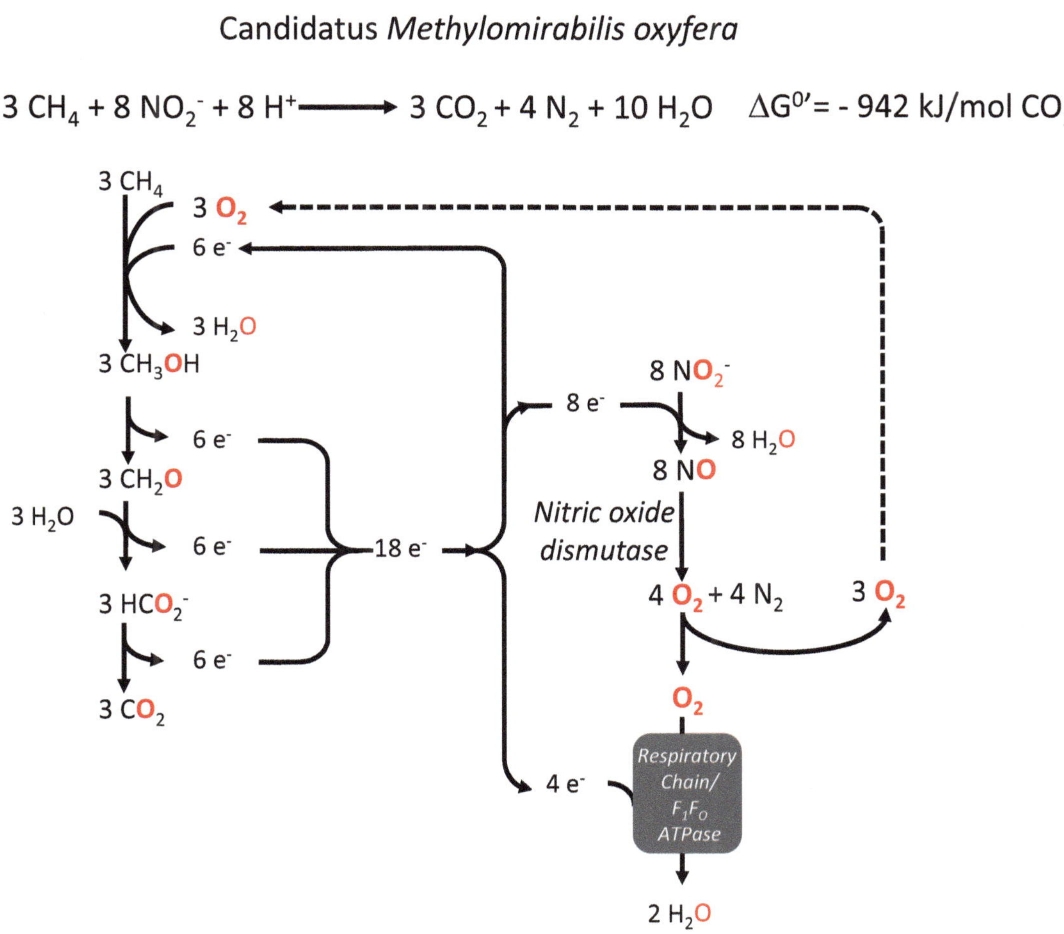

Fig. 10.24 Proposed pathway for nitrite-driven "anaerobic" methane oxidation by oxygenic "*Candidatus* Methylomirabilis oxyfera." Molecular oxygen for *methane monooxygenase* is released from NO by a *nitric oxide dismutase*. Of the 18 electrons released from oxidation of 3 methanol to 3 CO_2, six are used as co-substrate by *methane monooxygenase*, while eight are used by *nitrite reductase*. Disproportionation of 2 NO to O_2 and N_2 is exergonic by −43 kJ/mol NO. Four electrons are consumed in an aerobic respiratory chain, presumably coupled to energy conservation. See text for details

to N_2 and O_2. In the overall nitrite-dependent oxidation of 3 CH_4, three of the four O_2 formed from NO are used in the *methane monooxygenase* reaction, while the fourth O_2 serves as electron acceptor (Fig. 10.24). This enzymatic formation of O_2 from nitrite via NO emphasizes that under apparent anoxic conditions, NO_2^--dependent respirations can still involve molecular oxygen!

Further Readings

Mand TD, Metcalf WW (2019) Energy conservation and hydrogenase function in methanogenic archaea, in particular the genus *Methanosarcina*. Microbiol Mol Biol Rev 83(4):e00020–e00019

Schuchmann K, Müller V (2016) Energetics and application of heterotrophy in acetogenic bacteria. Appl Environ Microbiol 82(14):4056–4069

Thauer RK, Kaster AK, Seedorf H, Buckel W, Hedderich R (2008) Methanogenic archaea: ecologically relevant differences in energy conservation. Nat Rev. Microbiol 6:579–591

Welte C, Deppenmeier U (2014) Bioenergetics and anaerobic respiratory chains of aceticlastic methanogens. Biochim Biophys Acta 1837:1130–1147

Zhuo Z, Zhang C et al (2022) Non-syntrophic methanogenic hydrocarbon degradation by an archaeal species. Nature 601:257–262

Ettwig KF et al (2010) Nitrite-driven anaerobic methane oxidation by anoxygenic bacteria. Nature 464:543–548

Correction to: Principles of Microbial Metabolism and Metabolic Ecology

Correction to:
A. M. Spormann, *Principles of Microbial Metabolism and Metabolic Ecology*, https://doi.org/10.1007/978-3-031-28218-8

The original version of the book was inadvertently published with errors in Chapters 3, 7, 8, and 10. Figure 3.8, Figure 7.18, Figure 8.7, Figure 8.9, Figure 8.10, Figure 10.11, and Figure 10.17 have been updated with the corrected version. In addition, some text corrections have been made to Section 10.2, Section 10.3, Section 10.3.1, Section 10.3.2, and Section 10.4.2.3. These chapters have now been corrected and approved by the author.

The updated versions of these chapters can be found at
https://doi.org/10.1007/978-3-031-28218-8_3
https://doi.org/10.1007/978-3-031-28218-8_7
https://doi.org/10.1007/978-3-031-28218-8_8
https://doi.org/10.1007/978-3-031-28218-8_10

A. M. Spormann, *Principles of Microbial Metabolism and Metabolic Ecology*,
https://doi.org/10.1007/978-3-031-28218-8_11

Appendix

Table A.1 Molal Gibbs free energies of formation from the elements for compounds of microbiological interest at 25 °C, 2 °C, 70 °C, and 100 °C. The physical state for each substance is indicated in the column headed "State" as crystalline solid (c), liquid (liq), or gaseous (g). Solutions in water with solutes at 1 molal activity are listed as aqueous (aq). P_{SAT} denotes saturation pressures for H_2O (P-T boiling curve for water). p*K* values are for 25 °C. Temperature dependent ΔGf^0 are from Amend and Shock (2001). FEMS Microbiol. Rev. 25, 175–243. All others are from Thauer, Jungermann, and Decker, K (1977) Bacteriol. Rev. 41, 100–180 and from Alberty (1998a, b) Arch. Biochem. Biophys. 353, 116–130 and 358, 25–39

Substance	State	$-\Delta Gf^0$ kJ/mol at P_{SAT} 25 °C 2 °C 70 °C 100 °C			
H_2	g	0	−2.98	5.97	10.05
H_2	aq	−17.72	−18.89	−14.62	−12.08
H^+	aq	0	0	0	0
H^+ (neutral)	aq	39.95	39.13	42.07	43.78
H_2O	liq	237.18	235.64	240.57	243.08
HO^-	aq	157.30	157.39	156.44	155.54
H_2O_2	aq	134.02	130.74	140.64	145.21
O_2	g	0	−4.69	9.33	15.64
O_2	aq	−16.54	−18.82	−10.95	−6.56
C graphite	c	0			
CO	g	137.17	132.65	146.16	152.25
CO	aq	120.01	117.01	127.52	129.76
CO_2	g	394.359	389.48	404.10	410.73
CO_2	aq	385.98	383.51	399.91	396.66
H_2CO_3	aq	623.16			
HCO_3^-	aq	586.94	584.63	591.29	594.11
CO_3^{2-}	aq	527.98	528.83	524.91	522.07
Hydrocarbons					
CH_4	g	50.72	46.47	59.22	65.02
CH_4	aq	34.46	32.71	39.23	43.17
C_2H_6	g	32.89			
C_2H_6	aq	16.26	14.01	22.23	27.42
C_2H_4	g	−68.12			
C_2H_2	g	−209.2			
Propane	aq	8.22	5.38	15.75	22.04

(continued)

A. M. Spormann, *Principles of Microbial Metabolism and Metabolic Ecology*,
https://doi.org/10.1007/978-3-031-28218-8

Table A.1 (continued)

Substance	State	$-\Delta Gf^0$ kJ/mol at P_{SAT} 25 °C 2 °C 70 °C 100 °C			
Butane	aq	−0.14	−3.43	9.02	16.61
Hexadecane	liq	−49.44	−63.24	−19.85	1.54
Alcohols					
Methanol	aq	175.94	172.98	182.53	187.47
Ethanol	aq	181.30	178.08	188.90	194.83
n-propanol	aq	175.36	171.70	184.28	191.38
Iso-propanol	aq	185.56	182.87	193.62	200.21
n-butanol	aq	162.51	158.41	172.74	180.96
Ethylene glycol	aq	[330.51]			
Glycerol	aq	497.48			
Mannitol	aq	942.61			
Sorbitol	aq	942.70			
Aldehydes and ketones					
Formaldehyde	aq	121.5			
Formaldehyde	g	102.53			
Acetaldehyde	aq	139.9			
Acetaldehyde	g	128.91			
Acetone	aq	159.70			
Dihydroxyacetone	aq	−451.00			
Monocarboxylic acids					
Formate$^-$	aq	350.88	348.69	354.78	357.26
Formic acid (pK = 3.7)	aq	372.30	368.64	379.96	385.44
Acetate$^-$	aq	369.33	367.36	373.31	376.08
Acetic acid (pK = 4.7)	aq	396.48	392.52	405.08	411.40
Propionate$^-$	aq	363.09	360.66	368.55	372.66
Propionic acid (pK = 4.8)	aq	391.00	386.48	401.15	408.80
Butyrate$^-$	aq	354.18	351.28	360.79	365.84
Butyric acid (pK = 4.8)	aq	381.63	376.55	393.26	402.09
Valerate$^-$ (pK = 4.8)	aq	345.73	342.31	353.91	360.37
Caproate$^-$ (pK = 4.9)	aq	[335.96]			
Palmitate$^-$	aq	259.4			
Benzoate$^-$ (pK = 4.2)	aq	210.88	207.85	218.41	224.17
Benzoic acid	aq	234.86	229.90	246.40	255.29
Acrylate$^-$ (pK = 4.2)	aq	[286.19]			
Crotonate$^-$ (pK = 4.7)	aq	277.4			
Hydroxy/oxocarboxylic acids					
Glycolate$^-$ (pK = 3.8)	aq	506.98	504.47	512.03	515.51
Glycolic acid	aq	528.86	524.89	537.60	544.07
Lactate$^-$	aq	512.67	509.69	519.06	523.72
Lactic acid (pK = 3.9)	aq	534.72	530.19	544.99	552.74
β-Hydroxypropionate$^-$ (pK = 4.2)	aq	518.4			
Glyoxylate$^-$ (pK = 3.3)	aq	468.6			
Pyruvate$^-$ (pK = 2.5)	aq	472.27			
β-Ketobutyrate$^-$ (pK = 3.6)	aq	493.7			
Dicarboxylic acids					
Oxalate^{1-} (pK_2 = 4.2)	aq	698.34	694.80	704.94	709.18
Oxalate^{2-}	aq	674.05	672.66	675.18	675.09
Oxalic acid	aq	705.59	701.44	714.21	720.32

(continued)

Table A.1 (continued)

Substance	State	$-\Delta Gf^0$ kJ/mol at P_{SAT} 25 °C 2 °C 70 °C 100 °C			
Malonate^{1-} (pK_2 = 5.7)	aq	717.69	713.55	725.68	730.96
Malonate^{2-}	aq	685.18	683.68	668.72	686.83
Malonic acid	aq	733.97	728.75	744.98	752.88
Succinate^{1-} (pK_2 = 5.6)	aq	719.91	715.58	728.56	734.51
Succinate^{2-}	aq	687.77	685.69	690.78	692.07
Succinic acid	aq	743.92	738.13	756.37	765.41
Glutarate^{1-} (pK_2 = 5.4)	aq	714.89	709.86	725.14	732.30
Glutarate^{2-}	aq	683.97	681.28	688.38	690.76
Glutaric acid	aq	739.66	732.86	754.30	764.92
Fumaric acid	aq	647.14			
Fumarate^{2-} (pK_2 = 4.4)	aq	604.21			
L-Malate^{2-} (pK_2 = 5.2)	aq	845.08			
Oxalacetate^{2-} (pK_2 = 4.3)	aq	793.18			
α-ketoglutarate^{2-} (pK_2 = 4.3)	aq	797.53			
Tricarboxylic acids					
Citrate^{3-} (pK_3 = 5.4)	aq	1162.69			
Isocitrate^{3-}	aq	1156.04			
Cis-Aconitate^{3-}	aq	917.13			
Carbohydrates					
Glyceraldehyde	aq	[437.65]			
Dihydroxyacetone	aq	[445.18]			
D-Erythrose	aq	[598.3]			
D-ribose	aq	738.79			
D-ribulose	aq	735.94			
D-xylose	aq	750.49			
D-Xylulose	aq	746.15			
α-D-glucose	aq	915.90			
α-D-galactose	aq	908.93			
D-fructose	aq	915.51			
Lactose	aq	1567.33			
β-Maltose	aq	1574.69			
Sucrose	aq	1564.70			
Glycogen (per unit of glucose)	aq	662.33			
Amino acids					
L-alanine	aq	371.59	367.97	379.38	385.08
L-arginine	aq	240.01	232.53	256.96	267.84
L-asparagine	aq	525.06	519.57	536.47	544.71
L-aspartate$^-$	aq	699.44	695.24	707.74	713.51
L-cysteine	aq	336.11	331.80	345.50	352.51
L-Cystine	aq	666.51			
L-Glutamate$^-$	aq	699.62	695.38	708.08	714.03
L-Glutamine	aq	528.36	522.50	540.81	549.92
Glycine	aq	380.54	376.78	388.10	393.36
L-Histidine	aq	202.60	196.45	215.87	225.68
L-Isoleucine	aq	343.1	338.63	353.36	361.89
L-Leucine	aq	352.30	347.82	363.05	371.48
L-Lysine	aq	337.58	332.23	349.39	358.27
L-Methionine	aq	502.59	496.85	515.29	524.75

(continued)

Table A.1 (continued)

Substance	State	$-\Delta Gf^0$ kJ/mol at P_{SAT} 25 °C 2 °C 70 °C 100 °C			
L-Phenylalanine	aq	206.83	201.72	218.76	228.09
L-Proline	aq	307.78	303.13	317.75	325.04
L-Serine	aq	518.49	514.03	527.84	534.63
L-Threonine	aq	501.38	497.29	510.44	517.31
L-Tryptophan	aq	112.23	106.79	125.05	135.34
L-Tyrosine	aq	384.10	378.64	396.30	405.54
L-Valine	aq	357.03	352.91	366.59	373.96
Glycylglycine	aq	520.20			
Purines					
Adenine	aq	−312.29			
Hypoxanthine	aq	−89.5			
$Urate^-$	aq	25.9			
Uric acid	aq	356.9			
Other N-containing compounds					
Urea	aq	202.80			
Creatine	aq	259.2			
Creatinine	aq	23.14			
Allantoin	c	446.098			
NH_2-CH_2-CH_2-SO_3^-	aq	509.4			
CH_3NH_2	aq	−21.08	−23.88	−14.83	−10.12
$CH_3NH_3^+$	aq	39.86			
$(CH_3)_2NH_2^+$	aq	3.3			
$(CH_3)_3NH^+$	aq	−37.2			
Aromatic compounds					
Benzene	liq	−124.5			
Toluene	liq	−118.11	−122.76	−108.23	−101.08
Toluene	aq	−126.60	−130.42	−116.96	−109.15
Ethylbenzene	liq	−119.83	−125.55	−107.72	−98.99
Ethylbenzene	aq	−135.72	−140.05	−124.80	−115.98
Phenol	c	47.6			
Phenol	aq	52.1			
Benzoate	aq	210.88	207.85	218.41	224.17
Benzoic acid (p*K* = 4.2)	aq	234.86	229.90	246.40	255.29
o-Hydroxybenzoic acid (p*K* = 4.4)	c	421.33			
m-Hydroxybenzoic acid	c	422.58			
p-Hydroxybenzoic acid	c	423.00			
Benzyl alcohol	liq	31.25			
Benzyl alcohol	aq	27.6			
Indole	aq	−223.80			
N compounds					
N_2	g	0	−4.38	8.72	14.63
N_2	aq	−18.18	−20.15	−13.18	−9.19
NH_3	aq	26.71	24.30	31.82	35.50
NH_4^+	aq	79.45	76.96	84.68	88.37
NO	g	−86.57	−91.39	−76.99	−70.50
NO	aq	−102.06	−104.56	−95.98	−91.25
NO_2^-	aq	32.22	29.28	37.49	40.77
NO_3^-	aq	110.91	107.45	117.35	121.49

(continued)

Table A.1 (continued)

Substance	State	$-\Delta Gf^0$ kJ/mol at P_{SAT} 25 °C 2 °C 70 °C 100 °C			
N_2O	g	−104.20	−109.22	−94.18	−87.36
N_2O	aq	−113.38	−115.84	−107.27	−102.41
N_2H_4	aq	−128.03			
P compounds					
$H_2PO_4^-$	aq	1130.27	1128.15	1134.27	1136.87
HPO_4^{2-}	aq	1089.14	1089.65	1086.94	1084.81
$H_2PO_3^-$	aq	846.63	843.60	852.69	856.80
HPO_3^{2-}	aq	811.70	811.13	811.98	811.70
S compounds					
S rhombic	c	0	−0.71	1.51	2.59
S^{2-}	aq	−85.8			
SH^-	aq	−11.97	−13.63	−9.16	−7.55
SH_2	g	33.56	28.86	42.93	49.30
SH_2	aq	27.92	25.21	34.12	38.76
S_5^{2-}	aq	−65.68	−69.11	−59.87	−56.49
SO_3^{2-}	aq	486.60	486.98	484.48	482.30
HSO_3^-	aq	527.73	524.50	534.02	538.21
SO_4^{2-}	aq	744.63	743.74	744.56	743.94
SO_4H^-	aq	755.76	752.89	761.49	765.38
$S_2O_3^{2-}$	aq	522.59	520.79	524.92	525.83
$S_2O_4^{2-}$	aq	600.41	598.07	603.96	605.76
$S_3O_6^{2-}$	aq	958.14	954.77	963.84	967.15
$S_4O_6^{2-}$	aq	1040.60	1034.50	1050.80	1059.10
Transition metals					
MnO_2	s	464.4			
Mn^{2+}	aq	230.54	232.07	227.45	225.37
Fe^{3+}	aq	17.23	23.53	4.54	−4.14
$FeOH^{2+}$	aq	241.83	244.24	236.97	233.65
FeO^+	aq	222.17	223.02	219.52	217.28
FeO_2^-	aq	368.19	366.92	369.53	369.83
Fe^{2+}	aq	91.50	93.90	86.66	83.36
$FeOH^+$	aq	275.52	276.53	273.83	272.90
FeO	aq	212.21	213.17	210.35	209.16
$HFeO_2^-$	aq	399.15	400.68	396.61	395.14
Chlorine compounds					
Cl^-	aq	131.29	129.86	133.48	134.58
ClO^-	aq	36.82	35.73	38.38	39.14
ClO_2^-	aq	−17.15	−19.57	−12.81	−10.11
ClO_3^-	aq	7.95	4.16	15.13	19.80
ClO_4^-	aq	8.54	4.32	16.69	22.13

Table A.2 Catabolically important redox couples and their standard redox potentials E^{0}' at 25 °C and pH 7. H_2, CO_2, CH_4, N_2, NO, N_2O, and O_2 in the gaseous state at 1 bar (10^5 Pa) and all other compounds in aqueous solution at 1 molal activity. S°, FeS_2, Fe^0, Fe^{3+}, and MnO_2 are considered to be in the solid state. In situ E′ can deviate from E^{0}'considerably (see Sect. 3.3.4). Estimated maximal deviations are given for electron/hydride carriers and for those redox couples, in which the concentration of c_{ox} or c_{red} is constant (H_2O, H^+, S°, FeS_2, Fe^0, Fe^{3+}, and MnO_2). All other c_{ox}/c_{red} deviate from 1 by an estimated factor of maximally $\pm 10^3$ (±90 mV when $n = 2$). Abbreviations: CoM-SH, coenzyme M; CoB-SH, coenzyme B; H_4MPT, tetrahydromethanopterin; H_4F, tetrahydrofolate; MFR, methanofuran; APS, Adenosine-5′-phosphosulfate

Redox couple	Number e-transferred	E^{0}' [mV]	Maximal deviation from E^{0}'in situ [mV]
Electron/hydride carriers			
Ferredoxins: Fd_{ox}/Fd_{red}^-	1	−500 to −350	±120
Flavodoxins: Fld_{ox}/Fld^r_e	1	−450 to −300	±120
Rubredoxins: Rb_{ox}/Rb_{red}	1	−90 to +40	±120
$F_{420}/F_{420}H_2$	2	−360	+ 60
NAD^+/NADH	2	−320	+ 60
$NADP^+$/NADPH	2	−320	−60
CoM-S-S-CoB/CoM-SH + CoB-SH	2	−140	+ 60
FAD/$FADH_2$, FMN/$FMNH_2$ [a]	2	−220[d]	±200[a]
Flavin/FlavinH$^\bullet$ [e]	1	−310[d]	
FlavinH$^\bullet$/FlavinH_2	1	−120[d]	
Menaquinone (MK)/Menaquinol [b]	2	−70	±60
MK/MKH$^\bullet$ [e]	1	−260[d]	
Ubiquinone (UQ)/Ubiquinol [b]	2	+ 100	±60
UQ/UQH$^\bullet$ [e]	1	−163[d]	
Cytochromes b ox/red [c]	1	−400 to +200	±120
Cytochromes *c* ox/red [c]	1	−400 to +300	±120
H_2 and O_2			±90
$2H^+/H_2$	2	−420	+ 240
$O_2/O_2^{\bullet-}$	1	−330	
$O_2^{\bullet-}/H_2O_2$	1	+ 890	
O_2/H_2O_2	2	+280	
$O_2/2H_2O$	4	+ 820	−240
$H_2O_2/2H_2O$	2	+ 1350	−240
CO_2			±90
CO_2/CO	2	−520	
CO_2/Formate$^-$	2	−430	
CO_2 + MFR/formyl-MFR	2	−500	±60
CO_2 + acetyl-CoA/pyruvate	2	−500	±60
6 CO_2/glucose	24	−410	
CO_2/acetate$^-$	8	−290	
CO_2/CH_4	8	−240	
Reduced C_1-units			±90
Formate/formaldehyde	2	−580	
Formaldehyde/methanol	2	−140	
Methanol/methane	2	+ 170	
Methenyl-H_4MPT/methylene-H_4MPT	2	−390	±60
MethyleneH_4MPT/methyl-H_4MPT	2	−320	±60
Methenyl-H_4F/methylene-H_4F	2	−300	±60
Methylene-H_4F/methyl-H_4F	2	−200	±60

(continued)

Table A.2 (continued)

Redox couple	Number e-trans-ferred	$E^{0\prime}$ [mV]	Maximal deviation from $E^{0\prime}$in situ [mV]
C_2-units			
Acetate$^-$/acetaldehyde	2	−580	
Acetyl-CoA/acetaldehyde	2	−390	±60
Acetyl-CoA + CO_2/pyruvate	2	−500	±60
Acetaldehyde/ethanol	2	−200	
>C_2-units			**±90**
1,3-Bishosphoglycerate/GAP	2	−350	
Gluconate-6-P/Glucose-6-P	2	<−420	
Ribulose-5-P + CO_2/Gluconate-6-P	2	−350	
2-ketoglutarate+CO_2/Isocitrate	2	−380	
Acetone/isopropanol	2	−290	
Acetoacetyl-CoA/3-HO-butyryl-CoA	2	−250	
Pyruvate/lactate	2	−190	
Dihydroxyacetone-P/glycerol-P	2	−190	
Acetone/2,3-Butanediol	2	−190	
Oxaloacetate/malate	2	−170	
Crotonyl-CoA/Butyryl-CoA	2	−10	
Glycine/acetate + NH_3	2	−10	
Fumarate/succinate	2	+ 30	
Acrylyl-CoA/Propionyl-CoA	2	+ 70	
Halogenated compounds			±90
Vinyl chloride/Ethene	2	+450	
Trichloroacetate/Dichloroacetate	2	+470	
Trichloroethene/cis-Dichloroethene	2	+540	
1,1,2-Trichloroethane/1,1-Dichloroethane	2	+540	
1,1,1-Trichloroethane/1,1-Dichloroethane	2	+560	
Tetrachloroethene/Trichloroethene	2	+580	
CCl_4/$CHCl_3$	2	+670	
1,1,2-Tetrachloroethane/vinyl chloride	2	+830	
1,2-Dichloroethane/ethane	2	+740	
S compounds			±90
SO_4^{2-}/HSO_3^-	2	−520	
HSO_3^-/HS^-	6	−120	
SO_4^{2-}/HS-	8	−220	
S^0/HS^-	2	−270	+ 180
$S_2O_3^{2-}$/HS^- + HSO_3^-	2	−400	
APS/AMP + HSO_3^-	2	−60	
Dimethylsulfoxide/Dimethylsulfide	2	+160	
N compounds			±90
N_2/2NH_3	6	−320	
NO_3^-/NO_2^-	2	+ 430	
NO_2^-/NO	1	+ 350	
NO_2^-/NH_3	6	+ 330	
NO_3-/NH_3	8	+ 360	
2NO/N_2O	2	+ 1175	
N_2O/N_2	2	+ 1360	
2NO_3^-/N_2	10	+ 760	

(continued)

Table A.2 (continued)

Redox couple	Number e-trans-ferred	$E^{0'}$ [mV]	Maximal deviation from $E^{0'}$in situ [mV]
Trimethylamine N-oxide/Trimethyl-amine	2	+ 130	
Transition metals			
FeS_2 (pyrite)/$FeS + H_2S$	2	−610	+ 180
Fe^{2+}/Fe^0	2	−450	−180
$Fe(OH)_3 + HCO^-_3/FeCO_3$	1	+ 200	+ 180
MnO_2/Mn^{2+}	2	+ 420	+ 180

[a] Mostly enzyme bound
[b] Dissolved in the lipid phase of the cytoplasmic membrane
[c] Associated with the cytoplasmic membrane
[d] In water
[e] Crossed over redox potential meaning that the first redox potential in a 2 × 1 electron reduction reaction is more negative than the second one

Table A.3 Overview of the most common prototypic and auxiliary reactions and enzymes in microbial catabolism

		Examples of Enzymes	Major Pathways	Environments	Chapter
Oxidations/reductions	Alcohol oxidation 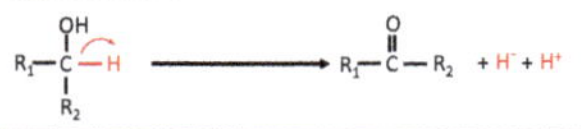	*Ethanol dehydrogenase* *Malate dehydrogenase* *Isocitrate dehydrogenase* *Lactate dehydrogenase*	TCA cycle β-oxidation Alcohol fermentations Hydrocarbon degradation	oxic anoxic	7.2.1
	Hemiacetal oxidation 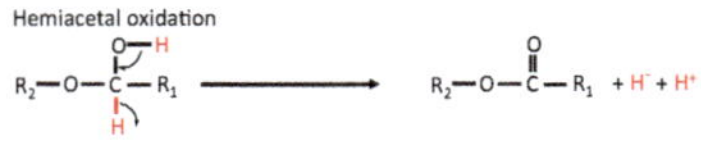	*Glucose-6-phosphate Dehydrogenase* *2-hydroxy-4-carboxymuconate semialdehyde dehydrogenase*	Pentose phosphate pathway Heterofermentative lactic acid fermentation, Meta cleavage pathway	oxic anoxic	7.2.2
	Aldehyde oxidation 	*Glyceraldehyde-3-phosphate dehydrogenase* *Acetaldehyde dehydrogenase*	Glycolysis TCA cycle	oxic anoxic	7.2.3
	Acyl-CoA oxidation 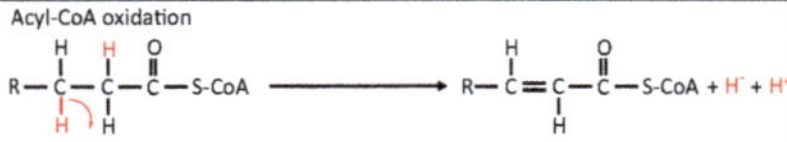	*Butyryl-CoA dehydrogenase* *Succinate dehydrogenase* *Pimelyl-CoA dehydrogenase* *3-Methyl butyryl-CoA dehydrogenase*	TCA cycle β-oxidation Anaerobic hydrocarbon degradation	oxic anoxic	7.2.4
	Hydrogen oxidation/H^+ reduction $H_2 \rightleftharpoons 2e^- + 2H^+$	*Fd-dependent H_2ase* *NAD^+-dependent H_2ase* *Formate:H_2 lyase* *H_2-dependent CO_2 reductase* *Heterodisulfide reductase*	Many anaerobic fermentations, acetogenesis and methanogenesis, Sulfate-reduction	mostly anoxic	7.2.5
C-C cleavage/ condensations	Aldol cleavage 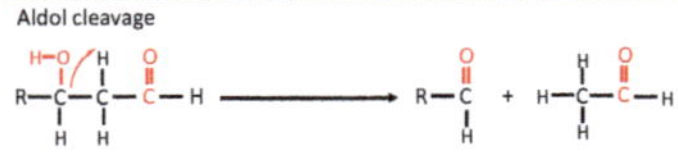	*Fructose-1,6-bisphosphate aldolase* *2-Keto-3-deoxy-6-phosphogluconate aldolase* *β-Ketothiolase* *Citrate lyase*	Glycolysis, Entner Doudoroff, various carbohydrate, amino acid degradations	oxic anoxic	7.3.1
	Ketol cleavage 	*Xylulose-5-phosphate phosphoketolase* *Fructose-6-phosphate ketolase* *Sedoheptulose-7-phosphate ketolase* *Pyruvate decarboxylase*	Phosphoketolase pathway Bifidobacterium shunt Pentosephosphate pathway	oxic anoxic	7.3.2
Auxiliary reactions	Mono- and dioxygenases 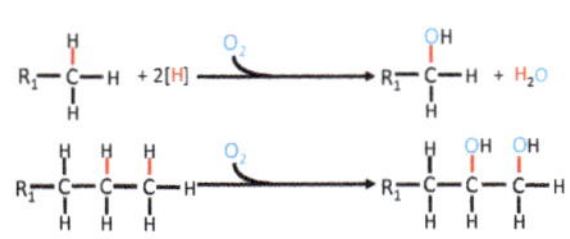	*Methane monooxygenase* *Toluene monooxygenase* *Phenol monooxygenase* *Ammonia monooxygenase* *Benzene 1,2-dioxygenase* *Catechol dioxygenase*	Aerobic degradation of aliphatic and substituted aromatic hydrocarbons, of aromatic compounds, of amino acids Aerobic degradation of aromatic hydrocarbons and aromatic compounds, ortho/meta cleavage pathways	oxic	7.4.1
	Fumarate additions 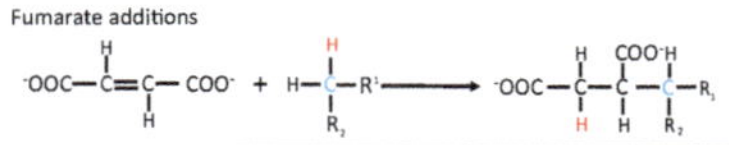	*Benzylsuccinate synthase* *Alkylsuccinate synthase*	Anaerobic degradation of aliphatic and aromatic hydrocarbons	anoxic	7.4.2
	Benzoyl-CoA reduction (dearomatizing) 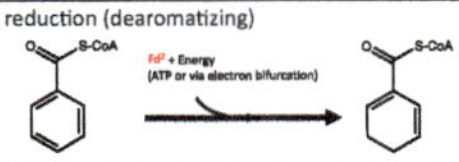	*Benzoyl-CoA reductase (Class I or Class II)*	Anaerobic degradation of aromatic compounds	anoxic	7.4.3
	Carbon-carbon rearrangements 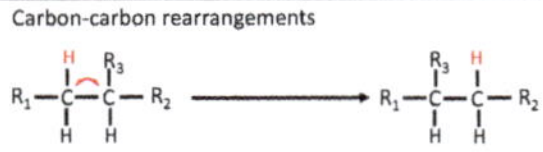	*Glutamate mutase* *Methylmalonyl-CoA mutase* *Ethanolamine ammonia lyase* *Glycerol dehydratase*	Anaerobic fermentations of lactate, amino acids ethanolamine, glycerol	anoxic	7.4.4
	Nucleophilic addition/elimination 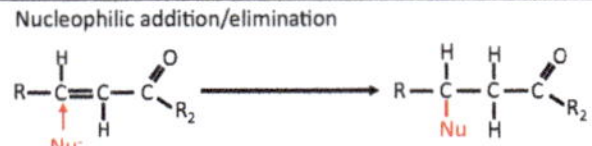	*Succinate dehydrogenase* *Fumarase* *Fumarate reductase* *Butyryl-CoA dehydrogenase* *Crotonase*	TCA cycle, butyrate fermentation Anaerobic alkane and aromatic hydrocarbon degradation	oxic anoxic	7.4.5
	α-Hydroxy/amino elimination 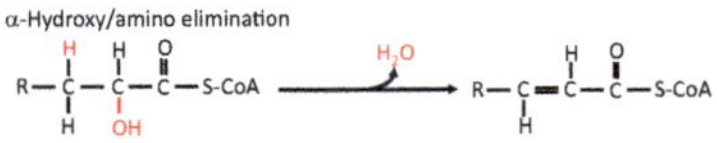	*Lactoyl-CoA dehydratase* *Glutaconyl-CoA dehydrogenase* *4-Hydroxybutyryl-CoA dehydratase* *Phenyllactoyl-CoA dehydratase*	Anaerobic fermentation of amino acids, propionate fermentation	anoxic	7.4.6

Table A.4 Recurrent modules in microbial catabolism

Acylate↔acyl-CoA module : Acylate + mATP (+ 1 HS-CoA) ⟶ Acyl-CoA

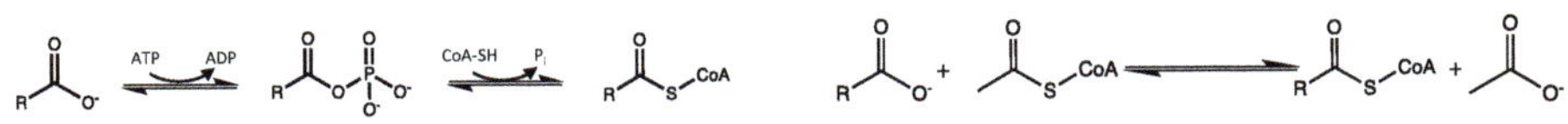

β-oxidation module : $Acyl_{(n)}$-CoA ⟶ $Acyl_{(n-2)}$-CoA + Acetyl-CoA + 4 [H]

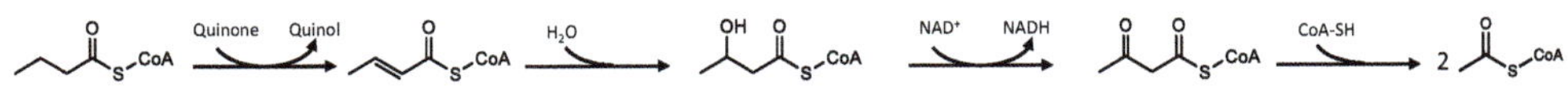

Succinate ↔oxaloacetate module : Succinate ⟶ Oxaloacetate + 4 [H]

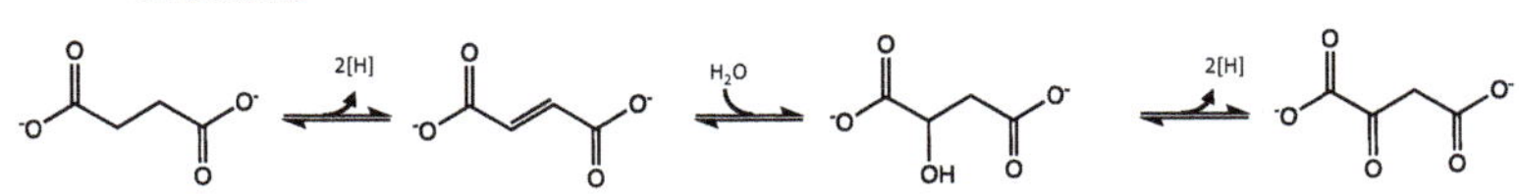

Acyl-CoA mutase module : Acyl-CoA ⟶ Methyl-$acyl_{(n-1)}$ -CoA

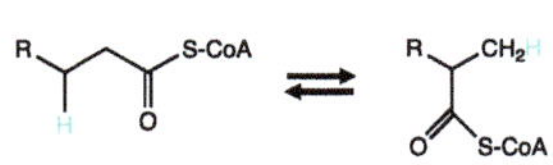

Methylmalonyl-CoA module : Propionyl-CoA + ATP + CO_2 ⟶ Succinyl-CoA

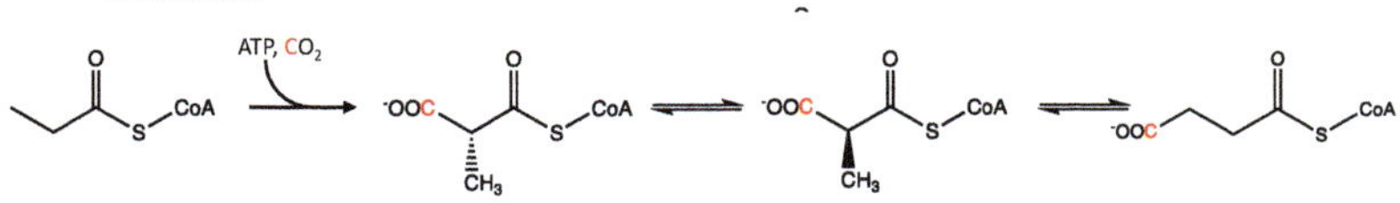

Glyceraldehyde-3-phosphate Dehydrogenase Module (NAD$^+$ dependent, phosphorylating)* : Glyceraldehyde-3-phosphate → Pyruvate + 2[H] + 2 ATP

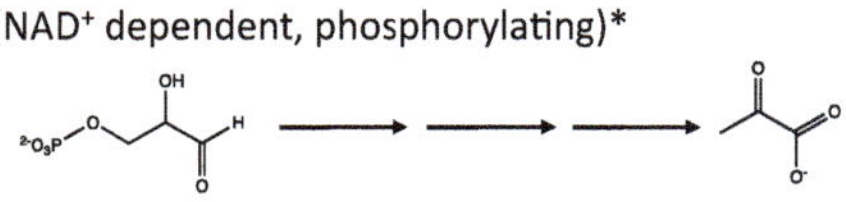

Table A.5 Recurrent catabolic pathways and the modules involved

Substrate	Pathway and overall metabolic reaction*	Modules and prototypic enzymes involved						Environmental conditions for pathway operation
Organic acids, hydrocarbons	***n*-Alkane Oxidation (e.g., octane)** Octane + O_2 → 4 Acetyl-CoA + 8 [H]	Oxygenase	Alcohol dehydrogenase	Aldehyde dehydrogenase	β-oxidation module			Oxic
	Alkylsuccinate Pathway (e.g., octane) Octane → 4 Acetyl-CoA + 12 [H]	Fumarate addition	AK/PTA module	Acyl-CoA mutase module	β-oxidation module			Anoxic
	Benzylsuccinate Pathway (e.g., toluene) Toluene → Benzoyl-CoA + 6 [H]	Fumarate addition	AK/PTA module	β-oxidation module				Anoxic
	Reductive Benzoyl-CoA Pathway Benzoate/Benzoyl-CoA → Acetyl-CoA + Glutaryl-CoA + 2 [H]	Kinase	Benzoyl-CoA reductase	β-oxidation module				Anoxic
	Citric Acid Cycle $Acetate^-$/Acetyl-CoA → 2 CO_2 + 8 [H]	AK/PTA module	Aldolase	Addition/ elimination	Alcohol dehydrogenase	Ketolase	β-oxidation module	Oxic/anoxic
	Methylcitrate Cycle $Propionate^-$/Propionyl-CoA → $Pyruvate^-$ + 4[H]	AK/PTA module	Aldolase	Addition/ elimination	Aldolase	β-oxidation module		Oxic/anoxic
	Glyoxylate Shunt 2 Acetyl-CoA → 1 Oxaloacetate + 6 [H]	AK/PTA module	Aldolase	Addition/ elimination	Aldolase	Aldolase	Alcohol dehydrogenase	Oxic/anoxic
Carbohydrate	**Embden-Meyerhof (glycolytic) Pathway** Glucose → 2 Pyruvate + 4[H] + 1-2 ATP	Kinase	Kinase	Aldolase	GAP-DH module			Oxic/anoxic
	Entner-Doudoroff Pathway Glucose → 2 Pyruvate + 4[H] + 0-1 ATP	Kinase	Alcohol de-hydrogenase	β-elimination	Aldolase	GAP-DH module		Oxic/anoxic
	***Phosphoketolase* Pathway** Pentose → 1 Pyruvate + 1 Acetate + 2[H] + 2 ATP	Kinase	Ketolase	GAP-DH module	AK/PTA module			Anoxic
	Pentosephosphate Pathway 2.5 Hexoses ↔ 3 Pentoses	Kinase	Ketolase	Aldolase	Ketolase			Oxic/anoxic
	Oxidative Pentose Phosphate Pathway Glucose → 6 CO_2 + 24 [H]	Kinase	Alcohol dehydrogenase	Ketolase	Aldolase	Ketolase		Oxic

Table A.6 Recurrent reductive pathways of anaerobic catabolism and the modules and prototypic enzymes involved

Substrate of reductive pathway	Reductive fermentative pathway and overall metabolic reaction*		Modules and prototypic enzymes involved in reductive pathway				Environmental conditions for pathway operation
Pyruvate	**Lactate-forming fermentation** Pyruvate + 2 [H] → D,L Lactate		Alcohol dehydrogenase				Anoxic
	Propionate-forming fermentation Pyruvate/lactate + 4/2 [H] → 1 Propionate	Methylmalonyl-CoA Pathway	Aldolase	Succinate↔oxaloacetate module	Acyl-CoA mutase module	AK/PTA module	Anoxic
	Propionate-forming fermentation Lactate + 2 [H] → 1 Propionate	Acryloyl-CoA Pathway	Alcohol dehydrogenase	α-Hydroxy dehydratase	AK/PTA module		Anoxic
	2,3 Butanediol-forming fermentation 2 Pyruvate + 2 [H] → 2,3-Butanediol + 2 CO_2		Ketolase	Aldolase	Alcohol dehydrogenase		Anoxic
	Ethanol-forming fermentation Pyruvate + 2 [H] → Ethanol + 1 CO_2		Ketolase	Alcohol dehydrogenase			Anoxic
	Formate-forming fermentation Pyruvate → Acetyl-CoA + 1 Formate		Pyruvate:formate lyase				Anoxic
Acetyl-CoA	**Ethanol-forming fermentation** Acetyl-CoA + 4 [H] → Ethanol		Alcohol/aldehyde dehydrogenase				Anoxic
	Butyrate-forming fermentation 2 Acetyl-CoA + 4 [H] → Butyrate + 1-1.5 ATP		β-oxidation module	AK/PTA module			Anoxic
	Butanol-forming fermentation 2 Acetyl-CoA + 8 [H] → Butanol		β-oxidation module	Alcohol/aldehyde dehydrogenase			Anoxic
	Isopropanol-forming fermentation 2 Acetyl-CoA + 2 [H] → Isopropanol + 1 CO_2		β-ketothiolase	Aldolase	Alcohol dehydrogenase		Anoxic
Citrate	**Citrolactate fermentation** Citrate + 2 [H] → 1 Lactate + 1 Acetate + 1 CO_2		Aldolase	Aldolase	Alcohol dehydrogenase		Anoxic
	Citrate-consuming fermentation Citrate + 2 [H] → 1 Succinate + 1 Acetate + 1 CO_2		Aldolase	Succinate↔oxaloacetate module			Anoxic
Glycerol	**Glycerol-consuming fermentation** 4 Glycerol → Acetate + 3 1,3-Propanediol + 1 CO_2		1,3 Propanediol module				Anoxic
CO_2	**Wood-Ljungdahl Pathway** 2CO_2 + 8 [H] → 1 Acetate		Hydrogenase	Formate↔methyl module	CO dehydrogenase/ acetyl-CoA synthase	AK/PTA module	Anoxic
	Methanogenesis Pathway CO_2 + 8 [H] → 1 Methane		Hydrogenase	Formate↔methyl module	Methyl-CoM reductase	Heterodisulfide reductase	Anoxic

Index

A. M. Spormann, *Principles of Microbial Metabolism and Metabolic Ecology*,
https://doi.org/10.1007/978-3-031-28218-8

D

Q

R

GPSR Compliance

The European Union's (EU) General Product Safety Regulation (GPSR) is a set of rules that requires consumer products to be safe and our obligations to ensure this.

If you have any concerns about our products, you can contact us on ProductSafety@springernature.com

In case Publisher is established outside the EU, the EU authorized representative is:

Springer Nature Customer Service Center GmbH
Europaplatz 3
69115 Heidelberg, Germany

Batch number: 10263223

Printed by Printforce, the Netherlands